TEN CHAPTERS IN TURBULENCE

Turbulence is ubiquitous in science, technology and daily life and yet, despite years of research, our understanding of its fundamental nature is tentative and incomplete. More generally, the tools required for a deep understanding of strongly interacting many-body systems remain underdeveloped.

Inspired by a research programme held at the Newton Institute in Cambridge, this book contains reviews by leading experts that summarize our current understanding of the nature of turbulence from theoretical, experimental, observational and computational points of view. The articles cover a wide range of topics, including the scaling and organized motion in wall turbulence; small scale structure; dynamics and statistics of homogeneous turbulence, turbulent transport and mixing; and effects of rotation, stratification and magnetohydrodynamics, as well as superfluid turbulence.

The book will be useful to researchers and graduate students interested in the fundamental nature of turbulence at high Reynolds numbers.

PETER A. DAVIDSON is a Professor in the Department of Engineering at the University of Cambridge.

YUKIO KANEDA is now an Emeritus Professor at Nagoya University and currently a Professor at Aichi Institute of Technology, Japan.

KATEPALLI R. SREENIVASAN is a Professor in the Department of Physics and in the Courant Institute for Mathematical Sciences at New York University.

TEN CHAPTERS IN TURBULENCE

Edited by

PETER A. DAVIDSON
University of Cambridge

YUKIO KANEDA
Aichi Institute of Technology, Japan

KATEPALLI R. SREENIVASAN
New York University

BOWLING GREEN STATE
UNIVERSITY LIBRARIES

CAMBRIDGE UNIVERSITY PRESS
Cambridge, New York, Melbourne, Madrid, Cape Town,
Singapore, São Paulo, Delhi, Mexico City

Cambridge University Press
The Edinburgh Building, Cambridge CB2 8RU, UK

Published in the United States of America by Cambridge University Press, New York

www.cambridge.org
Information on this title: www.cambridge.org/9780521769440

© Cambridge University Press 2013

This publication is in copyright. Subject to statutory exception
and to the provisions of relevant collective licensing agreements,
no reproduction of any part may take place without the written
permission of Cambridge University Press.

First published 2013

Printed and bound in the United Kingdom by the MPG Books Group

A catalogue record for this publication is available from the British Library

ISBN 978-0-521-76944-0 Hardback

Cambridge University Press has no responsibility for the persistence or
accuracy of URLs for external or third-party internet websites referred to
in this publication, and does not guarantee that any content on such
websites is, or will remain, accurate or appropriate.

Contents

Preface		*page* ix
Contributors		xi
1	**Small-Scale Statistics and Structure of Turbulence – in the Light of High Resolution Direct Numerical Simulation**	1
	Yukio Kaneda and Koji Morishita	
	1.1 Introduction	1
	1.2 Background supporting the idea of universality	2
	1.3 Examination of the ideas underlying the 4/5 law	7
	1.4 Intermittency of dissipation rate and velocity gradients	14
	1.5 Local structure	21
	1.6 Inertial subrange	28
	1.7 Concluding remarks	34
	References	35
2	**Structure and Dynamics of Vorticity in Turbulence**	43
	Jörg Schumacher, Robert M. Kerr and Kiyosi Horiuti	
	2.1 Introduction	43
	2.2 Basic relations	44
	2.3 Temporal growth of vorticity	52
	2.4 Spatial structure of the turbulent vorticity field	60
	2.5 Vorticity statistics in turbulence	73
	References	80
3	**Passive Scalar Transport in Turbulence: A Computational Perspective**	87
	T. Gotoh and P.K. Yeung	
	3.1 Introduction	87
	3.2 Computational perspective	89

v

	3.3	Background theory	93
	3.4	Approach to low-order asymptotic state	95
	3.5	High-order statistics: fine-scale structure and intermittency	108
	3.6	Concluding remarks	121
	References		124
4	**A Lagrangian View of Turbulent Dispersion and Mixing**		132
	Brian L. Sawford and Jean-François Pinton		
	4.1	Introduction	132
	4.2	Single particle motion and absolute dispersion	135
	4.3	Two particle motion and relative dispersion	153
	4.4	n-particle statistics	162
	4.5	Conclusions	165
	References		167
5	**The Eddies and Scales of Wall Turbulence**		176
	Ivan Marusic and Ronald J. Adrian		
	5.1	Introduction	176
	5.2	Background	179
	5.3	Scales of coherent structures in wall turbulence	190
	5.4	Relationship between statistical fine-scales and eddy scales	206
	5.5	Summary and conclusions	209
	References		214
6	**Dynamics of Wall-Bounded Turbulence**		221
	J. Jiménez and G. Kawahara		
	6.1	Introduction	221
	6.2	The classical theory of wall-bounded turbulence	227
	6.3	The dynamics of the near-wall region	231
	6.4	The logarithmic and outer layers	241
	6.5	Coherent structures and dynamical systems	249
	6.6	Conclusions	260
	References		261
7	**Recent Progress in Stratified Turbulence**		269
	James J. Riley and Erik Lindborg		
	7.1	Introduction	269
	7.2	Scaling, cascade and spectra	271
	7.3	Numerical simulations	282
	7.4	Laboratory experiments	292
	7.5	Field data	297

	7.6 Conclusions	309
	Appendix	311
	References	312
8	**Rapidly-Rotating Turbulence: An Experimental Perspective**	318
	P.A. Davidson	
	8.1 The evidence of the early experiments	318
	8.2 Background: inertial waves and the formation of Taylor columns	321
	8.3 The spontaneous growth of Taylor columns from compact eddies at low Ro	325
	8.4 Anisotropic structuring via nonlinear wave interactions: resonant triads	332
	8.5 Recent experimental evidence on inertial waves and columnar vortex formation	337
	8.6 The cyclone–anticyclone asymmetry: speculative cartoons	343
	8.7 The rate of energy decay	345
	8.8 Concluding remarks	347
	References	348
9	**MHD Dynamos and Turbulence**	351
	S.M. Tobias, F. Cattaneo and S. Boldyrev	
	9.1 Introduction	351
	9.2 Dynamo	356
	9.3 Mean field	374
	9.4 Conclusions	394
	References	397
10	**How Similar is Quantum Turbulence to Classical Turbulence?**	405
	Ladislav Skrbek and Katepalli R. Sreenivasan	
	10.1 Introduction	405
	10.2 Preliminary remarks on decaying QT	410
	10.3 Comparisons between QT and HIT: energy spectrum	417
	10.4 Decaying vorticity	422
	10.5 Decay of HIT when the shape of the energy spectra matters	425
	10.6 Effective viscosity	429
	10.7 Conclusions	431
	References	432

Preface

Vorticity fields that are not overly damped develop extremely complex spatial structures exhibiting a wide range of scales. These structures wax and wane in coherence; some are intense and most of them weak; and they interact nonlinearly. Their evolution is strongly influenced by the presence of boundaries, shear, rotation, stratification and magnetic fields. We label the multitude of phenomena associated with these fields as *turbulence*[1] and the challenge of predicting the statistical behaviour of such flows has engaged some of the finest minds in twentieth century science.

The progress has been famously slow. This slowness is in part because of the bewildering variety of turbulent flows, from the ideal laboratory creations on a small scale to heterogeneous flows on the dazzling scale of cosmos. Philip Saffman (*Structure and Mechanisms of Turbulence II*, Lecture Notes in Physics **76**, Springer, 1978, p. 273) commented: "... we should not altogether neglect the possibility that there is no such thing as 'turbulence'. That is to say, it is not meaningful to talk about the properties of a turbulent flow independently of the physical situation in which it arises. In searching for a theory of turbulence, perhaps we are looking for a chimera ... Perhaps there is no 'real turbulence problem', but a large number of turbulent flows and our problem is the self-imposed and possibly impossible task of fitting many phenomena into the Procrustean bed of a universal turbulence theory."[2]

More than forty years have passed since Steven Orszag (*J. Fluid Mech.* **41**, 363, 1970) made the whimsical comment, as if with an air of resignation: "It must be admitted that the principal result of fifty years of turbulence research is the recognition of the profound difficulties of the subject". The hope he expressed that "This is not meant to imply that a fully satisfactory theory is beyond hope", appearing almost as an afterthought in his paper, is still unrealized but progress has been made. In recent years one of the driving forces behind this progress has been the ever increasing power of computer simulations.[3] These simulations, in conjunction with ever more ambitious laboratory and field experiments, have helped us understand the

---

[1] We will not use the term here in its more general connotation of complex behaviour in an array of many-dimensional systems.
[2] Saffman was merely pointing out that he was open to this multiplicity, not declaring that research in turbulence is tantamount to taxonomy.
[3] The progress in numerical simulations of turbulence is a tribute to Orszag's untiring efforts. Sadly, he passed away during the preparation of this book.

role of organised motion in near-wall turbulence, and the structure and dynamics of small scales in statistically homogeneous turbulence and other turbulent flows. Together, experiments and simulations have enhanced our understanding of turbulent mixing and dispersion, allowing us to probe the validity and refinement of many classical scaling predictions, and have successfully complemented each other. In strongly stratified turbulence, for example, we have learnt that we are not free to prescribe the vertical Froude number; rather, nature dictates that it is of order unity. This understanding has been crucial to developing a self-consistent scaling theory of stratified turbulence. Similarly, in rapidly rotating turbulence, simulations have supplemented laboratory experiments and there is now a vigorous debate as to what precise role is played by inertial waves in forming the long-lived columnar vortices evident in both simulations and experiments. Yet another area in which simulations and experiments have admirably supplemented each other is turbulence in superfluids, which displays some of the same macroscopic phenomena as classical turbulence even though the microscopic physics in the two instances is quite different.

This book was conceived during an Isaac Newton programme on turbulence held in Cambridge in the Fall of 2008, and in particular after the workshop on *Inertial-Range Dynamics and Mixing* organised by the editors. The chapters, which take the form of reviews, are written by leading experts in the field and should appeal to specialists and non-specialists alike. They cover topics from small-scale turbulence in velocity and passive scalar fields to organized motion in wall flows, from dispersion and mixing to quantum turbulence as well as rotating, stratified and magnetic flows. They reflect the breadth, the nature and the features of turbulence already mentioned. The chapters progress from the those dealing with more general issues, such as homogeneous turbulence and passive scalars through to wall flows, ending with chapters dealing with stratification, rapidly rotating flows, the effect of electromagnetic fields and quantum turbulence. Each is intended to be a comprehensive account of lasting value that will help to open up new lines of enquiry. Yet, the title of the book conforms to the pragmatic style of the articles, rather than to a grand vision which promises more than it delivers.

The editors wish to thank the director and staff of the Isaac Newton Institute for their constant support during the 2008 turbulence programme, and Peter Bartello, David Dritschel and Rich Kerswell who co-organised the programme with great enthusiasm. They wish to thank CUP for the professionalism in preparing the book, and all the authors for their hard work and patience.

Peter Davidson
Yukio Kaneda
Katepalli Sreenivasan

Contributors

Ronald J. Adrian *Department of Mechanical and Aerospace Engineering, Arizona State University, Tempe, AZ 85287, USA*

S. Boldyrev *Department of Physics, University of Wisconsin – Madison, 1150 University Avenue, Madison, WI 53706 USA*

F. Cattaneo *Department of Astronomy and Astrophysics,* and *The Computation Institute, University of Chicago, Chicago, IL 60637, USA*

P.A. Davidson *Department of Engineering, University of Cambridge, Trumpington St., Cambridge CB2 1PZ, UK*

T. Gotoh *Department of Scientific and Engineering Simulation, Graduate School of Engineering, Nagoya Institute of Technology, Gokiso, Showa-ku, Nagoya, Japan*

Kiyosi Horiuti *Department of Mechano-Aerospace Engineering, I1-64, Tokyo Institute of Technology, 2-12-1, O-okayama, Meguro-ku, Tokyo, 152-8552, Japan*

J. Jiménez *School of Aeronautics, Universidad Politécnica, 28040 Madrid, Spain;* and *Centre for Turbulence Research, Stanford University, Stanford, CA 94305, USA*

Yukio Kaneda *Center for General Education Aichi Institute of Technology, Yakusa-cho, Toyota, 470–0392, Japan*

G. Kawahara *Graduate School of Engineering Science, Osaka University, Tayonaka, Osaka 560-8531, Japan*

Robert M. Kerr *Mathematics Institute and School of Engineering, University of Warwick, Coventry CV4 7AL, UK*

Erik Lindborg *Department of Mechanics, KTH, SE-100 44 Stockholm, Sweden*

Ivan Marusic *Department of Mechanical Engineering, University of Melbourne, Victoria, 3010, Australia*

Koji Morishita *Department of Computational Science, Graduate School of System Informatics, Kobe University, 1–1, Rokkodai-cho, Nada-ku, Kobe, 657–8501, Japan*

Jean-François Pinton *École Normale Supérieure de Lyon, Laboratoire de Physique, UMR CNRS 5672, 46, allée d'Italie, 69364 Lyon Cedex 07, France*

James J. Riley *Department of Mechanical Engineering, University of Washington, Seattle, WA 98195, USA*

Brian L. Sawford *Department of Mechanical and Aerospace Engineering, Monash University Clayton Campus, Wellington Road, Clayton, VIC 3800, Australia*

Jörg Schumacher *Institute of Thermodynamics and Fluid Mechanics, Ilmenau University of Technology, P.O. Box 100565, D-98684 Ilmenau, Germany*

Ladislav Skrbek *Department of Mechanical Engineering, Charles University, Prague, Czech Republic*

Katepalli R. Sreenivasan *Department of Physics* and *Courant Institute of Mathematical Sciences, New York University, New York, NY 10012, USA*

S.M. Tobias *Department of Applied Mathematics, University of Leeds, Leeds LS2 9JT, UK*

P.K. Yeung *Schools of Aerospace Engineering, Computational Science and Engineering, and Mechanical Engineering, Georgia Institute of Technology, Atlanta, GA 30332, USA*

1

Small-Scale Statistics and Structure of Turbulence – in the Light of High Resolution Direct Numerical Simulation

Yukio Kaneda and Koji Morishita

1.1 Introduction

Fully developed turbulence is a phenomenon involving huge numbers of degrees of dynamical freedom. The motions of a turbulent fluid are sensitive to small differences in flow conditions, so though the latter are seemingly identical they may give rise to large differences in the motions.[1] It is difficult to predict them in full detail.

This difficulty is similar, in a sense, to the one we face in treating systems consisting of an Avogadro number of molecules, in which it is impossible to predict the motions of them all. It is known, however, that certain relations, such as the ideal gas laws, between a few number of variables such as pressure, volume, and temperature are insensitive to differences in the motions, shapes, collision processes, etc. of the molecules.

Given this, it is natural to ask whether there is any such relation in turbulence. In this regard, we recall that fluid motion is determined by flow conditions, such as boundary conditions and forcing. It is unlikely that the motion would be insensitive to the difference in these conditions, especially at large scales. It is also tempting, however, to assume that, in the statistics at sufficiently small scales in fully developed turbulence at sufficiently high Reynolds number, and away from the flow boundaries, there exist certain kinds of relation which are universal in the sense that they are insensitive to the detail of large-scale flow conditions. In fact, this idea underlies Kolmogorov's theory (Kolmogorov, 1941a, hereafter referred as K41), and has been at the heart of many modern studies of turbulence. Hereafter, universality in this sense is referred to as universality in the sense of K41.

Although most of the energy in turbulence resides at large scales, most of

---

[a] This work was undertaken while both authors were at Nagoya University.
[1] This does not prevent satisfactory averages being measured, at least those belonging to small scales.

the degrees of dynamical freedom resides in the small scales. In Fourier space, for example, most of the Fourier modes are in the high-wavenumber range. Hence properly understanding the nature of turbulence at small scales is interesting, not only from the theoretical, but also from the practical point of view, because such an understanding can be expected to be useful for developing models of turbulence to properly reduce the degrees of freedom to be treated.

This chapter will review studies of the nature of turbulence at small scales. Of course, more intensive studies have been performed on this interesting subject than we can cover here; in addition, we cannot review all of the issues related to each study that we do cover. We present a review of a few topics in the light of recent progress in high resolution direct numerical simulation (DNS) of turbulence. An analysis is also made on elongated local eddy structure and statistics. An emphasis is placed upon the Reynolds number dependence of the statistics and on the difference between active and non-active regions in turbulence.

1.2 Background supporting the idea of universality

1.2.1 Kolmogorov's 4/5 law

The existence of universality in the sense of K41 has not yet been proven rigorously, but there is evidence supporting it. Among this is Kolmogorov's 4/5 law (Kolmogorov, 1941c), which is derived as a consequence of the Navier–Stokes (NS) equation governing fluid motion.

Let $\mathbf{u} = \mathbf{u}(\mathbf{x}, t)$ be an incompressible turbulent velocity field obeying the Navier–Stokes equation,

$$\frac{\partial}{\partial t}\mathbf{u} - (\mathbf{u} \cdot \nabla)\mathbf{u} - \frac{1}{\rho}\nabla p + \nu \nabla^2 \mathbf{u} + \mathbf{f}, \quad (1.1)$$

and the incompressibility condition,

$$\nabla \cdot \mathbf{u} = \mathbf{0}, \quad (1.2)$$

where ν is the kinematic viscosity, p the pressure, $\mathbf{f}$ the external force per unit mass, and ρ the fluid density.

For homogeneous and isotropic (HI) turbulence, the NS equation with the incompressibility condition (1.2) yields the Kármán–Howarth equation (Kármán and Howarth, 1938)

$$B_3^L(r) = -\frac{4}{5}\langle\epsilon\rangle r + 6\nu \frac{\partial B_2^L(r)}{\partial r} + F(r) - \frac{3}{r^4}\int_0^r \frac{\partial B_2^L(\tilde{r})}{\partial t}\tilde{r}^4 d\tilde{r}, \quad (1.3)$$

where $\langle\epsilon\rangle$ is the average of the rate of energy dissipation ϵ per unit mass, and $B_n^L(r)$ is the nth order structure function of the longitudinal velocity difference δu^L defined as

$$B_n^L(r) \equiv \left\langle \left[\delta u^L(r)\right]^n \right\rangle, \quad \delta u^L(r) \equiv [\mathbf{u}(\mathbf{x}+r\mathbf{e}) - \mathbf{u}(\mathbf{x})] \cdot \mathbf{e}, \qquad (1.4)$$

in which $\mathbf{e}$ is an arbitrary unit vector. In (1.3), F is expressed in terms of the correlation $g(r) \equiv \langle [\mathbf{u}(\mathbf{r}+\mathbf{x}) - \mathbf{u}(\mathbf{x})] \cdot [\mathbf{f}(\mathbf{r}+\mathbf{x}) - \mathbf{f}(\mathbf{x})] \rangle$. It is shown by simple algebra that

$$F(r) = \frac{6}{r^4} \int_0^r \tilde{r} G(\tilde{r}) d\tilde{r}, \quad G(r) = \int_0^r \tilde{r}^2 g(\tilde{r}) d\tilde{r}.$$

If the forcing $\mathbf{f}$ is confined only to large scales, say $\sim L$ (where the symbol $\sim$ denotes an equality up to a coefficient of order unity), and the viscosity ν is very small, then it is plausible to assume that in (1.3),

(i) the forcing term $F(r)$ is negligible at $r \ll L$,
(ii) the viscosity term works only at small scales, say $\sim \eta$, so that it is negligible at $r \gg \eta$, and
(iii) the statistics is almost stationary at small scales, so that the last term is negligible at $r \ll L$.

Under these assumptions, (1.3) yields the 4/5 law,

$$B_3^L(r) = -\frac{4}{5} \langle \epsilon \rangle r, \qquad (1.5)$$

for $L \gg r \gg \eta$.

Note that the 4/5 law (1.5) applies not only to the stationary but also to the freely-decaying case, as long as one may assume (iii), in addition to (i) and (ii), where L is to be understood appropriately, e.g., as the characteristic length scale of the energy containing eddies.

The relation (1.5) asserts that $B_3^L(r)$ is specified only by $\langle\epsilon\rangle$ and r. It holds independently of the shapes, internal structures, deformations, positions, alignments, interactions, collision and reconnection processes, etc. of small-scale eddies, however the term 'eddies' may be defined, and also of the forcing and boundary conditions outside the range $r \ll L$, as long as (i), (ii) and (iii) hold. (This doesn't mean that $B_3^L(r)$ is independent of these factors and conditions, as they may still affect $\langle\epsilon\rangle$: rather, the relation (1.5) means that their influence, if any, is only through $\langle\epsilon\rangle$.)

The relation (1.5) holds independently of these factors, just as the ideal gas laws hold independently of the shapes, internal structures, interactions, collision processes etc. of the molecules comprising the gas, and independently of the shape of the container of the gas. The relation is in this sense

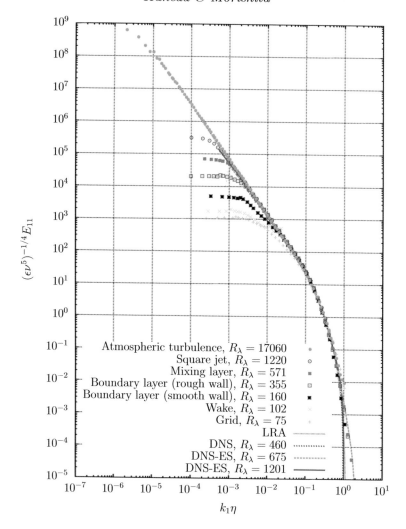

Figure 1.1 Normalized longitudinal one-dimensional energy spectrum. The data except those by DNS-ES are re-plotted from Tsuji (2009).

universal, and supports the idea of existence of universality in the sense of K41.

1.2.2 Energy spectrum

More support for the existence of universality in the sense of K41 is given by the second-order two-point velocity correlations, or, equivalently, the velocity correlation spectra observed in experiments and DNS. If the second-order moments of $\mathbf{u}(\mathbf{x} + \mathbf{r}) - \mathbf{u}(\mathbf{x})$ are universal in a certain sense at small scale,

$r \ll L$, then so are their spectra, i.e., their Fourier transforms with respect to **r** at large $k \gg 1/L$. The converse is also true. Here L is the characteristic length scale of the energy containing eddies. The universality of these statistics is at the heart of Kolmogorov's theory, K41 (see below). Experimental and DNS data have been accumulated, including those of the well-known tidal channel observation by Grant *et al.* (1962), according to which the spectra under different flow conditions overlap well at high wavenumbers under appropriate normalization (see, e.g., Monin and Yaglom, 1975).

Figure 1.1 illustrates collections of energy spectra. It shows the longitudinal one-dimensional energy spectrum E_{11} vs. $k_1\eta$ in turbulent mixing layers, boundary layers and atmospheric turbulence, at Taylor micro-scale Reynolds numbers R_λ up to $\approx 17,000$. Here E_{11} satisfies

$$\langle u_1^2 \rangle = \int_0^\infty E_{11}(k_1) dk_1,$$

in which $\langle u_1^2 \rangle$ and k_1 are respectively the mean-square fluctuation velocity and the wavenumber component in the longitudinal direction, $\eta \equiv (\nu^3/\langle \epsilon \rangle)^{1/4}$ is the Kolmogorov length scale, and $R_\lambda \equiv u'\lambda/\nu$ with $3u'^2/2 = E$ being the total kinetic energy of the fluctuating velocity per unit mass, and $\lambda \equiv (15\nu u'^2/\langle \epsilon \rangle)^{1/2}$ is the Taylor micro-scale. It also shows the spectra by DNS ($R_\lambda = 460$) by Gotoh *et al.* (2002) and a series of DNS performed on the Earth Simulator, hereafter referred to as DNS-ES, with R_λ and the number of grid points up to approximately 1,200 and 4096^3, respectively (Yokokawa *et al.*, 2002; Kaneda *et al.*, 2003). DNS with grid points as large as 4096^3 has been also used in studies of turbulence fields, see for example Donzis and Sreenivasan (2010) and Donzis *et al.* (2010).

In spite of the different flow conditions, the spectra are seen to overlap well at large $k_1\eta$, at least to the extent visible in the figure. This supports the idea that there may be certain kinds of relations which are insensitive to the details of flow conditions at large scales.

The overlap is in agreement with Kolmogorov's K41 theory, according to which $E_{11}(k_1)$ is of the form

$$E_{11}(k_1)/(\langle \epsilon \rangle \nu^5)^{1/4} = \phi_{11}(k_1\eta), \qquad (1.6)$$

in the wavenumber range $k \gg k_L \equiv 1/L$, and in particular

$$E_{11}(k_1) \approx C_K \langle \epsilon \rangle^{2/3} k_1^{-5/3}, \qquad (1.7)$$

in the inertial subrange $k_L \ll k_1 \ll k_d$, where $k_d \equiv 1/\eta$, the universal function ϕ_{11} depends only on $k_1\eta$, and where C_K is a dimensionless universal constant. In terms of the three-dimensional energy spectrum $E(k)$, (1.7) is equivalent to

$$E(k) \approx K_\text{o} \langle \epsilon \rangle^{2/3} k^{-5/3}, \quad \text{with } K_\text{o} = \frac{55}{18} C_\text{K}, \tag{1.8}$$

where k is the three-dimensional wavenumber. In physical space, (1.7) is equivalent to

$$B_2^L(r) \approx C_\text{R} \langle \epsilon \rangle^{2/3} r^{2/3}, \quad \text{with } C_\text{R} = \frac{3}{2} \Gamma\left(\frac{1}{3}\right) C_\text{K}, \tag{1.9}$$

in the range $L \gg r \gg \eta$. The scaling $r^{2/3}$ in (1.9), as well as $k^{-5/3}$ in (1.7) and (1.8), can be derived from the 4/5 law and by assuming the skewness

$$\langle [\delta u^L(r)]^3 \rangle / \langle [\delta u^L(r)]^2 \rangle^{3/2}$$

to be constant in the inertial subrange (Kolmogorov, 1941c).

The experimental and DNS data reported thus far suggest that $C_\text{K} \approx 0.5 \pm 0.05$ (see, e.g., Sreenivasan, 1995; Sreenivasan and Antonia, 1997), although a close inspection of the spectra shows that they don't agree with (1.7) in a strict sense. Readers may refer to e.g., Tsuji (2009) for a review of studies on this discrepancy.

One can consider at least two kinds of origins for the discrepancy:

(a) the inertial range intermittency;
(b) the behavior of the spectrum around $k\eta \sim 1$, where the compensated spectrum $E(k)/(\langle \epsilon \rangle^{2/3} k^{-5/3})$ shows a "bump", known as bottleneck (see, e.g., the review in Donzis and Sreenivasan, 2010).

Such a bump is known to be more noticeable in the normalized spectrum of $E(k)$ than $E_{11}(k_1)$ (see, e.g., Kaneda and Ishihara, 2007). The bottleneck does not contradict (1.6), (1.7), or (1.8) under the conditions of K41. A simple closure, Obukhov's constant skewness model (Obukhov, 1949), captures the bottleneck (see p. 404 in Monin and Yaglom, 1975, or Fig. 6.19 in Davidson, 2004).

By using data from DNS with the number of grid points up to 4096^3 and $R_\lambda \approx 1{,}000$, Donzis and Sreenivasan (2010) addressed the difficulty in estimating the constants K_o, and proposed a procedure for determining the constant by taking into account the effects (a) and (b), which yields $K_\text{o} \approx 1.58$.

The experimental value $C_K \approx 0.5 \pm 0.05$, (i.e. $K_\text{o} \approx 1.53 \pm 0.15$) is not far from the one derived by Lagrangian closure theories, the Abridged Lagrangian History Direct Interaction Approximation, ALDHIA, (Kraichnan,

1965, 1966), Strain based-LHDIA (Kraichnan and Herring, 1978; Herring and Kraichnan, 1979) and the Lagrangian Renormalized Approximation, LRA, (Kaneda, 1981; Gotoh et al., 1988), which are fully consistent with K41, and can be obtained by the lowest-order truncations of systematic renormalized perturbative expansions without introducing any ad hoc adjusting parameter. They give $K_o \approx 1.77, 2.0$ and 1.72, respectively. The spectrum for $R_\lambda \to \infty$ by the LRA is also plotted in Fig. 1.1.

1.3 Examination of the ideas underlying the 4/5 law

1.3.1 Energy dissipation rate at $\nu \to 0$

Although plausible in itself, Kolmogorov's 4/5 law is derived on the basis of several assumptions. Among them is one concerned with the average $\langle \epsilon \rangle$ in the limit $\nu \to 0$, or equivalently in the limit of the Reynolds number $Re \equiv u'L/\nu \to \infty$. It is assumed that $\langle \epsilon \rangle$ is not so small that the first term is dominant on the right hand side of (1.3) for $L \gg r \gg \eta$, in the limit $\nu \to 0$. This assumption is concerned with a fundamental question on the smoothness of the turbulent field:

(a) does $\langle \epsilon \rangle$ remain non-zero finite;
(b) or does $\langle \epsilon \rangle \to 0$ in the limit?

If (a) is true, we may safely assume that the first term is dominant on the right-hand side of (1.3) under the conditions (i), (ii) and (iii) used in deriving (1.5). Since $\langle \epsilon \rangle = 2\nu \langle e_{ij} e_{ij} \rangle$, (a) implies that $\langle e_{ij} e_{ij} \rangle \to \infty$, i.e., the mean square of at least one component of the velocity gradient tensor diverges as $\nu \to 0$, where $e_{ij} \equiv (\partial u_i/\partial x_j + \partial u_j/\partial x_i)/2$, and we use the summation convention for repeated indices. The condition (a) also implies that the dissipation in the limit $\nu \to 0$ is different from that in the ideal fluid with $\nu = 0$, in which $\langle \epsilon \rangle$ must be zero, and in this sense the limit is singular. Such a singularity is well known for flow past a solid body: neglecting it gives rise to D'Alembert's paradox.

Let D_ϵ be the normalized dissipation rate defined by $D_\epsilon \equiv L \langle \epsilon \rangle / u'^3$. Rigorous upper bounds for D_ϵ have been derived; for example, it has been shown that, in body-forced turbulence,

$$D_\epsilon \leq \frac{a}{Re} + b, \tag{1.10}$$

where the prefactors a and b depend only on the functional shape of the body force and not on its magnitude or any other length scales in the force, the domain or the flow, and $\langle \epsilon \rangle$ and L are to be understood as the time-average of ϵ and the longest length scale in the applied forcing function, respectively (Doering and Foias, 2002; Doering, 2009). As regards the lower band, an application of Poincaré's inequality gives $D_\epsilon \geq 4\pi^2/(\alpha^2 Re)$, as noted by Doering and Foias (2002), where $\alpha = L'/L$ is the ratio of the domain size L' to L. However, these inequalities are not sufficient to answer the question of whether (a) is true or not.

Studies have been made to address this question through DNS. The DNS data with R_λ up to approximately 200 compiled by Sreenivasan (1998), which include both decaying and forced turbulence data, shows that D_ϵ decreases with R_λ for $R_\lambda < 200$ or so, while adding the DNS-ES data up to $R_\lambda \approx 1200$ strongly suggests that D_ϵ remains finite and independent of Re in the limit of $Re \to \infty$ (Kaneda et al., 2003).

The right-hand side of (1.10) decreases rapidly with Re at small Re, then decreases slowly at larger Re, and approaches a constant as $Re \to \infty$. The latter is consistent with the expectation that $D_\epsilon \to$ constant as $Re \to \infty$, and the former, or the first term in (1.10), is the result valid for asymptotically small Re, (Sreenivasan, 1984; Doering and Foias, 2002). By using the rigorous relation $Re = D_\epsilon R_\lambda^2/15$ (see (1.12) below), (1.10) can be cast into the form

$$D_\epsilon \leq \frac{b}{2}\left(1 + \sqrt{1 + \frac{4a}{b^2}\frac{1}{R_\lambda'^2}}\right), \qquad (1.11)$$

where $R_\lambda' \equiv R_\lambda/\sqrt{15}$. Donzis et al. (2005) showed that the relation (1.11), with the inequality replaced by equality, provides a good fit for the Reynolds number dependence of D_ϵ in DNS. Another expression of the Reynolds number dependence of D_ϵ was proposed by Lohse (1994) on the basis of physical assumptions. Readers may refer to Tsinober (2009) and references cited therein for more DNS and experimental studies on D_ϵ.

Simple algebra gives

$$\frac{L}{\eta} = D_\epsilon^{1/2} Re^{3/4} = 15^{-3/4} D_\epsilon^{5/4} R_\lambda^{3/2} \propto Re^{3/4}, \qquad (1.12)$$

where the last proportionality is satisfied if D_ϵ is independent of Re.

1.3.2 Influence of finite Re, L/r and r/η

The 4/5 law assumes that $Re, L/r$ and r/η are sufficiently large. The same is true in K41, which assumes $Re, L/r \to \infty$. Even if these theories are correct, they themselves do not provide any quantitative answer to the question of how large $Re, L/r$ and r/η must be for the theories to achieve any required accuracy. In this respect (1.3) is interesting, because it is an exact relation (subject to the homogeneity and isotropy conditions), and may give a quantitative answer to such a question.

As regards the forcing term in (1.3), it can be shown by simple algebra that $F(r)/(\langle\epsilon\rangle r) \approx C_{\mathrm{f}}(r/L)^2$, provided that the forcing is exerted only at low wavenumber modes, as in DNS-ES, where C_{f} is a dimensionless constant of order unity which may depend on the method of forcing (see, for example (Gotoh et al., 2002) and (Kaneda et al., 2008)).

The viscosity term is not easy to evaluate rigorously, but a simple model based on (1.8) may be applicable as a first approximation (Lindborg, 1999; Moisy et al., 1999; Qian, 1999; Lundgren, 2003; Davidson, 2004; Kaneda et al., 2008, the last of which is referred to below as KYI). Substituting (1.9) into the viscosity term of (1.3) and assuming statistical stationarity gives

$$\Delta(r) \equiv \frac{B_3^L(r)}{\langle\epsilon\rangle r} + \frac{4}{5} = C_{\mathrm{v}}\left(\frac{r}{\eta}\right)^{-4/3} + C_{\mathrm{f}}\left(\frac{r}{L}\right)^2, \qquad (1.13)$$

where $C_{\mathrm{v}} = (44/9)C_{\mathrm{R}}$. Although this model is not rigorous, especially outside the inertial subrange, it was confirmed to be in good agreement with DNS-ES at large R_λ (> 400 or so) and $L \gg r \gg \eta$ (KYI; Ishihara et al., 2009).

Equation (1.13) implies that the difference $\Delta(r)$ takes its minimum $\Delta_{\min}$ as function of r at $r = r_m$, where

$$\Delta_{\min} \propto R_\lambda^{-6/5}, \quad \frac{r_m}{\eta} \propto \left(\frac{L}{\eta}\right)^{3/5} \propto R_\lambda^{9/10}, \qquad (1.14)$$

and the last proportionality is satisfied when (1.12) holds (see, for example, (Moisy et al., 1999; Qian, 1999) and KYI).

For decaying turbulence in the absence of external forcing $\mathbf{f}$, one may insert $F = 0$ in (1.3). But in this case its last term is not zero because turbulence cannot be stationary without external forcing. Lindborg (1999) showed that a simple model based on (1.9) and K–ϵ type modeling à la Kolmogorov (1941b) gives

$$\Delta(r) = C_{\mathrm{v}}\left(\frac{r}{\eta}\right)^{-4/3} + C_n R_\lambda^{-1}\left(\frac{r}{\eta}\right)^{2/3}, \qquad (1.15)$$

(C_n is a constant), so that $\Delta(r)$ takes its minimum $\Delta_{\min}$ at $r = r_{\mathrm{m}}$, where

$$\Delta_{\min} \propto R_\lambda^{-2/3}, \quad \frac{r_{\mathrm{m}}}{\eta} \propto R_\lambda^{1/2}. \qquad (1.16)$$

The approximation (1.15) can be derived also by a method of matched asymptotic expansion (Lundgren, 2003).

The decay of $\Delta_{\min}$ in proportion to $R_\lambda^{-6/5}$ in (1.14) for forced turbulence, and in proportion to $R_\lambda^{-2/3}$ in (1.16) for decaying turbulence, is in agreement with the data compiled by Antonia and Burattini (2006). The estimates (1.15) and (1.16) are consistent with those by Qian (1999). Decay in proportion to $R_\lambda^{-6/5}$ in forced turbulence was also confirmed to be in agreement with the DNS-ES data (KYI). For experimental or DNS data of $B_3^L(r)$, or for the comparison of (1.13) or (1.15) with experimental or DNS data, readers may refer to, for example, (Watanabe and Gotoh, 2004; Antonia and Burattini, 2006) and references cited therein, in addition to KYI.

Sreenivasan and Bershadskii (2006) proposed an approximation for $\Delta(r)$ based on an expansion of $\Delta(r)$ in powers of $\log(r/r_m)$, and argued that experimental and DNS data fit well to $r_m/\eta \propto R_\lambda^\mu$ with $\mu = 0.73 \pm 0.05$. The difference between this and the exponents in (1.14) and (1.16) is not surprising in view of the fact that the data set used for fitting includes data both from decaying turbulent flows and from forced turbulent flows.

1.3.3 Spectral space

The NS equation for homoegeneous and isotropic (HI) turbulence gives the spectral equation,

$$\frac{\partial}{\partial t}E(k) = T_E(k) - 2\nu k^2 E(k) + \hat{F}(k), \tag{1.17}$$

where $E(k)$ is the energy spectrum, $T_E(k)$ is the energy transfer function due to the nonlinear interaction, and $\hat{F}(k)$ represents the energy input due to the forcing **f**.

Integrating (1.17) with respect to k from K to ∞ and replacing K with k gives

$$\Pi(k) = \langle \epsilon \rangle - 2\nu \int_0^k p^2 E(p)dp - \int_k^\infty \hat{F}(p)dp + \int_k^\infty \frac{\partial}{\partial t}E(p)dp, \tag{1.18}$$

where $\Pi(k)$ is the energy flux across the wavenumber k, and is defined by

$$\Pi(k) \equiv \int_0^k T_E(p)dp,$$

where we have used

$$\int_0^\infty T_E(k)dk = 0, \quad \langle \epsilon \rangle = 2\nu \int_0^\infty k^2 E(k)dk.$$

Under the assumptions similar to those used in deriving (1.5), i.e., (i), (ii) and (iii), above, (1.18) gives

$$\Pi(k) = \langle \epsilon \rangle, \qquad (1.19)$$

for $1/L \ll k \ll 1/\eta$. The relation (1.19) is equivalent to the 4/5 law (1.5). Readers may refer to Frisch (1995) for some details on the relation between (1.5) and (1.19).

An estimate similar to (1.13) may be obtained by substituting (1.8) into the viscosity term of (1.18). As regards the forcing term, in DNS such as DNS-ES where the forcing is only at wavenumbers $\leq k_\mathrm{C}$, the third term on the right-hand side for $k > k_\mathrm{C}$ may be neglected. In this case, (1.18) yields

$$\Delta_k(k) \equiv \frac{\Pi(k)}{\langle \epsilon \rangle} - 1 = -\frac{3}{2} K_\mathrm{o} (k\eta)^{4/3}, \qquad (1.20)$$

for $k_\mathrm{C} < k \ll 1/\eta$ in stationary turbulence. Equation (1.20) has been confirmed to be in good agreement with DNS-ES at large R_λ and $k_\mathrm{C} < k < k_\mathrm{m}$ with $k_\mathrm{m} \approx 0.1/\eta$ (Ishihara et al., 2009).

1.3.4 Where is the inertial subrange situated?

The existence of a sufficiently wide range satisfying (1.19) is a basic prerequisite of various theories or models of energy cascade or intermittency in the inertial subrange of turbulence at high Re. From the viewpoint of these theories, it is legitimate to require that $\Delta_k(k)$, or its equivalent $\Delta(r)$, be sufficiently small within that range, i.e., its smallness may be regarded as a necessary condition for identifying a range as "the inertial subrange".

In comparing theories with statistics derived from experimental and/or DNS data, such as the discrepancy of the so-called anomalous scaling from K41, one shouldn't confuse statistics or discrepancies due to the finiteness of Re, L/r, r/η, hereafter referred to as Type A, with those due to something else such as intermittency, hereafter referred to as Type B; unlike Type A, Type B should remain finite in the limit $Re \to \infty$. If $\Delta(r)$ is not small enough, this implies that Type A is not small enough, and so might therefore mask Type B, i.e., it may be difficult to distinguish Type B from Type A. Similar but slightly different concerns have been posed elsewhere, e.g., by Qian (1997, 1998, 1999) and Lindborg (1999) .

From this viewpoint, it seems reasonable to define the inertial subrange as the range, say the δ-subrange, in which $\Delta(r)$ or $\Delta_k(k)$ is sufficiently small – that is, smaller than a certain threshold, say δ. In the framework of (1.13) for forced turbulence, to the leading order for small η/L, this requirement is equivalent to insisting that both the viscous and forcing terms are less than δ. The latter and (1.11) imply that the δ-subrange is identified as the range of r such that $r_{\min} < r < r_{\max}$, where

$$r_{\min} \equiv \eta C_{\mathrm{v}}^{3/4} \delta^{-3/4}, \quad r_{\max} \equiv L C_{\mathrm{f}}^{-1/2} \delta^{1/2},$$

so that

$$\frac{r_{\max}}{r_{\min}} = C \frac{L}{\eta} \delta^{5/4}, \quad \text{(where } C = C_{\mathrm{f}}^{-1/2} C_{\mathrm{v}}^{-3/4}\text{).} \tag{1.21}$$

This implies that the ratio $r_{\max}/r_{\min}$ decreases in proportion to $\delta^{5/4}$. If one uses $C_{\mathrm{f}} \approx 0.42$, $C_{\mathrm{v}} \approx 9.9$ and $D \approx 0.41$ from the data reported in KYI, then (1.12) and (1.21) give

$$\frac{r_{\max}}{r_{\min}} \approx 0.12 \frac{L}{\eta} \delta^{5/4} \approx 5.0 \times 10^{-3} R_{\lambda}^{3/2} \delta^{5/4}.$$

Suppose, for example, that $\delta = 0.1$. In this case, the ratio $r_{\max}/r_{\min}$ is approximately $1/150$ of L/η. If one requires not only $\delta = 0.1$ but also $r_{\max}/r_{\min} > 10$, then one has $R_{\lambda} > 1100$.

As regards decaying turbulence, simple algebra based on (1.15) shows that the δ-subrange is given by $r_{\min} < r < r_{\max}$, where

$$\frac{r_{\max}}{r_{\min}} = C' R_{\lambda}^{3/2} \delta^{9/4}, \quad \text{(where } C' = C_{\mathrm{n}}^{-3/2} C_{\mathrm{v}}^{-3/4}\text{).} \tag{1.22}$$

A comparison of (1.14) and (1.21) with (1.16) and (1.22) shows that the realization of the inertial range through experiments or DNS is more difficult in decaying turbulence than in forced turbulence, not only because the decrease of the difference $\Delta_{\min}$ with R_{λ} is slower in decaying turbulence, as noted after (1.16), but also because the decrease of the ratio $r_{\max}/r_{\min}$ with δ is faster.

In spectral space, if we define the range $k_{\min} < k < k_{\max}$ as that where $\Delta_k(k)$ is less than a small threshold, then (1.20) gives

$$\frac{k_{\max}}{k_{\min}} \propto R_{\lambda}^{3/2}, \tag{1.23}$$

for forced turbulence, provided that (1.12) holds. It is not difficult to apply a similar idea to decaying turbulence and to show that (1.23) holds also for decaying turbulence.

The exponent $3/2$ in (1.23) is consistent with (1.21) and (1.22), provided that (1.12) holds. It is larger, however, than either of the exponents for the inertial range of the energy spectrum $E_{11}(k_1)$ suggested by experiments, namely, ≈ 1 by Donnelly and Sreenivasan (1998) and ≈ 0.8 by Tsuji (2009).

As regards the origin of the difference, as well as the possibility of inaccuracy of the theory and/or experimental measurements, and the difference of flow conditions such as the forcing conditions, one must also consider the possibility that the influence of finite Re, L/r and/or r/η may depend on the quantity to be measured, i.e., the influence on B_3^L may be different from that on E_{11}, so that the width and position of the "inertial subrange" of B_3^L may be different from those of E_{11}.

It is worthwhile mentioning here a few experimental studies. By measuring velocity in the atmospheric boundary layer at R_λ between 10,000 and 20,000, Sreenivasan and Dhruva (1998) argued that even at these high Reynolds numbers, the structure functions do not scale unambiguously, and concluded that it is difficult to discuss the scaling effectively without first understanding quantitatively the effects of finite shear and finite Reynolds number. Kurien and Sreenivasan (2000) presented a method of extraction of anisotropic exponents that avoids mixing with the isotropic counterparts, and argued that the anisotropic effects diminish with decreasing scale, although much more slowly than previously thought. On the basis of the analysis of high Reynolds number data in field experiments, Kholmyansky and Tsinober (2009) argued that there are a substantial number of strong events contributing significantly to the statistics in the nominally-defined inertial subrange, for which viscosity (dissipation) is of utmost importance at high Reynolds number, and that the conventionally-defined inertial subrange is not a well-defined concept.

1.3.5 Anisotropy, non-stationarity, inhomogeneity

The 4/5 law also assumes that the turbulence is isotropic. However, the boundary conditions and forcing generally do not allow the field to be statistically isotropic at large scales. This naturally raises the question of how strongly or weakly, the anisotropy at large scales penetrates into small scales. A data analysis of the DNS-ES field, which is generated under periodic boundary conditions with forcing at low wavenumber modes, shows that the anisotropy decreases with the scale, but the decrease may be slow. For example, the difference

$$\delta_i(r) \equiv B_3^L(r\mathbf{e}_i) - \langle B_3^L(\mathbf{r}) \rangle_r$$

decays with r/L, but this decay is slow, and approximately only algebraic in the inertial subrange, where $B_3^L(r\mathbf{e})$ is defined by (1.4) but with r in $B_n^L(r)$ and $\delta u^L(r)$ replaced by $r\mathbf{e}$, $\mathbf{e}_i$ is the unit vector in the ith direction, i.e., $\mathbf{e}_1 = (1, 0, 0)$ etc., and $\langle \ \rangle_r$ denotes the average over all the directions of

r with given r, so that $\delta_i(r)$ must be zero in strictly isotropic turbulence (KYI).

There have been extensive studies on the anisotropy, not only on $B_3^L(r)$ but also on the other statistics. Readers may refer, for example, to the review by Biferale and Procaccia (2005), especially on the anisotropy scaling, as well as the review by Tsinober (2009) and references cited therein.

The 4/5 law also assumes that the non-stationarity of turbulence is negligible. In forced turbulence, the statistics are approximately but not exactly stationary. For example, although the total kinetic energy is kept almost constant independent of time, the other statistics in DNS-ES are not exactly stationary even after an initial transient stage. According to KYI, the influence of non-stationarity on the last term in the Kármán–Howarth equation (1.3) is not large, yet it may be not that small, and it has non-simple scale dependence. Another assumption incorporated into the 4/5 law is homogeneity; studies have been undertaken, e.g., by Danaila *et al.* (2002), on the influence of inhomogeneity, but much work on this subject remains to be done.

The questions discussed above on the influence of anisotropy, non-stationarity and inhomogeneity are concerned with statistics in a "local" sense. In a global sense, however, their influence may be small. In forced turbulence, for example, if one considers time averages over a sufficiently long time interval rather than temporal statistics, then the influence of non-stationarity must be weak. For DNS fields under periodic boundary conditions, one need not be worried about the inhomogeneity if one takes the volume average over the fundamental periodic domain. Similarly, if one considers the average over all directions, then the influence of anisotropy can be weak (see, e.g., Nie and Tanveer, 1999; Taylor *et al.*, 2003, KYI). The constraints of homogeneity and isotropy may be relaxed by considering averages over space, orientation and time. Readers may refer to Duchon and Robert (2000) and Eyink (2003) for discussion of the 4/5 law in a locally-averaged sense.

1.4 Intermittency of dissipation rate and velocity gradients

1.4.1 Intermittency – Landau's remark

Although energy spectra observed in experiments and DNS generally support the prediction by K41, as Fig. 1.1 shows, all is nevertheless not well with K41. Aside from the differences between K41's predictions and the results of experiments and DNS, particularly regarding high-order moments, K41 also has a problem in its logic, as pointed out by Landau (Landau and

Lifshitz, 1987, the first English edition was published in 1959). In view of Landau's comments, one may question the universality of ϕ_{11} in (1.6), because $\langle f(\epsilon)\rangle$ is not in general equal to $f(\langle\epsilon\rangle)$, (Kraichnan, 1974; Monin and Yaglom, 1975, the latter was originally published in Russian in 1965). The 4/5 law is free from this question, because $\langle\epsilon\rangle^n = \langle\epsilon^n\rangle$ for $n = 1$.

Landau's remark suggests the importance of taking into account the intermittency, or fluctuations in time and space, of the dissipation ϵ. This remark and related considerations on the small-scale statistics led Kolmogorov (1962) and Obukhov (1962) to modify K41 and to propose the log-normal theory. The hypotheses underlying the theory have, however, also been questioned (see, e.g., Novikov, 1971; Mandelbrot, 1974; Kraichnan, 1974), and alternative models or theories such as She & Leveque's model (She and Leveque, 1994) and fractal and multi-fractal models have been proposed by other authors. Many of the later intermittency models share much in common with the log-normal model. The link between these models and the NS dynamics governing fluid motion appears to still be missing, however. It is outside the scope of this chapter to provide a review of these models. Readers may refer, for example, to Nelkin (1994), Frisch (1995), Sreenivasan and Antonia (1997) and Davidson (2004), for reviews of these models and comparison of their predictions with experiments and DNS.

1.4.2 Statistics of energy dissipation rate

It has long been known that turbulence exhibits strong intermittency at small scales. Among the quantities showing intermittency is the energy dissipation rate ϵ.

Figure 1.2 presents an example showing the statistics of ϵ. It shows the probability function (PDF) of $\log \epsilon$ in DNS-ES at $R_\lambda = 429$. Note that the PDF is based on direct measurement of the "true" ϵ rather than on surrogates; in contrast to those obtained through experiments and field observations, this PDF is free from Taylor's frozen hypothesis. It is remarkable that, though questions may be raised concerning the log-normal hypothesis, the PDF is very close to log-normal distribution, for $|\log \epsilon - m_\epsilon|/\sigma_\epsilon$ up to 4, where m_ϵ and σ_ϵ are the mean and the standard deviation of $\log \epsilon$. This is consistent with experiments which show that the PDF fits well to the log-normal theory unless $|\log \epsilon - m_\epsilon|/\sigma_\epsilon$ is not very large (see, e.g., Monin and Yaglom, 1975). Yeung et al. (2006) noted that the PDFs of $\log \epsilon$, $\log \zeta$ and $\log \phi$, where $\zeta \equiv \nu\omega^2$, ($\omega \equiv |\boldsymbol{\omega}|$, $\boldsymbol{\omega} \equiv \nabla \times \mathbf{u}$), $\phi \equiv \nu(\partial u_i/\partial x_j)(\partial u_i/\partial x_j)$, in their high resolution DNS with 2048^3 grid points and $R_\lambda = 680$, display departures from the symmetry between positive and negative $\log X - \langle \log X \rangle$,

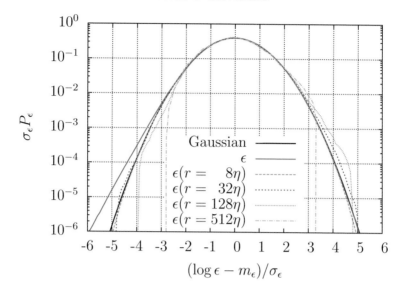

Figure 1.2 PDF of $\log \epsilon$ vs. $\log \epsilon / \sigma_\epsilon$, and PDF of $\log \epsilon(r)$ vs. $\log \epsilon(r)/\sigma_r$ for $r/\eta = 8, 32, 128, 512$ in DNS-ES at $R_\lambda = 429$, where m_ϵ and σ_ϵ are the mean and the standard deviations of $\log \epsilon$ or $\log \epsilon(r)$, respectively.

where $X = \epsilon$, ζ or ϕ. Such an asymmetry is also seen in Fig. 1.2, in particular at large $|\log \epsilon - m_\epsilon|/\sigma_\epsilon$. They also argued that the departures from log-normality become weaker at increasing Reynolds number.

Figure 1.2 also presents the PDF of $\log \epsilon(r)$ for $r/\eta = 8$–512, which plays a key role in the log-normal theory. Here $\epsilon(r) \equiv \epsilon(r, \mathbf{x}, t)$ is the local average at time t of ϵ over a sphere of radius r centred at $\mathbf{x}$: we will omit writing $\mathbf{x}$ and t on occasion. The PDF is seen to have wide tails that are wider for smaller r, i.e., the intermittency is stronger for smaller r.

Among the representative models of small-scale intermittency are those based on the idea of self-similar multiplicative random process (see, e.g., the reviews by Meneveau and Sreenivasan, 1991; Nelkin, 1994; Sreenivasan and Stolovitzky, 1995; Frisch, 1995; Sreenivasan and Antonia, 1997; Jiménez, 2000, 2001; Jiménez et al., 2001). This idea was first applied to turbulence by Gurvich and Yaglom (1967). Let ϵ_n be the average of ϵ over a volume of size $r_n = r_0 a^n$ with $0 < a < 1$, and α_n be the breakdown coefficients defined by $\alpha_n \equiv \epsilon_{n+1}/\epsilon_n$. Gurvich & Yaglom assumed that a range exists where

(1) the statistics of the coefficient α_n are independent of n, i.e., the scale (scale similarity), and

(2) α_n are statistically independent of each other.

The PDFs of the breakdown coefficients of coarse-grained surrogate dissipation were measured by Van Atta and Yeh (1975) and by Chhabra and Sreenivasan (1992). These authors argued that the PDFs are universal in the inertial subrange, in agreement with (1). On the other hand, on the basis of experimental data, Sreenivasan and Stolovitzky (1995) argued that while (1) holds to a good approximation, (2) is questionable. The breakdown coefficients were also measured by Pedrizzetti *et al.* (1996), who showed that the scale similarity is approximately valid, but a logarithmic correction is depicted in their data. Jiménez (2001) and Jiménez *et al.* (2001) argued on the basis of original measurements as well as those published in the two papers (Van Atta and Yeh, 1975; Chhabra and Sreenivasan, 1992) that the PDFs vary with scale, in contrast to (1). The idea of multiplicative random processes has been extended to the velocity field by e.g., Naert *et al.* (1997), who have developed intermittency models based on Fokker–Planck equations.

An analysis of the data of $\epsilon_n \equiv \epsilon(r_n)$ based on direct measurements of $\epsilon(r)$ rather than on surrogates was made by using the DNS-ES data with R_λ up to 429 (Kaneda and Morishita, 2007). The analysis shows that:

(i) the PDFs of the coefficients vary with scale, in contrast to (1);
(ii) the statistical dependence between the α_n's is not negligible, in contrast to (2), at least within the analysis with $R_\lambda < 732$ and $a > 2^{1/4}$.

The development of a theory consistent with these observations is awaited with interest.

The degree of intermittency can also be seen in the concentration of ϵ. According to Hosokawa and Yamamoto (1989), only 0.82% of the total volume accounts for 14% of the energy dissipation in DNS at $R_\lambda \approx 100$. Figure 1.3 shows the integrated PDFs

$$P_1(x) \equiv \int_x^\infty P(\epsilon)d\epsilon, \quad P_2(x) \equiv \frac{1}{\langle\epsilon\rangle}\int_x^\infty \epsilon' P(\epsilon')d\epsilon', \qquad (1.24)$$

vs. $x = \epsilon/\sigma_\epsilon$ in DNS-ES with R_λ up to 429, where P is the PDF of ϵ, and σ_ϵ is the standard deviation of ϵ. It is seen, for example, that at $R_\lambda = 429$, $P_1(2) \approx 0.05$ and $P_2(2) \approx 0.33$. This implies that only 5% of the total volume accounts for 33% of the energy dissipation. The figure also shows that the concentration of the energy dissipation is stronger at larger R_λ.

1.4.3 Statistics of velocity gradients

Intermittency at small scales in turbulence is reflected not only in the statistics of ϵ but also in those of velocity gradients. Among the statistics are the

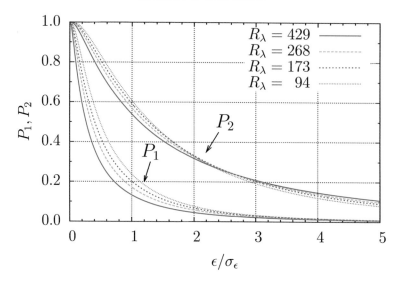

Figure 1.3 Integrated PDFs P_1 and P_2 defined by (1.24).

skewness and flatness factors of the longitudinal velocity gradients, defined as

$$S \equiv \left\langle \left(\frac{\partial u_1}{\partial x_1}\right)^3 \right\rangle \bigg/ \left\langle \left(\frac{\partial u_1}{\partial x_1}\right)^2 \right\rangle^{3/2}, \quad F \equiv \left\langle \left(\frac{\partial u_1}{\partial x_1}\right)^4 \right\rangle \bigg/ \left\langle \left(\frac{\partial u_1}{\partial x_1}\right)^2 \right\rangle^2.$$

As regards the skewness S, it is in general negative in fully developed turbulence. The data up to $R_\lambda \approx 200$ show a trend suggesting that S would tend to a constant independent of R_λ as R_λ increases (e.g., the data compiled by Sreenivasan and Antonia, 1997 and Ishihara et al., 2007). This trend cannot be extrapolated to higher R_λ, however: experiments as well as the DNS-ES data for larger $R_\lambda (> 200)$ show that S has a weak dependence on R_λ at larger R_λ. A least-square fit of the DNS-ES data gives $S \propto R_\lambda^\mu$ with $\mu = 0.11 \pm 0.01$, which is in agreement with the estimate $\mu \approx 0.11$ proposed by Hill (2002) on the basis of the experimental data for $R_\lambda > 400$ of Antonia et al. (1981), and is not far from the estimate $\mu = 0.09$, proposed by Gylfason et al. (2004).

In incompressible HI turbulence, any third-order moment of the velocity gradient tensor $g_{ij} = (\partial u_i/\partial x_j)$ can be expressed in terms of $\langle (\partial u_1/\partial x_1)^3 \rangle$; in particular, the average of the enstrophy production $\alpha \equiv e_{ij}\omega_i\omega_j$ is given by

$$\langle \alpha \rangle = -4 \langle abc \rangle = -\frac{35}{2} \left\langle \left(\frac{\partial u_1}{\partial x_1} \right)^3 \right\rangle, \qquad (1.25)$$

where a, b and c are the eigenvalues of the tensor e_{ij} satisfying $a+b+c=0$ (Townsend, 1951; Betchov, 1956; Champagne, 1978). Hence the negativity of S is equivalent to the positivity of the average of enstrophy production. Since $S = \langle \alpha \rangle = 0$, if the PDF of the velocity field is Gaussian or if the flow is two-dimensional, the negativity of S is a manifestation of the non-normality of the PDF and the three-dimensionality of the flow. The enstrophy production α plays key roles in vortex dynamics (see, e.g., Davidson, 2004; Tsinober, 2009).

As regards the flatness factor F, experimental data compiled by Sreenivasan and Antonia (1997) and the DNS-ES data (Ishihara et al., 2007) both show that F has a weak dependence on R_λ. Experiments by Tabeling et al. (1996) show a transition in its R_λ-dependence at $R_\lambda \approx 700$. Such a transition is not observed in the DNS-ES data (Ishihara et al., 2007), however, nor was it observed in experiments by Gylfason et al. (2004). The DNS-ES data agree well with atmospheric measurements of Antonia et al. (1981). A least-square fitting of the DNS-ES data gives $F \propto R_\lambda^\mu$ with $\mu = 0.34 \pm 0.03$, which is in agreement with the estimate $\mu \approx 0.31$ proposed by Hill (2002), and is not far from the experimental result of Gylfason et al. (2004) that $\mu = 0.39$ for the range $100 \leq R_\lambda < 10,000$ (Ishihara et al., 2007).

In incompressible HI turbulence, any fourth-order moment of the velocity gradients g_{ij} is expressed in terms of the following four rotational invariants (Siggia, 1981; Hierro and Dopazo, 2003):

$$I_1 \equiv \langle s^4 \rangle, \quad I_2 \equiv \langle s^2 \omega^2 \rangle, \quad I_3 \equiv \langle \omega_i e_{ij} \omega_k e_{kj} \rangle, \quad I_4 \equiv \langle \omega^4 \rangle,$$

where $s^2 \equiv e_{ij}e_{ij}$. The R_λ-dependence of these invariants was studied through DNS with R_λ up to ≈ 83 (Kerr, 1985) and through experiments with R_λ up to ≈ 100 (Zhou and Antonia, 2000). The former suggested that the invariants scale with R_λ in different ways, while the latter suggested that they scale similarly. Data obtained through DNS-ES with R_λ up to ≈ 680, on the other hand, suggests that they scale similarly when $R_\lambda > 100$, but not when $R_\lambda < 100$, (Ishihara et al., 2007). Thus the scalings provide another example in which a trend observed at lower R_λ cannot be extended to higher R_λ.

The results discussed in this subsection imply that the PDFs of the velocity gradients normalized by their standard deviations do not converge to a single curve, as $Re \to \infty$.

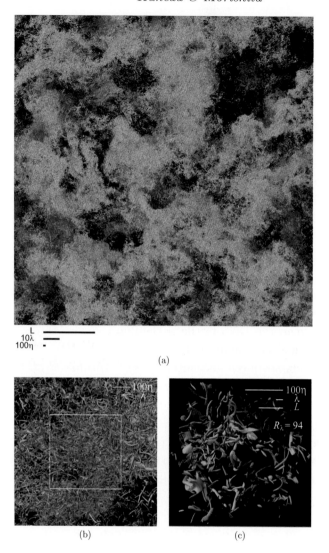

Figure 1.4 Equivorticity surface showing the intense vorticity regions with $\omega > \langle\omega\rangle + 4\sigma_\omega$ in DNS-ES, where $\langle\omega\rangle$ and σ_ω are the mean and the standard deviation of ω, respectively. (a) In DNS-ES with $R_\lambda = 1131$, $N^3 = 4096^3$ grid points, and $k_{\max}\eta \approx 1$. (b) A zoomed-in view of a region in DNS-ES with $R_\lambda = 675$, $N^3 = 4096^3$, and $k_{\max}\eta \approx 2$. (c) In DNS-ES with $R_\lambda = 94$, $N^3 = 256^3$, and $k_{\max}\eta \approx 2$. The scales L, 10λ and 100η are shown by solid lines. (a) and (b) are from Ishihara *et al.* (2009) and (c) is from Ishihara *et al.* (2007).

1.4.4 Visualization—worms/sinews/clusters

Visualization of the field through DNS helps us get some intuitive idea of the turbulent field underlying the statistics. One of the striking features of the structure of the turbulence observed in many DNS (e.g., Siggia, 1981; Kerr, 1985; Hosokawa and Yamamoto, 1989; She *et al.*, 1990; Vincent and Meneguzzi, 1991; Kida, 1993; Jiménez *et al.*, 1993; Miyauchi and Tanahashi, 2001; Kaneda and Ishihara, 2006; Ishihara *et al.*, 2007) is the emergence of a high-enstrophy region concentrated in tube-like structures. The tubes have been reported to have a cross-sectional radius of approximately 5η–10η. Yamamoto and Hosokawa (1988) suggested the term "worms" for these tubes, while the alternative term "sinews" has been suggested by Küchemann (1965), Saffman and Baker (1979) and Moffatt *et al.* (1994).

Figures 1.4(a) and (b) present two such visualizations and show regions of high enstrophy in DNS-ES at $R_\lambda = 1131$ and 675, respectively, with $N^3 = 3096^3$ grid points. They are to be compared with Fig. 1.4(c), which shows a DNS field with lower R_λ, $R_\lambda = 94$ and $N^3 = 256^3$. (Note that a high threshold has been set for ω in the figures.) Fig. 1.4(c) might give the impression that the field is dominated by relatively few strong eddies, some of which span the whole domain. However, Fig. 1.4(a) shows that this is not the case at high R_λ: rather, the field consists of a huge number of small eddies. It is also evident that the field consists of voids and clusters of eddies at a wide scale range. The characteristic length scale of each of these clusters is much larger than 100η, and their structures are distinctly different from those of the small eddies.

1.5 Local structure

1.5.1 Radii of curvatures of stream and vorticity lines

Figure 1.4 shows that vorticity is high in some regions and low in others. It would not be surprising if low-vorticity and high-vorticity regions were different in terms of their structure and statistics. This can be confirmed intuitively: see, for example, Figs. 1.5(a) and 1.5(b), which show stream lines near $\mathbf{x}_s$ and $\mathbf{x}_w$ in a DNS-ES run with $R_\lambda = 263$, where $\mathbf{x}_s$ and $\mathbf{x}_w$ are respectively the positions at which ω^2 reaches its maximum and minimum in the DNS field.

A simple quantitative measure characterizing the difference observed between Figs.1.5(a) and (b) is given by R_u, the radius of curvature of the stream lines. It can be shown by simple algebra (Moffatt, private communi-

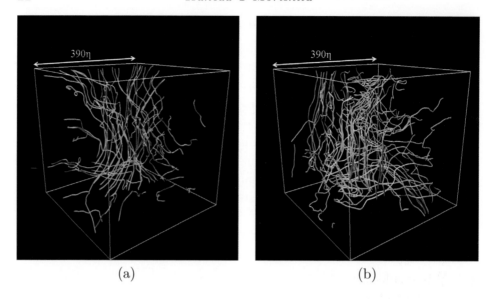

Figure 1.5 Snapshots of stream lines (a) near $\mathbf{x}_s$ and (b) near $\mathbf{x}_w$, where $\mathbf{x}_s$ and $\mathbf{x}_w$ are the positions at which ω^2 reaches its maximum and minimum in the field, respectively. $R_\lambda = 263$.

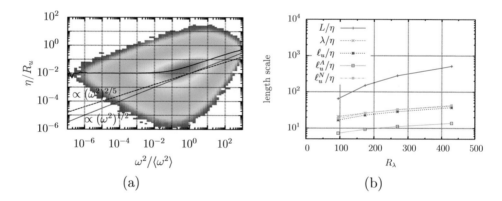

Figure 1.6 (a) Joint PDF of (ω^2, R_u), at $R_\lambda = 429$, and (b) R_λ-dependence of $\ell_u \equiv 1/\langle 1/R_u \rangle$ normalized by η. The solid line in (a) shows the conditional average $\langle \eta/R_u | \omega^2 \rangle$ of η/R_u for given ω^2. The dotted lines show the slopes $\propto (\omega^2)^{2/5}$ and $(\omega^2)^{1/2}$.

cation) that

$$\frac{1}{R_u^2} = \frac{1}{u^4}\left\{a^2 - \frac{(\mathbf{u}\cdot\mathbf{a})^2}{u^2}\right\}, \quad \mathbf{a} \equiv (\mathbf{u}\cdot\nabla)\mathbf{u}. \tag{1.26}$$

Figure 1.6(a) shows the joint PDF of $(\omega^2, \eta/R_u)$. It is seen that

(i) the conditional average $\langle \eta/R_u |\, \omega^2 \rangle$ for given ω^2 is larger for larger ω^2, and

(ii) $\langle \eta/R_u |\, \omega^2 \rangle$ is almost independent of ω^2 and is approximately 0.01 at $c_\omega \equiv \omega^2/\langle \omega^2 \rangle < 0.01$, while $\langle \eta/R_u |\, \omega^2 \rangle \approx (\omega^2)^\gamma$ with $\gamma \approx 2/5$ at $c_\omega > 1$.

Figure 1.6(b) shows the R_λ-dependence of the length scale ℓ_u defined by $1/\ell_u \equiv \langle 1/R_u \rangle$. It also shows the length scales ℓ_u^A and ℓ_u^N defined by $1/\ell_u^A \equiv \langle 1/R_u \rangle_A$ and $1/\ell_u^N \equiv \langle 1/R_u \rangle_N$, where the brackets $\langle .. \rangle_A$ and $\langle .. \rangle_N$ denote the averages over the regions satisfying $c_\omega > 3$ and $c_\omega \leq 3$, respectively. These two regions are hereafter referred to as the active and non-active regions, respectively. It is seen that

(iii) ℓ_u is close to λ,
(iv) $\ell_u^N \approx \ell_u$,
(v) ℓ_u^A is smaller than ℓ_u^N (by a factor of 4, approximately), but larger than η (by a factor of 10, approximately), and
(vi) ℓ_u as well as ℓ_u^A and ℓ_u^N scale with R_λ in a manner similar to λ.

The smallness of ℓ_u^A compared with ℓ_u^N is in agreement with our intuition that the influence of vorticity must be stronger, so that the stream lines are more curly in active rather than in non-active regions.

The estimate in (iii) agrees with the well-known hypothesis concerning the weakness of the statistical interdependence of large and small eddies, i.e., $\mathbf{u}$ and $\nabla \mathbf{u}$, which allows us to write $|\mathbf{a}| = |(\mathbf{u} \cdot \nabla)\mathbf{u}| \sim u' v_\eta / \eta$ and $u^2 \sim u'^2$, so that

$$\left\langle \frac{1}{R_u^2} \right\rangle \sim \frac{(v_\eta/\eta)^2}{u'^2} \sim \frac{1}{\lambda^2},$$

where v_η is the characteristic difference between the velocities at two points separated by the distance η. We have used $v_\eta \sim (\epsilon \eta)^{1/3}$.

The same analysis can be performed also for the radius R_ω of curvature c_ω of the vortex line given by (1.26) with $\mathbf{u}$ and u replaced by $\boldsymbol{\omega}$ and ω, respectively. Analysis of the DNS-ES data at $R_\lambda = 429$ shows the following.

(i) In contrast to $\langle \eta/R_u |\, \omega^2 \rangle$, $\langle \eta/R_\omega |\, \omega^2 \rangle$ is smaller for larger ω^2. This is consistent with the observation that the curvature c_ω decreases with the magnitude of $\boldsymbol{\omega}$, noted in Tsinober (2009),
(ii) $\langle \eta/R_\omega |\, \omega^2 \rangle$ is almost independent of c_ω at $c_\omega > 1$, and $\langle \eta/R_\omega |\, \omega^2 \rangle \propto (\omega^2)^\gamma$ with $\gamma \approx -1/2$ at $c_\omega < 0.1$.

A comparison of the data for different R_λ (less than 429) shows that:

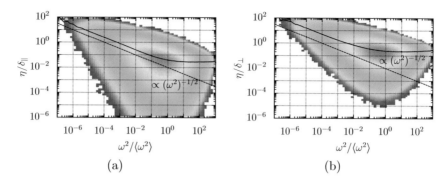

Figure 1.7 (a), (b) Joint PDF's of $(\omega^2, 1/\delta_\parallel)$, and $(\omega^2, 1/\delta_\perp)$, at $R_\lambda = 429$. The dotted lines show the slope $\propto (\omega^2)^{-1/2}$.

(iii) the dependence of $\ell_\omega \equiv 1/\left\langle \frac{1}{R_\omega} \right\rangle$, $\ell_\omega^A \equiv 1/\left\langle \frac{1}{R_\omega} \right\rangle_A$, and $\ell_\omega^N \equiv 1/\left\langle \frac{1}{R_\omega} \right\rangle_N$ on R_λ is similar to that of η, rather than that of λ;

(iv) nevertheless, their values are closer to λ than to η, and $\eta < \ell_\omega < \lambda$, within the R_λ range used in the analysis.

It is worthwhile commenting here on the resolution in DNS. In DNS that are based on the Fourier spectral method, the resolution is represented by the maximum wavenumber $k_{\max}$ of the modes retained in the DNS. It has been confirmed through a comparison of data with $k_{\max}\eta = 1$ and data with $k_{\max}\eta = 2$ that, in DNS-ES, the influence of the resolution is not significant for low-order statistics dominated by modes with $k\eta$ approximately < 1, provided that $k_{max}\eta = 2$ or so. However, this may be not the case for high-order statistics or for statistics associated with very high velocity gradients or ω (Schumacher et al., 2007; Donzis et al., 2008). The results presented in this section are from runs with $k_{\max}\eta = 2$. The possibility of the influence on the statistics, especially those dominated by high ω regions, remains to be examined.

1.5.2 Local Anisotropy

Figure 1.4 suggests that the high enstrophy regions are tube-like or sinew-like, i.e., they are elongated or anisotropic. The elongation/anisotropy is reflected in the gradient of the enstrophy. Let $\delta_\parallel$ and $\delta_\perp$ stand for the local scales parallel and perpendicular to $\boldsymbol{\omega}$, respectively, defined as

$$\frac{1}{\delta_\parallel} = \frac{|\mathbf{A} \cdot \boldsymbol{\omega}|}{\omega}, \quad \frac{1}{\delta_\perp} = \frac{|\mathbf{A} \times \boldsymbol{\omega}|}{\omega}, \quad \mathbf{A} \equiv \frac{\nabla \omega}{\omega}.$$

1: Small-Scale Statistics and Structure of Turbulence 25

Figures 1.7(a) and (b) show the joint PDFs of $(1/\delta_\|, \omega^2)$ and $(1/\delta_\perp, \omega^2)$ respectively. It is seen that:

(i) both $\langle 1/\delta_\| | \, \omega^2 \rangle$ and $\langle 1/\delta_\perp | \, \omega^2 \rangle$ are smaller for larger ω^2;
(ii) they are almost independent of ω^2 and

$$\langle 1/\delta_\| | \, \omega^2 \rangle \approx 0.03/\eta \ll \langle 1/\delta_\perp | \, \omega^2 \rangle \approx 0.2/\eta \text{ at } c_\omega > 10;$$

(iii) they are $\approx (\omega^2)^\gamma$ with $\gamma \approx -1/2$ at $c_\omega < 0.1$.

An analysis of the R_λ-dependence of the averages of the length scales defined by $1/\ell_\| \equiv \langle 1/\delta_\| \rangle, 1/\ell_\perp \equiv \langle 1/\delta_\perp \rangle, 1/\ell_\|^A \equiv \langle 1/\delta_\| \rangle_A$, etc. shows that:

(iv) $\ell_\|^A > \ell_\| \approx \ell_\|^N > \ell_\perp^A > \ell_\perp \approx \ell_\perp^N$;
(v) the R_λ-dependence of all the length scales in (iv) normalized by η is weak.

According to the DNS-ES data at $R_\lambda = 429$, we have:

(vi) $\ell_\|^A \approx 32.3\eta, \quad \ell_\| \approx \ell_\|^N \approx 12.3\eta, \quad \ell_\perp^A \approx 4.49\eta, \quad \ell_\perp \approx \ell_\perp^N \approx 2.96\eta$;
(vii) the average $\langle \delta_\|/\delta_\perp \rangle$ is approximately 6.13, and the conditional average for a given ω^2 is approximately 12.2 and 2.41 at large and small values of ω^2, respectively.

The estimates in (vi) implies, for example, that $\ell_\|$ is larger than $\ell_\perp$ by a factor of approximately $12.3/2.96 \approx 4.2$, and (vii) means that the elongation is stronger in the active region.

1.5.3 Dissipation range spectrum

Among the various forms proposed thus far for the spectrum $E(k)$ in the dissipation range where k is similar to or larger than $1/\eta$, the most widely accepted form is

$$\frac{E(k)}{(\langle \epsilon \rangle \nu^5)^{1/4}} = C(k\eta)^\alpha \exp(-\beta(k\eta)^n), \quad (1.27)$$

where α, β and C are constants independent of k (see, e.g., the review by Martinez *et al.*, 1997 and Sreenivasan and Antonia, 1997).

As regards n, a class of spectral closure theories give $n = 1$ (e.g., Kraichnan, 1959; Orszag, 1977; Tatsumi, 1980; Kaneda, 1993). Orszag (1977) argued that if the energy spectrum is of the form of (1.27), it is unlikely that $n > 1$, because the wavevector interactions involving the wavevectors of order $\frac{1}{2}\mathbf{k}$ ($\frac{1}{2}\mathbf{k} + \frac{1}{2}\mathbf{k} \to \mathbf{k}$) would drive too large a response in $\hat{\mathbf{u}}(\mathbf{k})$; if $n > 1$, then $|\hat{\mathbf{u}}(\mathbf{k}/2)\hat{\mathbf{u}}(-\mathbf{k}/2)| \gg c|\hat{\mathbf{u}}(\mathbf{k})|$ for $k\eta \gg 1$, where c is an appropriate constant, and $\hat{\mathbf{u}}(\mathbf{k})$ is the Fourier transform of $\mathbf{u}(\mathbf{x})$. Foias et al. (1990) showed that $n \geq 1$, under certain assumptions of the smoothness of the velocity field in a finite box. Experimentally, Sreenivasan (1985) found that $n = 1$ for $0.5 < k\eta < 1.5$. He also showed that if the spectrum falls off as rapidly as $\exp(k^n)$ for $n = 2$, then it will lead to unrealistic results in bandpass-filtered signals. DNS studies by Kida and Murakami (1987), Sanada (1992), and Kida et al. (1992) also support that $n = 1$ for $k\eta \approx 1$. She and Jackson (1993) showed that at $k\eta \approx 1$, experimental data fit well to $E(k)$ given by a superposition of two terms of the form of the right-hand side of (1.27) with $n = 1$.

Chen et al. (1993) showed that the form (1.27) fits the DNS data well in the range of wavenumbers as large as $5 < k\eta < 10$, but at relatively low Reynolds number $R_\lambda \approx 15$. Martinez et al. (1997) studied the R_λ-dependence of α, β and C, and showed that (1.27) also fits their DNS data well for $k\eta \approx 1$ with R_λ up to ≈ 90.

The form (1.27) was further confirmed to agree well with the DNS-ES data at $k\eta \approx 1$ with R_λ up to as large as 675. The data show that α, β and C depend on R_λ at low to moderate R_λ. A curve fit to the DNS-ES data suggests that, as $R_\lambda \to \infty$, these variables approach constants, say $(\alpha_\infty, \beta_\infty, C_\infty)$, independent of R_λ, in accordance with K41, where

$$(\alpha_\infty, \beta_\infty, C_\infty) \approx (-2.9, 0.62, 0.044).$$

The approach to these limits is very slow, however. Even when R_λ is as large as $R_\lambda = 10,000$, for example, the difference $\beta - \beta_\infty$ is still as large as 261% of β_∞ (Ishihara et al., 2005).

Although it is not easy to measure the spectrum $E(k)$ at very large $k\eta \gg 1$ in experiments, because of the weakness of the signal at large k, and, in DNS, because of the difficulty in achieving the required high accuracy and resolution, Schumacher (2007) examined the possible dependence of the statistics on the cut-off wavenumber k_{max} in DNS, and showed that (1.27) agrees well with DNS with $k\eta$ as large as 10, but with moderate R_λ up to 80.

The observed exponential dependence of $E(k) \propto \exp(-k)$ on k is in conflict with the spectra given by models of local structure of turbulence, such as those by Burgers (1948), Townsend (1951) and Lundgren (1982), which yield $E(k) \propto \exp(-k^2)$ (here we have omitted prefactors and constants). If one decomposes the flow field $\mathbf{u}$ into large- and small-scale components $\mathbf{u}^L$ and $\mathbf{u}^S$, i.e., $\mathbf{u} = \mathbf{u}^L + \mathbf{u}^S$, then in these models the large-scale component, $\mathbf{u}^L$, can be locally approximated by a linear function of the position vector

x, and the nonlinear interaction between the small-scale components $\mathbf{u}^S$ is exactly zero or assumed to be negligible.

Davidson (2004) proposed a model and argued that the observed $\exp(-k)$ spectrum and the $\exp(-k^2)$ spectrum predicted by the local models may be kinematically compatible. In his model, the total field consists of eddies with local structure giving $\exp[-(sk)^2]$ where s is eddy size and random, and an appropriate collection or PDF of the size s results in the total energy spectrum $\propto \exp(-k)$.

The local spectrum $\propto \exp(-k^2)$ may be questioned from a dynamical point of view, however. The question is concerned with interactions involving the wavevectors of order $\mathbf{k}/2$ or the inequality $|\hat{\mathbf{u}}(\mathbf{k}/2)\hat{\mathbf{u}}(-\mathbf{k}/2)| \gg c|\hat{\mathbf{u}}(\mathbf{k})|$ for $n > 1$, as noted above.

This inequality, however, does not exclude the possibility that the non-local (in scale) interactions between $\mathbf{u}^L$ and $\mathbf{u}^S$ may be dominant in certain cases, e.g., when $\mathbf{u}^L$ is strong and persistent, as it is in turbulent shear flow. Neither does this inequality exclude a priori the possibility that certain terms drop out identically: the vortex stretching term $(\boldsymbol{\omega}^S \cdot \nabla)\mathbf{u}^S$, where $\boldsymbol{\omega}^S \equiv \nabla \times \mathbf{u}^S$, for example, is zero if $\mathbf{u}^S$ is two-dimensional, and the vortex convection term $(\mathbf{u}^S \cdot \nabla)\boldsymbol{\omega}^S$ is also zero if $\mathbf{u}^S$ is axisymmetric. In fact, this is the case for the Burgers vortex.

The question here is whether such symmetry is generic, or what would be the generic spectrum if such symmetry were perturbed so that the nonlinear interaction between small-scale components $\mathbf{u}^S$ were enabled. Such a perturbation was analyzed by Moffatt et al. (1994), where $\mathbf{u}^S$ is expanded as $\mathbf{u}^S = \mathbf{u}_0 + (1/R_\Gamma)\mathbf{u}_1 + \cdots$, in which $\mathbf{u}_0$ is the Burgers model field and R_Γ is a Reynolds number based on the circulation. Whether the expansion is uniformly valid at large k remains to be examined.

In this respect, it is interesting to note that a simple model of local shear with nonlinear interaction between small-scale components (Ishihara and Kaneda, 1998) and a field consisting of interacting anti-parallel Euler vorticity (Bustamante and Kerr, 2008) both yield a one-dimensional spectrum approximately proportional to $\exp(-k_1)$.

It is also worth noting that, according to Chen et al. (1993), not only the spectrum $E(k)$, but also the local spectra defined by the **x**-space filter $\propto \exp[?|\mathbf{x}?\mathbf{x}_c|^2/(\Delta)^2]$ in the subregions of the periodic box (of size 2π) agree well with (1.27), where $\mathbf{x}_c$ is the subregion center and $\Delta = \pi/8$. A preliminary analysis of the DNS-ES data with $R_\lambda = 268$ also suggests that $n = 1$ for the local spectra in the subregions with $\mathbf{x}_c = \mathbf{x}_s$ and $\mathbf{x}_c = \mathbf{x}_w$ and for $\Delta/\eta \approx 160, 80, 40$, where the fundamental periodic box size in the DNS is $(2\pi)^3 \approx (1560\eta)^3$.

1.6 Inertial subrange

1.6.1 Examination of hypotheses of K62

Activity may influence not only the statistics of the dissipation range but also the statistics of the inertial subrange. In fact, this idea underlies the hypotheses of Kolmogorov's refined similarity hypotheses (Kolmogorov, 1962, hereafter called K62), where it is assumed that:

(1) the PDF of $\ln \epsilon(r)$ is Gaussian with the variance given by

$$\sigma_r^2 = A + \mu \ln\left(\frac{L}{r}\right),$$

where A and μ are constants;

(2) the moments of the velocity differences $\langle (\delta u(r))^n | \epsilon_r \rangle$ conditioned on $\epsilon(r)$ to ϵ_r are given by

$$\langle (\delta u(r))^n | \epsilon_r \rangle = (\epsilon_r r)^{n/3} g_n(R_{\epsilon r}), \quad \text{for} \quad r \ll L,$$

where $g_n(R_{\epsilon r})$ are universal functions of $R_{\epsilon r} \equiv \epsilon^{1/3} r^{4/3}/\nu$;

(3) when $R_{\epsilon r} \gg 1$, $g_n(R_{\epsilon r}) \sim D_n$, where D_n are constants.

These assumptions have been tested by experiments and DNS; see, for example, the review by Wang *et al.* (1996). They compared these assumptions with DNS data with R_λ up to 195 and the number of grid points N^3 up to 512^3, as well as with LES data. The comparison showed that:

(i) the PDF of $\ln \epsilon(r)$ is nearly Gaussian, except for the tails;
(ii) μ is in the range 0.20–0.28 in good agreement with experimental measurements using surrogate dissipation and numerical simulations (see the references cited in Wang *et al.*, 1996);
(iii) the normalized moments $\langle (\delta u(r))^n | \epsilon_r \rangle / (\epsilon_r r)^{n/3}$ depend on $R_{\epsilon r} \gg 1$ in a manner consistent with (2).

The analysis was based on the use of surrogates, which are the averages of the dissipation values $\epsilon(r)$ over lines rather than over volumes (spheres). An examination based on the use of volume averages rather than surrogates was made by using the DNS-ES data with R_λ up to 429 and N^3 as large as 2048^3 in unpublished work at Nagoya University (by Kutsuna, Ishihara and Kaneda). This study confirmed (i) for the volume-averaged $\epsilon(r)$, but also showed that:

(iv) the width of the range where μ is insensitive to r does not increase with R_λ;

(v) in the range defined in (iv), $\mu \approx 0.35$ if the volume average $\epsilon(r)$ is used, while $\mu \approx 0.28$ if the surrogate of line average is used;

(vi) for large $R_{\epsilon r}$, the moments scale with ϵ and r in a manner consistent with (2), and $g_2 \to 1.2$.

This value 1.2 agrees with that reported by Wang et al. (1996), but one difference must be noted: in the DNS-ES data, each of the curves of

$$\langle (\delta u(r))^n | \epsilon_r \rangle / (\epsilon_r r)^{n/3} \text{ vs. } R_{\epsilon r}$$

for various ϵ_r tends to 1.2 for large $R_{\epsilon r}$, while in Wang et al. (1996), it is the envelope of the curves, not the curves themselves, that tends to 1.2 for large $R_{\epsilon r}$. The estimates $\mu \approx 0.35$ and $\mu \approx 0.28$ are a little larger than $\mu \approx 0.25$ estimated by Yeung et al. (2006) from high resolution DNS data of $\log \epsilon$ with R_λ up to 680.

In connection with K62, it is appropriate to mention that by using high resolution DNS data, Schumacher et al. (2007) studied the relation between the inertial and dissipative regions within the framework of K62 and a multifractal formalism (Yakhot and Sreenivasan, 2004; Yakhot, 2004). They argued that the asymptotic state of turbulence is attained for the velocity gradients at far lower Reynolds numbers than those required for the inertial subrange to appear.

1.6.2 Energy transfer

An essence of turbulence lies in the energy transfer between different scales by the nonlinear NS dynamics. A quantitative measure characterizing this transfer is given by

$$T \equiv -\tau_{ij} \bar{e}_{ij},$$

where

$$\tau_{ij} = (\overline{u_i u_j} - \overline{u_i}\, \overline{u_j}) - \frac{2}{3}\delta_{ij} q, \quad q = \frac{1}{2}\left(\overline{u_k u_k} - \overline{u_k}\, \overline{u_k}\right), \quad \bar{e}_{ij} = \frac{1}{2}\left(\frac{\partial \bar{u}_i}{\partial x_j} + \frac{\partial \bar{u}_j}{\partial x_i}\right),$$

in which the bar $^-$ denotes the filtering operation defined by $\hat{\bar{f}}(\mathbf{k}) = G(\mathbf{k})\hat{f}(\mathbf{k})$, and $\hat{f}(\mathbf{k})$ and $G(\mathbf{k})$ are the Fourier transforms of f and the filtering function, respectively. In Large Eddy Simulation, $\bar{f}$ and $f - \bar{f}$ are called the grid scale and subgrid scale components, respectively, of f.

Here, we consider the so-called spectral cut filter by which all the Fourier modes of f with wavenumbers k larger than the cut-off wavenumber k_c are removed, i.e., $G(\mathbf{k}) = 1$ for $k < k_c$ and $= 0$ for $k \geq k_c$. In this case, $T = T(k_c, \mathbf{x})$ is a function of the position $\mathbf{x}$ as well as of the cut-off wavenumber

k_c, and may be regarded as the energy transfer due to the nonlinear NS dynamics across the wavenumber k_c at position $\mathbf{x}$. This is an analogue of the transfer discussed by Kraichnan (1974), and studies based on DNS have been made on the statistics of T (e.g., Piomelli *et al.*, 1991; Domaradzki, 1992; Cerutti and Meneveau, 1998).

Using a spectral cut-off filter, we have $\langle T(k) \rangle = \Pi(k)$ and, in the inertial subrange,

$$\langle T(k) \rangle = \langle \epsilon \rangle = \langle \epsilon(r) \rangle,$$

where $T(k) \equiv T(k, \mathbf{x})$ and $kr = \pi$. This does not mean, however, that the statistics of $T(k)$ are similar to those of ϵ or $\epsilon(r)$. One should note that:

(a) $T(k)$ can be negative, in contrast to ϵ and $\epsilon(r)$;

(b) $\epsilon(r)$ is not a genuine inertial subrange quantity, even if r is in the inertial subrange; it is an integral of the dissipation range quantity ϵ (Kraichnan, 1974).

An analysis of the DNS–ES data by Aoyama *et al.* (2005) shows that:

(i) the PDF of $T(k)$ is far from Gaussian, and it is more intermittent for larger k;

(ii) the volume ratio of the region where $T(k) < 0$ (backward transfer region) is as large as approximately 40% in the inertial subrange;

(iii) the skewness S and flatness factor F of the fluctuation $\Delta T \equiv T - \langle T \rangle$ increases with k, and fits well $S \propto (kL)^\alpha$, $F \propto (kL)^\beta$ in the inertial subrange, where $\alpha \approx 2/3$ and $\beta \approx 1$;

(iv) the decrease of the PDF of $T(k)$ with $\Delta T / \sigma_T$ at large $\Delta T / \sigma_T$ is similar to that of $\epsilon(r)$ with $\epsilon(r)/\sigma_\epsilon$ at large $\epsilon(r)/\sigma_\epsilon$, where σ_T and σ_ϵ are the standard deviations of $T(k)$ and $\epsilon(r)$, respectively, and $kr = \pi$.

(v) the correlation between $\epsilon(r)$ and $T(\pi/r)$ is not high: indeed, it is as low as 0.25.

The nonlinear NS dynamics transfers not only energy but also enstrophy. Unlike energy, enstrophy can be generated in a turbulent flow, so that one may divide the effects of the nonlinear dynamics into two distinct processes: the generation, $G^L(r)$, of enstrophy above a given scale r; and the flux, $F(r)$, of enstrophy across scale r to smaller scales. Under the same conditions for the 4/5 law to hold, and as $Re \to \infty$, we have $G^L(r) \to F(r) \sim \langle \epsilon \rangle / r^2$ in the inertial subrange (Davidson *et al.*, 2008).

1.6.3 Linear response theory

Kolmogorov introduced the idea of a local equilibrium state in formulating K41. Although not all of the hypotheses in K41 are sound, it is attractive to assume the existence of a certain kind of equilibrium state at small scales, though we don't know how to accurately specify such a state. If one accepts the assumption that such a state exists, one may consider a framework similar to that of the linear response theory for thermal equilibrium systems.

Linear response theory for thermal equilibrium systems

Let us first briefly review the linear response theory for thermal equilibrium systems, following Kubo (1966). If an external force $X(t)$ is applied to a system in a state of thermal equilibrium, then the distribution function or the density matrix ρ representing the statistical ensemble is given by

$$\rho = \rho_e + \Delta\rho + \cdots, \quad (1.28)$$

where ρ_e is the distribution function or the density matrix in the absence of the force X, and $\Delta\rho$ is the change of ρ due to X and is first order in X. The response of the system to the force is observed in the change of a certain physical quantity, say B, which may be written, in a symbolic notation, as

$$\langle B \rangle = \langle B \rangle_e + \Delta\langle B \rangle + \cdots, \quad (1.29)$$

where $\langle B \rangle_e$ is the average over the equilibrium distribution ρ_e,

$$\Delta\langle B \rangle = cX, \quad (1.30)$$

in which c is a constant, and the possible time dependence of X is ignored.

An example of (1.30) is the relation between X and the so-called generalized flux J,

$$J = CX, \quad (1.31)$$

where (J, X) may stand for (mean density gradient, mass flux), (mean electric field, electric current), etc., and the coefficient C satisfies certain reciprocal relations reflecting the symmetry in the equilibrium state (Onsager, 1931).

In the context of fluid mechanics, the relation between the stress tensor τ_{ij} and the rate of strain tensor $\partial u_i/\partial x_j$ for a Newtonian fluid,

$$\tau_{ij} = c_{ijmn}\frac{\partial u_m}{\partial x_n},$$

is another example of (1.31) or (1.30), where c_{ijmn} is a fourth-order isotropic

tensor, i.e.
$$c_{ijmn} = \lambda^{(1)}\delta_{im}\delta_{jn} + \lambda^{(2)}\delta_{in}\delta_{jm} + \lambda^{(3)}\delta_{ij}\delta_{mn}, \tag{1.32}$$
and $\lambda^{(\alpha)}$, $\alpha = 1,2,3$, are constants.

Turbulent Shear Flow

Let $\mathbf{u}_0 \equiv \mathbf{u}(\mathbf{x}_0, t_0)$ be the velocity at a point $\mathbf{x}_0$ and time t_0 in a turbulent shear flow, and D be the local space-time domain such that $D = \{(\mathbf{r}, \tau) | r \ll L, |\tau| \ll T\}$, where $\mathbf{r} \equiv \mathbf{x} - (\mathbf{x}_0 + \mathbf{u}_0\tau)$, $\tau \equiv t - t_0$, and $L = \min(L_e, L_S)$ and $T = \min(T_e, T_S)$, in which L_e, L_S and T_e, T_S are the characteristic length and time scales of energy-containing eddies and the mean flow $\mathbf{U} \equiv \langle \mathbf{u} \rangle$, respectively.

Here we assume for the sake of simplicity that the rate of strain tensor $S_{ij} \equiv \partial U_i/\partial x_j$ of the mean flow $\mathbf{U}$ may be regarded to be constant independent of time in D, and that there is no external force, so that $\mathbf{f} = 0$ in (1.1). Then simple dimensional considerations suggest that the statistics of turbulence in D is dominated by the Navier–Stokes (NS) dynamics without mean flow $\mathbf{U}$, if

$$\delta_\ell \equiv \frac{S\ell}{v_\ell} \ll 1, \tag{1.33}$$

where v_ℓ is the characteristic magnitude of the velocity difference $\mathbf{v}(\mathbf{r}) \equiv \mathbf{u}(\mathbf{x}+\mathbf{r}) - \mathbf{u}(\mathbf{x})$ for $|\mathbf{r}| = \ell$, and S is an appropriate norm of the tensor S_{ij}. This implies that the influence of $\mathbf{U}$ may be treated as a small perturbation added to the equilibrium state which is determined by the inherent NS dynamics in the absence of $\mathbf{U}$ (Kaneda and Yoshida, 2004).

Let ρ be the probability distribution of $\mathbf{v}$ in D. The above argument yields ρ as (1.28), where ρ_e stands for the distribution in the absence of the mean shear, and $\Delta\rho$ represents the change of ρ in response to the shear and linear in S_{ij}. Corresponding to (1.29) and (1.30), we have

$$\langle v_i(\mathbf{r})v_j(\mathbf{r})\rangle = \langle v_i(\mathbf{r})v_j(\mathbf{r})\rangle_e + \Delta\langle v_i(\mathbf{r})v_j(\mathbf{r})\rangle + \cdots, \tag{1.34}$$

where

$$\Delta\langle v_i(\mathbf{r})v_j(\mathbf{r})\rangle = C_{ijmn}(\mathbf{r})S_{mn}, \tag{1.35}$$

$\langle .. \rangle_e$ is the average over the equilibrium distribution ρ_e, and C_{ijmn} is a fourth-order tensor satisfying $C_{ijmn} = C_{jimn}$ and compatible with the incompressibility condition of $\mathbf{v}$.

So far, it has not been necessary to specify the equilibrium distribution ρ_e. If one assumes that the statistics of the equilibrium state in the absence of the mean shear are locally homogeneous and isotropic in D, then C_{ijmn}

1: Small-Scale Statistics and Structure of Turbulence

must be a fourth-order isotropic tensor independent of **x**. In the following we assume this to be the case.

When the incompressibility condition has to be taken into account, it is convenient to work in wavevector space. Let $Q_{ij}(\mathbf{k})$ be the Fourier transform of $\langle u_i(\mathbf{x}+\mathbf{r})u_j(\mathbf{r})\rangle$ with respect to $\mathbf{r}$. Then (1.34) and (1.35) give

$$Q_{ij}(\mathbf{k}) = \langle Q_{ij}(\mathbf{k})\rangle_e + \Delta Q_{ij}(\mathbf{k}) + \cdots, \qquad (1.36)$$

where $\langle Q_{ij}(\mathbf{k})\rangle_e$ is the equilibrium spectrum in the absence of the mean flow, and

$$\Delta Q_{ij}(\mathbf{k}) = C'_{ijmn}(\mathbf{k})S_{mn}, \qquad (1.37)$$

in which C'_{ijmn} is a fourth-order isotropic tensor depending on $\mathbf{k}$ satisfying $k_i C'_{ijmn} = k_j C'_{ijmn} = 0$ and $C'_{ijmn} = C'_{jimn}$. It may be therefore written as

$$C'_{ijmn}(\mathbf{k}) = q^{(1)}(k)\left[P_{im}(\mathbf{k})P_{jn}(\mathbf{k}) + P_{jm}(\mathbf{k})P_{in}(\mathbf{k})\right]$$
$$+ q^{(2)}(k)P_{ij}(\mathbf{k})\frac{k_m k_n}{k^2} + q^{(3)}(k)P_{ij}(\mathbf{k})\delta_{mn}, \qquad (1.38)$$

where $P_{ij}(\mathbf{k}) = \delta_{ij} - (k_i k_j)/k^2$, and $q^{(\alpha)}(k)$, $\alpha = 1, 2, 3$, depends on $\mathbf{k}$ only through $k = |\mathbf{k}|$. The $q^{(3)}$-term may be neglected without loss of generality, because $S_{mm} = 0$.

The expression (1.38) is to be compared with (1.32). Both of these are fourth-order isotropic tensors, but, in contrast to c_{ijmn}, C'_{ijmn} depends on $\mathbf{k}$ and (1.38) holds only in the domain D where (1.33) is satisfied.

It is tempting to assume that the equilibrium state in the inertial subrange may be approximated as in K41, in which the subrange is characterized only by ϵ and $\mathbf{r}$ or $\mathbf{k}$. This assumption is consistent with (1.33) in the inertial subrange, because, according to K41, $v_\ell \sim (\epsilon\ell)^{1/3}$, which gives $\delta_\ell \sim S\ell^{3/2}/\epsilon^{1/3}$ in the range $L \gg \ell \gg \eta$, and the condition $\delta_\ell \ll 1$ is therefore satisfied for sufficiently small values of ℓ such that $\ell \ll (\epsilon^{1/3}/S)^{2/3}$. Note that the inequality (1.33) may be violated outside the inertial subrange.

Under these assumptions, we have

$$\langle Q_{ij}(\mathbf{k})\rangle_e = \frac{K_o}{2\pi}\langle\epsilon\rangle^{2/3} k^{-11/3} P_{ij}(\mathbf{k}), \qquad (1.39)$$

and (1.37), (1.38) but with

$$q_1(k) = A\langle\epsilon\rangle^{1/3} k^{-13/3}, \quad q_2(k) = B\langle\epsilon\rangle^{1/3} k^{-13/3}, \qquad (1.40)$$

for $1/L \ll k \ll 1/\eta$, provided that $S/(\epsilon k^2)^{1/3} \ll 1$, where A and B are dimensionless universal constants (Ishihara et al., 2002). Equation (1.40) implies that $\Delta Q_{ij}(\mathbf{k}) \propto^{-13/3}$. This scaling is consistent with that by Lumley

(1967), and also with those by Leslie (1973) and Yoshizawa (1998), both of which were derived by a spectral closure approaches. The spectra given by (1.36)–(1.40) are consistent with those in experiments by Saddougghi and Veeravalli (1994) and in DNS by Ishihara et al. (2002). A spectral closure analysis suggests that for accurate estimation by DNS, or experiments, of the universal constants A, B that determine the anisotropic component of the spectrum $Q_{ij}(\mathbf{k})$, the Reynolds number must be much higher than that required for the estimation of K_o which determines the isotropic counterpart (Yoshida et al., 2003).

Stably stratified turbulence and MHD turbulence

Cinsiderations similar to those already discussed are applicable not only to turbulent shear flow but also to strongly stably stratified incompressible turbulence obeying the Boussinesq approximation (Kaneda and Yoshida, 2004) as well as to magneto-hydrodynamic turbulence of incompressible fluid under a stationary strong uniform magnetic field at small magnetic Reynolds numbers (Ishida and Kaneda, 2007). Results from these applications were confirmed to be in good agreement with those of DNS.

1.7 Concluding remarks

Due in part to space constraints and even more to the authors' lack of expertise, many interesting and challenging subjects had to be left out of this chapter. These include vortex dynamics, anomalous scaling, wavelet analysis, and statistics and structure in four dimensions (space + time).

DNS of turbulence provides us with detailed information on turbulence without the experimental uncertainties that occur even under well-controlled conditions. It must be noted, however, that universality in the sense of K41 is expected to exist only at sufficiently high Reynolds numbers Re, and that properties and trends observed at lower Re should not necessarily be extrapolated to higher Re, as the examples reviewed in this chapter have shown. Different conditions must not be confused. To that end, performing DNS at sufficiently large Re is expected to contribute much to our understanding of the nature of turbulence at high Re.

Fortunately, it has recently become possible to perform high resolution DNS at values of Re comparable to the highest ones achievable in laboratory experiments for canonical flows under simple boundary conditions such as periodic boundary conditions. It is unlikely, however, that it will become possible in the near future to perform DNS to solve practical flow problems under complex geometric conditions and physical situations, such as those in environmental studies. Such problems require certain kinds of simplified models in order to have any validity. Improving our understanding of the nature of turbulence at small scales is expected to help us develop sound models with a firm basis in fact.

With respect to this goal, it is instructive to recall the history of statistical mechanics of systems in thermal equilibrium. It tells us that universal phenomenological laws such as the three laws of thermodynamics had been established prior to the construction of statistical mechanics, and that they played a guiding role in its construction. Unfortunately, no phenomenological lawa as universal and solid as thermodynamics have yet been established for turbulence at high Re. However, the history of statistical mechanics for systems in thermal equilibrium suggests the importance of the search for, and the establishment of, universal phenomenological laws of turbulence. It is hoped that "extensive, but well-planned computational efforts" (von Neumann, 1949) exploiting the recent rapid developments in both hardware and software will contribute to this search and establishment, and so "break the deadlock" of turbulence research.

Acknowledgements The contents of Section 1.5 of this chapter owe a great deal to our stimulating discussion with K. Moffatt, which started during participation of Moffatt and YK in the Turbulence Program at the Isaac Newton Institute in 2008. This chapter could not have been written without the warm encouragement of P. Davidson and K.R. Sreenivasan, to whom the authors would like to express their thanks. They are also grateful to them, R. Rubinstein, J. Schumacher, M. Tanahashi, P.K. Yeung and T. Ishihara for their valuable comments, and to T. Tsuji and T. Gotoh for providing us with the data for Fig. 1.1. Figure 1.5 was made by using Vapor. This work was supported partly by Grant-in-Aids for Scientific Research (B)20340099 and (S)20224013 from the Japan Society for the Promotion of Science.

References

Antonia, R. A., Chambers A. J. and Satyaprakash B. R. 1981. Reynolds number dependence of high-order moments of the streamwise turbulent velocity derivative. *Boundary-Layer Met.*, **21**, 159–171.

Antonia, R. A. and Burattini, P. 2006. Approach to the 4/5 law in homogeneous isotropic turbulence. *J. Fluid Mech.*, **550**, 175–184.

Aoyama, T., Ishihara, T., Kaneda, Y., Yokokawa, M., Itakura, K. and Uno, A. 2005. Statistics of energy transfer in high-resolution direct numerical simulation of turbulence in a periodic box. *J. Phys. Soc. Jpn.*, **74**, 3202–3212.

Betchov, R. 1956. An inequality concerning the production of vorticity in isotropic turbulence. *J. Fluid Mech.*, **1**, 497–504.

Biferale, L. and Procaccia, I. 2005. Anisotropy in turbulent flows and in turbulent transport. *Physics Reports*, **414**, 43–164.

Burgers, J. M. 1948. A mathematical model illustrating the theory of turbulence. *Adv. in Appl. Mech.*, **1**, 171–199.

Bustamante, M. D. and Kerr, R. M. 2008. 3D Euler about a 2D symmetry plane. *Physica D*, **237**, 1912–1920.

Cerutti, S. and Meneveau, C. 1998. Intermittency and relative scaling of subgrid-scale energy dissipation in isotropic turbulence. *Phys. Fluids*, **10**, 928–937.

Chhabra, A. B. and Sreenivasan, K. R. 1992. Scale-invariant multiplier distributions in turbulence. *Phys. Rev. Lett.*, **68**, 2762–2765.

Champagne, F. H. 1978. The fine-scale structure of the turbulent velocity field. *J. Fluid Mech.*, **86**, 67–108.

Chen, S., Doolen, G., Herring, J. R., Kraichnan, R. H., Orszag, S. A. and She, Z. -S. 1993. Far-dissipation range of turbulence. *Phys. Rev. Lett.*, **70**, 3051–3054.

Danaila, L., Anselmet, F. and Antonia, R. A. 2002. An overview of the effect of large-scale inhomogeneities on small-scale turbulence. *Phys. Fluids*, **14**, 2475–2484.

Davidson, P. A. 2004. *Turbulence: An Introduction for Scientists and Engineers.* Oxford University Press.

Davidson, P. A., Morishita, K. and Kaneda, Y. 2008. On the generation and flux of enstrophy in isotropic turbulence. *J. Turbulence*, **9**, 42.

Doering, C. R. and Foias, C. 2002. Energy dissipation in body-forced turbulence. *J. Fluid Mech.*, **467**, 289–306.

Doering, C. R. 2009. The 3D Navier–Stokes problem. *Ann. Rev. Fluid. Mech.*, **41**, 109–128.

Domaradzki, J. A. 1992. Nonlocal triad interactions and the dissipation range of isotropic turbulence. *Phys. Fluids A*, **4**, 2037–2045.

Donnelly, R. J. and Sreenivasan, K. R. 1998. *Flow at Ultra-High Reynolds and Rayleigh Numbers: A Status Report.* Springer-Verlag.

Donzis, D. A., Sreenivasan, K. R. and Yeung, P. K. 2005. Scalar dissipation rate and dissipative anomaly in isotropic turbulence. *J. Fluid Mech.*, **532**, 199–216.

Donzis, D. A., Yeung, P. K. and Sreenivasan, K. R. 2008. Dissipation and enstrophy in isotropic turbulence: Resolution effects and scaling in direct numerical simulations. *Phys. Fluids*, **20**, 045108.

Donzis, D. A. and Sreenivasan, K. R. 2010. The bottleneck effect and the Kolmogorov constant in isotropic turbulence. *J. Fluid Mech.*, **657**, 171–188.

Donzis, D. A., Sreenivasan, K. R. and Yeung, P. K. 2010. The Batchelor spectrum for mixing of passive scalars in isotropic turbulence. *Flow Turbulence Combust*, **85**, 549–566.

Duchon, J. and Robert, R. 2000. Inertial energy dissipation for weak solutions of incompressible Euler and Navier–Stokes equations. *Nonlinearity*, **13**, 249–255.

Eyink, G. L. 2003. Local 4/5–law and energy dissipation anomaly in turbulence. *Nonlinearity*, **16**, 137–145.

Frisch, U. 1995. *Turbulence: the Legacy of A. N. Kolmogorov.* Cambridge University Press.

Foias, C., Manley, O. and Sirovich, L. 1990. Empirical and Stokes eigenfunctions and the far-dissipative turbulent spectrum. *Phys. Fluids A*, **2**, 464–467.

Gotoh, T., Kaneda, Y. and Bekki, N. 1988. Numerical-integration of the Lagrangian renormalized approximation. *J. Phys. Soc. Jpn.*, **57**, 866–880.

Gotoh, T., Fukayama, D. and Nakano, T. 2002. Velocity field statistics in homogeneous steady turbulence obtained using a high-resolution direct numerical simulation. *Phys. Fluids*, **14**, 1065–1081.

Grant, H. L., Stewart, R. W. and Moilliet, A. 1962. Turbulence spectra in a tidal channel. *J. Fluid Mech.*, **12**, 241–268.

Gurvich, A. S. and Yaglom, A. M. 1967. Breakdown of eddies and probability distributions for small-scale turbulence. *Phys. Fluids*, **10**, S59–S65.

Gylfason, A., Ayyalasomayajula, S. and Warhaft, Z. 2004. Intermittency, pressure and acceleration statistics from hot-wire measurements in wind-tunnel turbulence. *J. Fluid Mech.*, **501**, 213–229.

Herring, J. R. and Kraichnan, R. H. 1979. A numerical comparison of velocity-based and strain-based Lagrangian-history turbulence approximations. *J. Fluid Mech.*, **91**, 581–597.

Hill, R. J. 2002. Possible alternative to R-lambda-scaling of small-scale turbulence statistics. *J. Fluid Mech.*, **463**, 403–412.

Hierro, J. and Dopazo, C. 2003. Fourth-order statistical moments of the velocity gradient tensor in homogeneous, isotropic turbulence. *Phys. Fluids*, **15**, 3434–3442.

Hosokawa, I., and Yamamoto, K. 1989. Fine structure of a directly simulated isotropic turbulence. *J. Phys. Soc. Jpn.*, **58**, 20–23.

Ishida, T. and Kaneda, Y. 2007. Small-scale anisotropy in magnetohydrodynamic turbulence under a strong uniform magnetic field. *Phys. Fluids*, **19**, 075104.

Ishihara, T. and Kaneda, Y. 1998. Fine-scale structure of thin vortical layers. *J. Fluid Mech.*, **364**, 297–318.

Ishihara, T., Yoshida, K. and Kaneda, Y. 2002. Anisotropic velocity correlation spectrum at small scales in a homogeneous turbulent shear flow. *Phys. Rev. Lett.*, **88**, 154501.

Ishihara, T., Kaneda, Y., Yokokawa, M., Itakura, K. and Uno, A. 2005. Energy spectrum in the near dissipation range of high resolution direct numerical simulation of turbulence. *J. Phys. Soc. Jpn.*, **74**, 1464–1471.

Ishihara, T., Kaneda, Y., Yokokawa, M., Itakura, K. and Uno, A. 2007. Small-scale statistics in high-resolution direct numerical simulation of turbulence: Reynolds number dependence of one-point velocity gradient statistics. *J. Fluid Mech.*, **592**, 335–366.

Ishihara, T., Gotoh, T. and Kaneda, Y. 2009. Study of high-Reynolds number isotropic turbulence by direct numerical simulation. *Ann. Rev. Fluid Mech.*, **41**, 165–180.

Jiménez, J. 2000. Intermittency and cascades. *J. Fluid Mech.*, **409**, 99–120.

Jiménez, J. 2001. Self-similarity and coherence in turbulent cascade. In *IUTAM Symposium on Geometry and Statistics of Turbulence*, T. Kambe, T. Nakano and T. Miyauchi (eds.), 57–66. Kluwer Academic Publishers.

Jiménez, J., Wray, A. A., Saffman, P. G. and Rogallo, R. S. 1993. The structure of intense vorticity in isotropic turbulence. *J. Fluid Mech.*, **255**, 65–90.

Jiménez, J., Moisy, F., Tabeling, P. and Willaime, H. 2001. Scaling and structure in isotropic turbulence. In *Intermittency in Turbulent Flows*, J. C. Vassilicos (ed.), Cambridge University Press.

Kármán, T. and Howarth, L. 1938. On the statistical theory of isotropic turbulence. *Proc. R. Soc. Lond, Ser. A*, **164**, 192–215.

Kaneda, Y. 1981. Renormalized expansions in the theory of turbulence with the use of the Lagrangian position function. *J. Fluid Mech.*, **107**, 131–145.

Kaneda, Y. 1993. Lagrangian and Eulerian time correlations in turbulence. *Phys. Fluids A*, **5**, 2835–2845.

Kaneda, Y. and Ishihara, T. 2006. High-resolution direct numerical simulation of turbulence. *J. Turbulence*, **7**, 20.

Kaneda, Y. and Ishihara, T. 2007. Attempts at a computational science of turbulence. *Nagare*, **26**, 375–383. (*in Japanese*)

Kaneda, Y. and Morishita, K. 2007. Intermittency of energy dissipation in high-resolution direct numerical simulation of turbulence. *J. Phys. Soc. Jpn.*, **76**, 073401.

Kaneda, Y. and Yoshida, K. 2004. Small-scale anisotropy in stably stratified turbulence. *New Journal of Physics*, **6**, 34.

Kaneda, Y., Ishihara, T., Yokokawa, M., Itakura, K. and Uno, A. 2003. Energy dissipation rate and energy spectrum in high resolution direct numerical simulations of turbulence in a periodic box. *Phys. Fluids*, **15**, L21–L24.

Kaneda, Y., Yoshino, J. and Ishihara, T. 2008. Examination of Kolmogorov's 4/5 law by high-resolution direct numerical simulation data of turbulence. *J. Phys. Soc. Jpn.*, **77**, 064401.

Kerr, R. M. 1985. Higher-order derivative correlations and the alignment of small-scale structures in isotropic numerical turbulence. *J. Fluid Mech.*, **153**, 31–58.

Kholmyansky, M. and Tsinober, A. 2009. On an alternative explanation of anomalous scaling and how well-defined is the concept of inertial range. *Physics Lett. A*, **373**, 2364–2367.

Kida, S. 1993. Tube-like structures in Turbulence. *Lecture Notes in Numerical Applied Analysis*, **12**, 137–159.

Kida, S. and Murakami, Y. 1987. Kolmogorov similarity in freely decaying turbulence. *Phys. Fluids*, **30**, 2030–2039.

Kida, S., Kraichnan, R. H., Rogallo, R. S., Waleffe, F. and Zhou, Y. 1992. Triad Interactions in the Dissipation Range. *Proc. CTR Summer Program 1992*, 83–99.

Kolmogorov, A. N. 1941a. The local structure of turbulence in incompressible viscous fluid for very large Reynolds numbers. *Dokl. Akad. Nauk SSSR*, **30**, 301–305.

Kolmogorov, A. N. 1941b. On degeneration of isotropic turbulence in an incompressible viscous liquid. *Dokl. Akad. Nauk SSSR*, **31**, 538–540.

Kolmogorov, A. N. 1941c. Dissipation of energy in the locally isotropic turbulence. *Dokl. Akad. Nauk SSSR*, **32**, 16–18.

Kolmogorov, A. N. 1962. A refinement of previous hypotheses concerning the local structure of turbulence in a viscous incompressible fluid at high Reynolds number. *J. Fluid Mech.*, **13**, 82–85.

Kraichnan, R. H. 1959. The structure of isotropic turbulence at very high Reynolds numbers. *J. Fluid Mech.*, **5**, 497–543.

Kraichnan, R. H. 1965. Lagrangian-history closure approximation for turbulence. *Phys. Fluids*, **8**, 575–598.

Kraichnan, R. H. 1966. Isotropic turbulence and inertial-range structure. *Phys. Fluids*, **9**, 1728–1752.

Kraichnan, R. H. 1974. On Kolmogorov's inertial-range theories. *J. Fluid Mech.*, **62**, 305–330.

Kraichnan, R. H. 1978. A strain-based Lagrangian-history turbulence theory. *J. Fluid Mech.*, **88**, 355–367.

Kubo, R. 1966. The fluctuation–dissipation theorem. *Rep. Progr. Phys.*, **29**, 255–284.

Küchemann, D. 1965. Reports on IUTAM Symposium on concentrated vortex motions in fluids. *J. Fluid Mech.*, **21**, 1–20.

Kurien, S. and Sreenivasan K. R. 2000. Anisotropic scaling contributions to high-order structure functions in high-Reynolds-number turbulence. *Phys. Rev. E*, **62**, 2206–2212.

Landau, L. D. and Lifshitz, E. M. 1987. *Fluid Mechanics, 2nd ed.*, Pergamon Press.

Leslie, D. C. 1973. *Developments in the Theory of Turbulence*. Clarendon Press.

Lindborg, E. 1999. Correction to the four-fifths law due to variations of the dissipation. *Phys. Fluids*, **11**, 510–512.

Lohse, D. 1994. Crossover from high to low Reynolds number turbulence. *Phys. Rev. Lett.*, **73**, 3223–3226.

Lumley, J. L. 1967. Similarity and the turbulent energy spectrum. *Phys. Fluids*, **10**, 855–858.

Lundgren, T. S. 1982. Strained spiral vortex model for turbulent fine structure. *Phys. Fluids*, **25**, 2193–2203.

Lundgren, T. S. 2003. Linearly forced isotropic turbulence. *CTR Annual Research Briefs 2003*, 461–473.

Mandelbrot, B. B. 1974. Intermittent turbulence in self-similar cascades: divergence of high moments and dimension of the carrier. *J. Fluid Mech.*, **62**, 331–358.

Martínez, D. O., Chen, S., Doolen, G. D., Kraichnan, R. H., Wang, L.-P. and Zhou, Y. 1997. Energy spectrum in the dissipation range of fluid turbulence. *J. Plasma Phys.*, **57**, 195–201.

Meneveau, C. and Sreenivasan, K. R. 1991. The multifractal nature of turbulent energy dissipation. *J. Fluid Mech.*, **224**, 429–484.

Miyauchi, T. and Tanahashi, M. 2001. Coherent fine scale structure in turbulence. In *IUTAM Symp. Geometry and Statistics of Turbulence*, T. Kambe, T. Nakano and T. Miyauchi (eds.), 67–76, Kluwer Academoic Publishers.

Moisy F., Tabeling P. and Willaime, H. 1999. Kolmogorov equation in a fully developed turbulence experiment. *Phys. Rev.Lett.*, **82**, 3993–3997.

Moffatt, H. K., Kida, S. and Ohkitani, K. 1994. Stretched vortices - the sinews of turbulence; large-Reynolds-number asymptotics. *J. Fluid Mech.*, **259**, 241–264.

Monin, A. S. and Yaglom, A. M. 1975. *Statistical Fluid Mechanics: Mechanics of Turbulence*, **2**, MIT Press.

Nelkin, M. 1994. Universality and scaling in fully developed turbulence. *Adv. Physics*, **43**, 143–181.

Novikov, E. A. 1971. Intermittency and scale similarity of the structure of turbulent flow. *Prikl. Math. Mekh.*, **35**, 266–277.

Obukhov, A. M. 1949. Local structure of atmospheric turbulence. *Dokl. Acad. Nauk SSSR*, **67**, 643–646.

Obukhov, A. M. 1962. Some specific features of atmospheric turbulence. *J. Fluid Mech.*, **13**, 77–81.

Naert, A., Friedrich, R. and Peinke, J. 1997. Fokker–Planck equation for the energy cascade in turbulence *Phys. Rev. E*, **56**, 6719–6722.

Nie, Q. and Tanveer, S. 1999. A note on third-order structure functions in turbulence. *Proc. R. Soc. Lond. A*, **455**, 1615–1635.

Onsager, L. 1931. Reciprocal relations in irreversible processes. I. *Phys. Rev.*, **37**, 405–426.

Orszag, S. A. 1977. Lectures on the statistical theory of turbulence. In *Fluid Dynamics*, Balian, R. and Peube, J.-L. (eds.), 235–374. Gordon and Breach Science Publishers.

Pedrizzetti, G., Novikov, E. A. and Praskovsky, A. A. 1996. Self-similarity and probability distributions of turbulent intermittency. *Phys. Rev. E*, **53**, 475–484.

Qian, J. 1997. Inertial range and the finite Reynolds number effect of turbulence. *Phys. Rev. E*, **55**, 337–342.

Qian, J. 1998. Normal and anomalous scaling of turbulence. *Phys. Rev. E*, **58**, 7325–7329.

Qian, J. 1999. Slow decay of the finite Reynolds number effect of turbulence. *Phys. Rev. E*, **60**, 3409–3412.

Piomelli, U., Cabot, W. H., Moin, P. and Lee, S. 1991. Subgrid-scale backscatter in turbulent and transitional flows. *Phys. Fluids A*, **3**, 1766–1771.

Saddoughi, S. G. and Veeravalli, S. V. 1994. Local isotropy in turbulent boundary layers at high Reynolds number. *J. Fluid Mech.*, **268**, 333–372.

Saffman, P. G. and Baker, G. R. 1979. Vortex interactions. *Ann. Rev. Fluid Mech.*, **11**, 95–122.

Sanada, T. 1992. Comment on the dissipation-range spectrum in turbulent flows. *Phys. Fluids A*, **4**, 1086–1087.

Schumacher, J. 2007. Sub-Kolmogorov-scale fluctuations in fluid turbulence. *EPL*, **80**, 54001.

Schumacher, J., Sreenivasan, K. R. and Yakhot, V. 2007. Asymptotic exponents from low-Reynolds-number flows. *New Journal of Physics*, **9**, 89.

She, Z. -S. and Jackson, E. 1993. On the universal form of energy spectra in fully developed turbulence. *Phys. Fluids A*, **5**, 1526–1528.

She, Z. -S. and Leveque, E. 1994. Universal scaling laws in fully developed turbulence. *Phys. Rev. Lett.*, **72**, 336–339.

She, Z. -S., Jackson, E. and Orszag, S. A. 1990. Intermittent vortex structures in homogeneous isotropic turbulence. *Nature*, **344**, 226–228.

Siggia, E. D. 1981. Numerical study of small-scale intermittency in three-dimensional turbulence. *J. Fluid Mech.*, **107**, 375–406.

Sreenivasan, K. R. 1984. On the scaling of the turbulence energy dissipation rate. *Phys. Fluids*, **27**, 1048–1951.

Sreenivasan, K. R. 1985. On the fine-scale intermittency of turbulence. *J. Fluid Mech.*, **151**, 81–103.

Sreenivasan, K. R. 1995. On the universality of Kolmogorov constant. *Phys. Fluids*, **7**, 2778–2784.

Sreenivasan, K. R. 1998. An update on the energy dissipation rate in isotropic turbulence. *Phys. Fluids*, **10**, 528–529.

Sreenivasan, K. R. and Antonia, R. A. 1997. The phenomenology of small-scale turbulence. *Ann. Rev. Fluid Mech.*, **29**, 435–472.

Sreenivasan, K. R. and Bershadskii, A. 2006. Finite-Reynolds-number effects in turbulence using logarithmic expansions. *J. Fluid Mech.*, **554**, 477–498.

Sreenivasan, K. R. and Dhruva, B. 1998. Is there scaling in high-Reynolds-number turbulence? *Prog. Theor. Phys. Supplement*, **130**, 103–120.

Sreenivasan, K. R. and Stolovitzky, G. 1995. Turbulent cascades. *J. Stat. Phys.*, **78**, 311–333.

Tabeling, P., Zocchi, G, Belin, F., Maurer, J. and Willaime, H. 1996. Probability density functions, skewness, and flatness in large Reynolds number turbulence. *Phys. Rev. E*, **53**, 1613–1621.

Tatsumi, T. 1980. Theory of homogeneous turbulence. *Adv. Appl. Mech.*, **20**, 39–133.

Taylor, M. A., Kurien, S. and Eyink, G. L. 2003. Recovering isotropic statistics in turbulence simulations: the Kolmogorov 4/5th law. *Phys. Rev. E*, **68**, 026310.

Townsend, A. A. 1951. On the Fine-Scale Structure of Turbulence. *Proc. R. Soc. Lond. A*, **208**, 534–542.

Tsinober, A. 2009. *An Informal Conceptual Introduction to Turbulence, 2nd ed.* Springer.

Tsuji, Y. 2009. High-Reynolds-number experiments: the challenge of understanding universality in turbulence. *Fluid Dyn. Res.*, **41**, 064003.

Van Atta, C. W. and Yeh, T. T. 1975. Evidence for scale similarity of internal intermittency in turbulent flows at large Reynolds numbers. *J. Fluid Mech.*, **71**, 417–440.

Vincent, A. and Meneguzzi, M. 1991. The spatial structure and statistical properties of homogeneous turbulence. *J. Fluid Mech.*, **225**, 1–20.

von Neumann, J. 1949. In *Collected Works,* Vol.6: *Theories of Games, Astrophysics, Hydrodynamics and Meteorology,* A. H. Taub (ed.), 437–472 (1963). Pergamon Press.

Wang, L. -P., Chen, S., Brasseur, J. G. and Wyngaard, J. C. 1996. Examination of hypotheses in the Kolmogorov refined turbulence theory through high-resolution simulations. Part 1. Velocity field. *J. Fluid Mech.*, **309**, 113–156.

Watanabe T. and Gotoh T. 2004. Statistics of a passive scalar in homogeneous turbulence. *New Journal of Physics*, **6**, 40.

Yamamoto, K. and Hosokawa, I. 1988. A decaying isotropic turbulence pursued by spectral method. *J. Phys. Soc. Jpn.*, **57**, 1532–1535.

Yakhot V. and Sreenivasan K. R. 2004. Towards a dynamical theory of multifractals in turbulence. *Physica A*, **343**, 147–155.

Yakhot V. 2006. Probability densities in strong turbulence. *Physica D*, **215**, 166–174.

Yeung, P. K., Pope, S. B., Lamorgese, A. G. and Donzis, D. A. 2006. Acceleration and dissipation statitics of numerically simulated isotropic turbulence. *Phys. Fluids*, **18**, 065103.

Yokokawa, M., Itakura, K., Uno, A., Ishihara, T. and Kaneda, Y. 2002. 16.4-Tflops direct numerical simulation of turbulence by a Fourier spectral method on the Earth Simulator. *Proc. IEEE/ACM SC2002 Conf., Baltimore, 2002*, http://www.sc-2002.org/paperPDFs/pap.pap273.PDF.

Yoshida, K., Ishihara, T. and Kaneda, Y. 2003. Anisotropic spectrum of homogeneous turbulent shear flow in a Lagrangian renormalized approximation. *Phys. Fluids*, **15**, 2385–2397.

Yoshizawa, A. 1998. *Hydrodynamic and Magnetohydrodynamic Turbulent Flow.* Kluwer Academic Publishers.

Zhou, T. and Antonia, R. A. 2000. Reynolds number dependence of the small-scale structure of grid turbulence. *J. Fluid Mech.*, **406**, 81–107.

2
Structure and Dynamics of Vorticity in Turbulence

Jörg Schumacher, Robert M. Kerr and Kiyosi Horiuti

2.1 Introduction

Ancient depictions of fluids, going back to the Minoans, envisaged waves and moving streams. They missed what we would call vortices and turbulence. The first artist to depict the rotational properties of fluids, vortical motion and turbulent flows was da Vinci (1506 to 1510). He would recognize the term vortical motion as it comes from the Latin vortere or vertere: to turn, meaning that vorticity is where a gas or liquid is rapidly turning or spiraling. Mathematically, one represents this effect as twists in the velocity derivative, that is the curl or the anti-symmetric component of the velocity gradient tensor. If the velocity field is **u**, then for the vorticity is $\boldsymbol{\omega} = \nabla \times \mathbf{u}$.

The aspect of turbulence which this chapter will focus upon is the structure, dynamics and evolution of vorticity in idealized turbulence – either the products of homogeneous, isotropic, statistically stationary states in forced, periodic simulations, or flows using idealized initial conditions designed to let us understand those states. The isotropic state is often viewed as a tangle of vorticity (at least when the amplitudes are large), an example of which is given in Fig. 2.1. This visualization shows isosurfaces of the magnitude of the vorticity, and similar techniques have been discussed before (see e.g. Pullin and Saffman, 1998; Ishihara et al., 2009; Tsinober, 2009). The goal of this chapter is to relate these graphics to basic relations between the vorticity and strain, to how this subject has evolved to using vorticity as a measure of regularity, then focus on the structure and dynamics of vorticity in turbulence, in experiments and numerical investigations, before considering theoretical explanations. Our discussions will focus upon three-dimensional turbulence.

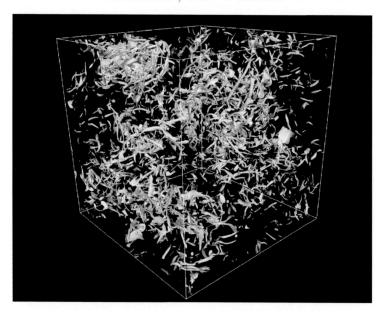

Figure 2.1 Isosurface plot of the squared vorticity (also denoted as local enstrophy) at the level of $10\langle\omega_i^2\rangle$. Data are from a direct numerical simulation of homogeneous isotropic turbulence in a periodic cube with sidelength 2π which is resolved with 2048^3 grid points (Schumacher et al., 2007; Schumacher, 2007).

2.2 Basic relations

The time evolution of an incompressible turbulent flow is governed by the Navier–Stokes equations for the velocity vector **u**

$$\frac{\partial \mathbf{u}}{\partial t} + (\mathbf{u}\cdot\nabla)\mathbf{u} = \rho^{-1}\nabla p + \nu\triangle\mathbf{u} + \mathbf{f} \qquad (2.1)$$

where ρ is the mass density, p the kinematic pressure and **f** the forcing which sustains turbulence. The symbol $\triangle$, represents the three-dimensional Laplacian, sometimes written ∇^2 or $(\partial/\partial x_i)^2$.

For a compressible flow one needs to add equations for mass and energy conservation. This chapter will focus upon incompressible flows, where these relations can be removed by using the divergence-free condition:

$$\nabla\cdot\mathbf{u} = 0\,. \qquad (2.2)$$

Vorticity $\boldsymbol{\omega}$ is the anti-symmetric component of the velocity gradient, written as a vector or with indices as follows:

$$\boldsymbol{\omega} = \nabla\times\mathbf{u} \quad \text{or} \quad \omega_i = \epsilon_{ijk}M_{jk} \quad \text{where} \quad M_{ij} = \frac{\partial u_i}{\partial x_j} \qquad (2.3)$$

Here ϵ_{ijk} is the fully anti-symmetric Levi-Civita tensor and summation over indices is implied using the Einstein summation convention. The antisymmetric part of M_{ij} is the vorticity tensor: $\Omega_{ij} = -\frac{1}{2}\epsilon_{ijk}\omega_k = \frac{1}{2}(M_{ij} - M_{ji})$. The symmetric component of M_{ij} is the strain S_{ij}, which governs the growth and twisting of vorticity,

$$S_{ij} = \frac{1}{2}(M_{ij} + M_{ji}) \quad \text{so that} \quad M_{ij} = S_{ij} + \Omega_{ij}. \tag{2.4}$$

The vortical form of the Navier–Stokes equations are obtained by taking the curl of the usual velocity equations. Thus, for an incompressible flow where $\nabla \cdot \mathbf{u} = 0$, one has

$$\nabla \times [2.1] \quad \Rightarrow \quad \frac{\partial \boldsymbol{\omega}}{\partial t} + (\mathbf{u} \cdot \nabla)\boldsymbol{\omega} = (\boldsymbol{\omega} \cdot \nabla)\mathbf{u} + \nu \triangle \boldsymbol{\omega} + \nabla \times \mathbf{f}. \tag{2.5}$$

The velocity field can be recovered from the vorticity field using a Biot-Savart integral (2.6) over space

$$u_i(\mathbf{x}, t) = \frac{1}{4\pi} \int_{V'} \epsilon_{ijk}\, \omega_j(\mathbf{x}', t) \frac{x_k - x'_k}{|\mathbf{x} - \mathbf{x}'|^3}\, dV', \tag{2.6}$$

or an inverse curl operator

$$\mathbf{u}(\mathbf{x}, t) = \nabla \times \triangle^{-1} \boldsymbol{\omega}. \tag{2.7}$$

The vorticity equation (2.5) written in indices (for $\mathbf{f} = 0$) is given by

$$\frac{D\omega_i}{Dt} = S_{ik}\omega_k + \nu \frac{\partial^2 \omega_i}{\partial x_k^2}, \tag{2.8}$$

and the strain equation is

$$\frac{DS_{ij}}{Dt} = -S_{ik}S_{kj} - \frac{1}{4}(\omega_i\omega_j - \delta_{ij}\boldsymbol{\omega}^2) - \rho^{-1}P_{ij} + \nu \triangle S_{ij}. \tag{2.9}$$

where

$$P_{ij} = \frac{\partial^2 p}{\partial x_i \partial x_j} \tag{2.10}$$

is known as the pressure Hessian.

2.2.1 Conservation laws

Equations (2.1) are known as the Euler equations for $\nu = 0$ (Euler, 1761). They conserve two quadratic invariants, depending upon dimension. In two-dimensions, these are the kinetic energy $E(t)$ and the enstrophy $\Omega(t)$:

$$E(t) = \frac{1}{2}\int_A \mathbf{u}^2(\mathbf{x}, t)\, dA \quad \text{and} \quad \Omega(t) = \frac{1}{2}\int_A \boldsymbol{\omega}^2(\mathbf{x}, t)\, dA, \tag{2.11}$$

where A is the two-dimensional flow plane. The quantity ω^2 will be refered to as the enstrophy density or local enstrophy.

In three-dimensions, the conserved quantities are the kinetic energy $E(t)$ and the helicity $H(t)$:

$$E(t) = \frac{1}{2}\int_V \mathbf{u}^2(\mathbf{x},t)\,\mathrm{d}V \quad \text{and} \quad H(t) = \int_V \mathbf{u}(\mathbf{x},t)\cdot\boldsymbol{\omega}(\mathbf{x},t)\,\mathrm{d}V\,. \qquad (2.12)$$

The helicity H is not Galilean invariant and the three-dimensional equations do not conserve enstrophy due to the vortex stretching term $(\boldsymbol{\omega}\cdot\nabla)\mathbf{u}$ of (2.5), which is identically zero in two-dimensions. Attempts to understand the stretching term form the basis for much of the mathematical and numerical analysis discussed here.

The ideal equations also conserve the circulation about Lagrangian patches. That is for a closed curve $\mathcal{C}(s,t)$ of points $\mathbf{x}(s,t)$ in space, the circulation Γ, where

$$\Gamma = \oint_\mathcal{C} \mathbf{u}(\mathbf{x},t)\cdot\mathrm{d}\mathbf{x}\,. \qquad (2.13)$$

is conserved.

2.2.2 Invariants related to strain and vorticity

Rotationally-invariant combinations of the vorticity and strain have been determined numerically and used in models. Two underlying sets of equations are the Cayley-Hamilton relations between different orders and the time-dependent restricted Euler equations. Invariants only up to order four appear in the restricted Euler equations.

The vorticity squared, or enstrophy density ω^2, and the strain squared $|\mathbf{S}|^2 = S_{ij}^2$ are the two local second-order rotational invariants of the stress tensor. The strain squared is directly related to the local kinetic energy dissipation rate $\epsilon(\mathbf{x},t)$ and for periodic boundary conditions, or for a patch of turbulence isolated from boundaries, when integrated, the vorticity squared and integral of strain squared are related to the globally averaged rate of energy dissipation $\langle\epsilon\rangle$. Given

$$\epsilon(\mathbf{x},t) = 2\nu\mathbf{S}^2\,, \qquad (2.14)$$

we have

$$\langle\epsilon\rangle = 2\nu\langle\mathbf{S}^2\rangle = \nu\langle\omega^2\rangle = 2\nu\Omega,\,. \qquad (2.15)$$

The third-order rotational invariants, first noted by Betchov (1956), are:

$$R_\omega = \omega_i S_{ij}\omega_j \quad \text{and} \quad R_S = S_{ij}S_{jk}S_{ki}\,. \qquad (2.16)$$

2: Structure and Dynamics of Vorticity in Turbulence

In three-dimensions, these are globally related (Betchov, 1956) by

$$\langle R_\omega \rangle = -\frac{4}{3}\langle R_S \rangle. \tag{2.17}$$

There are four fourth-order rotational invariants (Siggia, 1981a)

$$I_1 = |\mathbf{S}|^4, \tag{2.18}$$
$$I_2 = \omega^2|\mathbf{S}|^2, \tag{2.19}$$
$$I_3 = \omega_i S_{ij} S_{jk} \omega_k = (\mathbf{S}\boldsymbol{\omega})^2 = \omega^2 \chi^2, \tag{2.20}$$
$$I_4 = \omega^4. \tag{2.21}$$

Their global values are linearly independent and I_3 will have a role when the 'restricted' Euler equations are introduced.

The Cayley-Hamilton theorem is that any second-order tensor $\mathbf{M}$ will satisfy the relation

$$\mathbf{M}^3 - \mathcal{P}\mathbf{M}^2 + \mathcal{Q}\mathbf{M} - \mathcal{R}\mathbf{I} = \mathbf{O}, \tag{2.22}$$

where $\mathbf{I}$ is the identity matrix and $\mathbf{O}$ is the zero matrix. $\mathcal{P}$, $\mathcal{Q}$ and $\mathcal{R}$ are the rotationally invariant first, second and third invariants for the velocity gradient tensor $\mathbf{M}$:

$$\mathcal{P} = \nabla \cdot \mathbf{u}, \quad \mathcal{Q} = -\frac{1}{2}\mathbf{S}^2 + \frac{1}{4}\omega^2, \quad \mathcal{R} = -\frac{1}{3}R_S - \frac{1}{4}R_\omega. \tag{2.23}$$

The first-invariant $\mathcal{P}$ is zero in an incompressible flow. The second-invariant $\mathcal{Q}$ has a natural role in determining the pressure for the incompressible equations.

The strain S_{ij} can be decomposed into three eigenvectors

$$\{\mathbf{e}_-, \mathbf{e}_r, \mathbf{e}_+\} \quad \text{with eigenvalues} \quad \sigma_- > \sigma_r > \sigma_+ \tag{2.24}$$

where $\sigma_\pm = -\sigma_r/2 \pm \sigma_i$ and incompressibility implies $\sigma_- + \sigma_r + \sigma_+ = 0$. It was during a study of the statistics of the terms above (Kerr, 1985) that it was first noted that the eigenvector $\mathbf{e}_r$, the one with the smallest positive eigenvalue σ_r, is statistically aligned with the vorticity.

This inspired a study of the statistics of the normalized middle eigenvalue β_N in Ashurst et al. (1987). Further related statistics are: Kerr (1987) noted that there is a one-to-one mapping from the β_N statistics to the statistics of R_S normalized, S_N^3. Given:

$$\beta_N = \sqrt{6}\frac{\sigma_r}{\mathbf{S}^{1/2}}, \quad \alpha_N = \sqrt{6}\frac{R_\omega}{\omega^2 \mathbf{S}^{1/2}} \quad \text{and} \quad S_N^3 = \sqrt{6}\frac{R_S}{\mathbf{S}^{3/2}} \tag{2.25}$$

then at every point

$$S_N^3 = -\frac{1}{2}\beta_N(3 - \beta_N^2). \tag{2.26}$$

α_N is related to the von Kármán-Howarth equation discussed below. The limits of β_N, $\alpha_N = \sqrt{6}$ and S_N^3 are:

$$-1 \leq (\beta_N, \alpha_N, S_N^3) \leq 1. \tag{2.27}$$

By taking a divergence of the velocity equation (2.1) one obtains:

$$\rho^{-1}\triangle p = -\nabla \cdot (\mathbf{u} \cdot \nabla)\mathbf{u} = -u_{i,j}u_{j,i} = -\mathbf{S}^2 + \frac{1}{2}\boldsymbol{\omega}^2 = 2Q \tag{2.28}$$

Related to the Cayley-Hamilton relation is the following symmetric tensor

$$\Lambda_{ij} = S_{ik}S_{kj} + \Omega_{ik}\Omega_{kj}. \tag{2.29}$$

The use of eigenvalues and eigenvectors to identify structures is discussed below.

2.2.3 Hierarchies of the S_{ij} equation

Starting with the inviscid equations for the vorticity and the strain, two equivalent hierarchies have been described. One leads to the 'restricted' Euler equations, which is closed by an isotropic assumption upon the form of $\triangle P$. The other is the quaternionic formulation for the evolution of components of S_{ij} aligned with the direction of vorticity $\hat{\boldsymbol{\omega}} = \boldsymbol{\omega}/|\boldsymbol{\omega}|$.

After the equations for $\boldsymbol{\omega}^2$ and $\mathbf{S}^2$, the next equations in the hierarchy are those for R_ω and R_S. Using

$$H_{ij} = -P_{ij} + \frac{\delta_{ij}}{3}P_{kk} \tag{2.30}$$

they are given by

$$\frac{D}{Dt}R_\omega = I_3 - \frac{2}{3}\omega^2 Q + \omega_i H_{ij}\omega_j \tag{2.31}$$

$$\frac{D}{Dt}R_S = -\frac{3}{4}I_3 + S^2 Q + 3S_{ij}H_{jk}S_{ki} \tag{2.32}$$

Finally one needs an equation for $I_3 = (\omega_i S_{ij})^2$ which is given by

$$\frac{D}{Dt}I_3 = -\frac{4}{3}QR_\omega + 2\omega_i H_{ij}S_{jk}\omega_k. \tag{2.33}$$

If one assumes that the pressure Hessian is isotropic, or $H_{ij} \equiv 0$, one gets a closed system. This version of the 'restricted' Euler equations was first recognized in the PhD thesis of J. Leorat (1975, unpublished) and then developed by Vieillefosse (1982, 1984), whose name is often attached to this approximation.

Next consider the invariants Q and R of the Cayley-Hamilton equations:

$$\frac{D}{Dt}Q = -3R - H_{ij}S_{ij} \tag{2.34}$$

and

$$\frac{D}{Dt}R = \frac{2}{3}Q^2 - S_{ij}H_{ij}S_{ki} - \frac{1}{4}\omega_i H_{ij}\omega_j. \tag{2.35}$$

This version of the 'restricted' Euler equations was first noted by Vieillefosse, and applied by Cantwell (1992). There have been numerous studies, applications and modifications since, discussed below in Section 2.5.3.

The other hierarchy to consider are the equations for two components of S_{ij} contracted with the direction of vorticity. With $\omega = |\boldsymbol{\omega}|$ one defines

$$\boldsymbol{\xi} = \frac{\boldsymbol{\omega}}{\omega}, \quad \alpha = \boldsymbol{\xi} \cdot (\mathbf{S} \cdot \boldsymbol{\xi}) \quad \text{and} \quad \boldsymbol{\chi} = \boldsymbol{\xi} \times (\mathbf{S} \cdot \boldsymbol{\xi}). \tag{2.36}$$

The evolution equations for the scalar α and vector $\boldsymbol{\chi}$ are

$$\frac{D}{Dt}\alpha = -\alpha^2 + \chi^2 - P_\alpha \quad \text{and} \quad \frac{D}{Dt}\boldsymbol{\chi} = -2\alpha\boldsymbol{\chi} - \mathbf{P}_\chi \tag{2.37}$$

where $P_\alpha = \boldsymbol{\xi} \cdot (\mathbf{P} \cdot \boldsymbol{\xi})$ and $\mathbf{P}_\chi = \boldsymbol{\xi} \times (\mathbf{P} \cdot \boldsymbol{\xi})$ where $\mathbf{P}$ is again the pressure Hessian. This hierarchy has been continued (Gibbon et al., 2006), but cannot be closed. However, analysis of the next order can be used to impose an additional initial condition that the Euler equations must obey in order to have possible singular behavior.

2.2.4 Balance and production equations

Besides looking at the structures and statistics related to their alignment, several measures of the intensity of growth have been proposed. For comparison let us first consider the balance equation for kinetic energy, which in Navier–Stokes turbulence can only decrease. The equation is obtained by taking scalar product of Eq. (2.1) with u_i, integrating in the absence of forcing over V and then averaging

$$\frac{dE}{dt} = -\oint_{\partial V}\left[u_j\left(p + \frac{u_i^2}{2}\right) - \frac{\nu}{2}\frac{\partial u_i^2}{\partial x_j}\right]dA_j - \nu\int_V\left(\frac{\partial u_i}{\partial x_j}\right)^2 dV \tag{2.38}$$

By applying the same procedure to the vorticity equation Eq. (2.8), one can then get the enstrophy balance equation:

$$\frac{d\Omega}{dt} = \int_V \omega_i S_{ij}\omega_j\, dV - \nu\int_V\left(\frac{\partial \omega_i}{\partial x_j}\right)^2 dV + \nu\oint_{\partial V}\omega_i\frac{\partial \omega_i}{\partial x_j}\, dA_j. \tag{2.39}$$

The first term of (2.39) is an enstrophy production term due to vortex stretching, the second term is the enstrophy dissipation. For simple configurations such as an infinitely extended domain or a periodic box which is frequently used in simulations, the surface integral contribution in (2.39) disappears.

Some standard turbulent length scales need to be defined. These are the Kolmogorov length scale

$$\eta_K = \left(\frac{\nu^3}{\langle\epsilon\rangle}\right)^{1/4}, \qquad (2.40)$$

and the Taylor microscale with (2.12)

$$\lambda = \left(\frac{E/(3V)}{\langle\epsilon\rangle/15\nu}\right)^{1/2}. \qquad (2.41)$$

The von Kármán–Howarth equation (von Kármán and Howarth, 1938) is a one-dimensional, cumulant expansion of the enstrophy equation for homogeneous isotropic turbulence:

$$\frac{\partial}{\partial t}(\langle u_x^2\rangle f) = \langle u_x^2\rangle^{3/2}\left(\frac{\partial k}{\partial r} + 4\frac{k}{r}\right) - 2\nu\langle u_x^2\rangle\left(\frac{\partial^2 f}{\partial r^2} + \frac{4}{r}\frac{\partial f}{\partial r}\right). \qquad (2.42)$$

This equation formed much of the basis of early turbulence work, especially in the interpretation of hot-wire probes which are inherently one-dimensional. We expand the functions which determine the longitudinal second-order and third-order correlations in homogeneous isotropic turbulence as:

$$f(r) = 1 + \frac{1}{2!}f_0''r^2 + \frac{1}{4!}f_0''''r^4 + \cdots, \qquad (2.43)$$

$$k(r) = \frac{1}{3!}f_0'''r^3 + \cdots, \qquad (2.44)$$

Inserting (2.43) and (2.44) into (2.42) one gets equations for each even power of r. The equation for r^0 is the turbulent kinetic energy balance. The enstrophy balance follows for $\mathcal{O}(r^2)$ and is given by (Batchelor, 1953; Rotta, 1997)

$$\frac{d}{dt}\left(\langle u_x^2\rangle f_0''\right) = \frac{7}{6}\langle u_x^2\rangle^{3/2}k_0''' + \frac{7}{3}\nu\langle u_x^2\rangle f_0''''. \qquad (2.45)$$

Assuming homogeneous isotropic turbulence, the energy dissipation rate can be determined from one-dimensional data using: $\langle\epsilon\rangle = 15\nu\langle(\partial u_x/\partial x)^2\rangle$.

The enstrophy production and triple strain correlation can be determined from the one-dimensional velocity-derivative skewness by

$$S = k_0''' = \frac{\langle(\partial u_x/\partial x)^3\rangle}{\langle(\partial u_x/\partial x)^2\rangle^{3/2}} = -\frac{6}{7}\frac{\langle R_\omega\rangle}{(\langle\epsilon\rangle/15\nu)^{3/2}} = \frac{24}{21}\frac{\langle R_S\rangle}{(\langle\epsilon\rangle/15\nu)^{3/2}}, \quad (2.46)$$

and it follows from (2.45) that

$$\frac{d\langle\omega^2\rangle}{dt} = -\frac{7S}{3\sqrt{15}}\langle\omega^2\rangle^{3/2} - 70\nu\left\langle\left(\partial^2 u_x/\partial x^2\right)^2\right\rangle. \quad (2.47)$$

As a consequence of the negative value of S, the first term on the r.h.s. of (2.47) is positive and denotes the local enstrophy production due to vortex stretching. The second term is the enstrophy dissipation.

This equation has served to connect hot-wire measurements to theory and numerical simulations. In this context, simulations originally meant spectral closure calculations and now refers to direct numerical simulations (DNS). In numerical simulations of isotropic turbulence, as the Reynolds number goes to zero, so does the skewness. For $R_\lambda \geq 20$ (Chen et al., 1993; Herring and Kerr, 1982), $S \approx -0.5$. The numerical value of S has only increased slowly as resolution and the Reynolds number have increased. Starting with a value of -0.4 in the first DNS (Orszag and Patterson, 1972) for $R_\lambda = 35$, until the largest Reynolds number of $R_\lambda = 1130$ with $S \approx -0.64$ on machines currently available (Ishihara et al., 2007).

Spectral analysis suggests there is an upper bound to S. This could be related to the distributions of α_N and β_N (2.25) from Kerr (1987), which seem to be universal (see Figure 2.12). In Section 2.5.3 the relationship of these statistics to the alignment of vorticity and strain responsible for vortex tubes is discussed.

Reaching an appropriate value of S has been a valuable tool in DNS of isotropic, homogeneous turbulence for identifying when the flow is fully turbulent. From random initial conditions with $S = 0$, its magnitude increases rapidly, overshooting the asymptotic value. The overshoot could be related to transient structures where, in the infinite Reynolds number limit, the growth of enstrophy could be unbounded. This possibility is discussed below in the context of rigorous mathematics for bounds on the growth rate of enstrophy in the Navier–Stokes equations and possible singularities of the Euler equation.

Once S has reached its asymptotic value, most agree that the resulting flow represents a homogeneous and isotropic state of turbulence and this is the regime from which most analysis and visualizations are taken. This is sometimes called the von Kármán-Howarth test and was introduced for DNS by Siggia and Patterson (1978).

2.3 Temporal growth of vorticity

2.3.1 Blowup in three-dimensional Euler flows

Vorticity plays a central role in the fundamentally important and still open questions of the existence and smoothness of three-dimensional solutions of the incompressible Navier–Stokes equation (2.1,2.2) and its inviscid ($\nu = 0$) version, the Euler equation (Euler, 1761). Loosely speaking, the open questions are whether the velocity and pressure (u_i, p) for any smooth initial condition remains smooth while evolving for all time, i.e. $u_i, p \in C^\infty(\mathbb{R}^3, [0, \infty))$, with no higher derivatives blowing up in a finite time. More mathematical details for the Navier–Stokes case are found in Fefferman (2000) and Doering (2009) and for the Euler case in Gibbon (2008) or Bardos and Titi (2007).

Most of the known mathematics about the Euler equation and possible blow-up derives from the theorem of Beale, Kato and Majda (1984), which states that if the time integral of the supremum of vorticity is bounded from above as

$$\int_0^T \left\{ \sup_{\mathbf{x} \in \mathbb{R}^3} |\boldsymbol{\omega}(\mathbf{x}, t)| \right\} dt \leq \infty, \qquad (2.48)$$

then no other quanity can blow up. In Beale et al. (1984), the L_∞ Sobelev norm is taken as the supremum, that is $\sup |\boldsymbol{\omega}| = \|\boldsymbol{\omega}\|_\infty$. This result has also been proven using the bounded mean oscillation norm to identify the supremum (Kozono and Taniuchi, 2000).

Due to this and related mathematical results, shortly after publication of Beale et al. (1984), there was a surge in attempts by the numerical community to address the Euler singularity problem because it finally gave them the tools needed to identify a blowup of the Euler equation. That is, to identify a singularity, all one needs to demonstrate is that vorticity increases in a manner consistent with exceeding this bound. No higher order derivatives of velocity need to be calculated. Furthermore, the integral condition implies a bound on power law growth of the maximum of vorticity against which calculations can be compared. That is if T_c is a projected singular time and it is assumed that

$$\sup |\boldsymbol{\omega}| \approx \frac{C}{(T_c - t)^\gamma} \quad \text{then} \quad \gamma \geq 1 \qquad (2.49)$$

is required if there is a singularity. While, in principle, satisfying (2.48) is all that is needed to show singular growth, in practice several additional conditions should be satisfied. The reason is that even if growth looks like a power law it could just as easily be super-exponential. The strongest complementary test follows from a corrolary to (2.48) that the time integral of the supremum of strain also bounds any singularity of the Euler equation (Ponce, 1985). One component of the strain is the stretching along the vorticity, usually denoted as α (see Eq. (2.37)). It is also the growth rate for vorticity. From this, a sensitive test, where possible, would be to monitor the stretching at the point of maximum vorticity, denoted here α_∞. If this can be monitored, and if $\sup|\boldsymbol{\omega}|$ grows as a power law, then α_∞ should grow as

$$\alpha_\infty = \frac{\gamma}{(T_c - t)} \tag{2.50}$$

where γ is the exponent of the vorticity power law behavior (2.49). No calculation reporting singular growth has so far been able to perform this α_∞ test satisfactorily. The closest is the pseudospectral simulation of Kerr (1993) which obtained growth of $\max(\alpha)$ that had the $(T_c - t)^{-1}$ scaling. Three-dimensional isosurfaces of vorticity modulus and a cross-section through the symmetry plane are displayed in Figs. (2.2) and (2.3), respectively.

One problem in using large simulations to extract sensitive mathematical or physical properties is the difficulty of getting two different codes, using similar initial conditions, to agree. In the case of Euler calculations, Hou and Li (2006, 2007) have claimed that by starting with an initial condition which qualitatively resembled the initial condition of Kerr (1993), that singular behavior is eventually suppressed. Whether or not the conclusions of Hou and Li are premature, as suggested by Bustamante and Kerr (2008), one must conclude that a more robust initialization for this problem is needed. If this can be resolved, then the study of a so-called Kida–Pelz flow (Kida and Murakami, 1987; Boratav and Pelz, 1994) by Grafke et al. (2008) suggests that the resolution of the vorticity will eventually prove to be more important than the order of the numerical scheme.

Further diagnostics were needed if numerical analysis was to play a role. Three categories have been suggested and attempted. These are: (1) Different local measures such as velocity; (2) The collapse of relevant length scales; (3) Growth of enstrophy, a global measure. An example of what we would like to achieve is given by the two-dimensional surface geostrophic equations, where it can be shown that in addition to an integral constraint similar to (2.48), that the time integrals of the squares of the velocity and the equivalent of curvature of vortex lines also bound any possible singularities. Essentially

$$\int_0^T \left\{ \sup_{\mathbf{x} \in \mathbb{R}^2} |\mathbf{u}(\mathbf{x},t)|^2 \right\} dt \leq \infty, \tag{2.51}$$

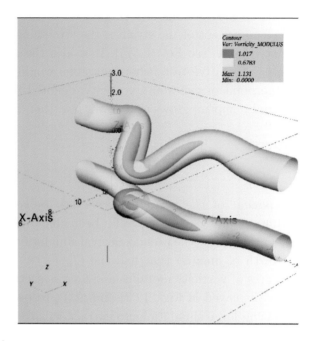

Figure 2.2 Euler anti-parallel vortices in full periodic domain near $t = 2.51$. taken from Bustamante and Kerr (2008). Yellow tubes are isosurface contours of vorticity modulus corresponding to 60% of the instantaneous maximum of vorticity modulus. Red elongated blobs are isosurfaces corresponding to 90% of the maximum of vorticity modulus.

and

$$\int_0^T \left\{ \sup_{\mathbf{x} \in \mathbb{R}^2} |\kappa(\mathbf{x}, t)|^2 \right\} \mathrm{d}t \leq \infty \,. \tag{2.52}$$

The same analysis has been attempted for the Euler equations (Constantin et al., 1996), and the method illustrates a number of additional properties.

First, the Biot–Savart integral for the stretching α in terms of the vorticity is given by (Constantin, 1994)

$$\alpha(\mathbf{x}, t) = \frac{3}{4\pi} P.V. \int_V \frac{(\hat{y}_m \xi_m)\, \epsilon_{ijk} \hat{y}_i \xi_j (\mathbf{x} + \mathbf{y}) \xi_k(\mathbf{x}) \omega(\mathbf{x} + \mathbf{y})}{y^3} \mathrm{d}^3 y \,, \tag{2.53}$$

where $y = (y_i^2)^{1/2}$ and $\hat{y}_i = y_i/y$. The source terms in this integral is then broken into two inner regions around the position of sup $|\boldsymbol{\omega}|$. The philosophy is similar to how (2.48) was proven (Beale et al., 1984), where only one length scale was necessary. The elusive step in the proof for curvature and velocity bounds is the necessity to bound both collapsing inner length scales. To close the proof a conditional assumption was needed (Constantin, 2001).

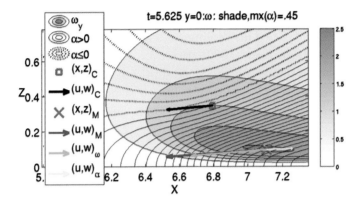

Figure 2.3 Cross-section through the x–z symmetry plane from the above simulation at $t = 5.625$. Vorticity (gray-scale contours) is flattened in z. The maximum vorticity $\sup |\omega|$ is at $(x, z)_M$ (red cross). The maximum of the stretching rate (lines: $\alpha > 0$, dots: $\alpha \leq 0$) is in the lower right corner. Overall motion (upper black arrow) is to the left. At $(x, z)_M$, both the Lagrangian motion $(u, w)_M$ (blue) and the motion of $(x, z)_M$ (green) are shown.

However, what we learn from this is that if three-dimensional Euler case is in fact singular, we should not be looking for ideal, single length scale collapse. There are probably two lengths involved. This would be consistent with an energy argument from Constantin (1994). Numerical evidence is discussed below.

Despite current ambiguities in the mathematics of curvature and velocity, further analysis of their growth properties should be attempted from the perspective that secondary properties might not have singular growth in the immediate vicinity of the maximum of vorticity, as numerics have suggested for the velocity (Kerr, 2005). This could be consistent with new constraints based upon integrals along finite lengths of the vortex lines (Deng et al., 2005). Another approach is to consider the evolution of complex time singularities, which were introduced by Brachet et al. (1983). There has been more recent analysis of power series by Cichowlas and Brachet (2005) for inviscid Kida–Pelz and Taylor–Green flows, and a Kida–Pelz power series expansion by Gulak and Pelz (2005). We should also start thinking about the pressure, either for how it could regularise the solutions as in Pelz and

Ohkitani (2005), or the consideration of additional constraints related to the pressure Hessian as in Gibbon et al. (2006). It was originally expected, based upon a variety of scaling arguments, including how vortex filament calculations collapse, that the innermost length scale for the Euler case would scale as $R \sim (T_c - t)^{1/2}$. This scaling does appear in the re-analysis (Kerr, 2005) of the data from Kerr (1993). However, the stronger conclusion from Kerr (1993) is that many positions of $\sup(|\partial u_i/\partial x_j|)$, including vorticity, collapse to a common point with the length scale going as $R \sim (T_c - t)$. This conclusion is being supported by ongoing analysis of the new data set (Bustamante and Kerr, 2008) depicted in Fig. 2.3.

If a test based upon a global measure such as enstrophy (2.11) could be proven, then this would be the easiest type of test to implement. These would be less sensitive to local resolution than pointwise measures such as $\sup |\boldsymbol{\omega}|$. No analytic result exists for such a test, but there is an empirical result. Bustamante and Kerr (2008) have found that by simultaneously tracking the growth of enstrophy and enstrophy production (2.39) that

$$\Omega \sim (T_c - t)^{1/2} . \tag{2.54}$$

Further work has suggested that this is a result of the following relation between enstrophy production $\dot{\Omega}$, enstrophy Ω and the circulation of the initial vortex Γ.

$$\dot{\Omega} \sim \frac{\Omega^3}{\Gamma^3} \quad \text{with} \quad \Omega \sim a \frac{\Gamma^{3/2}}{(T_c - t)^{1/2}} . \tag{2.55}$$

This relation bears an uncanny resemblance to the upper bound for the growth of enstrophy in the Navier–Stokes equations discussed in the next subsection. Similar conditions do not exist that could bound and test numerical solutions of the Navier–Stokes equation, which is our ultimate interest. Only tentative steps have been taken in simultaneously including the effects of curvature in the analysis of the viscous term and stretching (Constantin and Fefferman, 1993). Most analysis of the Navier–Stokes case is instead based upon definitions of measure and intermittency, which is reviewed by Gibbon (2008) or Bardos and Titi (2007).

With advances in computational power and numerical algorithms, one might have expected that by now a new level of calculations of the three-dimensional Euler case would be providing us with new insight (Bustamante and Kerr, 2008; Grafke et al., 2008; Hou and Li, 2008). However, the current conclusion is that even though computational power has increased, a good initial condition is still elusive. The safest type of initial condition is still a small set of Fourier modes (Grafke et al., 2008), but this does not get one very far into the singular regime. The conclusion is that aside from the empirical, global enstrophy test discussed above, we are still far from being able to satisfy the sup $|\boldsymbol{\omega}|$, see (2.48), and related tests conclusively.

2.3.2 Identifying growth in three-dimensional Navier–Stokes flow

This section will focus on the evidence for intense enstrophy growth in the Navier–Stokes equation that could be relevant to the numerical search for the most strongly dissipating structures in a turbulent flow. This section will not discuss the current state of the mathematics for possible singularities of the Navier–Stokes equations, which has recently been reviewed by Doering (2009).

The first step in this search is to identify numerical tools that do not presuppose any particular type of structure, in particular are not prejudiced by the most intense structures found in simulation of the Euler equations. The importance of such a test is that the existing evidence (Holm and Kerr, 2007; Schumacher et al., 2010) that the most intense structures of Navier–Stokes flows are not consistent with anti-parallel Euler structures discussed in the previous section. This suggests that finite viscosity has a role that begins in the early stages of evolution.

What is known about the role of viscosity in the transition from a nearly inviscid flow to Navier–Stokes turbulence? If the initial condition is prescribed in form of vortices, the first obvious viscous effect is reconnection, or the cut-and-connect of vortices and thus a change in topology. The topological change is from one Lagrangian frame of reference to another. For anti-parallel configurations, this manifests itself as an abrupt change in the circulation from the original symmetry plane the vortices pass through, to the one containing the reconnected vortices (Kerr et al., 1990; Virk et al., 1995).

To understand the role of viscosity in more general flows, initial conditions using only a few Fourier modes are preferred. The expectation was that this will lead to more general vortical configurations at the small scales. However, the symmetric Taylor–Green initial condition (Brachet et al., 1983) evolves into a flow dominated by vortex sheets, which are less general than using prescribed vortex tubes as the initial condition. Eventually reconnection does occur and the topological coordinate change has been illustrated by evolving Weber–Clebsch variables in addition to the velocity and vorticity (see Cartes et al. (2007)). The Kida–Pelz symmetric initial condition leads more directly to a flow with small-scale interacting vortices (Kida and

Murakami, 1987; Boratav and Pelz, 1994; Grafke et al., 2008), followed by reconnection.

Enstrophy growth occurs in Navier–Stokes as well as Euler flows, but is expected to saturate once viscous effects dominate. Some of the earliest numerical searches for possibly singular behavior of the Navier–Stokes and Euler equations (Brachet et al., 1983) used fits of the high-wavenumber kinetic energy spectra as their tool. Results have proven inconclusive due to the difficulty in obtaining sufficiently smooth spectra.

What was clear from these results was that enstrophy growth was much weaker than spectral closure predictions, which is also the prediction of the von Kármán–Howarth theory (2.47) with $\nu = 0$. The latter predicts that a scaling of $\Omega \sim (T_c - t)^{-2}$ would be found. The problem is that the von Kármán–Howarth theory (2.47) implicitly assumes that the stretching $\alpha \sim \omega$. This is true for homogeneous turbulence once a statistical equilibrium is reached, but is in general not true for arbitrary initial conditions similar to those used for Euler equations.

However, there is now an analytic upper bound to what the enstrophy growth rate could be for an instant in time that avoids assumptions on the stretching with respect to the vorticity. Following Lu and Doering (2008), it is given by

$$\frac{\mathrm{d}\Omega(t)}{\mathrm{d}t} \leq \frac{27c^3}{4\nu^3}\Omega(t)^3 \,. \tag{2.56}$$

The proof depends upon the realization that while enstrophy production via stretching is increasing in a Navier–Stokes flow, simultaneously the dissipation is also increasing. This moderates any possible increase in enstrophy.

Can any configuration of vortex structures achieve this upper bound? If so, the configuration does not have to persist in time, but could still give guidance as to the most strongly dissipating structures in a turbulent flow. An answer to this question has been provided by variational numerics. Lu and Doering (2008) first found solutions identified with interacting Burgers vortices, for which $\dot{\Omega} \sim \Omega^{7/4}$. Then in a refined search they found a solution whose growth rate matched the analytic bound with $c = \sqrt{2/\pi}$. The structure consists of colliding, axially symmetric vortex rings, as shown in Fig. 2.4.

Could solutions such as this exist in isotropic turbulence? For this to happen, the upper bound on the growth rate (2.56) would have to persist. The power law growth in enstrophy follows then to

$$\Omega \sim \frac{\nu^{3/2}}{(T_c - t)^{1/2}}, \tag{2.57}$$

2: *Structure and Dynamics of Vorticity in Turbulence*   59

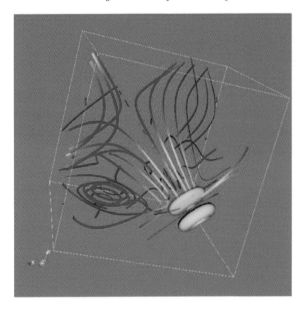

Figure 2.4 The highly symmetric configuration of two colliding vortex rings saturates the upper bound on enstrophy growth (2.56). Displayed are field lines of the vorticity field and an isosurface of the vorticity magnitude. The image is taken from Lu and Doering (2008).

which is identical in form to the empirically determined relation from an Euler calculation in relation (2.55) once the viscosity ν is substituted by the circulation Γ.

This type of short-term enstrophy growth was not seen in two recent Navier–Stokes calculations (Holm and Kerr, 2007; Schumacher et al., 2010). However, Schumacher et al. (2010) have identified small regions where for short times the inviscid case prediction of the von Kármán–Howarth theory with $\dot{\Omega} \sim \Omega^{3/2}$ is obeyed. This growth is not associated with the $(T_c - t)^{-2}$ growth because there is strong viscous dissipation of enstrophy during these periods. The study of the local enstrophy growth is done in fluid subvolumes $V_\ell = \ell^3$ with $\ell = 17\eta_K$ which are attached to Lagrangian tracers that are advected through the turbulent flow. These comoving laboratory systems allowed vorticity averages and their growth to be determined locally in Navier–Stokes turbulence, not on the integral scales of the flow.

Individual growth histories of enstrophy in these subvolumes, defined by

$$\Omega_\ell(t) = \frac{1}{2} \int_{V_\ell} \omega^2(\mathbf{x}, t) \mathrm{d}V, \qquad (2.58)$$

can vary strongly in time. Schumacher et al. (2010) find that these are bounded from above by the scaling $d\Omega_\ell/dt < \Omega_\ell^{3/2}$.

In both of these calculations (Holm and Kerr, 2007; Schumacher et al., 2010) the vorticity configurations associated with the maxima of enstrophy growth could more accurately be described as orthogonal or transverse vorticity rather than as anti-parallel vortices or colliding vortex rings. Siggia (1981b) was the first paper to identify vortex structures in a numerically generated three-dimensional simulation. In that and other past papers that have tried to identify the most intense, and possibly most dissipative, structures, the transverse configuration is usually identified. Recently, the data from Holm and Kerr (2007) were replotted in the same manner suggested by Schumacher et al. (2010), and it was found that $\dot{\Omega} \sim \Omega^{3/2}$ is obeyed. Surprisingly from the time vortex sheets form until well after vortex reconnection has begun and an energy cascade has started to form.

2.4 Spatial structure of the turbulent vorticity field

2.4.1 Exact vortex solutions of the Navier–Stokes equations

Since the very beginning of turbulence research, efforts were taken to understand some aspects of the complex fluid motion from exact solutions of the Navier–Stokes equations. They were thought as basic building blocks that can be found repeatedly on different spatial and temporal scales. For a comprehensive review of the history of these developments in turbulence research we refer the reader to articles by Pullin and Saffman (1998) and Rossi (2000).

Probably the most famous of such exact solutions is the Burgers vortex solution (Burgers, 1948). The problem is formulated in cylindrical coordinates (r, θ, z). An axisymmetric vortex is stretched in an outer irrotational straining field. The total velocity field is thus composed of two contributions, the straining field with a rate a_0 and an azimuthal induced part

$$\mathbf{u}(\mathbf{x}) = -\frac{a_0}{2} r \mathbf{e}_r + a_0 z \mathbf{e}_z + u_\theta^{(\omega)}(r) \mathbf{e}_\theta . \qquad (2.59)$$

The evolution for the axial vorticity component $\omega_z(r,t)$ follows from (2.8) and is given by

$$\frac{\partial \omega_z}{\partial t} = \frac{a_0}{2r} \frac{\partial}{\partial r}(r^2 \omega_z) + \frac{\nu}{r} \frac{\partial}{\partial r}\left(r \frac{\partial \omega_z}{\partial r}\right), \qquad (2.60)$$

which tends then to a steady solution for $t \to \infty$ and takes the form

$$\omega_z(r) = \frac{\Gamma}{\pi r_B^2} \exp(-r^2/r_B^2), \tag{2.61}$$

$$u_\theta^{(\omega)}(r) = \frac{\Gamma}{2\pi r_B} \frac{1 - \exp(-r^2/r_B^2)}{r/r_B}, \tag{2.62}$$

with the circulation Γ and the Burgers radius $r_B = \sqrt{4\nu/a_0}$. Generalizations with a time-dependent strain rate $a(t)$ are possible and result in an additional equation for $r_B(t)$ which is given by

$$\frac{\mathrm{d} r_B^2}{\mathrm{d}t} + a(t) r_B^2 = 4\nu. \tag{2.63}$$

Solving (2.63) and taking the limit $t \to \infty$ brings us back to a_0 given that $a(t) \to a_0$. Another extension to a nonuniformly stretched vortex has been formulated by Rossi et al. (2004). The spatially constant stretching rate a_0 was generalized to a stretching rate $a(r,z)$ localized in the vortex core with $a(r = \infty, z = 0) = a_0$. This extension of Burgers classical model was in good agreement with an experimentally generated roll-up vortex in a separated boundary layer behind an obstacle (Cuypers et al., 2003). The structure and evolution of the roll-up vortex is shown in Fig. 2.5.

A special case of the Burgers vortex model was put forward by Lundgren (1982). This model considers passive perturbations about the central vortex. Their evolution yields a time-averaged $k^{-5/3}$ energy spectrum due to a combination of two-dimensional spiral winding and stretching. The general case is discussed by Saffman (1997). The result is important for two reasons. First, this is the only analytic calculation of a $-5/3$ spectrum that is not based on statistics or equations that are severe spectral truncations of the Navier–Stokes equation. This is possible because there is an energy cascade generated by the Burgers vortex, not by statistics, as the fluctuations are pulled into the center of the flow. Second, the $-5/3$ spectrum only appears as a time average. Therefore we should not expect to see instantaneous $-5/3$ spectra in turbulent flows.

Given this insight, Pullin and Saffman (1998) examined higher order derivative statistics in this model once the one free parameter was set. For low-order statistics the results were suggestive, but higher-order statistics were not consistent with observations and simulations, which suggest that these statistics are dominated by events with steeper gradients than axisymmetric vortices (Saffman and Pullin, 1996). And when vortices of roughly equal magnitude interact in calculations of homogeneous turbulence, and eventually generate a $-5/3$ spectrum, spirals are not observed (Holm and

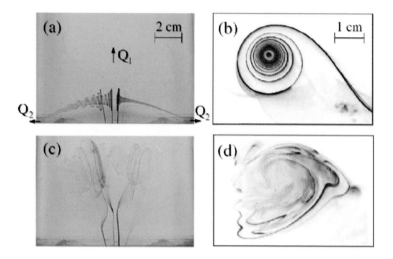

Figure 2.5 Vortex roll-up as generated behind an obstacle in a laminar boundary layer flow experiment. Q_1 is the downstream flowrate and Q_2 is the suction flowrate in a slot behind the ramp. The left figures provide a top view and the right figures a side view. Figures (a) and (b) show the initial stage of the evolution and figures (c) and (d) a later stage. The image is taken from Cuypers et al. (2003).

Kerr (2007)). Nonetheless, the Lundgren model offers an intuitive explanation of experimental observations of spiral vortices such as those by Cuypers et al. (2003) and new insight into how a cascade could form.

2.4.2 Separation of high-amplitude vorticity regions

It has been observed in many direct numerical simulations that the vorticity field is indeed concentrated in tube-like regions of high vorticity as shown in Fig. 2.1 and as suggested by the Burgers vortex solution. Numerical experiments by Siggia (1981b), Kerr (1985) or Vincent and Meneguzzi (1991) were among the first that visualized high-amplitude vorticity events. In laboratory experiments of von Kármán swirling flows, Douady et al. (1991) used the cavitation of bubbles to visualize vortex filaments. The existence of several criteria, which have been suggested in the past to separate coherent vortex filaments from the turbulent background, illustrates that there is no unique definition of what determines a coherent vortex filament. Here, we list just some examples:

- The simplest extraction is done by the cut-off criterium. Vortex filaments

are all space-time points with $|\omega_i|(\mathbf{x},t) \geq \omega_c$. The criterion was used in Fig. 2.1.
- The so-called Q-criterium in which coherent vortices are defined as space-time points with a positive second invariant $Q > 0$ of the velocity gradient tensor M_{ij} from (2.23). The regions of large Q or intense vorticity corresponds to local minima of the pressure.
- Vortical flow regions have local topologies of elliptical fixed points which correspond to the eigenvalues of M_{ij}. Recalling the eigenvalues from (2.24), $(\sigma_\pm = -\sigma_r/2 \pm \sigma_i, \sigma_r)$, then the sign of σ_r determines if it is an in- or outgoing spiral and isosurfaces of σ_i can be used to display vortex filaments.
- An identification method for the vortex tube alternative to the Q-criterion is the λ_2-method (Jeong and Hussain, 1995), in which the eigenvalues, λ_i of Λ_{ij} given by (2.29) were utilized. These eigenvalues were ordered in the descending order as $\lambda_1 \geq \lambda_2 \geq \lambda_3$. The corresponding eigenvectors are denoted as $\mathbf{e}^{(1)}$, $\mathbf{e}^{(2)}$ and $\mathbf{e}^{(3)}$. The vortex core was defined as the region with $\lambda_2 < 0$. Note that $Q = -(\lambda_1 + \lambda_2 + \lambda_3)/2$.
- The vorticity-dominated region similar to a core region of the vortex tube can be extracted as the region in which the condition $0 > \lambda_+ \geq \lambda_-$ holds (Horiuti, 2001), where the eigenvalues are reordered according to the degree of alignment of the corresponding eigenvectors with the vorticity vector. In detail the three eigenvalues λ_1, λ_2 and λ_3 of Λ_{ij} are reordered according to the alignment of the corresponding eigenvectors with the vorticity vector. The one with $\cos(e_k^{(i)}, \omega_k)$ maximal is denoted as λ_s followed by λ_+ and λ_-. This ordering is adopted to eliminate the mixture of the stretching and intermediate eigenvalues which occurs due to the interplay between the background straining and the field induced by the vortex itself (Andreotti, 1997) (see also Section 2.5.3).

Further criteria can be found for example in a recent review by Wallace (2009). The applicability of most of these criteria and their comparison has been intensively discussed for the detection of growing coherent vortices in the buffer layer of wall-bounded flows. These details will not be discussed here.

2.4.3 Vorticity measurement

The measurement of vorticity in a turbulent flow is a challenge since one has to determine the velocity simultaneously at two or more space points which are very close to each other, at least of the order of the Kolmogorov scale η_K. Optical techniques have turned out to successfully determine the

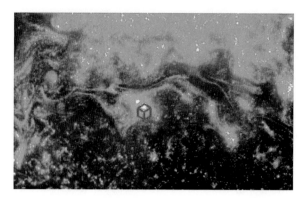

Figure 2.6 Turbulent grid flow illuminated by a laser sheet. The cube in the middle of the figure indicates the velocity gradient measurement volume. The image is taken from Lanterman et al. (2004).

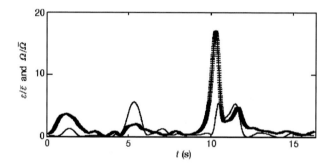

Figure 2.7 Time series of local enstrophy ω_i^2 and energy dissipation rate ϵ for an intense event. Both quantities are given in units of the long-time averages. The image is taken from Zeff et al. (2003).

components of the velocity gradient tensor and thus of the vorticity. Single plane particle image velocimetry (PIV) is able to reproduce the four components of M_{ij} which are the interrogation plane e.g. $i, j = 1, 2$ (Tropea et al., 2007; Wallace and Vukoslavčević, 2010). This allows to determine the three diagonal elements of M_{ij} for incompressible flows. If we extend the single plane to a dual plane stereo setup, the third velocity component u_z can be determined and thus $\partial u_z/\partial x$ and $\partial u_z/\partial y$. The extension of the dual plane setup to the two-frequency dual plane stereo PIV provides eventually all nine components of M_{ij}. They are measured simultaneously and independently in stereo PIV arrangements with laser light illumination by different frequencies. Both planes are differentially spaced such that gradients out of plane can be evaluated. The resolution that can be achieved goes down to four Kolmogorov length scales for a jet flow with an outer-scale Reynolds number of $Re = 6000$ (Mullin and Dahm, 2006a,b).

All nine components of the velocity gradient tensor could also be directly determined in a three-orthogonal-plane setup of PIV (Zeff et al., 2003; Lanterman et al., 2004). Three laser light sheets illuminate three faces of a small cube of sidelength of $1.8\eta_K$. The homogeneous isotropic turbulence was generated by two oscillating grids and a Taylor microscale Reynolds number of $R_\lambda = 54$. Figure 2.6 shows an instantaneous snapshot of the grid turbulence flow as illuminated by the laser sheet. The velocity gradient is determined in the small cube that is added to the figure. The apparatus made the temporal tracking of high-amplitude dissipation and local enstrophy events possible as given in Fig 2.7. Alternatively, holographic PIV techniques and thermal anemometers arrays with up to 20 hot wires have been used as summarized in the review by Wallace and Vukoslavčević (2010).

Direct use of the Lagrangian frame of reference is made by particle tracking velocimetry (PTV). If the number of particle tracks is sufficiently large the vorticity field (and velocity gradient field) can be reconstructed (Lüthi et al., 2005). This is done by assuming that the following relation

$$u_i(\mathbf{x}_0) = u_{i,0} + \frac{\partial u_i}{\partial x_1}x_1 + \frac{\partial u_i}{\partial x_2}x_2 + \frac{\partial u_i}{\partial x_3}x_3 , \qquad (2.64)$$

holds locally. Lagrangian tracks are obtained in a saturated copper sulfate solution that is driven electromagnetically and was to a good degree homogeneously isotropic turbulence at $R_\lambda \approx 50$. The particle seeding density in the reported experiment is about 180 cm^{-3}. Very recently, particle tracking was combined with digital holographic microscopy by Sheng et al. (2009) to study hairpin vortices in the buffer layer. This method requires a rather dense suspension of tracer particles. Particle velocity and gradients follow from the displacement in successive frames.

To summarize, different methods (especially optical methods) have been applied in both, the Eulerian and Lagrangian frames, and are able to resolve the space-time structure of the vorticity field almost down to the Kolmogorov length, and thus the velocity gradient field. All measurements techniques are limited to a small observation area and to moderate Reynolds numbers. One of the main bottlenecks is the fast streaming of the data from the digital cameras to the data disks. Given the progress which has been made in the past decade in this field, one can be reasonably certain that many of these limitations will be overcome in the future.

2.4.4 Coherent vortex extraction by wavelets

Wavelets have one big advantage to a standard Fourier analysis of turbulent fields. In contrast to the Fourier case, a structure or feature of the field to be analysed which is localized in physical space remains localized in the signal or wavelet space. This is exactly the reason why wavelets have been used to study vorticity fields in turbulence (see e.g. Farge (1992)). Wavelets form an orthonormal basis $\{\psi_\lambda(\mathbf{x})\}_{\lambda \in \Lambda}$ of the Hilbert space $L^2(\mathbb{R}^3)$ by successive dilation and translation of a mother wavelet. One thus can represent any square-integrable function $f(\mathbf{x})$, such as the components of the vorticity field, as

$$f(\mathbf{x}) = \sum_{\lambda \in \Lambda} w_\lambda \psi_\lambda(\mathbf{x}), \qquad (2.65)$$

where the wavelet coefficient w_λ is given by

$$w_\lambda := \int_{\mathbb{R}^3} f(\mathbf{x}) \psi_\lambda^*(\mathbf{x}) \, dV. \qquad (2.66)$$

The multi-index set $\Lambda = \{j, i_n, \mu\}$ is used where $j = 0, \ldots, J-1$ denotes the dilation and $i_n = 0, \ldots, 2^j - 1$ with $n = 1, 2, 3$ the translations. The largest scale is 2^0 and the smallest one 2^{-J+1}. Finally, the index $\mu = 1, \ldots, 7$ is required to construct an orthogonal system in the three-dimensional case which is done by combining one-dimensional wavelets $\psi_{j,i_n}(x_n)$ with scaling functions $\phi_{j,i_m}(x_m)$ in tensorial products, e.g. for $\mu = 1$ one has $\psi^1_{j,i_1,i_2,i_3}(\mathbf{x}) = \psi_{j,i_1}(x) \otimes \phi_{j,i_2}(y) \otimes \phi_{j,i_3}(z)$. Okamoto et al. (2007) used the Coiflet 12 wavelet for the coherent vortex extraction in DNS data sets.

To set up the formalism, one has to ask now which criterion has to be applied in order to identify coherent vortex structures in the wavelet spectrum. The main idea which was suggested by Farge (1992) is to decompose the vorticity field into coherent structures and noise

$$\omega_i(\mathbf{x}) = \omega_i^{(C)}(\mathbf{x}) + \omega_i^{(N)}(\mathbf{x}). \qquad (2.67)$$

All wavelet coefficients that remain after denoising, i.e. after removing the wavelet coefficients that do not exceed a certain threshold, compose the coherent part of the vorticity field, $\omega_i^{(C)}(\mathbf{x})$. In practice it turned out that the criterium by Donoho and Johnstone (1994) was optimal. All coefficients with magnitude

$$|w_{i,\lambda}| < \sqrt{\frac{2}{3}\sigma^2 \log_e(N)} \qquad (2.68)$$

will be considered as noise where σ^2 is the variance of the incoherent vorticity part. Here, $w_{i,\lambda}$ are the wavelet coefficients of ω_i. This variance has to be evaluated iteratively since it is unknown at the beginning of the analysis. Here, N is the number of degrees of freedom of the DNS of box turbulence. From the classical Kolmogorov analysis follows

$$\frac{L}{\eta_K} = Re^{3/4} \quad \Rightarrow \quad N = \left(\frac{L}{\eta_K}\right)^3 = Re^{9/4} \sim R_\lambda^{9/2}. \qquad (2.69)$$

Okamoto et al. (2007) performed this analysis for box turbulence data records at Taylor microscale Reynolds numbers between 167 and 732. The total number of wavelet coefficients (which equals the number of degrees of freedom in the analysis) was found to grow as $N \sim R_\lambda^{4.11}$. The number of degrees of freedom of the coherent part of the vorticity field, which make up only 2.6% of N for the highest Reynolds number, yielded a slightly smaller exponent with $N_C \sim R_\lambda^{3.90}$. The analysis suggests that the essential nonlinear information, which is contained in the subset $\omega_i^{(C)}(\mathbf{x})$, puts a weaker constraint on the resolution in a simulation which is based directly on a wavelet decomposition. A next challenging step is to evolve the fields in time by using a wavelet basis. A further analysis that should probably be made is to study the sensitivity of the coherent vortex extraction with respect to the resolution of the data.

2.4.5 Combined vorticity and strain analysis by curvelets

Curvelets extend the multiscale representation of turbulent flow structures which were just discussed in Section 2.4.4 by including orientation. The basis functions of the curvelet transform are localized in scale (Fourier space), position (physical space) and orientation. For a discrete curvelet transform, a family of curvelets at a scale 2^{-j} is obtained from a mother wavelet (at scale 2^0) by successive rotations about an equispaced sequence of angles $0 \le \theta_\ell < 2\pi$ and translations $i_n \in \mathbb{Z}^3$ with $n = 1, 2, 3$. Indices (j, i_n, ℓ) are a compact notation for the dilation, translation and rotation, respectively (Ying et al., 2005). Recall that the latter operation is missing in the wavelet case. The curvelet coefficient $c(j, i_n, \ell)$ is defined as the inner product between the function $f \in L^2(\mathbb{R}^3)$ and the curvelet $\varphi_{j,i_n,\ell}(\mathbf{x})$

$$c(j, i_n, \ell) := \int_{\mathbb{R}^3} f(\mathbf{x}) \varphi^*_{j,i_n,\ell}(\mathbf{x}) \, dV, \qquad (2.70)$$

where the asterisk denotes the conjugate complex curvelet. The data analysis occurs in Fourier space. By applying Plancherel's theorem (see e.g. Mallat (1999)), we can thus calculate the curvelet coefficients alternatively via the

Fourier transforms

$$c(j, i_n, \ell) := \frac{1}{(2\pi)^3} \int_{\mathbb{R}^3} \hat{f}(\mathbf{k}) \, \hat{\varphi}^*_{j,i_n,\ell}(\mathbf{k}) \, \mathrm{d}^3 k \,. \tag{2.71}$$

The curvelets $\hat{\varphi}_{j,i_n,\ell}(\mathbf{k})$ are defined in Fourier space by

$$\hat{\varphi}_{j,i_n,\ell}(\mathbf{k}) = \tilde{U}_{j,\ell}(\mathbf{k}) \exp\left(\frac{-2\pi i \sum_{n=1}^{3} i_n k_n / L_{n,j,\ell}}{\sqrt{\prod_{i=1}^{3} L_{i,j,\ell}}}\right). \tag{2.72}$$

The frequency space is smoothly windowed in radial and angular spherical coordinates, providing the decomposition in different scales and orientations, respectively. $\tilde{U}_{j,\ell}(\mathbf{k})$ is the wavenumber window $\tilde{U}_{j,\ell}(\mathbf{k}) = \tilde{W}_j(\mathbf{k}) \tilde{V}_{j,\ell}(\mathbf{k})$, with $\tilde{W}_j(\mathbf{k})$ and $\tilde{V}_{j,\ell}(\mathbf{k})$ the radial and angular windows.

This analysis was applied to scalar fields obtained from DNS of three-dimensional turbulence on a uniform Cartesian grid for a periodic box $V = (2\pi)^3$ (Bermejo-Moreno and Pullin, 2008; Bermejo-Moreno et al., 2009). There $\tilde{U}_{j,\ell}$ is supported in a rectangular box of integer size $L_{1,j,\ell} \times L_{2,j,\ell} \times L_{3,j,\ell}$. For a given scale, j, the radial window smoothly extracts the frequency contents near the dyadic corona $[2^{j-1}, 2^{j+1}]$. A low-pass radial window is introduced for the coarsest scale, j_0. The unit sphere representing all directions is partitioned, for each scale $j > j_0$, into smooth angular windows whose numbers are $O(2^j)$, each with a disk-like support of radius $O(2^{-j/2})$ as

$$\tilde{W}_{j_0}(\mathbf{k}) = \Phi_{j_0}(\mathbf{k}) \tag{2.73}$$

$$\tilde{W}_j(\mathbf{k}) = \sqrt{\Phi^2_{j+1}(\mathbf{k}) - \Phi^2_j(\mathbf{k})} \quad \text{for} \quad j > j_0, \tag{2.74}$$

where $\Phi_j(k_1, k_2, k_3) = \phi(2^{-j} k_1) \phi(2^{-j} k_2) \phi(2^{-j} k_3)$, and ϕ is a smooth function such that $0 \le \phi \le 1$; it is equal to 1 on $[-1, 1]$ and zero outside $[-2, 2]$. The last scale is given by $j_e = \log_2(n/2)$, and a set of $j_e - j_0 + 1$ filtered scalar fields results from the original field. The angular window for the ℓth direction is defined as

$$\tilde{V}_{j,\ell}(\mathbf{k}) = \tilde{V}\left(2^{j/2} \frac{k_2 - \alpha_\ell k_1}{k_1}\right) \tilde{V}\left(2^{j/2} \frac{k_3 - \beta_\ell k_1}{k_1}\right), \tag{2.75}$$

where $(1, \alpha_\ell, \beta_\ell)$ is the direction of the center line of the wedge defining the center slope for the ℓth wedge.

Although applications of wavelets to turbulent flows have become increasingly popular, traditional wavelets perform well only at representing point singularities since they ignore the geometric properties of structures and do not exploit the regularity of edges. Therefore, wavelet-based compression and structure extraction becomes computationally inefficient for geometric features with line and surface singularities. Discrete wavelet thresholding could lead to oscillations along the edges of vortices. The curvelet transform allows a multiscale decomposition by filtering in curvelet space the different scales of interest, individually or in groups. Unlike the isotropic elements of wavelets, curvelet transform uses basis functions that have a position and scale and orientation. In particular, fine-scale basis functions are long skinny ridges. Thereby, the curvelet transform possesses very high directional sensitivity and anisotropy, and is very efficient in representing vortex edges (Jianwei et al., 2009). Curvelets are suited for detecting, organizing, or providing a compact representation of intermediate multi-dimensional structures. Besides, the longitudinal and cross-sectional dimensions of curvelet basis functions in physical space follow the relation width $\approx$ length^2 (parabolic scaling), and curvelets are an optimally sparse basis for representing surface-like singularities of codimension one as shown by Candès and Donoho (2002).

Based on the curvelet transform, a multi-scale methodology which enables the analysis of nonlocal geometrical structures of the turbulence was developed by combining with statistical and clustering techniques (Bermejo-Moreno and Pullin, 2008). It consists of extraction, characterization and classification of structures. In the extraction step, multiscale decomposition of the field associated is done by means of the curvelet transform. After this multi-scale analysis, the structures of interest associated to each relevant range of scales are deduced. These structures of interest are defined as the individual disconnected surfaces obtained by isocontouring each filtered field at particular contour values. The characterization of the extracted structures is done in terms of the area-based joint probability density function (jpdf) of two differential-geometry properties: the shape index, S and the curvedness, C, which are polar coordinates in the Cartesian space of the surface principal curvatures (κ_1, κ_2),

$$S = \left| -\frac{2}{\pi} \arctan\left(\frac{\kappa_1 + \kappa_2}{\kappa_1 - \kappa_2}\right) \right|, \tag{2.76}$$

$$C = \mu \sqrt{\frac{\kappa_1^2 + \kappa_2^2}{2}}, \tag{2.77}$$

where $\mu = 3V/A$, V and A are the volume and area, respectively, of the closed surface. Note that the absolute value is taken for the shape index and the curvedness is made dimensionless (Koenderink and van Doorn (1992)). Furthermore, a dimensionless stretching parameter

$$\lambda = \frac{\sqrt[3]{36\pi V^2}}{A} \tag{2.78}$$

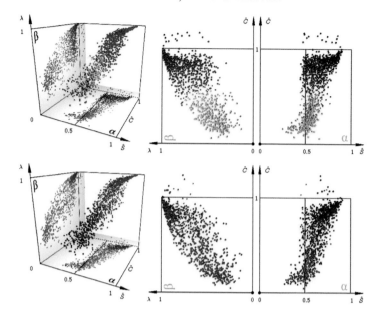

Figure 2.8 Three-dimensional view (left) and lateral (centre) and top (right) projections with glyphs (spheres) representing the optimum clusters of structures educed from the stratified random sample with optimum allocation of the sets of ω_i^2 (top) and $S_{ij}S_{ji}$ (bottom) structures (Bermejo-Moreno et al., 2009).

is defined. The jpdf, $P(S,C)$, and parameter λ define the geometrical signature of the structure. The classification step uses the feature centre $(\hat{S},\hat{C})$, the upper and lower distances in terms of its first- and second-order moments, and λ. A three-dimensional visualization space with coordinates $(\hat{S},\hat{C},\lambda)$ represents the geometry of individual structures by means of glyphs. Classification of each structure is followed by clustering of the structures which is carried out using learning-based clustering techniques and a locally scaled spectral partitional algorithm. For each member of a set of structures at a particular scale, $\mathcal{L}$, the contracted computational signature, $p[k]$, is defined using the moments of pdfs of S, C and λ. $p[k]$ is used to measure dissimilarity between the two elements of $\mathcal{L}$, and a locally scaled affinity matrix, L, is constructed. The elements of $\mathcal{L}$ are mapped onto a different eigenspace of L where clusters can be better identified, and the number of clusters is optimized and automatically determined.

Figure 2.8 shows the results of the clustering algorithm applied to ω^2 and $S_{ij}S_{ji}$ structures obtained from the decaying isotropic turbulence DNS data (Bermejo-Moreno et al., 2009). In the coordinates $(\hat{S},\hat{C},\lambda)$, blob-like structures are located near the region $(\hat{S},\hat{C},\lambda) \approx (1,1,1)$; tube-like structures concentrate near the axis $(\hat{S}, \hat{C}, \lambda) \approx (1/2, 1, \lambda)$ (more stretched for smaller values of λ), while sheet-like structures present small values of both $\hat{C}$ and λ. Here, glyphs are chosen as spheres with centre given by $(\hat{S}, \hat{C}, \lambda)$, and scaled by the normalized silhouette value, which indicates the degree of membership of that element to the assigned cluster. It can be seen that tube-like structures are dominant at intermediate scales of ω^2, which is consistent with the presence of worms (She et al., 1991; Vincent and Meneguzzi, 1991; Jiménez and Wray, 1998). Tube-like structures also appear at intermediate scales of $S_{ij}S_{ji}$ but in less proportion than for ω^2. At all scales analysed, $S_{ij}S_{ji}$ shows, on average, more planar geometries than ω^2, and the transition to sheet-like structures occurs in larger scale for $S_{ij}S_{ji}$. At the smallest scale, both fields show a clear dominance of sheet-like structures, and they are highly stretched.

2.4.6 Vortex sheets in turbulence

In Section 2.4.2, the methods to detect vortex tubes were presented. As for structures which are responsible for dissipation of turbulent energy, estimates of fractal dimension, D, for the set where energy dissipation takes place yielded the values of $D \simeq 2.7$ (Sreenivasan and Antonia, 1997), and $D \simeq 1.7$ (Moisy and Jiménez, 2004). These dimensions consistently seem not to be related to filamentary tube-like structures, but rather to (wrinkled) vortex sheets.

An identification method for vortex sheets is presented now. Most prominent characteristic feature of vortex sheet is that the strain rate and vorticity are highly correlated and their magnitudes are comparably large. By taking advantage of this feature, Tanaka and Kida (1993) proposed an identification method for vortex sheets by imposing restrictions on their magnitudes when educing the isosurfaces of vorticity. Classification of the structures can be done using the reordered eigenvalues of Λ_{ik}, as shown in Section 2.4.2. A region in which vorticity and strain are comparably large is classified by imposing the condition as $\lambda_+ \geq 0 \geq \lambda_-$, a strain-dominated region by $\lambda_+ \geq \lambda_- > 0$ (Horiuti, 2001). This method, however, is not suitable for selective extraction of the vortex sheet.

To take into account of the correlation between S_{ij} and Ω_{ij}, a symmetric second-order velocity gradient tensor which consists of S_{ij} and Ω_{ij} and is denoted by

$$A_{ij} \equiv (S_{ik}\Omega_{kj} + S_{jk}\Omega_{ki}), \tag{2.79}$$

is considered in Horiuti and Takagi (2005). Since $A_{ii} = 0$, the eigenvalues of

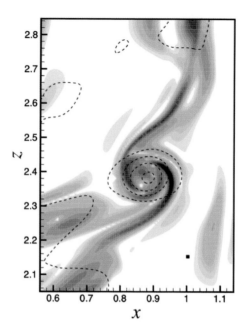

Figure 2.9 Isocontours of $[A_{ij}]_+$ (grey scale) [100 to 2250] and p [−7.0 to −1.0] (dashed line) in the x–z plane. The black square at $x = 1.0, y = 2.15$ represents the computational grid cell in x–z plane. The sizes of the plotted domain are $72.5\eta_K$ and $100\eta_K$ in the x- and z-directions, respectively (Horiuti and Fujisawa, 2008).

A_{ij} are obtained from the following characteristic equation

$$\lambda^3 - \frac{1}{2}(A_{ij}A_{ji})\lambda - \frac{1}{3}(A_{ij}A_{jk}A_{ki}) = 0. \qquad (2.80)$$

Evaluation of the terms $A_{ij}A_{ji}$ and $A_{ij}A_{jk}A_{ki}$ terms using the DNS data reveals that the $(A_{ij}A_{ji})^3$ term is larger than the $(A_{ij}A_{jk}A_{ki})^2$ term by two to three orders of magnitude, thus the solutions are approximated as $[A_{ij}]_\pm \simeq \pm\sqrt{A_{ij}A_{ji}/2}$. Since $[A_{ij}]_s \simeq 0$, $[A_{ij}]_+$ is suitable for extracting the two-dimensional planar structure such as the vortex sheet. (For definition of the reordered eigenvalues of $[A_{ij}]$, see Section 2.4.2.) As an example of application of the $[A_{ij}]_+$ method, the isocontours of $[A_{ij}]_+$ and the pressure p obtained using the decaying isotropic turbulence DNS data is shown in Figure 2.9 (Horiuti and Fujisawa, 2008). The lower and upper sheets are stretched and entrained by the tube, causing the sheets to form a spiral. This spiral tightens and forms spiral turns, which is attributed to differential rotation. By tightening of the spiral turns, spiral sheets are stretched to extreme lengths. The cascade of velocity fluctuations to smaller scale and

2: Structure and Dynamics of Vorticity in Turbulence

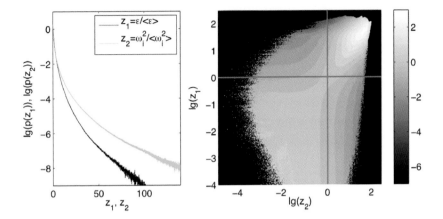

Figure 2.10 Statistics of local enstrophy and energy dissipation rate. Left: Probability density functions of local enstrophy, ω_i^2, and energy dissipation rate, ϵ, given in units of their means, respectively. Right: Joint probability density function of local enstrophy and energy dissipation rate. The distribution is normalized by both single quantity distributions, $p(z_1, z_2)/[p(z_1)p(z_2)]$, in order to highlight the statistical correlations between $z_1 = \epsilon/\langle\epsilon\rangle$ and $z_2 = \omega^2/\langle\omega^2\rangle$. Color coding is in decadic logarithm. Data are from Schumacher et al. (2010).

intense dissipation take place along spiral sheets (Lundgren, 1982). With regard to the vorticity alignment along two spiral sheets and the vortex tube in the core region, a sheet whose vorticity is either parallel or perpendicular to a vortex tube is wrapped into a spiral (Pullin and Lundgren, 2001), or one branch of the spiral has vorticity parallel to the tube and the other one perpendicular to the tube (Horiuti and Fujisawa, 2008).

2.5 Vorticity statistics in turbulence

2.5.1 Joint vorticity and strain statistics

Although the ensemble averaged values of energy dissipation and squared vorticity (or local enstrophy) are related by $\langle\epsilon\rangle = \nu\langle\omega_i^2\rangle$, this relation does not apply to their instantaneous and local values. Two tools for characterizing the joint statistics of vorticity and strain are distributions and the invariants (2.14), (2.16) and (2.18) to (2.21) of the velocity gradient tensor M_{ij}. These invariants appear in their own equations, describe alignments (2.25) and show up in the Cayley–Hamilton theorem (2.22). This subsection will use the probability density function (pdf) to characterize the joint statistics of the second-order quantities, the local energy dissipation and squared vorticity.

A common feature of these pdfs are stretched exponential tails, which indicate strong small-scale intermittency instead of Gaussian behavior. The tail is more extended for $\boldsymbol{\omega}^2$ than for ϵ, in agreement with the observations in Zeff et al. (2003) and first reported by Chen et al. (1997). This difference between both tails seems to decrease when the Reynolds number of the flow is increased as indicated by the simulations of Donzis et al. (2008). Fat tails imply that large amplitude events are significantly more probable than for a Gaussian distributed signal with the same second moment. The statistical correlation between high-amplitude local enstrophy and energy dissipation events can be studied by the following joint pdf (jpdf) which is normalized by the single-quantity pdfs

$$\Pi(\epsilon/\langle\epsilon\rangle, \boldsymbol{\omega}^2/\langle\boldsymbol{\omega}^2\rangle) := \frac{p(\epsilon/\langle\epsilon\rangle, \boldsymbol{\omega}^2/\langle\boldsymbol{\omega}^2\rangle)}{p(\epsilon/\langle\epsilon\rangle)p(\boldsymbol{\omega}^2/\langle\boldsymbol{\omega}^2\rangle)}. \qquad (2.81)$$

Π is shown in Figure 2.10(right). The maximum values appear in the upper right of the support where the largest amplitudes for both are present. High-amplitude fluctuations in energy dissipation and local enstrophy density are thus strongly correlated and found very close together, in both, space and time. The latter point is demonstrated in Figure 2.11. Time series of energy dissipation rate and enstrophy, which are accumulated in boxes of size $V_\ell = \ell^3$ (see Sections 2.54 and (2.58)) are displayed as a function of time.

2.5.2 Single-point vorticity probability density function

We have seen in the last section already that the statistics of the local enstrophy is highly intermittent. The same can thus be expected for the single-point pdf of the vorticity components. In the past decades several attempts have been made to derive equations of motion for the pdfs directly from the corresponding equations of motion (Lundgren, 1967; Monin, 1967; Novikov, 1968). It is clear that one faces the same closure problem as in the original equations of motion. Direct numerical simulations however allow us to gain a deeper understanding of the dynamical origin of the non-Gaussianity of the pdfs as recently done by Wilczek and Friedrich (2009). Starting point is the fine-grained pdf which is given by

$$\hat{p}(\boldsymbol{\omega}; \mathbf{x}, t) = \delta[\tilde{\boldsymbol{\omega}}(\mathbf{x}, t) - \boldsymbol{\omega}], \qquad (2.82)$$

where $\tilde{\boldsymbol{\omega}}(\mathbf{x}, t)$ is a realization of the (turbulent) vorticity field and $\boldsymbol{\omega}$ a sample variable. The single-point pdf follows by

$$p(\boldsymbol{\omega}; \mathbf{x}, \mathbf{t}) = \langle\delta[\tilde{\boldsymbol{\omega}}(\mathbf{x}, \mathbf{t}) - \boldsymbol{\omega}]\rangle. \qquad (2.83)$$

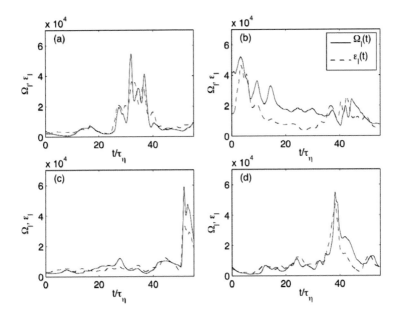

Figure 2.11 Temporal behaviour of local enstrophy $\Omega_\ell(t)$ as defined in (2.58) and energy dissipation rate ϵ_ℓ monitored in different subvolumes V_ℓ which are attached to Lagrangian tracers. Data are from Schumacher et al. (2010).

The equation for the pdf follows by taking first temporal and spatial derivatives of the fine-grained pdf $\hat{p}$ and substituting the corresponding terms of the vorticity equation (2.5). The second step is an ensemble averaging where unclosed terms are expressed as conditional averages, e.g. $\langle \mathbf{u}\hat{p}\rangle = \langle \mathbf{u}|\boldsymbol{\omega}\rangle p$. This results to

$$\frac{\partial p}{\partial t} + \nabla \cdot [\langle \mathbf{u}|\boldsymbol{\omega}\rangle p] = -\nabla_\omega \cdot [\langle \mathbf{S}\cdot\tilde{\boldsymbol{\omega}} + \nu\Delta\tilde{\boldsymbol{\omega}} + \nabla\times\mathbf{f}|\boldsymbol{\omega}\rangle p]. \qquad (2.84)$$

The right hand side of Eq. (2.84) obeys terms due to vortex stretching, vorticity diffusion and the large-scale forcing. The latter can be neglected for the discussion at the small scales of the turbulent flow. A further simplification arises due to homogeneity and isotropy. Using that $\partial^2 p/\partial x_i^2 = 0$ allows to substitute the vorticity diffusion by an enstrophy dissipation term in (2.84). For a single component of the vorticity the following equation is obtained

$$\frac{\partial p}{\partial t} = -\frac{\partial}{\partial \omega_x}\langle S_{xj}\tilde{\omega}_j|\omega_x\rangle p - \frac{\partial^2}{\partial \omega_x^2}\langle \nu(\nabla\tilde{\omega}_x)^2|\omega_x\rangle p. \qquad (2.85)$$

which yields a formal stationary solution (Wilczek and Friedrich, 2009)

$$p(\omega_x) = \frac{C}{\langle \nu(\nabla\tilde{\omega}_x)^2|\omega_x\rangle} \exp\left(-\int_{-\infty}^{\omega_x} \frac{\langle S_{xj}\tilde{\omega}_j|\omega_x'\rangle}{\langle \nu(\nabla\tilde{\omega}_x)^2|\omega_x'\rangle}\,d\omega_x'\right), \qquad (2.86)$$

where C is a normalization constant. The necessary input for both conditional moments in (2.86) can be obtained from DNS of homogeneous isotropic turbulence. In addition the dynamical origin of the highly non-Gaussian statistics of the vorticity components can be understood within this framework when starting with a Gaussian vorticity distribution. Wilczek and Friedrich (2009) could show that in the course of such an evolution of the turbulent flow towards a statistically stationary state the developing fat tails can be traced back to intense vortex-stretching events while the core of pdf remains dominated by vorticity diffusion.

2.5.3 Alignment of vorticity and strain

Our modern view that the small scales of turbulence are generated by the stretching of eddies was introduced by Taylor (1938). This paper followed upon Taylor and Green (1937), which introduced the Taylor–Green vortex to show how small eddies could be produced from larger ones – what we now call the energy cascade. These ideas were developed further by Burgers (1948), Betchov (1956) and others. The picture is that the small-scale vortex dynamics and vortex stretching that creates the small scales of turbulence are controlled by an interplay between vorticity and strain.

Over two decades ago, several direct numerical simulations (see e.g. Kerr, 1985; Ashurst et al., 1987; She et al., 1991; Nomura and Post, 1998; Ishihara et al., 2007; Donzis et al., 2008) provided more details of this relationship between vorticity and strain, including identifying a preferred alignment of the vorticity with a positive intermediate eigenvector of the rate of strain tensor. This result was unexpected by the statistical physics community, but had been anticipated by vortex models such as Tennekes (1968), Lundgren (1982) and Vieillefosse (1982, 1984). This has since been confirmed by multi-component experimental studies of the fine-scale structure (Tsinober et al., 1992; Zeff et al., 2003; Lüthi et al., 2005; Mullin and Dahm, 2006b).

The first two models listed that address the alignment assume a cylindrical geometry consistent with a Burgers vortex and alignment of vorticity and the middle eigenvalue. Several observed statistical features of a turbulent flow can be represented in this way. This includes the generation of a $k^{-5/3}$ energy spectrum for vorticity fluctuations about the central vortex in Lundgren (1982). The papers by Vieillefosse (1982, 1984) demonstrate how this configuration can arise by a combination of local twisting and stretching from almost any initial three-dimensional configuration of vorticity and strain. This is done using the restricted Euler equations, that is (2.30) and (2.33) with $H_{ij} = 0$. While showing that the vorticity tries to align with the middle eigenvalue of strain, these models cannot derive the relative magnitudes of the eigenvalues. However, numerical simulations and some experiments can give these values, and show that the dominant strain alignment has eigenvalues with the ratios $\sigma_- : \sigma_r : \sigma_+ = -4 : 1 : 3$ or similar (see e.g. Ashurst et al., 1987).

Since then, there have been numerous studies of the statistics of strain and vorticity in numerical turbulence, usually done by comparing with variations upon the restricted Euler equations. The basis of the studies conducted using numerical data was the identification of a characteristic curve in an R–Q space that Lagrangian points could follow (Cantwell, 1992). This was followed by numerical evidence that points clustered around this curve (Cantwell, 1993), and the evidence that the trajectories follow the predicted paths (Martin et al., 1998). More recently, extensions to the original restricted model have been proposed that are designed to represent some of the $H_{ij} \neq 0$ effects using tetrads (Chertkov et al., 1999) and stochastic terms (Chevillard and Meneveau, 2006).

However, none of these studies has answered some basic questions arising from the early studies. These questions include why the alignment noted above seems to dominate, whether the structures containing this alignment are vortex sheets, tubes or something else, and what length scales control this alignment.

The statistics usually used to examine these issues are distributions of β_N and S_N^3 as given by (2.25) and the alignment of vorticity direction with all three principal components of the strain. Another useful distribution examines the local vortex stretching, i.e., the normalized α_N as defined in (2.25).

There has been some argument about the relative value of examining at β_N or S_N^3, although it is shown in Kerr (1987) that the two measures are equivalent (2.26). The issue is that when one looks at the distributions of β_N, the value of $\beta_N = 1/2$ suggests vortex tubes. When one looks at S_N^3, one finds that this distribution is clustered around $S_N^3 = -1$, which would suggest vortex sheets (Lundgren and Rogers, 1994). These separate features are contrasted in Fig. 2.12. This dichotomy between tubes and sheets is partially resolved by considering the Lundgren model which evolves into a strong central vortex tube surrounded by vortex sheets.

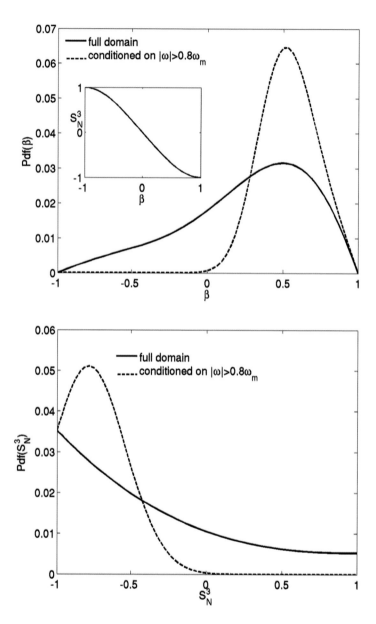

Figure 2.12 Distributions of alignments of the strain eigenvalues calculated from Ashurst et al. (1987). Top: Distributions of β_N for the full domain and conditioned on where $\omega > 0.8\max(\omega)$. The inset shows the transformation from β_N to S_N^3. Bottom: Distributions of S_N^3 for the same data.

These considerations suggest the question: How do the particular ratios of principal strain rates arise? In a stretched vortex tube picture, σ_i is the order of the vorticity, but σ_r should be related to the ratio of the diameter of small-scale vortex tubes to a characteristic length or radius of curvature. Since the tube diameter is known to be the order of the Kolmogorov length scale η_K, the other length scale must be some large multiple of η_K, or perhaps an inertial range scale such as the Taylor microscale λ.

How does the distribution of β_N change if all the small scales are filtered out? In this case, both β_N and S_N^3 are symmetrically distributed about $\beta_N = 0$ and $S_N^3 = 0$ (Ashurst et al., 1987). But this does not tell us what happens if the strain is filtered while leaving the vorticity field unchanged.

Therefore, one can consider the distributions of $\mathbf{e}_i \cdot \boldsymbol{\xi}$ where $\boldsymbol{\xi}$ is given by (2.36). This supports a strong alignment between $\mathbf{e}_r$ and $\boldsymbol{\xi}$ where $\mathbf{e}_r$ is given by (2.24). However a different alignment is obtained by comparing the alignment of an unfiltered vorticity field with the eigenvalues $\tilde{\mathbf{e}}_i$ of the filtered strain. In this case, $\boldsymbol{\xi}$ aligns more strongly with the eigenvector with the largest positive eigenvalue, $\tilde{\mathbf{e}}_+$. Therefore the correlation responsible for the typical value of $\beta_N = 0.5$ comes from the alignment with the component of the strain resulting from this, or a larger, filter size.

The filter is applied using the following application of the Biot–Savart law (2.6) to the strain (Ohkitani, 1994):

$$S_{ij}(\mathbf{x}) = \frac{3}{8\pi} \int_{V'} (\epsilon_{ikl} r_j + r_i \epsilon_{jkl}) \frac{r_k}{r^5} \omega_l(\mathbf{x}') \, dV', \qquad (2.87)$$

where $r = |\mathbf{x} - \mathbf{x}'|$ and $r_i = x_i - x_i'$. Hamlington et al. (2008a,b) made direct use of this relation in order to calculate the background strain field from high-resolution DNS data. In these data, the Kolmogorov scale η_K was resolved to within three grid cells so that the integral (which has to be understood as a principal value) could be calculated numerically.

The decomposition of the strain is

$$S_{ij}(\mathbf{x}) = S_{ij}^B(\mathbf{x}) + S_{ij}^R(\mathbf{x}), \qquad (2.88)$$

where S_{ij}^R is the local strain induced by a vortical structure in its neighboring vicinity with a characterisitc (yet undetermined) scale R, and S_{ij}^B is the nonlocal background strain. The vortical structure would then be expected to align with the principal axis corresponding to the most extensional eigenvalue of the background rate of strain tensor $S_{ij}^B(\mathbf{x})$.

One combines (2.88) and (2.87) to get

$$S_{ij}^B(\mathbf{x}) = S_{ij}(\mathbf{x}) - \frac{3}{8\pi} \int_{B_R(\mathbf{x}')} (\epsilon_{ikl} r_j + r_i \epsilon_{jkl}) \frac{r_k}{r^5} \omega_l(\mathbf{x}') \, dV'. \qquad (2.89)$$

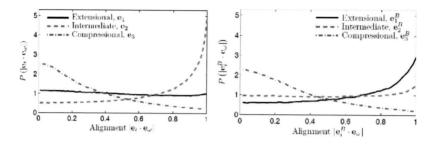

Figure 2.13 Distributions of vorticity vector alignment cosines. Left: with the total strain S_{ij}. Right: with the background strain S_{ij}^B which is calculated by (2.89). Images are taken from Hamlington et al. (2008a).

Here, $B_R(\mathbf{x}')$ denotes a sphere of radius R. Clearly R is a free parameter which separates the two contributions. A possible physical choice for R is the characteristic decorrelation length of the spatial vorticity correlations $C_{\omega\omega}(\mathbf{r}) = \langle \omega_i(\mathbf{x})\omega_i(\mathbf{x}+\mathbf{r}) \rangle$. For the DNS data with $R_\lambda = 107$ (Schumacher et al., 2007), the scale R is at least $R \approx 12\eta$. The results of this analysis are shown in Fig. 2.13 for alignments with the total strain and with the background strain. They show that indeed the filtering results in the preferential alignment of the extensional strain with the most extensional background strain.

In this chapter, we have discussed the central role of vorticity in gaining a deeper understanding of fundamental aspects of the mathematical foundations of the equations of fluid motion of ideal and real fluids as well as of the connection of statistical and structural properties of turbulence. We have shown that the vorticity field plays a central role not only in its own, but also together with the strain, the symmetric counterpart of the velocity gradient tensor. The new noninvasive experimental measurement techniques as well as the larger computational ressources give us new possibilities which will allow us to study further aspects of the variety of aspects of the turbulent vorticity field in turbulence. It was also shown that particularly elementary dynamical processes and instabilities of the vorticity field can be thought as fundamental building blocks which can be successively observed in a turbulent flow, probably even in more complex settings than the one discussed mostly here.

References

Andreotti, B. 1997. Studying Burgers models to investigate the physical meaning of the alignments statistically observed in turbulence. *Phys. Fluids*, **9**, 735–742.

Ashurst, W. T., R., Kerstein A., Kerr, R. M., and Gibson, C. H. 1987. Alignment of vorticity and scalar gradient with strain rate in simulated Navier–Stokes turbulence. *Phys. Fluids*, **30**, 2343–2353.

Bardos, C., and Titi, E. 2007. Euler equations for incompressible ideal fluids. *Russian Math. Surveys*, **62**, 409–451.

Batchelor, G. K. 1953. *Theory of Homogeneous Turbulence*. Cambridge University Press, Cambridge.

Beale, J. T., T., Kato, and Majda, A. 1984. Remarks on the breakdown of smooth solutions for the 3D Euler equations. *Commun. Math. Phys.*, **94**, 61–66.

Bermejo-Moreno, I., and Pullin, D. I. 2008. On the non-local geometry of turbulence. *J. Fluid Mech.*, **603**, 101–135.

Bermejo-Moreno, I., Pullin, D. I., and Horiuti, K. 2009. Geometry of enstrophy and dissipation, grid resolution effects and proximity issues in turbulence. *J. Fluid Mech.*, **620**, 121–166.

Betchov, R. 1956. An inequality concerning the production of vorticity in isotropic turbulence. *J. Fluid Mech.*, **1**, 497–504.

Boratav, O. N., and Pelz, R. B. 1994. Direct numerical simulation of transition to turbulence from a high-symmetry initial condition. *Phys. Fluids*, **6**, 2757–2784.

Brachet, M. E., Meiron, D. I., Orszag, S. A., Nickel, B. G., Morf, R. H., and Frisch, U. 1983. Small-scale structure of the Taylor-Green vortex. *J. Fluid Mech.*, **130**, 411–452.

Burgers, J. M. 1948. A mathematical model illustrating the theory of turbulence. *Adv. Appl. Math.*, **1**, 171–199.

Bustamante, M. D., and Kerr, R. M. 2008. 3D Euler about a 2D symmetry plane. *Physica D*, **237**, 1912–1920.

Candès, E.J., and Donoho, D. L. 2002. New tight frames of curvelets and optimal representations of objects with piecewise-C^2 singularities. *Comm. Pure Appl. Math.*, **57**, 219–266.

Cantwell, B. 1992. Exact solution of a restricted Euler equation for the velocity gradient tensor. *Phys. Fluids A*, **4**, 782–793.

Cantwell, B. 1993. On the behavior of velocity gradient tensor invariants in direct numerical simulations of turbulence. *Phys. Fluids A*, **5**, 2008–2013.

Cartes, C., Bustamante, M., and Brachet, M.E. 2007. Generalized Eulerian–Lagrangian description of Navier–Stokes dynamics. *Phys. Fluids*, **19**, 077101.

Chen, S., Doolen, G., Herring, J.R., Kraichnan, R.H., Orszag, S.A., and She, Z.S. 1993. Far-dissipation range of turbulence. *Phys. Rev. Lett.*, **70**, 3051–3054.

Chen, S., Sreenivasan, K.R., and Nelkin, M. 1997. Inertial range scalings of dissipation and enstrophy in isotropic turbulence. *Phys. Rev. Lett.*, **79**, 1253–1256.

Chertkov, M., Pumir, A., and Shraiman, B.I. 1999. Lagrangian tetrad dynamics and the phenomenology of turbulence. *Phys. Fluids*, **11**, 2394–2410.

Chevillard, L., and Meneveau, C. 2006. Lagrangian dynamics and statistical geometric structure of turbulence. *Phys. Rev. Lett.*, **97**, 174501.

Cichowlas, C., and Brachet, M.-E. 2005. Evolution of complex singularities in Kida–Pelz and Taylor–Green inviscid flows. *Fluid Dyn. Res.*, **36**, 239–248.

Constantin, P. 1994. Geometric statistics in turbulence. *SIAM Review*, **36**, 73–98.

Constantin, P. 2001. Three lectures on mathematical fluid mechanics. In: *From Finite to Infinite Dimensional Dynamical Systems*, Glendinning, P.A., and Robinson, J.C. (eds), Kluwer Academic, Dordrecht, The Netherlands, for NATO ASI workshop at the Isaac Newton Institute, August 1995.

Constantin, P., and Fefferman, C. 1993. Direction of vorticity and the problem of global regularity for the Navier–Stokes equations. *Indiana Univ. Math. Journal*, **42**, 775–789.

Constantin, P., Fefferman, C., and Majda, A.J. 1996. Geometric constraints on potentially singular solutions for the 3-D Euler equations. *Comm. in PDE*, **21**, 559–571.

Cuypers, Y., Maurel, A., and Petitjeans, P. 2003. Vortex burst as a source of turbulence. *Phys. Rev. Lett.*, **91**, 194502.

da Vinci, L. 1506 to 1510. *Codex Leicester.*

Deng, J., Hou, T.Y., and Yu, X. 2005. Geometric properties and non-blowup of 3D incompressible Euler Flow. *Commun. PDEs*, **30**, 225–243.

Doering, C. R. 2009. The 3D Navier–Stokes problem. *Ann. Rev. Fluid Mech.*, **41**, 109–128.

Donoho, D. L., and Johnstone, J. M. 1994. Ideal spatial adaptation by wavelet shrinkage. *Biometrika*, **81**, 425–455.

Donzis, D. A., Yeung, P. K., and Sreenivasan, K. R. 2008. Dissipation and enstrophy in isotropic turbulence: Resolution and scaling in direct numerical simulations. *Phys. Fluids*, **20**, 045108.

Douady, S., Couder, Y., and Brachet, M. E. 1991. Direct observation of the intermittency of intense vorticity filaments in turbulence. *Phys. Rev. Lett.*, **67**, 983–986.

Euler, L. 1761. Principia motus fluidorum. *Novi Commentarii Acad. Sci. Petropolitanae*, **6**, 271–311.

Farge, M. 1992. Wavelet transforms and their application to turbulence. *Annu. Rev. Fluid Mech.*, **24**, 395–457.

Fefferman, C. 2000. Existence and smoothness of the Navier–Stokes equation. Clay Millenium Prize description. *http://www.claymath.org/millenium*, 1–5.

Gibbon, J.D. 2008. The three-dimensional Euler equations: Where do we stand? *Physica D*, **237**, 1895–1904.

Gibbon, J.D., Holm, D.D., Kerr, R.M., and Roulstone, I. 2006. Quaternions and particle dynamics in the Euler fluid equations. *Nonlinearity*, **19**, 1969–1983.

Grafke, T., Homann, H., Dreher, J., and Grauer, R. 2008. Numerical simulations of possible finite time singularities in the incompressible Euler equations: comparison of numerical methods. *Physica D*, **237**, 1932–1936.

Gulak, Y., and Pelz, R. B. 2005. High-symmetry Kida flow: Time series analysis and resummation. *Fluid Dyn. Res.*, **36**, 211–220.

Hamlington, P. E., Schumacher, J., and Dahm, W. J. A. 2008a. Direct assessment of vorticity alignment with local and nonlocal strain rates in turbulent flows. *Phys. Fluids*, **20**, 111703.

Hamlington, P. E., Schumacher, J., and Dahm, W. J. A. 2008b. Local and nonlocal strain rate fields and vorticity alignment in turbulent flows. *Phys. Rev. E*, **77**, 026303.

Herring, J.R., and Kerr, R.M. 1982. Comparison of direct numerical simulation with prediction of two–point closures. *J. Fluid Mech.*, **118**, 205–219.

Holm, D. D., and Kerr, R. M. 2007. Helicity in the formation of turbulence. *Phys. Fluids*, **19**, 025101.

Horiuti, K. 2001. A classification method for vortex sheet and tube structures in turbulent flows. *Phys. Fluids*, **13**, 3756–3774.

Horiuti, K., and Fujisawa, T. 2008. Multi mode stretched spiral vortex in homogeneous isotropic turbulence. *J. Fluid Mech.*, **595**, 341–366.

Horiuti, K., and Takagi, Y. 2005. Identification method for vortex sheet structures in turbulent flows. *Phys. Fluids*, **17**, 121703.

Hou, T. Y., and Li, R. 2008. Blowup or no blowup? The interplay between theory and numerics. *Physica D*, **237**, 1937–1944.

Ishihara, T., Kaneda, Y., Yokokawa, M., Itakura, K., and Uno, A. 2007. Small-scale statistics in high-resolution direct numerical simulation of turbulence: Reynolds number dependence of one-point velocity gradient statistics. *J. Fluid Mech.*, **592**, 335–366.

Ishihara, T., Gotoh, T., and Kaneda, Y. 2009. Study of high-Reynolds number isotropic turbulence by direct numerical simulation. *Annu. Rev. Fluid Mech.*, **41**, 165–180.

Jeong, J., and Hussain, F. 1995. On the identification of a vortex. *J. Fluid Mech.*, **285**, 69–94.

Jianwei, M., Hussaini, M. Y., Vasilyev, O. V., and Le Dimet, F. 2009. Multiscale geometric analysis of turbulence by curvelets. *Phys. Fluids*, **21**, 075104.

Jiménez, J., and Wray, A. A. 1998. On the characteristics of vortex filaments in isotropic turbulence. *J. Fluid Mech.*, **373**, 255–285.

Kerr, R. M. 1985. Higher-order derivative correlations and the alignment of small-scale structures in isotropic numerical turbulence. *J. Fluid Mech.*, **153**, 31–58.

Kerr, R. M. 1987. Histograms of helicity and strain in numerical turbulence. *Phys. Rev. Lett.*, **59**, 783–786.

Kerr, R. M. 1993. Evidence for a singularity of the three-dimensional incompressible Euler equations. *Phys. Fluids A*, **5**, 1725–1746.

Kerr, R. M. 2005. Velocity and scaling of collapsing Euler vortices. *Phys. Fluids*, **17**, 075103.

Kerr, R.M., Virk, D., and Hussain, F. 1990. Effects of incompressible and compressible vortex reconnection. Page 1 of: *Topological Fluid Mechanics*, Moffatt, H.K., and Tsinober, A. (eds), Cambridge University Press, for IUTAM meeting at Cambridge University 1989.

Kida, S., and Murakami, Y. 1987. Kolmogorov similarity in freely decaying turbulence. *Phys. Fluids*, **30**, 2030–2039.

Koenderink, J.J., and van Doorn, A.J. 1992. Surface shape and curvature scales. *Image Vision Comput.*, **10**, 557–565.

Kozono, H., and Taniuchi, Y. 2000. Limiting case of the Sobolev inequality in BMO, with application to the Euler equations. *Commun. Math. Phys.*, **215**, 191–200.

Lanterman, D. D., Lathrop, D. P., Zeff, B. W., McAllister, R., Roy, R., and Kostellich, E. J. 2004. Characterizing intense rotation and dissipation in turbulent flows. *Chaos*, **14**, S8.

Lu, L., and Doering, C. R. 2008. Limits on enstrophy growth for solutions of the three-dimensional Navier–Stokes equations. *Indiana Univ. Math. J.*, **57**, 2693–2728.

Lundgren, T. S. 1967. Distribution functions in the statistical theory of turbulence. *Phys. Fluids*, **10**, 969–975.

Lundgren, T. S. 1982. Strained spiral vortex model for turbulent fine structure. *Phys. Fluids*, **25**, 2193–2202.

Lundgren, T. S., and Rogers, M. M. 1994. An improved measure of strain state probability in turbulent flows. *Phys. Fluids A*, **6**, 1838–1847.

Lüthi, B., Tsinober, A., and Kinzelbach, W. 2005. Lagrangian measurement of vorticity dynamics in turbulent flow. *J. Fluid Mech.*, **528**, 87–118.

Mallat, S. 1999. *A Wavelet Tour of Signal Processing*. Academic Press, San Diego.

Martin, J., Ooi, A., Chong, M.S., and Soria, J. 1998. Dynamics of the velocity gradient tensor invariants in isotropic turbulence. *Phys. Fluids*, **10**, 2336–2346.

Moisy, F., and Jiménez, J. 2004. Geometry and clustering of intense structures in isotropic turbulence. *J. Fluid Mech.*, **513**, 111–133.

Monin, A.S. 1967. Equations of turbulent motion. *Prikl. Mat. Mekh.*, **31**, 1057–1068.

Mullin, J. A., and Dahm, W. J. A. 2006a. Dual-plane stereo particle image velocimetry measurements of velocity gradient tensor fields in turbulent shear flow. I. Accuracy assessments. *Phys. Fluids*, **18**, 035101.

Mullin, J. A., and Dahm, W. J. A. 2006b. Dual-plane stereo particle image velocimetry measurements of velocity gradient tensor fields in turbulent shear flow. II. Experimental results. *Phys. Fluids*, **18**, 035102.

Nomura, K. K., and Post, G. K. 1998. The structure and dynamics of vorticity and rate of strain in incompressible homogeneous turbulence. *J. Fluid Mech.*, **377**, 65–97.

Novikov, E. A. 1968. Kinetic equations for a vortex field. *Sov. Phys. Dokl.*, **12**, 1006–1008.

Ohkitani, K. 1994. Kinematics of vorticity: Vorticity–strain conjugation in incompressible fluid flows. *Phys. Rev. E*, **50**, 5107–5110.

Okamoto, N., Yoshimatsu, K., Schneider, K., Farge, M., and Kaneda, Y. 2007. Coherent vortices in high-resolution direct numerical simulation of homogeneous isotropic turbulence: A wavelet viewpoint. *Phys. Fluids*, **19**, 115109.

Orszag, S.A., and Patterson, G.S. 1972. Numerical simulation of three-dimensional homogeneous isotropic turbulence. *Phys. Rev. Lett.*, **28**, 76–79.

Pelz, R. B., and Ohkitani, K. 2005. Linearly strained flows with and without boundaries-the regularizing effect of the pressure term. *Fluid Dyn. Res.*, **36**, 193–210.

Ponce, G. 1985. Remark on a paper by J.T. Beale, T. Kato and A. Majda. *Commun. Math. Phys.*, **98**, 349.

Pullin, D. I., and Lundgren, T. S. 2001. Axial motion and scalar transport in stretched spiral vortices. *Phys. Fluids*, **13**, 2553–2563.

Pullin, D. I., and Saffman, P. S. 1998. Vortex dynamics in turbulence. *Annu. Rev. Fluid Mech.*, **30**, 31–51.

Rossi, M. 2000. Of vortices and vortical layers: an overview. Pages 40–123 of: *Lectures Notes in Physics*, Maurel, A., and Petitjeans, P. (eds), vol. 555. Springer, Berlin and Heidelberg.

Rossi, M., Bottausci, F., Maurel, A., and Petitjeans, P. 2004. A nonuniformly stretched vortex. *Phys. Rev. Lett.*, **92**, 054504.

Rotta, J. C. 1997. *Turbulente Strömungen*. Teubner, Stuttgart.

Saffman, P. 1997. *Vortex Dynamics*. Cambridge University Press.

Saffman, P. G., and Pullin, D. I. 1996. Calculation of velocity structure functions for vortex models of isotropic turbulence. *Phys. Fluids*, **8**, 3072–3084.

Schumacher, J. 2007. Sub-Kolmogorov-scale fluctuations in fluid turbulence. *Europhys. Lett.*, **80**, 54001.

Schumacher, J., Sreenivasan, K. R., and Yakhot, V. 2007. Asymptotic exponents from low-Reynolds-number flows. *New J. Phys.*, **9**, 89.

Schumacher, J., Eckhardt, B., and Doering, C. R. 2010. Extremal vorticity growth in Navier–Stokes turbulence. *Phys. Lett. A*, **371**, 861–865.

She, Z.-S., Jackson, E., and Orszag, S. A. 1991. Structure and dynamics of homogeneous turbulence: Models and simulations. *Proc. R. Soc. London, Ser. A*, **434**, 101–124.

Sheng, J., Malkiel, E., and Katz, J. 2009. Buffer layer structures associated with extreme wall stress events in a smooth wall turbulent boundary layer. *J. Fluid Mech.*, **633**, 17–60.

Siggia, E. D. 1981a. Invariants for the one-point vorticity and strain rate correlation functions. *Phys. Fluids*, **24**, 1934–1936.

Siggia, E. D. 1981b. Numerical study of small-scale intermittency in three-dimensional turbulence. *J. Fluid Mech.*, **107**, 375–406.

Siggia, E. D., and Patterson, G.S. 1978. Intermittency effects in a numerical simulation of stationary three-dimensional turbulence. *J. Fluid Mech.*, **86**, 567–592.

Sreenivasan, K. R., and Antonia, R. A. 1997. The phenomenology of small-scale turbulence. *Ann. Rev. Fluid Mech.*, **29**, 435–472.

Tanaka, M., and Kida, S. 1993. Characterization of vortex tubes and sheets. *Phys. Fluids A*, **5**, 2079–2082.

Taylor, G.I. 1938. Production and dissipation of vorticity in a turbulent fluid. *Proc. R. Soc. London, Ser. A*, **164**, 15–23.

Taylor, G.I., and Green, A.E. 1937. Mechanism of the production of small eddies from large ones. *Proc. R. Soc. London, Ser. A*, **158**, 499–521.

Tennekes, H. 1968. Simple model for the small-scale structure of turbulence. *Phys. Fluids*, **11**, 669–671.

Tropea, C., Yarin, A. L., and Foss, J. F. 2007. *Handbook of Experimental Fluid Mechanics*. Springer, Berlin.

Tsinober, A. 2009. *An Informal Conceptual Introduction to Turbulence*. Springer, Berlin.

Tsinober, A., Kit, E., and Dracos, T. 1992. Experimental investigation of the field of velocity gradients in turbulent flows. *J. Fluid Mech.*, **242**, 169–192.

Vieillefosse, P. 1982. Local interaction between vorticity and shear in a perfect incompressible fluid. *J. Phys.*, **43**, 837–842.

Vieillefosse, P. 1984. Internal motion of a small element of fluid in an inviscid flow. *Physica A*, **125**, 150–162.

Vincent, A., and Meneguzzi, M. 1991. The spatial structure and statistical properties of homogeneous turbulence. *J. Fluid Mech.*, **225**, 1–20.

Virk, D., Kerr, R.M., and Hussain, F. 1995. Compressible vortex reconnection. *J. Fluid Mech.*, **304**, 47–86.

von Kármán, T., and Howarth, L. 1938. On the statistical theory of turbulence. *Proc. R. Soc. London, Ser. A*, **164**, 192–215.

Wallace, J. M. 2009. Twenty years of experimental and direct numerical simulation access to the velocity gradient tensor: What have we learned about turbulence? *Phys. Fluids*, **21**, 021301.

Wallace, J. M., and Vukoslavčević, P. V. 2010. Measurement of the velocity gradient tensor in turbulence. *Ann. Rev. Fluid Mech.*, **42**, 157–181.

Wilczek, M., and Friedrich, R. 2009. Dynamical origins for non-Gaussian vorticity distributions in turbulent flows. *Phys. Rev. E*, **80**, 016316.

Ying, L., Demanet, L., and Candès, E. J. 2005. 3D Discrete curvelet transform. *Tech. Rep. Applied and Computational Mathematics, California Institute of Technology*, 1–11.

Zeff, B. W., Lanterman, D. D., McAllister, R., Roy, R., Kostellich, E. J., and Lathrop, D. P. 2003. Measuring intense rotation and dissipation in turbulent flows. *Nature*, **421**, 146–149.

3
Passive Scalar Transport in Turbulence: A Computational Perspective

T. Gotoh and P.K. Yeung

3.1 Introduction

It is part of the nature of turbulence that local inhomogeneities of heat, chemical substances or other material properties are readily mixed with the surroundings. Turbulent mixing is important in a broad range of situations, such as combustion, pollutant dispersion and oceanic transport, as well as critical events such as volcanic eruptions, underwater oil spills, and accidental radioactive releases. The basic physics is represented by the transport of a passive scalar (Sreenivasan, 1991; Shraiman and Siggia, 2000; Warhaft, 2000), such as a small temperature fluctuation or low solute concentration, which does not affect the flow but has a finite molecular diffusivity. While practical applications are often more complex (Dimotakis, 2005) the mixing of passive scalars is the rate-limiting process in important phenomena such as nonpremixed combustion (Bilger, 1989). Studies of turbulent mixing also often provide useful diagnostics for the flow itself.

In this chapter we consider statistical and structural properties of passive scalar fields in turbulence, with emphasis on results from numerical simulations. Assuming Fickian diffusion with constant diffusivity (κ), the basic transport equation is

$$\partial \theta/\partial t + \mathbf{u} \cdot \nabla \theta = \kappa \nabla^2 \theta + f_\theta , \qquad (3.1)$$

which governs the evolution of the fluctuation θ of a passive scalar Θ (of mean $\langle \Theta \rangle$) in an incompressible fluid velocity field $\mathbf{u}$ that satisfies the Navier–Stokes equations. Here f_θ is a source term which may have a prescribed spectrum (e.g., Shraiman and Siggia (2000)) or may be represented by a mean scalar gradient, as $f_\theta = -\mathbf{u} \cdot \nabla \langle \Theta \rangle$ (e.g., Overholt and Pope (1996)). For simplicity, we shall assume $\nabla \langle \Theta \rangle$ (if present) to be spatially uniform, which is compatible with statistical homogeneity of fluctuations in space.

The efficiency of turbulent mixing is dependent on the process by which large scale inhomogeneities are broken down into smaller and smaller scales, through the advective term $\mathbf{u} \cdot \nabla \theta$ which in turn depends strongly on how velocity and scalar fluctuations at different scales interact. The range of scales in the velocity field, as the ratio of the integral length scale (L) to the Kolmogorov scale (η), is expressed by (a power of) the Reynolds number, while for scalars an additional parameter is the ratio of fluid kinematic viscosity (ν) to the molecular diffusivity, called the Schmidt number (Sc). The value of Sc varies widely, from $O(0.01)$ in liquid metals, to $O(1)$ in gas-phase combustion, to $O(10)$ for heat in water, and as high as $O(1000)$ for many organic liquids. Many differences between the moderately diffusive ($Sc \lesssim 1$; Corrsin (1951); Warhaft (2000)) and weakly diffusive ($Sc \gg 1$; Batchelor (1959); Antonia and Orlandi (2003a)) regimes are known. In particular, as discussed further in §3.3, for $Sc \lesssim 1$ the smallest scale is the Obukhov–Corrsin scale, $\eta_{OC} = \eta Sc^{-3/4}$, and in the so-called inertial–convective range at high Reynolds number the spectrum is expected to scale as $k^{-5/3}$ (where k is the wavenumber). However, for $Sc \gg 1$ the smallest scale is the Batchelor scale, $\eta_B = \eta Sc^{-1/2}$, the so-called viscous–convective range where a k^{-1} spectrum is of greatest interest. It is worth noting that experiments and computations at both very low or very large Schmidt numbers are more challenging, and current knowledge in those regimes is more limited.

In general, since molecular diffusion and chemical reaction occur primarily at the small scales, the study of scale similarity and small-scale structure in passive scalar mixing is very important (Sreenivasan and Antonia, 1997). Fundamental questions include, among others, the effects of the large scales expressed by the turbulent scalar flux, the status of local isotropy in relation to the imposed external mean gradient, and the intermittency of scalar gradients and fluctuations of the scalar dissipation rate. Usually the small scales are represented by higher-order statistical quantities, which are more difficult to resolve or sample adequately and are thus less well understood than lower-order quantities such as spectra and second-order moments. Likewise, the study of topological features of the scalar field requires consideration of two-point statistics and the use of advanced visualization techniques.

Most turbulence theories are asymptotic in nature, invoking an assumption of high Reynolds number, and perhaps the limits of very low or very high Schmidt number. Since data from both experiment and computation are typically obtained at adequate detail only at finite or modest values of these fundamental scaling parameters, an understanding of the rates at which various limiting behaviors are approached (if they are) becomes important. Consequently in this review we give considerable attention to the question of the approach to asymptotic states, for both low- and high-order statistics. At the same time, in some cases, simulations enabled by state-of-the-art computing resources are now close to the parameter ranges needed to evaluate fundamental theories that have long resisted detailed examination in the past. For this reason, and because of the background of the authors, but by no means under-estimating the importance of (and synergistic potential provided by) experiment, theory and modeling, the results discussed in detail in this paper are primarily computational.

The structure of this review is as follows. In §3.2 we wish to draw attention to potential opportunities, and the inherent challenges, for future progress through rapid advances in computing, which are expected to continue for the rest of this decade. In §3.3 we provide some further background on classical similarity theory, primarily in an Eulerian frame as a function of scale sizes in space. In §3.4 we consider low-order statistics, including Reynolds and Schmidt number scalings of the averaged rate of dissipation of the scalar fluctuations, spectra of scalar variance and scalar flux in wavenumber space, as well as structure functions in physical space. In §3.5 our focus shifts to higher-order quantities and small-scale behavior, including geometrical features, local isotropy, and the intermittency of the scalar dissipation rate fluctuations. Both one- and two-point statistics will be considered.

This review, even considering its emphasis on computation, is not meant to address all areas of active research in turbulent mixing. For example, while we focus on Eulerian scale similarity, a Lagrangian approach closely connected to turbulent dispersion from localized sources (as described by Sawford & Pinton in another chapter in this volume) is also very useful. In §3.6 we shall present an overall summary, make brief note of several less-studied aspects, and provide several suggestions for current research needs and future research directions in the field.

3.2 Computational perspective

As already mentioned, in this review we give special attention to results from direct numerical simulations (DNS), where the exact governing equations are solved numerically without modeling. The value of DNS for both physical understanding and model development is well known (Moin and Mahesh, 1998). However, careful thought needs to be given to numerical methods, resolution in space and time, and (increasingly) the technology of high-performance computing. We discuss the main points below.

3.2.1 Numerical methods

We consider flows in which the velocity field $\mathbf{u}$ in (3.1) is computed from the Navier–Stokes equations for incompressible flow, in the form

$$\partial \mathbf{u}/\partial t + \mathbf{u} \cdot \nabla \mathbf{u} = -(1/\rho)\nabla p + \nu \nabla^2 \mathbf{u} + \mathbf{f}, \qquad \nabla \cdot \mathbf{u} = 0, \qquad (3.2)$$

where **f** is a forcing term usually restricted to the large scales. The simplified geometry of homogeneous turbulence is convenient both physically and computationally. The dominant numerical approach is Fourier pseudo-spectral, where (3.1) and (3.2) are transformed to ordinary differential equations in wavenumber space, and nonlinear products are evaluated in physical space, with care taken to minimize any resulting aliasing errors that might affect the small scales. The fluctuating pressure field (p) is constrained by a Poisson equation. It is well known that, with CPU cost on an N^3 grid proportional to $N^3 \log_2 N$, discrete Fourier transforms at each time step are much more efficient than finite difference schemes at similar resolution.

A number of research groups (e.g., Overholt and Pope (1996); Wang et al. (1999); Yeung et al. (2002); Watanabe and Gotoh (2004); Schumacher and Sreenivasan (2005)) have performed DNS for the canonical problem of passive scalar mixing in forced, stationary isotropic turbulence. Variations in methodology are known in the choice of forcing schemes for the velocity field, isotropic versus mean-gradient forcing for the scalars, second- versus fourth-order time advance, as well as the treatment of aliasing errors. Nevertheless, as will be seen later, physical features observed in different simulations have been broadly consistent, while comparisons with wind-tunnel experiments (e.g. Mydlarski and Warhaft (1998)) are also encouraging.

The principal advantage of DNS is the ability to probe into massive datasets to extract physical information not otherwise available. While there are limitations, and stringent resolution criteria (discussed below) are often not met perfectly, this approach has proved to be of great value for discovering new knowledge, especially if advanced computing resources are available. An example of complex turbulence structure revealed through the immense degree of detail in DNS, which mimics that of the development of the telescope in astronomy, is that of clouds and voids in high-Reynolds-number turbulence in the chapter by of Kaneda & Morishita in this volume (see also Kaneda and Ishihara (2006); Ishihara et al. (2009)).

3.2.2 Resolution in space and time

A basic premise in DNS is that of faithful computation from basic principles, with all relevant scales resolved. This includes resolving small-scale features such as vortex tubes and ramp-cliff structures in scalar fields, and is, of course, more challenging at higher Reynolds number.

For simulations on a domain of linear size L_0 and N^3 grid points using standard de-aliasing error treatments (Rogallo, 1981) the smallest scale actually resolved is on the order of $1/k_{\max}$, where $k_{\max} = (\sqrt{2}/3) 2\pi N / L_0$.

Resolution of the small scales is often expressed by the parameter $k_{\max}\bar{\eta}$ (Eswaran and Pope, 1988), which gives the ratio of grid spacing (Δx) to Kolmogorov scale $(\nu^3/\bar{\epsilon})^{1/4}$ based on the averaged energy dissipation rate per unit mass, $\bar{\epsilon}$, as $(\sqrt{2}/3)(2\pi)/(k_{\max}\bar{\eta})$. Most simulations aimed at pushing the Reynolds number (Wang et al., 1999; Yeung et al., 2002; Gotoh et al., 2002; Kaneda et al., 2003) are in the range $1 \leq k_{\max}\bar{\eta} \leq 2$, corresponding to $3 \geq \Delta x/\bar{\eta} \geq 1.5$, and similarly for scalars in terms of the Batchelor scale η_B. However, while $\bar{\eta}$ is based on the mean energy dissipation rate ($\bar{\epsilon}$), the energy and scalar dissipation rates fluctuate, and so do $\eta(\mathbf{x},t) = (\nu^3/\epsilon(\mathbf{x},t))^{1/4}$ and $\eta_B(\mathbf{x},t) = Sc^{-1/2}\eta(\mathbf{x},t)$. As a result, the resolution criterion implied by a target value of $k_{\max}\bar{\eta}$ is broken locally, and high dissipation regions are under-resolved (Sreenivasan, 2004). Schumacher et al. (2005) found that even at very low Reynolds number there is significant probability for the local Batchelor scale η_B to be substantially smaller than $\bar{\eta}_B$. Watanabe and Gotoh (2007a) also reached similar conclusions at higher Reynolds number. These effects suggest a need to resolve small scales better than normally practiced, i.e. to use a higher $k_{\max}\bar{\eta}$ which however would imply lower values of Reynolds number and/or Schmidt number for a simulation of given size (Yakhot and Sreenivasan, 2005).

Pragmatically, since resolution in DNS will always be finite it becomes important to compare simulation results at different resolutions (but nominally the same Reynolds and Schmidt numbers), and to estimate the degree of resolution needed for a given type of statistics. Donzis et al. (2008a) and Donzis and Yeung (2010) addressed this issue for velocity and scalar fields respectively, with emphasis on structure functions and moments of dissipation rates at various orders. The conclusion is that $k_{\max}\bar{\eta} \approx 3$ and $k_{\max}\bar{\eta}_B \approx 3$ give adequate results for third and fourth moments of the dissipation rates, while simulations at "standard" resolution still give qualitatively correct behaviors for comparative quantities such as ratios of moments of dissipation to those of enstrophy. More importantly, local scaling exponents of the structure functions at intermediate (i.e. inertial–convective) scales appear to be insensitive to small-scale resolution (Watanabe and Gotoh, 2007a). This provides hope that inertial-range properties can still be studied reliably without huge extra costs in pursuit of finer small-scale resolution.

Improved resolution in space also leads to greater costs in time stepping. If the linear viscous and diffusive terms in the Fourier-transformed versions of (3.1) and (3.2) are integrated exactly via an integrating factor then the size of time step (Δt) is controlled by a convective Courant number condition for numerical stability. Essentially this means $\Delta t \propto \Delta x$, and hence a smaller Δx also leads to more time steps needed for a desired physical simulation time period (say, in terms of large-eddy time scales). Roughly speaking, when accounting for both space and time, a doubling of the parameter $k_{\max}\bar{\eta}$ can lead to a factor of 16 increase in overall CPU costs. Furthermore, at high Sc a smaller Δt may be needed to preserve numerical stability in the high-wavenumber scalar field.

3.2.3 High-performance computing

For DNS at high resolution it is critical to have parallel algorithms capable of utilizing advanced computer hardware effectively. The Fourier pseudo-spectral numerical approach is well suited for distributed-memory parallelism, which has been a dominant paradigm in high-performance computing (HPC) driven by machines of increasingly large processor counts. Essentially, workload and memory requirements are shared among a large number of processors, with each processor (or each unit of a multi-cored processor) operating on a part of the solution domain partitioned according to a suitable domain-decomposition scheme. However, significant overhead is incurred in interprocessor communication (usually handled by so-called message-passing software protocols), which is necessary for the task of re-dividing the domain along different coordinate directions (e.g., to perform Fourier transforms in each direction). Indeed, while algorithms with demonstrated success on $O(10^4$–$10^5)$ processor counts have been developed (Donzis et al., 2008b), communication has become an unavoidable bottleneck in the largest calculations attempted today. The development of advanced communication protocols and innovative programming models, as well as new-generation hardware, are thus very important areas of current HPC research. Alternative numerical schemes of a more local character in space and hence lesser requirements for interprocessor communication should also be explored.

Sources from the HPC community indicate that the past 25 years or so has been a period of sustained exponential growth in high-end computing power, which is projected to reach the Exaflops (10^{18} operations per second) level by 2018. This would be some 25,000 times faster than in 2002, when the first turbulence simulation at 4096^3 grid resolution was performed (Yokokawa et al., 2002). Clearly, an advance of this magnitude changes the line between what is feasible and what is not. While the actual design of future so-called Exascale computers is currently unknown, and daunting challenges lie ahead on the road to using them effectively, it is not difficult to compile a wish-list of problems in turbulent mixing worthy of the future line of machines. On the other hand, as larger problems become computable, future simulations are also likely to produce "avalanches" of data, underneath which the true

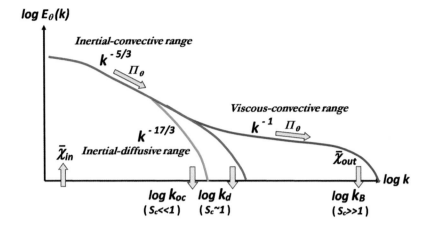

Figure 3.1 Schematic sketch of spectrum of passive scalar variance. Scalar fluctuations are injected at low wavenumbers at rate $\bar{\chi}_{in}$ and transferred to high wavenumbers with the mean flux Π_θ and then smeared by the molecular diffusivity at the rate $\bar{\chi}_{out}$. In a steady state $\bar{\chi}_{in} = \Pi_\theta = \bar{\chi}_{out}$. Note $k_{\rm OC} = k_d Sc^{3/4}$ and $k_{\rm B} = k_d Sc^{1/2}$, where $k_d = \bar{\eta}^{-1}$.

flow physics may be buried. Full exploitation of super-large simulations in the future will thus also require corresponding advances in the development of effective protocols for collaborative data analyses, involving participation by a much wider community than is usual at this time.

3.3 Background theory

The essence of similarity theory in turbulence is in characterizing the dependence of flow statistics on the range of scales present, while considering the effects of different physical processes acting at different scales in time and space. The distribution of spectral content as a function of scale size in three dimensions is a natural starting point. For the velocity field this gives the energy spectrum function, $E(k)$, whose integral in wavenumber space $\int_0^\infty E(k)\,dk$ is the turbulence kinetic energy, $K \equiv \frac{1}{2}\langle \mathbf{u} \cdot \mathbf{u} \rangle$ per unit mass. For turbulence at high Reynolds number, classical inertial range theory gives the form $E(k) = C_{\rm K} \bar{\epsilon}^{2/3} k^{-5/3}$ for $1/L \ll k \ll 1/\bar{\eta}$ where $\bar{\epsilon}$ is the mean rate of energy dissipation per unit mass, and $C_{\rm K}$ is called the Kolmogorov constant. For scalars, the form of the spectrum $E_\theta(k)$ (whose integral is the scalar variance, $\langle \theta^2 \rangle$) is sensitive to how the smallest (diffusive) scale compares with the scales of the velocity field. This leads to multiple regimes depending on the Schmidt number, as depicted in Fig. 3.1. We briefly review the classical theory here, as a basis for comparisons in the next subsection.

Obukhov (1949) and Corrsin (1951) were the first to propose extensions of Kolmogorov phenomenology to passive scalars. The time scale associated with the scalar spectrum at scale $1/k$ is $[kE_\theta(k)/\bar{\chi}]$, where $\bar{\chi} = 2\kappa\langle\nabla\theta\cdot\nabla\theta\rangle$ is the averaged scalar dissipation rate. If $Sc = O(1)$ then the range of scales in velocity and scalar fields will be similar, and their time scales should be the same as $\tau_k = (\bar{\epsilon}^{1/3}k^{2/3})^{-1}$. Equating $[kE_\theta(k)/\bar{\chi}]$ thus to τ_k gives

$$E_\theta(k) = C_{\text{OC}}\bar{\chi}\bar{\epsilon}^{-1/3}k^{-5/3}, \quad \text{for } 1/L \ll k \ll 1/\bar{\eta}_{\text{OC}}, \tag{3.3}$$

where C_{OC} is known as the Obukhov–Corrsin constant. The range of scales $L \gg 1/k \gg 1/\bar{\eta}_{\text{OC}}$, where neither viscosity nor molecular diffusion is important, is called the inertial–convective range.

In the case $Sc \gg 1$ since the scale $\bar{\eta}_{\text{B}} = \bar{\eta}Sc^{-1/2}$ is smaller than $\bar{\eta}$ a viscous–convective range may arise where $k\bar{\eta} \gg 1$ but $k\bar{\eta}_{\text{B}} \ll 1$, i.e., viscosity being important but not the diffusivity. Assuming $[kE_\theta(k)/\bar{\chi}]$ scales with $\tau_\eta = (\nu/\bar{\epsilon})^{1/2}$ yields the well-known k^{-1} Batchelor spectrum, i.e.

$$E_\theta(k) = C_{\text{B}}\bar{\chi}(\nu/\bar{\epsilon})^{1/2}k^{-1} \quad \text{for } 1/\bar{\eta} \ll k \ll 1/\bar{\eta}_{\text{B}}, \tag{3.4}$$

where C_{B} is called the Batchelor constant. Incidentally, Batchelor's original result (Batchelor, 1959) also predicts a rapid $\exp(-k^2)$ exponential decay in the viscous–diffusive range ($k\bar{\eta} \gg 1$ and $k\bar{\eta}_{\text{B}} \gg 1$), namely

$$E_\theta(k) = \bar{\chi}\gamma_{\text{eff}}k^{-1}\exp\left(-\kappa k^2/\gamma_{\text{eff}}\right), \quad \gamma_{\text{eff}}^{-1} = C_{\text{B}}(\nu/\bar{\epsilon})^{1/2}. \tag{3.5}$$

Batchelor assumed that isoscalar surfaces are subjected to an effective most-compressive principal strain rate proportional to $(\bar{\epsilon}/\nu)^{1/2}$. The exponential factor in (3.5) is equivalent to $\exp(-C_{\text{B}}(k\bar{\eta}_{\text{B}})^2)$ which clearly reverts to (3.4) in the limit $k\bar{\eta}_{\text{B}} \ll 1$. However since the strain rate does fluctuate, Batchelor's assumptions were not fully realistic. Kraichnan (1968) assumed that the velocity field changes rapidly in time and obtained

$$E_\theta(k) = C_{\text{B}}\bar{\chi}(\nu/\bar{\epsilon})^{1/2}k^{-1}\left(1+(6C_{\text{B}})^{1/2}\right)\exp\left(-(6C_{\text{B}})^{1/2}(k\bar{\eta}_{\text{B}})\right), \tag{3.6}$$

giving a milder viscous–diffusive decay but also predicts k^{-1} viscous–convective scaling. We shall compare both theories with the data available.

In both the inertial–convective and viscous–convective ranges considered above it is reasonable to assume spectral transfer at a constant rate (equal to $\bar{\chi}$, analogous to $\bar{\epsilon}$ for the energy spectrum) plays a dominant role. However this is not the case for scales smaller than the molecular cutoff where diffusivity becomes important. In the limit of extremely low Sc (such that $\bar{\eta}_{\text{OC}} \gg \bar{\eta}$) an inertial–diffusive (sometimes called inertial-conductive) range where $k\bar{\eta}_{\text{OC}} \gg 1$ but $k\bar{\eta} \ll 1$ may be possible. Two classical hypotheses are known. The first, by Batchelor et al. (1959), based on an assumed balance between convective and diffusive terms in the evolution equation for the scalar spectrum, gives a result of the form

$$E_\theta(k) = C_K/3 \bar{\chi} \bar{\epsilon}^{2/3} \kappa^{-3} k^{-17/3} . \tag{3.7}$$

The second, by Gibson (1968), based on consideration of zones of weak or zero scalar gradients, gives

$$E_\theta(k) = C_G \bar{\chi} \kappa^{-1} k^{-3}, \tag{3.8}$$

where C_G is a nondimensional constant. Unfortunately, the $Sc \ll 1$ regime, which occurs in liquid metals and astrophysics, is not readily accessible to laboratory measurement, and (to date) has not been addressed by DNS. Expressions (3.7) and (3.8) have thus remained largely untested.

In this section we have focused on classical spectral forms, which have played a fundamental role in theory development. In later sections we review current knowledge of the spectrum, and of many other important aspects, such as the scaling of mean scalar dissipation rate, structure functions in physical space, the scalar flux, and higher-order statistics which represent the effects of intermittency. We will also introduce additional results from classical theory as they arise in the discussions below.

3.4 Approach to low-order asymptotic state

We begin our review of results from the literature by focusing on low-order statistics such as the mean scalar dissipation rate and the scalar spectrum. An important goal is to examine how quickly are asymptotic results for high Reynolds and/or Schmidt number approached. Both single-point (spatially-averaged) and two-point (scale-dependent) quantities are relevant.

3.4.1 Mean dissipation rate and one-point statistics

It is well known that the mean energy dissipation rate $\bar{\epsilon}$ in high Reynolds number turbulence is determined by the large scales, and is (despite its definition as $2\nu \langle s_{ij} s_{ij} \rangle$, where s_{ij} denotes strain-rate fluctuations) asymptotically independent of the viscosity. (This concept is known as the dissipative anomaly.) An analysis by Doering (2009) of turbulence driven by a body force suggests the normalized energy dissipation is rigorously bounded from above by

$$f \equiv \bar{\epsilon} L / u'^3 \leq A + B/R_\lambda^2 , \tag{3.9}$$

where the constants A and B (in the limit $R_\lambda \to \infty$) depend only on the functional form of the body force. Surveys of data from DNS of forced isotropic turbulence (Sreenivasan, 1998; Ishihara et al., 2009) suggest that asymptotic constancy for f holds quite well for $R_\lambda > 100$, and that most forcing schemes give $A \approx 0.4$ although larger values are known in some cases.

For scalars, when a steady spectral cascade is sustained under the action of either isotropic injection or a mean gradient, extension of the dissipative anomaly concept above suggests the mean scalar dissipation rate $\bar{\chi}$ may be insensitive to molecular diffusivity. However since lower diffusivity also allows larger scalar fluctuations to build up, the normalized scalar dissipation $f' \equiv \bar{\chi} L / (\langle \theta^2 \rangle u')$ has some dependence on Sc. To help develop some theoretical estimates, Donzis et al. (2005) collected data from various sources in the literature. This is presented in Fig. 3.2(a), which includes data from simulations over a range of Schmidt numbers, and from measurements mainly of temperature fluctuations in air, at $Sc \approx 0.7$. While there is a spread at low and moderate Reynolds numbers, a trend towards a constant at high Reynolds number is also evident for scalars of $Sc = O(1)$. Data for $Sc < 1$ generally lie above data points for $Sc > 1$, since high Sc causes a slight increase in $\langle \theta^2 \rangle$ while $\bar{\chi}$ is maintained. A plot (not shown) of f' versus Péclet number (defined as $Pe_\lambda = R_\lambda Sc$) also supports a trend towards a constant for large Pe_λ when R_λ is low, while the scatter in the data remains. Full attainment of an asymptotic state of f' for high R_λ and Pe_λ with high Sc remains a challenge for both simulation and experiment.

In the literature on turbulent mixing models (Pope, 1985; Fox, 2003) the mechanical-to-scalar time scale ratio, $r_\theta = (K/\bar{\epsilon})/(\langle \theta^2 \rangle/\bar{\chi})$ is important, since knowledge of r_θ in a known velocity field allows the time scale of the mixing, $\langle \theta^2 \rangle/\bar{\chi}$, to be determined and used to relate $\langle \theta^2 \rangle$ and $\bar{\chi}$. The definition of r_θ differs from that of f' only via the use of different velocity time scales L/u' and $K/\bar{\epsilon}$, which are proportional to each other at $R_\lambda > O(200)$. By integrating the spectral forms of (3.3) and (3.4) from $k = 1/L$ to $1/\bar{\eta}_{\rm OC}$ (for $Sc \leq 1$), or from $k = 1/L$ to $1/\bar{\eta}$ followed by $1/\bar{\eta}$ to $1/\bar{\eta}_{\rm B}$ (for $Sc > 1$), Donzis et al. (2005) arrived at the result

$$(r_\theta C_{\rm OC})^{-1} = f^{2/3} - \frac{\sqrt{15}}{R_\lambda} \times \begin{cases} 1 - C_{\rm B}/(3C_{\rm OC})\ln(Sc) & (Sc > 1) \\ Sc^{-1/2} & (Sc < 1) \end{cases}. \quad (3.10)$$

This expression indicates that at fixed R_λ the time scale ratio decreases with Sc, but that dependence on Sc becomes weaker at high R_λ. It also suggests that in the limit $R_\lambda \to \infty$ (with $C_{\rm OC} \approx 0.6$ (Sreenivasan, 1996)), r_θ approaches an asymptotic value close to 3.0. A plot of DNS data as shown

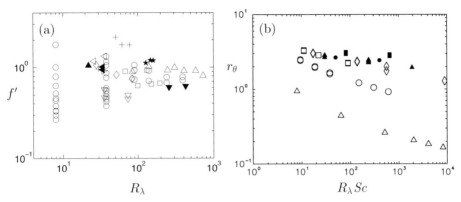

Figure 3.2 (a) Normalized scalar dissipation rate from both simulations and experiments: ○, Donzis et al. (2005); ▼, Watanabe and Gotoh (2004); □, Wang et al. (1999); ◇, Overholt and Pope (1996); △, Mydlarski and Warhaft (1998); ★, Tavoularis and Corrsin (1981); ◁, Sirivat and Warhaft (1983); ◀, Sreenivasan et al. (1980) ▷, Yeh and Atta (1973); ▶, Warhaft and Lumley (1978); ▲, Mills et al. (1958); +, Antonia et al. (2000); ▽, Bogucki et al. (1997). (b) Variation of mechanical-to-scalar time scale ratio in DNS over a range of R_λ, Sc and grid resolution from a single DNS database (Donzis and Yeung, 2010). Symbols denote simulations at different R_λ: △ (8), ○ (38), □ (90), ◇ (140), ▲ (240), • (390), ■ (650).

in Fig. 3.2(b) is in good agreement with all of these features. Since (3.10) involves an assumption of wide inertial–convective and viscous–convective ranges it is expected to become more accurate at higher R_λ and/or Sc.

In the case of scalar fluctuations maintained by a uniform mean gradient assuming homogeneity, the equation for scalar variance obtained from (3.1) can be written as

$$\partial \langle \theta^2 \rangle / \partial t = -2 \langle \mathbf{u}\theta \rangle \cdot \nabla \langle \Theta \rangle - 2\kappa \langle \nabla \theta \cdot \nabla \theta \rangle, \qquad (3.11)$$

where terms on the right hand side represent production (by a turbulent scalar flux) and dissipation (by molecular diffusivity). If the velocity field is isotropic the orientation of $\nabla \langle \Theta \rangle$ is arbitrary, and the resulting scalar flux becomes aligned with the mean gradient. Furthermore, since (3.1) is linear in the mean gradient, if the calculations are started with zero initial conditions then the resulting scalar fluctuations are proportional to the mean gradient everywhere and at all times. As a result, if we choose $\nabla \langle \Theta \rangle = (G, 0, 0)$ then $\langle \theta^2 \rangle$, $\langle u\theta \rangle$, $\bar{\chi}$ are proportional to G^2, G and G^2 respectively, while (in contrast) statistics of normalized quantities such as $\chi/\bar{\chi}$, which we consider later, are independent of the magnitude of G. Likewise, the Nusselt number, Nu, i.e. the ratio of turbulent to molecular fluxes, as $-\langle u\theta \rangle/(\kappa G)$ is also independent of G and instead mainly determined by the correlation between velocity and scalar fluctuations. In a steady state, DNS data indicate that

$\langle u\theta \rangle$ also depends little on Sc, so that Nu increases with Pe (Gotoh and Watanabe, 2012).

As an aside, we note that if the velocity field is not isotropic, as in the case of homogeneous shear flow, the alignment between scalar flux and mean gradient as assumed above does not hold – indeed Rogers et al. (1989) showed that a tensor eddy diffusivity is then needed to describe even the basic behavior. However, not many recent simulations of passive scalars in homogeneous turbulent shear flow are known.

3.4.2 Spectra of scalar variance and scalar flux

To isolate the behaviors of scalar fluctuations at different scale sizes it is useful to examine data from different sources in the literature on the spectral forms discussed in §3.3, as well as the spectrum of the scalar flux. However, while cross-checking between experiment, theory and simulation is clearly valuable, it is important to appreciate several types of uncertainty. First, both (3.3) and (3.4) require a wide scaling range, which may not be available especially in simulations at lower resolution in the past. Second, the precise determination of $\bar{\epsilon}$ and $\bar{\chi}$ can be problematic in experiments, where use of one-dimensional surrogates with or without the so-called Taylor's frozen-turbulence hypothesis is common. Third, since in experiments typically only 1D spectra are available, comparison with 3D spectra in DNS requires the use of various isotropy relations, which, because of sampling and domain size issues, may not hold sufficiently well in both DNS and experiment. Fourth, intermittency corrections to classical theory may be significant and at least make clean observations of scaling laws more difficult. Finally, bottleneck effects similar to those in the energy spectrum (Falkovich, 1994; Donzis and Sreenivasan, 2010) must be considered as well.

Experimental data for the inertial–convective range have mainly come from measurements of temperature fluctuations in air ($Sc \approx 0.7$), in the atmosphere (Aivalis et al., 2002), or the wind tunnel (Mydlarski and Warhaft, 1998; Danaila and Antonia, 2009), or in low-temperature helium gas ($Sc = 0.8$) laboratory devices (Moisy et al., 2001). Sreenivasan (1996) collected data available at the time and concluded that the Obukhov–Corrsin constant in the 1D scalar spectrum is about 0.4, which corresponds to $C_{OC} = (0.4)(5/3) = 0.67$ using isotropy relations. On the other hand smaller values have been derived from spectral closures by Kraichnan (1966) (0.208) and Kaneda (1986) (0.304). Measurements in an active-grid wind tunnel (Mydlarski and Warhaft, 1998) are in the range 0.75–0.92. Data from two sources of DNS of isotropic turbulence at high Reynolds number (Watanabe and Go-

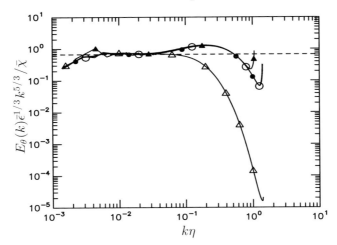

Figure 3.3 Scalar spectrum in Obukhov–Corrsin scaling, from two separate series of simulations of isotropic turbulence. Open symbols for data from Yeung et al. (2005) at $R_\lambda \approx 650$, 2048^3, $Sc = 1/8$ (triangles) and 1 (circles). Closed triangles for data from Watanabe and Gotoh (2004): $R_\lambda = 427$, 1024^3, $Sc = 1$; Closed circles for data from Watanabe and Gotoh (2007b): $R_\lambda = 585$, 2048^3, $Sc = 1$. The height of the dashed line gives the inferred value of Obukhov–Corrsin constant (0.67, for the 3D spectrum).

toh, 2004; Yeung et al., 2005) suggest, as shown in Fig. 3.3, close agreement with $C_{\rm OC} = 0.67$ suggested by Sreenivasan (1996). There is also a high and encouraging degree of consistency between these two series of simulations performed using different schemes for forcing, scalar injection, and numerical integration in time. A well-defined scaling range of about one decade is quite visible, especially in the case of $Sc = 1/8$. A spectral bump (or bottleneck effect) is apparent at $Sc = 1$ (making the scaling range narrower) but not at $Sc = 1/8$. This bottleneck effect may in part be a pre-cursor of Batchelor k^{-1} behavior which develops at higher Schmidt numbers. Incidentally, since (3.3) is linear in $\bar{\chi}$, intermittency effects are felt only through the variations of $\bar{\epsilon}$, independent of Schmidt number. Likewise, since the spectrum of a scalar of $Sc = 1/8$ drops rapidly at high wavenumber it is not affected much by limitations of finite resolution.

For the viscous–convective range, the need to resolve scales as small as $\bar{\eta}_{\rm B} = \bar{\eta} Sc^{-1/2}$ has generally constrained the highest Sc feasible in DNS, or forced a compromise between reaching high Reynolds number versus high Schmidt number. It seems reasonable to consider high Sc as more important than high R_λ for this type of scaling. If this is valid then less expensive simulations at moderate or low Reynolds number may be sufficient to reveal most of the important physics. Indeed, simulations by Bogucki et al. (1997); Yeung et al. (2002); Brethouwer et al. (2003) and Yeung et al. (2004) (with

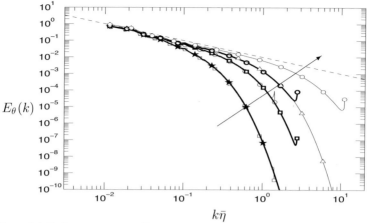

Figure 3.4 Scalar spectrum at $R_\lambda = 140$ in DNS data reported by Donzis et al. (2010), for $Sc = 1/8$, 1, 4 and 64 in the direction of the arrow, which approaches a dashed line of slope -1. Circles, triangles and squares denote data with resolution $k_{\max}\bar{\eta}_B \approx 1.4$, 2.9 and 5.6 respectively.

the latter at R_λ as low as 8 but Sc as high as 1024) have provided qualitative support for a trend towards k^{-1} scaling at high Sc (see Fig. 3.4). Furthermore, as seen in Fig. 3.5, the decay of $E_\theta(k)$ in the viscous–diffusive range is closer to the Kraichnan formula, (3.6), than Batchelor's, (3.5). The k^{-1} spectrum has also been observed in experiments at high Péclet and Schmidt numbers driven by electromagnetic forces (Jukkien et al., 2000). However, since mixing at very high Sc is dominated by straining by the small-scale motions a Reynolds number sufficiently large for small-scale universality is still important. More recent analyses in Donzis et al. (2010) supported by new data at higher R_λ show that the minimum Sc needed to observe viscous–convective scaling decreases systematically with increasing R_λ. Reduced sensitivity to Sc at higher R_λ is also consistent with the mathematical form of (3.10).

The exponential decay law for $E_\theta(k)$ in the far diffusive range was derived by Kraichnan (1968, 1974), who assumed the turbulent strain rate field to be white noise in time. While this assumption is not consistent with DNS (where the velocity field is smooth in time), the excellent agreement with exponential roll-off seen in Fig. 3.5 suggests that fluctuations in strain rates acting upon the scalar field are the key. In contrast, the work of Batchelor et al. (1959) can be considered as a mean field theory which assumes a constant effective strain rate everywhere and is thus less satisfactory in the viscous–diffusive range where strain-rate fluctuations become increasingly important.

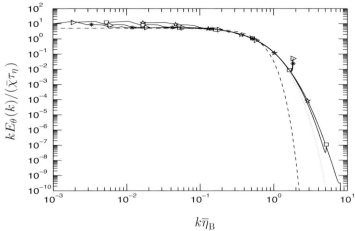

Figure 3.5 Spectrum of scalar variance normalized by Batchelor variables, in highest-resolution DNS data available at $R_\lambda = 240$: ☆, $Sc = 0.125$ (1024^3, $k_{\max}\bar{\eta} \approx 2.8$); □, $Sc = 1$ (2048^3, $k_{\max}\bar{\eta} \approx 5.1$); ▷, $Sc = 8$ (2048^3, $k_{\max}\bar{\eta} \approx 5.1$); ∗, $Sc = 32$ (4096^3, $k_{\max}\bar{\eta} \approx 10.4$). The data are compared with Batchelor's ((3.5), dashed line) and Kraichnan's (3.6), dotted line) formulas. [Courtesy D.A. Donzis]

C_B	Theory, EXP, DNS	
$\sqrt{3} < C_B < 2\sqrt{3}$		Gibson (1968)
$C_B < 0.9$	ALHDIA	Kraichnan (1968)
1.68	TFM	Newman and Herring (1979)
$\sqrt{10/3} \approx 1.83$	LRA	Gotoh (1989); Gotoh et al. (2000)
2		Batchelor (1959)
2	SBALHDIA	Herring and Kraichnan (1979)
$2\sqrt{5} \approx 4.47$		Qian (1995)
5		Borgas et al. (2004)
$2\sqrt{15} \approx 7.75$		Higgins et al. (2006)
3.9 ± 1.5	EXP	Grant et al. (1968)
3.7 ± 1.5	EXP	Oakey (1982)
3.4	DNS, decay, $R_\lambda = 38, Sc = 15$	Antonia and Orlandi (2003b)
5.26 ± 0.25	DNS, case R, $R_\lambda = 77, Sc = 7$	Bogucki et al. (1997)
5.5	DNS, case G, $R_\lambda = 38, Sc = 64$,	Yeung et al. (2002)
4.93	DNS, case G, $R_\lambda = 140, Sc = 64$	Donzis et al. (2010)

Table 3.1 Values of Batchelor's constant C_B from various sources, including theory, experiment, and DNS. Scalar fluctuations were either decaying or maintained by random forcing (Case R) or uniform mean gradient (Case G).

As noted by Donzis et al. (2010), despite strong evidence for k^{-1} scaling, the value of the (Batchelor) scaling constant, C_B, is much less settled. A list of values of C_B reported in the literature is given in Table 3.1; most values from DNS or experiment are between 3 and 6, while the theoretical predictions show more variation. Since the viscous–convective range observed (if any) is frequently narrow and subject to contamination by bottleneck effects, instead of using (3.4), C_B is usually estimated by fitting the spectrum to the Kraichnan form (3.6). The fact that almost all observed values of C_B are higher than the original 2.0 in Batchelor (1959) also implies that the correct effective strain rate γ_{eff} in (3.5) is less than $0.5\,(\tau_\eta)^{-1}$. Donzis et al. (2010) showed that this discrepancy can be partly resolved by adopting instead the most probable value of the most compressive principal strain rate. More rigorously, the theory of Kraichnan (1968) predicts that C_B is given by the time integral of the Lagrangian autocorrelation of the velocity gradients, as

$$C_B = (D+2)\left[\left(\frac{\nu}{\bar{\epsilon}}\right)^{1/2}\int_{-\infty}^{t}\langle A_{ij}(\mathbf{x},t|t)A_{ij}(\mathbf{x},t|s)\rangle\,\mathrm{d}s\right]^{-1}, \qquad (3.12)$$

where D is the number of spatial dimensions (3 for 3D turbulence) and $A_{ij}(\mathbf{x},t|s)$ is the velocity gradient tensor sampled at the position at time s of a fluid particle whose trajectory passes $\mathbf{x}$ at time t (Kraichnan, 1974; Herring and Kraichnan, 1979). At sufficiently high Reynolds number this Lagrangian correlation is likely to receive more contributions from motions at smaller scales, which suggests small-scale universality may lead to a trend of C_B becoming independent of R_λ. On the other hand, this expression for C_B can be further improved by noting that pure straining motion is primarily responsible for the deformation of a scalar blob and its Lagrangian correlation time is shorter than that of other velocity gradients (Yu and Meneveau, 2010; Watanabe and Gotoh, 2010). Indeed, calculations of C_B using pure strain-rate fluctuations in place of velocity gradient fluctuations in (3.12) produce results closer to DNS and experiments (Ishihara and Kaneda, 1992; Gotoh et al., 2000).

In the inertial–diffusive range, spectral closures invoked by Kraichnan (1968) and Gotoh (1989) both produced the same result as (3.7), while a different analysis by Qian (1986) suggested a larger scaling constant, $G_0 = 1.2\,C_K$ (instead of $G_0 = C_K/3$). Unfortunately, experiments in this regime are particularly difficult due to the toxic nature of fluids such as mercury, and available data are not conclusive (see Antonia and Orlandi (2003a)). Numerical simulation is also challenging because of the need for high Reynolds number and a concern for domain size effects as larger scales develop. To lessen the computational costs Chasnov et al. (1988) considered a frozen velocity field, which has some justification if the scalar evolves on time scales much shorter than those of the velocity. (This corresponds to f' and r_θ in Fig. 3.2 becoming very large as $Sc \to 0$.) Large-eddy simulations in Chasnov et al. (1988) based on this principle gave $E_\theta(k) \propto k^{-17/3}$, with

$G_0 \approx 0.33\, C_K$ for a Gaussian random velocity field with a prescribed Kolmogorov spectrum, and $G_0 \approx 0.39\, C_K$ for a velocity field taken from DNS with Navier–Stokes dynamics. However, the lowest Schmidt number in more recent high Reynolds number DNS is only 1/8 (Donzis et al., 2010) which is not sufficiently low for tests of inertial–diffusive scaling. Likewise, a channel flow calculation by Piller et al. (2002) showed behavior close to $k^{-17/3}$ but it was not conclusive.

In wavenumber space, the evolution of the scalar spectrum is the result of combined effects of nonlinear spectral transfer, production (if present), and dissipation. If a mean gradient is present, production is represented by the velocity–scalar co-spectrum $E_{u\theta}(k)$ (whose integral gives the scalar flux, $\langle u\theta \rangle$). Lumley (1964, 1967) postulated, on dimensional grounds,

$$E_{u\theta}(k) = C_{u\theta} G \bar{\epsilon}^{1/3} k^{-7/3} \tag{3.13}$$

in the inertial–convective rannge, where G is the uniform mean gradient, and $C_{u\theta}$ is a nondimensional constant of order unity. In wind-tunnel experiments up to $R_\lambda = O(700)$ Mydlarski and Warhaft (1998) and Mydlarski (2003) found the spectral exponent to be close to $-5/3$ and the approach to the asymptotic value $-7/3$ to be very slow. A spectral closure by Bos et al. (2005) also suggested $C_{u\theta}$ is indeed $O(1)$ but approach to a $-7/3$ exponent may require R_λ as high as 10^4. In DNS by Watanabe and Gotoh (2007b) up to $R_\lambda = 585$ and $Sc = 1$ at resolution 2048^3, as the Reynolds number increases the Lumley scaling range appears at lower wavenumbers, followed by a nearly k^{-2} range and finally an exponential roll off at high wavenumbers. Fig. 3.6 shows the shell-averaged cospectrum and the compensated spectra by multiplying $k^{7/3}$ and k^2 in the inset, respectively. The nearly k^{-2} range resembles a spectral bump. The computed value of $C_{u\theta}$ is 1.5 ± 0.08.

O'Gorman and Pullin (2005) showed that the Schwarz inequality gives an upper bound on the co-spectrum in terms of the energy and scalar spectra, in the form

$$E_{u\theta}(k) \leq \left(\frac{4}{3} E(k) E_\theta(k)\right)^{1/2}. \tag{3.14}$$

The relationship of $E_{u\theta}(k)$ to $E(k)$ and $E_\theta(k)$ can also be traced to the phase alignment between velocity and scalar modes in wavenumber space (Yeung, 1996), which is clearly sensitive to Schmidt number as well. The value of $C_{u\theta}$ in the inertial–convective range predicted by the spectral theory in O'Gorman and Pullin (2005) is more than twice that in experiments and numerical simulations. However the authors also studied Schmidt number effects in some detail. When $Sc \gg 1$, in the viscous–convective range since $E_\theta(k) \propto k^{-1}$ while $E(k)$ decays exponentially, $E_{u\theta}(k)$ must decay at least

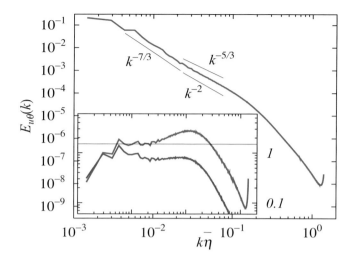

Figure 3.6 Spectrum of the scalar flux $E_{u\theta}(k)$ from DNS using 2048^3 grid points at $R_\lambda = 585$, $Sc = 1$, from Watanabe and Gotoh (2007b). Thin lines correspond to power laws for $k^{-7/3}$, k^{-2} and $k^{-5/3}$. The plot in the inset shows the compensated spectrum in Lumley's scaling, $G^{-1}\bar{\epsilon}^{-1/3}k^{7/3}E_{u\theta}(k)$ (upper curve) and $k^2 E_{u\theta}(k)$ (lower curve), respectively. The horizontal line in the inset is at the level of 1.50 obtained by averaging the compensated spectrum over the range $0.0085 \leq k\bar{\eta} \leq 0.023$.

exponentially as well. When $Sc \ll 1$, in the inertial–diffusive range, if we take $E_\theta(k) \propto k^{-17/3}$, then the upper bound above implies $E_{u\theta}$ decays like $k^{-11/3}$ or faster. Future simulations covering a wider range of Reynolds and Schmidt numbers are expected to provide a more complete picture.

A practical question common to discussions of energy, scalar and scalar flux spectra is the manner in which inertial-range scaling exponents approach their asymptotic values at increasing Reynolds number. In particular, if intermittency corrections are neglected, in experiments typically the exponents of 5/3 for $E(k)$ and $E_\theta(k)$ are approached from below (Mydlarski and Warhaft, 1998; Danaila and Antonia, 2009) while in DNS they are approached from above (Kaneda et al., 2003; Watanabe and Gotoh, 2007b). If intermittency is considered, then statistical realizability and the exact 4/5 law for third-order velocity structure functions imply that the exponent in the energy spectrum must be approached from above. However no similar constraints for the spectra of scalar variance and scalar flux are known.

3.4.3 Structure functions in physical space

A direct approach to describing spatial structure in turbulence is to use the statistics of spatial increments of velocity and scalar fluctuations over an interval of linear size r varied over a significant range. For homogeneous turbulence, a fundamental quantity is the mixed velocity-scalar structure function $\langle \delta u_j(\mathbf{r},t)(\delta\theta(\mathbf{r},t))^2 \rangle$, where $\delta u_j(\mathbf{r},t) = u_j(\mathbf{x}+\mathbf{r},t) - u_j(\mathbf{x},t)$ and $\delta\theta(\mathbf{r},t) = \theta(\mathbf{x}+\mathbf{r},t) - \theta(\mathbf{x},t)$ but $\mathbf{x}$ is statistically immaterial by homogeneity. From (3.1) one can derive an exact transport equation, as

$$\frac{\partial}{\partial r_j}\langle \delta u_j(\mathbf{r},t)(\delta\theta(\mathbf{r},t))^2 \rangle = -2\bar{\chi} + 2\kappa\frac{\partial^2 \langle(\delta\theta(\mathbf{r},t))^2\rangle}{\partial r_j \partial r_j} - \frac{\partial \langle(\delta\theta(\mathbf{r},t))^2\rangle}{\partial t} + 2F_\theta(\mathbf{r},t), \quad (3.15)$$

where the function $F_\theta(\mathbf{r},t) = \langle \delta\theta(\mathbf{r},t)\delta f_\theta(\mathbf{r},t)\rangle$, with $\delta f_\theta(\mathbf{r},t) = f_\theta(\mathbf{x}+\mathbf{r},t) - f_\theta(\mathbf{x},t)$, represents the external injection for the scalar. If isotropy applies then dependence on $\mathbf{r}$ reduces to dependence on $r = |\mathbf{r}|$ only, and only the longitudinal velocity increment $\delta u_L(r) = \delta u_j(\mathbf{r})r_j/r$ is important. Consequently, if there exists a range of scale size r where isotropy applies, and such that the diffusive, non-stationary and injection terms in Eq. (3.15) all become negligible, then the equation above becomes

$$\langle \delta u_L(r)(\delta\theta(r))^2 \rangle = -\frac{2}{3}\bar{\chi}r\,. \quad (3.16)$$

This is often known as Yaglom's relation or the two-thirds law (note Yaglom's definition of $\bar{\chi}$ was a factor of 2 smaller). It is noteworthy that, as one of the necessary conditions for Yaglom's relation, neglect of the diffusion term is appropriate in both the inertial–convective and viscous–convective ranges, if such ranges are indeed well formed. However, some care is needed, as below, to justify the neglect of non-stationary and injection terms.

If no forcing or injection mechanisms are present and initial conditions are isotropic, we can integrate all terms in (3.15) once in space to obtain

$$\langle \delta u_L(r,t)(\delta\theta(r,t))^2 \rangle = -\frac{2}{3}\bar{\chi}r + 2\kappa\frac{\partial}{\partial r}\langle(\delta\theta(r,t))^2\rangle - \frac{3}{r^2}\int_0^r r'^2 \frac{\partial}{\partial t}\langle(\delta\theta(r',t))^2\rangle\,dr', \quad (3.17)$$

(Yaglom, 1949; Danaila et al., 1999; Orlandi and Antonia, 2002; Hill, 2002) where the form of the non-stationary term in (3.17) is due to the use of spherical coordinates in the integration. For $\bar{\eta}_B \ll r \ll L_\theta$, where L_θ is the integral scale of the passive scalar, the diffusive term can be neglected. The non-stationary term decreases with decreasing r. However, if Sc is large, it follows, since this decrease may (due to small diffusivity) be slow, that the non-stationary term may be important (Danaila and Mydlarski, 2001; Orlandi and Antonia, 2002). Applicability of Yaglom's relation in these conditions thus rests on whether the initial conditions lead to substantial non-stationary effects at scales where molecular diffusion is weak.

When external injection is present and the scalar fluctuations attain a statistically homogeneous and stationary state, the effects of the initial condition vanish and we can neglect the non-stationary term. If the scalar is injected isotropically and only at large scales, $F_\theta(r)$ decreases rapidly with decreasing r and can be neglected at small scales, and thus Yaglom's relation holds. However, if the injection is applied via a mean uniform scalar gradient, we have

$$F_\theta(\mathbf{r}) = -G\left[2\langle\theta(\mathbf{x})u(\mathbf{x})\rangle - \langle\theta(\mathbf{x}+\mathbf{r})u(\mathbf{x})\rangle - \langle\theta(\mathbf{x})u(\mathbf{x}+\mathbf{r})\rangle\right].$$

Since $\langle\theta(\mathbf{x}+\mathbf{r})u(\mathbf{x})\rangle \to \langle\theta(\mathbf{x}+\mathbf{r})\rangle\langle u(\mathbf{x})\rangle = 0$ as $r \to \infty$, $F_\theta(\mathbf{r})$ is constant for large r and approaches monotonically to zero as r decreases (although the decay might be slow), so that the first term on the right of (3.15) dominates at small scales. At the same time, since injection by uniform mean gradient is not isotropic, the neglect of diffusive, non-stationary and injection contributions in (3.15) leads directly to

$$\partial\langle\delta u_j(\mathbf{r})(\delta\theta(\mathbf{r}))^2\rangle/\partial r_j = -2\bar{\chi} \ . \tag{3.18}$$

This expression is related to but not identical to Yaglom's relation (3.16). Yet, to complete the derivation, one may consider a volume integral over a sphere of radius r (Hill, 2002). Since the left-hand side of this equation is the divergence of a vector in $\mathbf{r}$-space, invoking Gauss's theorem converts the result to an integral over the surface of the sphere of the vector $\langle\delta\mathbf{u}(\mathbf{r})(\delta\theta(\mathbf{r}))^2\rangle$ projected onto $\mathbf{r}$, which is in the outward normal direction. The surface integral yields $\langle\delta u_L(r)(\delta\theta(r))^2\rangle(4\pi r^2)$ for the left of (3.18), while a straightforward volume integral gives $(-2\bar{\chi})(4\pi r^3/3)$ on the right. Equating these expressions allows us to reproduce Yaglom's relation: in other words, Yaglom's 2/3 law holds for the spherical average of $\langle\delta u_L(\mathbf{r})(\delta\theta(\mathbf{r}))^2\rangle$ when a uniform mean scalar gradient is present.

In classical similarity scaling of turbulent mixing Equation (3.16) has a central role analogous to that of Kolmogorov's exact relation for the third-order velocity structure function, i.e.

$$\langle(\delta u_L(r))^3\rangle = -\frac{4}{5}\bar{\epsilon}r \ . \tag{3.19}$$

While most attempts to test Yaglom's relation have focused on high Reynolds number for scalars of $Sc = O(1)$ (e.g. Mydlarski and Warhaft (1998); Watanabe and Gotoh (2004)), in the case of $Sc \gg 1$ this relation is uniformly valid at scales down to the Batchelor scale as well (Yeung et al., 2002). Furthermore, this equation can be transformed to wavenumber space (Frisch, 1995) where it corresponds to the statement that the spectral transfer flux ($\Pi_\theta(k)$,

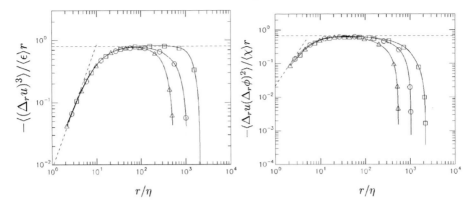

Figure 3.7 The third-order velocity structure functions (left) and the mixed velocity-scalar structure functions (right) scaled according to Kolmogorov's and Yaglom's relations at $R_\lambda = 240$ ($\triangle$), 400 ($\circ$), and 650 ($\square$), respectively. With the definition $\bar{\chi} = 2\kappa \langle \nabla \theta \cdot \nabla \theta \rangle$, the level of the plateau is 2/3. The dashed line of slope 2 gives the small r asymptote. Data are taken from Yeung et al. (2005) (where highest R_λ was quoted as 700, but upon averaging over a longer time period 650 is more accurate).

rate of spectral transfer from scales of wavenumbers lower than k to those higher than k) is constant and equal to $\bar{\chi}$ in the applicable scaling range. Specifically, one can write

$$\Pi_\theta(k) = -\frac{1}{4\pi^2} \int \frac{\sin(kr)}{r} \frac{\partial}{\partial r_l} \left(\frac{r_l}{r^2} \frac{\partial}{\partial r_j} \langle \delta u_j(\mathbf{r})(\delta \theta(\mathbf{r}))^2 \rangle \right) \, d\mathbf{r} = \bar{\chi} \,, \quad (3.20)$$

which, like (3.16), holds in both inertial–convective and viscous–convective ranges.

The approach to Yaglom's relation provides another measure for assessing how close the statistics of the passive scalar in turbulence are to an asymptotic state. Figure 3.7 shows a test of both (3.19) and (3.16) in DNS at several Reynolds numbers up to $R_\lambda = 650$, for a scalar of $Sc = 1$ with uniform mean gradient (Yeung et al., 2005). Closer examination suggests the plateau for Yaglom's relation (right) starts at a smaller r than that for Kolmogorov's (left). This is consistent with results from simulations where the scalars were injected isotropically (Watanabe and Gotoh, 2004). However as discussed in Yeung et al. (2002) there is also a significant degree of anisotropy at scale sizes in the reported scaling range. To clarify the interpretation of these results further in the future it would be useful to employ the spherical harmonics expansion which we discuss later.

3.5 High-order statistics: fine-scale structure and intermittency

Because of its close coupling with molecular processes, including chemical reaction, the structure of the smallest scales in turbulent mixing is often of the greatest importance. In addition to the spectrum (which is a second-order quantity), it thus becomes necessary to examine the geometry of the instantaneous scalar field and the property of intermittency represented by localized, intense fluctuations which are, in turn, expressed by high-order statistics of the scalar field. As we discuss below, high-order statistics are also subject to challenges in resolution and sampling.

3.5.1 Geometry of the scalar field

The statistical geometry of isoscalar surfaces has a direct impact on the rate of heat and mass transfer possible for a source or mean gradient field maintained at a given intensity. It is well known that scalar fluctuations driven by a mean gradient $\nabla\langle\Theta\rangle$ tend to develop a ramp-cliff structure (Sreenivasan et al., 1979; Antonia et al., 1986; Sreenivasan, 1991; Pumir, 1994; Holzer and Siggia, 1994; Warhaft, 2000; Celani et al., 2001). Computer visualization of large datasets can be very helpful. Figure 3.8 shows images of the total scalar $\Theta = Gx + \theta$ where x is the direction of $\nabla\langle\Theta\rangle$, and the scalar dissipation in orthogonal plane, from DNS at $R_\lambda = 468$ and $Sc = 1$ (Gotoh and Watanabe, 2012). A ramp (or plateau) and cliff structure is indeed evident, with a mix of regions of sharp changes and others of relative uniformity (e.g. large patches of red and green separated by narrow bands of yellow in the figure). In the plateau regions, which correspond to the ramp in θ, the total scalar (Θ) is nearly constant with only small fluctuations, meaning that the scalar is well mixed. In addition, Θ rises abruptly at the boundary of the plateau, as the cliff emerges and very large scalar dissipation occurs. Observations from images of different grid planes from multiple instantaneous snapshots suggest that the cliffs are preferentially aligned perpendicular to the mean gradient (Brethouwer et al., 2003). We can also infer that the linear dimension of the plateau is of the order of the integral scale, while large scalar dissipation occurs along cliffs of width about one Batchelor scale.

Since the instantaneous scalar dissipation χ is proportional to $|\nabla\theta|^2$ the bottom frame of Fig. 3.8 also provides information on scalar gradients, which are preferentially aligned with the direction of the most compressive principal strain rate (Ashurst et al., 1987; Vedula et al., 2001). It is clear that the scalar interface at the cliff becomes highly convoluted under the persistent straining, and that very high scalar dissipation occurs on a fractal

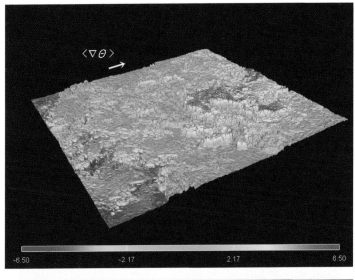

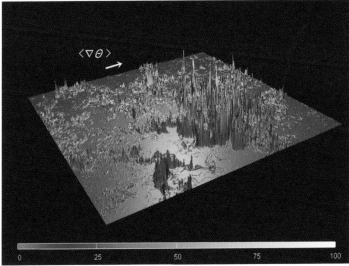

Figure 3.8 Color contour surfaces of the total scalar amplitude (top) and the scalar dissipation (bottom) in a selected plane at $R_\lambda = 460$ and $Sc = 1$ in 1024^3 DNS (Gotoh and Watanabe, 2012).

interface. Schumacher et al. (2005) computed the generalized dimension of the scalar dissipation as $D_q = 2.2$ for large q at $R_\lambda = 24$ and $Sc = 32$ using box-counting techniques. Sreenivasan and Prasad (1989) examined the generalized dimensions of the scalar interface in turbulent jet at $Re = 2000$ and Sc about 2000. They found two scaling ranges, with the dimension being 2.36 for $r > \bar\eta$ and 2.75 for $\bar\eta_B < r < \bar\eta$.

If isoscalar surfaces can be considered sheet-like in general, then the thickness and curvature become important attributes. Two possible measures of the thickness are the inverse of the gradient $h(\mathbf{x},t) \equiv |\partial \theta / \partial n|^{-1}$ of a normalized scalar θ, and the smallest scale of the scalar field, namely the Batchelor scale η_B. Aguirre and Catrakis (2005) studied the scalar interface thickness distribution across the width of a turbulent jet. While their results (within the resolution available) provide some support for a log-normal distribution for the scalar dissipation (χ), log-normal theory is also known to have a number of weaknesses (Frisch, 1995). Instead, DNS results by Schumacher and Sreenivasan (2005) and Donzis and Yeung (2010), together covering a relatively wide Reynolds number range, show that the probability density function (PDF) of χ is of stretched–exponential form, i.e. $P(\chi) \propto \exp(-c(\chi/\bar{\chi})^\alpha)$. Here both the exponent (α) and prefactor (c) depend on Reynolds number and Schmidt number, while the choice of mean gradient forcing versus Gaussian white noise forcing at large scales for the scalar field fluctuations may also have a secondary effect. Stronger intermittency at higher R_λ is represented by a wider tail with smaller values of α and c, and has important implications for spatial resolution requirements as well. Schumacher and Sreenivasan (2005) introduced the local Batchelor scale based on the molecular Schmidt number and the fluctuating energy dissipation rate, as $(\nu^3/\epsilon)^{1/4}/\sqrt{Sc}$, and examined the resolution needed to resolve this scale at every grid point in space. Subsequently Donzis and Yeung (2010) compared several results at different resolutions with R_λ and Sc fixed.

The curvature of a scalar isosurface can be quantified by two principal curvatures, K_1 and K_2, which can be computed from second derivatives of the scalar field. An elliptically convex or concave surface is indicated by K_1 and K_2 being of the same sign (with a positive Gauss curvature, $K_\mathrm{g} = K_1 K_2$); a surface is of saddle-like (hyperbolic) shape if they are of opposite signs ($K_\mathrm{g} < 0$). Dopazo et al. (2007) presented results on K_g and the mean curvature $K_\mathrm{m} = (K_1 + K_2)/2$ from a low Reynolds number but well-resolved simulation. The most probable state of surface curvature is close to that of a flattened sheet, with both $K_\mathrm{m} = 0$ and $K_\mathrm{g} = 0$ and a large scalar gradient normal to the surface. The build-up of such a large scalar gradient (and dissipation) is often associated with local compressive straining and high energy dissipation, although the overall correlation between energy and scalar dissipation rates is only modest (Vedula et al., 2001; Schumacher et al., 2005). An important issue is the coherency in time (Batchelor, 1959) of principal strain rates in the Lagrangian frame, whose time scale may be of order one Kolmogorov time scale (Brethouwer et al., 2003). This can be examined through Lagrangian velocity gradient time series in DNS (Yeung et al., 2006b).

3.5.2 PDF tails and high order statistics

The PDF of scalar fluctuations in turbulence is not as well understood as that for the velocity. In most data sources from DNS of 3D stationary isotropic turbulence, including Overholt and Pope (1996) and Schumacher and Sreenivasan (2005), scalar fluctuations are, like the velocity fluctuations, close to Gaussian, with a flatness factor roughly 2.6–2.8. If there is no mean gradient then, as the scalar fluctuations decay, the form of the PDF naturally depends on initial conditions (Jaberi et al., 1996). A functional form consisting of close-to-Gaussian cores with sub-Gaussian tails has been observed in both experiment (Thoroddsen and Van Atta, 1992) and DNS (Watanabe and Gotoh, 2004). Ferchichi and Tavoularis (2002) showed that the scalar PDF is almost Gaussian in sheared turbulence experiments for R_λ 180–253. However, in other experiments (Gollub et al., 1991; Jayesh and Warhaft, 1992) exponential tails have also been observed. Such tails are quite prominent in studies of turbulent convection at high Rayleigh number (Castaing et al., 1989; Ching, 1993) where temperature is an active scalar.

Several different approaches have been taken in the literature to explain or identify conditions for different forms for the passive scalar PDF. In the linear-eddy formulation by Kerstein (1991) an exponential form is obtained when the system size is much larger than the integral length scale. Pumir et al. (1991) and Shraiman and Siggia (1994) analyzed the PDF in terms of Lagrangian path integral method, and predicted an exponential tail arises if a fluid particle is able to carry scalar fluctuations over a long distance of several integral length scales in the velocity field. This is consistent with Kerstein's result. Likewise, a 2D calculation by Holzer and Siggia (1994) also produced exponential tails. Kimura and Kraichnan (1993) showed by using the mapping closure that the occurrence of non-Gaussian scalar PDFs depends on a subtle interplay between advective transport and molecular diffusion.

Although there is currently no consensus, studies suggest that the physical mechanisms generating non-Gaussian scalar PDFs are not unique (Jaberi et al., 1996). The presence of a mean scalar gradient is not a necessary condition. Common elements in the various arguments are that existence of multiple length scales of the scalar as well as the length scales of the velocity compared to those of the scalar field and to the linear size of the flow domain may be important. For example, Jaberi et al. (1996) suggested that the weak excitation of the scalar field simultaneously at both large and small scales tends to produce a wide tail in the scalar PDF, while Warhaft (2000) suggests that besides a requirement for high Reynolds number, the size of the apparatus or computational domain must be at least 8 integral length scales. A theory for the scalar PDFs which takes into account the lengths scales of the scalar and velocity and various parameters such as the Reynolds and Schmidt numbers is much desired. A comparison of results from DNS carried out on domains of different sizes is expected to be useful.

Interest in probability distributions extends to other quantities as well. Examples include the velocity-scalar joint PDF in PDF modeling (Pope, 2000) and the PDF of the product $q = u\theta$ (Thoroddsen and Van Atta, 1992; Jayesh and Warhaft, 1992; Mydlarski, 2003; Watanabe and Gotoh, 2006; Gotoh and Watanabe, 2012) whose average is the turbulent scalar flux. This latter PDF is found to develop exponential tails of the form $P(q) \propto |q|^{-1/2} \exp(-c_\pm |q|/\sigma_q)$ (where σ_q is the r.m.s. fluctuation of q), with $c_- < c_+$, which corresponds to a negative skewness. Thoroddsen and Van Atta explained theoretically how an exponential tail for q can arise despite u and θ being jointly Gaussian. They derived the exponential roll-off rate as $c_\pm \propto (1 \pm \rho_{u\theta})^{-1}$ where $\rho_{u\theta}(< 0)$ is the correlation coefficient between u and θ. Gotoh and Watanabe (2012) have shown that the Péclet number dependence of $c_\pm$ is very weak and $\rho_{u\theta}$ is determined by the time integral of the Lagrangian velocity autocorrelation function along the gradient.

Scalar gradient fluctuations are very non-Gaussian with PDFs of stretched-exponential form. When a mean gradient is present the component aligned with it, denoted by $\nabla_\| \theta$, is of greatest interest. Figure 3.9 shows the DNS data, organized to show the effects of Reynolds number (left) and Schmidt number (right) separately. In general this PDF is positively skewed and has tails of the form $P(\nabla_\| \theta) \propto \exp(-\alpha_\pm |\nabla_\| \theta|^{n_\pm})$ where $n_\pm \leq 1$ and α_+ and α_- are algebraic prefactors describing tails on the positive and negative sides respectively. We use data at highest resolution, in terms of $k_{\max}\bar{\eta}$ or $k_{\max}\bar{\eta}_B$ available for each combination of R_λ and Sc. While tails of these PDFs stretch well beyond 20 standard deviations, for the present purposes it is sufficient to focus on the data range shown which is statistically well-converged. It can be seen that at higher R_λ the negative tails become wider while the positive side changes little up to about 15 standard deviations (a trend does appear at yet larger $\nabla_\| \theta$ where sampling uncertainties also become substantial). This is consistent with a conclusion in support of reduced departures from local isotropy at high R_λ, although the improved resolution in Donzis and Yeung (2010) appears to be necessary to observe this trend. Increase of Sc has a similar effect on the negative tail but the

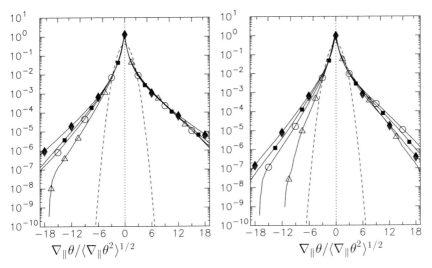

Figure 3.9 Standardized PDF of scalar gradient parallel to the mean gradient, based on DNS database (Donzis and Yeung, 2010) over a range of spatial resolutions. The *left* half shows data at fixed $Sc = 1$: △: $R_\lambda = 140$, $K_{\max}\bar{\eta} \approx 5.6$; ○: $R_\lambda = 240$, $K_{\max}\bar{\eta} \approx 5.6$; ■: $R_\lambda = 390$, $K_{\max}\bar{\eta} \approx 1.4$; ♦: $R_\lambda = 650$, $K_{\max}\bar{\eta} \approx 2.8$. The *right* half shows data at fixed $R_\lambda = 140$: △: $Sc = 1/8$, $K_{\max}\bar{\eta} \approx 2.8$; ○: $Sc = 1$, $K_{\max}\bar{\eta} \approx 5.6$; ■: $Sc = 4$, $K_{\max}\bar{\eta}_B \approx 5.6$; ♦: $Sc = 64$, $K_{\max}\bar{\eta}_B \approx 1.4$. In each case a dashed curve shows a standard Gaussian PDF for comparison.

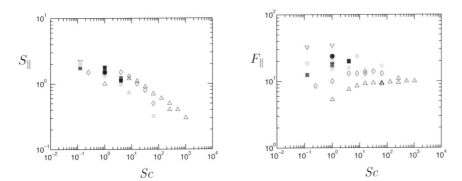

Figure 3.10 Variation of the skewness (left) and flatness (right) of the component of scalar gradient $\nabla_\| \theta$ along the mean gradient, with respect to Schmidt number. Open triangles: $R_\lambda = 8$ from Yeung et al. (2004); open diamonds: $R_\lambda = 38$ from Yeung et al. (2002). Squares and circles are for $R_\lambda = 140$ and 240 from Donzis and Yeung (2010), with light, middle and dark gray corresponding to $k_{\max}\bar{\eta}_B \approx 1.4$, 2.7 and 5.6 respectively. Inverted triangles: $R_\lambda \approx 650$, from Yeung et al. (2005).

positive tail shows a non-monotonic dependence – the tail actually becomes slightly narrower as Sc increases from 1 to 4 and further to 64.

The normalized third- and fourth-order moments of scalar gradient fluctuations $\nabla_\| \theta$ and $\nabla_\perp \theta$ in directions parallel and perpendicular to the mean

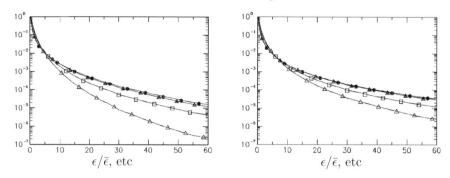

Figure 3.11 PDFs of the energy dissipation rate (△), enstrophy (□), and scalar dissipation rates at $Sc = 1/8$ (▲) and $Sc = 1$ (●). *Left*: 512^3 DNS at $R_\lambda = 240$. *Right*: 2048^3 DNS at $R_\lambda = 650$. All variables are normalized by the mean (Yeung et al., 2005).

gradient, say the skewness $(S_\parallel, S_\perp)$ and flatness factors $(F_\parallel, F_\perp)$ provide, respectively, important information on local isotropy and intermittency of the small scales. In general, $S_\parallel$ is positive and finite, while $S_\perp$ vanishes by symmetry in the plane perpendicular to the mean gradient (Mydlarski and Warhaft, 1998; Yeung et al., 2002; Antonia and Orlandi, 2003a). The magnitude of $S_\parallel$ is a measure of the departure from local isotropy at the small scales, which is an issue of long-standing interest (Sreenivasan, 1991). In most experiments (Mydlarski and Warhaft, 1998) and simulations (e.g. Yeung (1998)), $S_\parallel$ lies between 1.0 and 2.0, with no obvious trend towards smaller values at higher Reynolds number. However, as noted in our discussion of Fig. 3.9 above, data from highly-resolved simulations (Donzis and Yeung, 2010) suggest a trend of decrease at high Reynolds numbers which could be obscured in simulations at more limited resolution. An increase in Sc also leads to a systematic decrease of the gradient skewness, as seen in the left half of Fig. 3.10.

The flatness factors of $\nabla_\parallel \theta$ shown in the right half of Fig. 3.10 are typically higher than those of the velocity gradients at the same Reynolds number (see also Sreenivasan and Antonia (1997)). This is consistent with observations that scalar fluctuations at the small scales are more intermittent than the velocity. While in general $F_\parallel$ (slightly larger than $F_\perp$) increases with Reynolds number, simulations at high Sc also indicate saturation of intermittency beyond a high Sc threshold, which has a lower value if the Reynolds number is high (Donzis et al., 2010). Similar trends are also known for the PDF of scalar dissipation rate, which typically has wider tails than the PDFs of energy dissipation and enstrophy, as seen in Fig. 3.11 (Yeung et al., 2005).

In considering local isotropy and fine-scale intermittency of passive scalars it is important to ascertain whether simulation results are sensitive to the choice of Gaussian isotropic injection or uniform mean gradient as a source of scalar fluctuations. It is clear that if the scalar source and the underlying velocity field are both isotropic then no systematic departures from local isotropy will develop. In addition, while Gaussian isotropic forcing is usually confined to modes in a few low wavenumber spectral shells, in injection by uniform mean gradient the scalar field is excited at all wavenumbers. In the latter case, the source term $f_\theta(\mathbf{x}, t) = -Gu(\mathbf{x}, t)$ has a spatial correlation function of the form (Gotoh and Watanabe, 2012)

$$\langle f_\theta(\mathbf{x}+\mathbf{r},t) f_\theta(\mathbf{x},t) \rangle = G^2 u'^2 \left(\frac{1}{3r^2} \frac{d}{dr}(r^3 f_u) - \frac{r}{3} \frac{df_u(r)}{dr} P_2(\cos\phi) \right), \quad (3.21)$$

where ϕ is the angle between the direction of the mean gradient and $\mathbf{r}$, $f_u(r)$ is the longitudinal velocity correlation function, and P_2 is the Legendre polynomial of order 2. The term $P_2(\cos\phi)$ carries anisotropic contributions, which affect all scales from the integral length scale down to the Kolmogorov scale. Although this anisotropizing tendency may be opposed by turbulent scrambling, the passive scalar equation does not have a strongly isotropizing term analogous to pressure for the velocity field. As a result the skewness $S_\|$ remains finite for $Sc \lesssim 1$. When Sc is large enough, as inferred from Fig. 3.10, the anisotropy imprinted on the passive scalar at scale η is reduced by the straining motion which is constant in space but varies in time with time scale on the order of $(\nu/\bar{\epsilon})^{1/2}$. However it is not yet certain if $S_\|$ continues to decrease indefinitely with asymptotically large Sc (Schumacher et al., 2003) where the molecular diffusivity is very small but still finite. At the same time, detailed comparisons at $Sc = 1$ between simulations using isotropic forcing and mean gradient forcing (Watanabe and Gotoh, 2006) show that the latter leads to slightly stronger intermittency in the small scales.

3.5.3 High order statistics at two points

Two key issues in spatial structure as a function of scale size are the statistics of the spatial increment and the relationship between fluctuating scalar dissipation rates at distance r apart. We consider here the case of $Sc = O(1)$.

Figure 3.12 shows PDFs of the longitudinal velocity and scalar increments, i.e., $P_\|(\delta u, r)$ and $P_\|(\delta \theta, r)$, from DNS at $R_\lambda = 468$ and $Sc = 1$ (Watanabe and Gotoh, 2006), with δu taken in a direction aligned with vector separation $\mathbf{r}$, and r itself along the mean gradient for scalars. As expected, at large r both sets of PDFs are close to Gaussian but at small r they display wide tails which reflect the behavior of the small scales. The tails at small r are wider

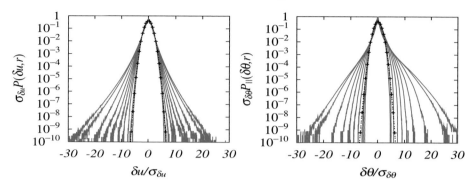

Figure 3.12 Normalized PDFs for the longitudinal velocity increment (left) and for the scalar increment parallel to the mean gradient (right), at $R_\lambda = 468$, $Sc = 1$ (Watanabe and Gotoh, 2006). The curves are for $r_n = 2^{n-1}(2\pi/N)$, $n = 1, 2, \ldots, 10$ and $N = 1024$ from the outermost. The black dotted line represents a standard Gaussian PDF for comparison.

for scalar increments than for velocity increments, which is consistent with stronger small-scale intermittency in the scalar field. The negative skewness of longitudinal velocity gradients (a property related to the energy cascade in incompressible turbulence) and positive skewness of the scalar gradient component $\nabla_\parallel \theta$ are reflected in wider tails on the negative and positive sides of these two PDFs. It can be seen also that the tails of $P_\parallel(\delta\theta, r)$ on the right have a different shape, being concave for low to moderate values of $\delta\theta$ but convex for larger $\delta\theta$ where a faster decay than in $P_\parallel(\delta u, r)$ is observed. These properties have been reproduced theoretically as well, e.g in a model by Li and Meneveau (2006).

If a robust inertial–convective range exists it is reasonable to expect the structure function to possess well-defined scaling exponents, i.e. $S_q(\mathbf{r}) \equiv \langle (\delta\theta(\mathbf{r}))^q \rangle \sim r^{\zeta_q}$ where ζ_q is, because of intermittency, smaller than the value $q/3$ suggested by classical Obukhov–Corrsin arguments. Indeed, for $\mathbf{r}$ in the direction of the mean gradient (unit vector $\mathbf{e}_\parallel$) we can write

$$S_q(r) = C_{\parallel,q} \bar\chi^{q/2} (\bar\epsilon)^{-q/6} r^{q/3} (r/L)^{\zeta_{\parallel,q} - q/3}, \qquad (3.22)$$

where $C_{\parallel,q}$ is a scaling coefficient, and Reynolds number dependence is expressed through the ratio r/L. The double subscript notation, as in $\zeta_{\parallel,q}$, is used to distinguish between analogous quantities based on whether $\mathbf{r}$ is parallel or perpendicular to the mean gradient. In practice, since the extent of the scaling range (if any) is not known *a priori*, a local exponent of the form $\zeta_q(r) = d\log[S_q(r)]/d\log r$ is usually computed and examined for evidence of a limited plateau within an applicable scaling range. The data of Watanabe and Gotoh (2004, 2006) suggest smaller scaling exponents for

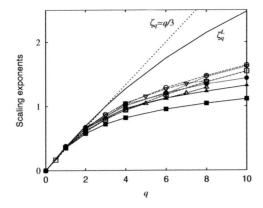

Figure 3.13 Comparison of passive scalar structure function scaling exponents ζ_q from various sources, with classical (non-intermittent) limit of $\zeta_q = q/3$ (dotted line) and ζ_q^L (unmarked solid line) for the longitudinal velocity (Gotoh et al., 2002). Open symbols are for random injection (Gaussian, white-noise large scale forcing): □: DNS, $R_\lambda = 220$ (Chen and Cao, 1997); ○: DNS, $R_\lambda = 427$ (Watanabe and Gotoh, 2004); △: DNS, $R_\lambda = 470$ (Ruiz-Chavarria et al., 1996); ▽: EXP., $R_\lambda = 650$ (Moisy et al., 2001). ◇: EXP., $R_\lambda = 850$ (Antonia et al., 1998). Filled symbols are for mean gradient: ■: DNS, $R_\lambda = 468$ (Watanabe and Gotoh, 2006); ●: EXP., $R_\lambda = 688$ (Gylfason and Warhaft, 2004); ▲: DNS, 2D turbulence (Celani et al., 2001).

injection by mean gradients but it is not clear if the difference also holds in the limit of very high Reynolds number.

Figure 3.13 shows the exponents ζ_q for the passive scalar obtained from multiple sources in the literature, including both experiments and computations. Despite a wider scatter than that seen in the velocity field the exponent ζ_q is clearly smaller than that for the longitudinal velocity. While the exponent increases with order, the rate of increase becomes smaller at higher order, and saturation may eventually occur. For example, ζ_q from a low temperature helium gas flow (Moisy et al., 2001) saturates at about $\zeta_\infty \approx 1.45$. This saturation of exponents is related to the ramp-and-cliff structures which dominate the high-order structure functions and enforce bounds on the PDF tails. Some support for this behavior can be found in theoretical work by Kraichnan (1994) for a velocity field delta-correlated in time, and in studies of 2D turbulence (Celani et al., 2000, 2001). However, for 3D Navier–Stokes turbulence future simulations at yet higher Reynolds numbers are still needed.

We may recall that in Fig. 3.12 the scalar increment PDFs at small r depend strongly on r, in contrast to classical predictions of small-scale universality. However, if the scaling exponent does saturate at high order (as discussed above), it is useful to ask if these PDFs satisfy a form of self-similarity, i.e. whether they will have the same form if the scalar increment is scaled by r^{ζ_∞} where ζ_∞ is the limiting scaling exponent at high order. Accordingly Watanabe and Gotoh (2006) considered the expression

$$P_\|(\delta\theta, r) = \theta_{\text{rms}}^{-1} (r/\bar{\eta})^{\zeta_{\|,\infty}} Q_\|(\delta\theta/\theta_{\text{rms}}) \,, \quad (3.23)$$

where $Q_\|(.)$ is a function independent of r, and analogous relations are assumed to apply to $P_\perp(\delta\theta, r)$ as well. Results at $R_\lambda = 468$ and $Sc = 1$ given in Watanabe and Gotoh (2006) (see also Celani et al. (2001)) show that the tails of the scalar increment PDF for r between 32 and 256 Kolmogorov scales collapse together very well according to this suggested scaling.

As noted earlier, when a uniform mean gradient is present, scalar fluctuations are injected anisotropically at all scales, so that local isotropy is not restored. However the structure functions are then expected to display statistical axisymmetry, such that $S_q(\mathbf{r})$ is a function only of the distance r and the angle (ϕ) between $\mathbf{r}$ and the unit vector along the mean gradient, with no dependence on the azimuthal angle (φ) in a spherical coordinate system. Odd-order structure functions for $\mathbf{r}$ in the orthogonal plane vanish by reflexion symmetry. To analyze the structure of the anisotropy here an expansion in terms of spherical erical harmonics (Kurien and Sreenivasan, 2001; Biferale and Procaccia, 2005) is useful. In this approach, under the symmetries noted above, the qth-order structure function (for $q = 2, 4, 6, 8, \ldots$) can be expanded in terms of even-order Legendre polynomials (P_{2l}) as

$$S_q(\mathbf{r}) = \sum_{l=0}^{\infty} S_q^{2l}(r) P_{2l}(\cos\phi) \,. \quad (3.24)$$

If perfect isotropy occurs at scale r then $S_q^0(r)$ is the only nonzero expansion coefficient; otherwise, $S_q^0(r)$ corresponds to a spherical average (which can be approximated by averaging over $\mathbf{e}_\|$ and then over two directions in the plane parallel to the mean gradient), while the ratios $S_q^{2l}(r)/S_q^0(r)$ (for $l \geq 1$) are measures of contributions from anisotropic sectors of different orders (Kurien et al., 2001). If inertial–convective scaling applies then one can write $S_q^l(r) = C_q^l (r/L)^{\zeta_q^l}$, with the behaviors of both the amplitude (C_q^l) and the exponent (ζ_q^l) being of interest. Results obtained by Celani et al. (2000, 2001) in 2D turbulence suggest the amplitude is influenced by the manner of forcing while the exponent is more universal. For even-order moments it is reasonable to expect that $\zeta_q^0 < \zeta_q^{2l}$ and ζ_q^{2l} is a monotonically increasing function of l, which means the isotropic sector $\zeta_q^0(r)$ has the lowest scaling exponent for a given order (Biferale et al., 2004; Biferale and Procaccia, 2005). However for odd-order moments $C_{2n+1}^0 = 0$ by reflexion symmetry, while $\zeta_{2n}^0 \leq \zeta_{2n+1}^1 \leq \zeta_{2n+2}^0$ inferred from the experiment of Gylfason and

Warhaft (2004), and $\zeta^1_{2n+1} \leq (2n+1)\zeta^0_2/2$ for the hyperskewnesses at various orders.

In general, while scaling exponents ζ_q at order q are expected to be monotonic nondecreasing functions of q, anisotropy effects lead to questions on whether these exponents are sensitive to coordinate direction. The wind-tunnel experiments of Mydlarski and Warhaft (1998), Gylfason and Warhaft (2004) as well as DNS by Watanabe and Gotoh (2006), at $R_\lambda = 566$ and 468 respectively, indicate $C_{\|,2n} = C_{\perp,2n}$ and $\zeta_{\|,2n} = \zeta_{\perp,2n}$ up to order $2n = 8$ for scalars driven by a mean gradient, while, as expected, $C_{\perp,2n+1} = 0$ but $C_{\|,2n+1} \neq 0$. Watanabe and Gotoh (2006) also found that even-order scaling exponents ζ^0_{2n} of the isotropic sector in the spherical harmonics expansion are virtually the same as the exponents deduced from structure functions in directions parallel or perpendicular to the mean gradient. Therefore, for 3D homogeneous isotropic turbulence with a mean gradient, the above observations imply that the scaling exponents of the leading term of the scalar structure function increase monotonically with the order regardless of whether the latter is odd or even. However, caution is still needed at high orders, since numerical resolution issues may prevent such exponents from being evaluated accurately, and the magnitude of the effect may also differ between the parallel and perpendicular directions (Donzis and Yeung, 2010). Furthermore, the inferences noted here are primarily based on data at $Sc = O(1)$, and no detailed data in the high Sc regime are yet available.

From the observations noted above (and additional information in the references cited) we may conclude that, for structure functions of the scalar field in homogeneous turbulence:

(1) the amplitude is not universal;
(2) the realizability condition $d\zeta_q/dq \geq 0$ holds for the scaling exponents of the leading term in the spherical harmonics expansion;
(3) $d\zeta_q/dq$ decreases with q (such that a plot of ζ_q versus q is concave) irrespective of scalar injection mechanisms in both 2D and 3D turbulence;
(4) ζ_q itself may have some dependence on injection mechanisms but the effect is weak;
(5) these scaling exponents are smaller (reflecting stronger intermittency) than those of the velocity field although there is substantial scatter among data from different sources. These suggest universality of the scalar scaling exponents in a very narrow sense.

However, theoretical work (Shraiman and Siggia, 1998; Chertkov et al., 1996) based on the Kraichnan model showed that the scalar exponents are affected by the time correlation and degrees of non-Gaussianity and symmetry of the

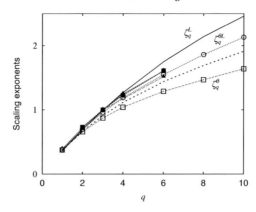

Figure 3.14 Scaling exponents of the mixed velocity-scalar structure functions ($\zeta_q^{\theta L}$). The unmarked dashed line is $(\zeta_q^L + 2\zeta_q)/3$ which provides the lower bound for $\zeta_q^{\theta L}$ according to (3.26) computed by using the DNS data of Watanabe and Gotoh (2004) for Gaussian, white-noise large scale isotropic injection (unmarked solid line for ζ_q^L, $\square$ for ζ_q, $\circ$ for $\zeta_q^{\theta L}$). Experimental data for $\zeta_q^{\theta L}$ measured at downstream of a circular cylinder without the mean scalar gradient from Lévêque et al. (1999) ($\blacksquare$) and the data with mean scalar gradient from Mydlarski (2003) ($\blacktriangle$) are also shown.

velocity field; this suggests non-universality of such exponents cannot yet be ruled out. Further study at high Reynolds numbers is necessary.

The form of the Yaglom relation (3.16) suggests the scaling of "mixed" structure functions, via the moments of $\tilde{\Pi}_\theta(r) = r^{-1}\delta u_L(r)(\delta\theta(r))^2$, is also important. An upper bound for the qth-order mixed structure function $J_q(r) = \langle (r\tilde{\Pi}_\theta(r))^q \rangle$ is given by the Hölder inequality

$$\left\langle \left(\delta u_L(r)(\delta\theta(r))^2\right)^{q/3} \right\rangle \leq \langle (\delta u_L(r))^q \rangle^{1/3} \langle (\delta\theta(r))^q \rangle^{2/3} . \quad (3.25)$$

It follows that the corresponding "mixed" scaling exponent $\zeta_q^{\theta L}$ must satisfy the inequality (Eyink, 1996; Boratav and Pelz, 1998)

$$\zeta_q^{\theta L} \geq (\zeta_q^L + 2\zeta_q)/3 > \zeta_q , \quad (3.26)$$

where ζ_q^L is the scaling exponent of the longitudinal velocity structure function, and we have used the fact that $\zeta_q < \zeta_q^L$ for $q > 3$ as seen in Fig. 3.13. Both experimental and numerical data are shown in Fig. 3.14, where the scatter for $\zeta_q^{\theta L}$ is less than that seen earlier for ζ_q (Fig. 3.13). The inequality (3.26) above is indeed satisfied. The scalar and velocity increments in the inertial–convective range are statistically dependent (Boratav and Pelz, 1998). The local scaling exponent $d\log J_q(r)/d\log r$ (not shown here) also typically shows a plateau that is better developed and wider than those seen for velocity and scalar structure functions separately. This is consistent with the fact that Yaglom's relation holds for both inertial–convective and viscous–convective ranges (Yeung et al., 2002; Watanabe and Gotoh, 2004).

In addition to statistics of spatial increments, important information on intermittency as a function of scale size r is also given by the correlation between scalar dissipation (χ) at two points of distance r apart, as well as the properties of χ averaged locally over a domain of linear size r. These quantities are fundamental in the formulation of refined similarity theory for passive scalars (e.g. Stolovitzky et al. (1995), Sreenivasan and Antonia (1997)) as a correction to classical Kolmogorov–Obukhov–Corrsin phenomenology. In particular the two-point correlator $\langle \chi(\mathbf{x}+\mathbf{r})\chi(\mathbf{x})\rangle$ as a function of $r = |\mathbf{r}|$ which, along with several analogous quantities (Sreenivasan and Kailasnath, 1993), may scale as $r^{-\mu_\chi}$ where μ_χ is called the intermittency exponent. The form of Yaglom's relation suggests a close connection between $\left\langle \left(\delta u_\mathrm{L}(\mathbf{r})(\delta\theta(\mathbf{r}))^2\right)^2\right\rangle$ and $r^2\langle\chi(\mathbf{x}+\mathbf{r})\chi(\mathbf{x})\rangle$, which then suggests $\mu_\chi = 2 - \zeta_6^{\theta\mathrm{L}}$.

Experimental data on the scalar dissipation intermittency exponent μ_χ include roughly 0.35 at $R_\lambda = 175$ in boundary layers (Antonia and Sreenivasan, 1977), 0.2–0.25 at $R_\lambda = 85$–731 in the wind tunnel (Sreenivasan and Antonia, 1997; Mydlarski and Warhaft, 1998; Mydlarski, 2003), and 0.25 ± 0.05 for atmospheric surface layer at $R_\lambda = 4300$–7800 (Chambers and Antonia, 1984). Despite high Reynolds number being clearly a requirement for inertial–convective scaling the observed Reynolds number dependency of μ_χ is very weak as long as $R_\lambda > 100$. In DNS Wang et al. (1999) reported $\mu_\chi = 0.77$ at $R_\lambda = 141$ and $Sc = 0.7$ from a 512^3 simulation with Gaussian forcing. Donzis and Yeung (2010) inferred 0.64 for $Sc = 1$ and 0.60 for $Sc = 4$ at almost the same R_λ (140) but with a mean gradient and at much higher (4096^3) resolution. High grid resolution is necessary for results at the highest Sc in a given simulation to be sufficiently accurate for a study Schmidt number trends. Donzis and Yeung (2010) observed a slow decrease of μ_χ with increasing Sc, which is consistent with earlier results at $R_\lambda = 38$ (Yeung et al., 2002). As in experiments, sensitivity of μ_χ to R_λ in DNS is relatively weak but finite resolution could have affected the results. Finally, DNS provides a great opportunity to clarify the absence or presence of any systematic effects of the use of one-dimensional surrogates (e.g. replacing χ by square of $\partial\theta/\partial x$, a common practice in experiments), and of averaging locally along a line instead of a full cube in 3D space.

3.6 Concluding remarks

In this review of limited scope we have focused on the present knowledge of Eulerian similarity scaling in the fundamental problem of passive scalar mixing in isotropic turbulence. Large-scale direct numerical simulations enabled by advances in high-performance computing have provided a large body of data at sufficiently high Reynolds or Schmidt numbers for observing the trends of the approach to asymptotic universality states, or unambiguous departures therefrom. We have considered both low- and high-order quantities as well as one-point and two-point statistics, with reference to experiments and both classical and contemporary theories constructed to incorporate the effects of intermittency.

Results at comparable Reynolds number taken from simulations differing in the use of Gaussian isotropic forcing versus uniform mean gradient as a source of scalar fluctuations are quite consistent, except that the latter introduces issues related to the turbulent scalar flux and departures from local isotropy, and perhaps slightly stronger small-scale intermittency. The mean scalar dissipation rate is insensitive to Schmidt number while the mixing time scale increases with Schmidt number but less so at higher Reynolds number. Inertial-convective scaling for the scalar spectrum is well achieved at Taylor-scale Reynolds number (R_λ) around 650 subject to a pronounced spectral bump for data at $Sc = 1$, while clear scaling in the scalar flux spectrum appears to require much higher Reynolds numbers. The approach to viscous–convective scaling at high Sc is quite definite although resolution requirements still limit the range of R_λ and Sc accessible using present capabilities. In contrast, theories for inertial–diffusive behavior have remained largely untested. In physical space Yaglom's relation for the mixed velocity-scalar structure function is well satisfied in both inertial–convective and viscous–convective scaling ranges. Scalar gradients have stretched-exponential PDFs whose tails become wider at increasing Reynolds number. However, higher Sc leads to greater symmetry and reduced skewness in the PDF of scalar gradient fluctuations parallel to the mean gradient. For higher-order structure functions, scaling exponents from various sources in the literature suggest a trend towards saturation at high Sc but this cannot be fully confirmed yet. Analyses of the anisotropy (due to mean scalar gradient) in the structure functions using spherical harmonics at different orders show that the leading term has a scaling exponent that increases monotonically with order. Finally we examined intermittency exponents inferred from two-point statistics of the scalar dissipation but caution is needed since these results are sensitive to the inevitable effects of finite resolution.

Many other important features of turbulent mixing are not addressed in this article. For example, from a theoretical point of view we have not discussed the relationship between the scaling properties of scalars transported by Navier–Stokes turbulence and the case of random advection (Kraichnan,

1994). From a modeling point of view, we have not discussed conditional statistics such as scalar flux and dissipation given the scalar (or including the velocity) needed for the development of stochastic models based on PDF transport equations where molecular mixing requires closure (Pope, 1985). Since stochastic models are often posed in a Lagrangian frame the Lagrangian statistics of passive scalars (Yeung, 2001; Fox and Yeung, 2003) are thus relevant. In reacting flow applications (where multiple species routinely occur) differential diffusion between scalars of different diffusivities (e.g. Fox (1999)) is also important. Subgrid scale modeling for passive scalars is also generally more difficult than for the velocity field, although the filtered-density-function approach (Colucci et al., 1998) offers some significant advantages in inhomogeneous flows. Finally many additional open questions arise when other physical processes, such as chemical reaction, density stratification, solid-body rotation, and electromagnetic forces must be considered. The first two of these leads to active-scalar problems (called Level II and Level III mixing by Dimotakis (2005)), which affect the velocity field dynamically and are unlikely to be well described by classical scaling theories (or simple modifications thereof).

To conclude this review, we emphasize that turbulent mixing is an important and challenging problem that has benefited from advances in high performance computing coupled to theory, modeling and experiment, and should continue to do so. However, since resources available are still finite, a choice of priorities becomes important. At present, the high Reynolds number behavior of scalars of Schmidt number close to unity is better understood than the physics of scalars of very low or very high Schmidt number, which deserve more attention in the future. One can also expect that, over time, the greatest needs for research will shift towards problems of yet greater complexity such as some of those outlined in the preceding paragraph.

As a final word, since the value of any massive computation is ultimately to be judged by its impact on research in the field, development of effective means of community-wide collaboration is also increasingly important. Clearly, the size of datasets involved adds further to this challenge, which also requires attention to the issues of data-intensive computing.

Acknowledgments The authors are indebted to many individuals, not least the Editors of this volume, for their encouragement and stimulating interactions during the preparation of this review. We thank D.A. Donzis and T. Watanabe for their contributions to much of the data presented in a number of figures; and we thank L. Biferale, R. J. Hill, S.B. Pope, R. Rubinstein, B.L. Sawford, K.R. Sreenivasan, and Z. Warhaft for their valuable comments and many thought-provoking discussions on the subjects addressed here. T.G's work was partially supported by Grant-in-Aid for Scientific Research No. 21360082 from the Ministry of Education, Culture, Sports, Science and Technology of Japan. Also TG would like to thank the Theory and Computer Simulation Center of the National Institute for Fusion Science (NIFS10KNSS008), and the Information Technology Center of Nagoya University for their support. PKY would like to acknowledge support from the National Science Foundation (NSF), USA and resource allocations over many years at major supercomputing centers supported by NSF and the U.S Department of Energy.

References

Aguirre, R. C., and Catrakis, H. 2005. On intermittency and the physical thickness of turbulent fluid interfaces. *J. Fluid Mech.*, **540**, 39–48.

Aivalis, K. G., Sreenivasan, K. R., Tsuji, Y, Klewicki, J. C., and Biltoft, C. A. 2002. Temperature structure functions for air flow over moderately heated ground. *Phys. Fluids*, **14**, 2439–2446.

Antonia, R. A., and Orlandi, P. 2003a. Effect of Schmidt number on passive scalar turbulence. *Appl. Mech. Rev.*, **56**, 615–632.

Antonia, R. A., and Orlandi, P. 2003b. On the Batchelor constant in decaying isotropic turbulence. *Phys. Fluids*, **15**, 2084–086.

Antonia, R. A., and Sreenivasan, S. R. 1977. Log-normality of temperature dissipation in a turbulent boundary layer. *Phys. Fluids*, **20**, 1800–1804.

Antonia, R. A., Chambers, A. J., Britz, D., and Browne, L. W. B. 1986. Organized structures in a turbulent plane jet: topology and contribution to momentum and heat transport. *J. Fluid Mech.*, **172**, 211–229.

Antonia, R. A., Hopfinger, E. J., Gagne, Y., and Anselmet, F. 1998. Temperature structure functions in turbulent shear flows. *Phys. Rev. A*, **30**, 2704–2707.

Antonia, R. A., Zhou, T., and Xu, G. 2000. Second-order temperature and velocity structure functions: Reynolds number dependence. *Phys. Fluids*, **12**, 1509–1517.

Ashurst, W. T., Kerstein, A. R., Kerr, R. M., and Gibson, C. H. 1987. Alignment of vorticity and scalar gradient with strain rate in simulated Navier-Stokes turbulence. *Phys. Fluids*, **30**, 2343–2353.

Batchelor, G. K. 1959. Small-scale variation of convected quantities like temperature in turbulent fluid. Part 1. General discussion and the case of small conductivity. *J. Fluid Mech.*, **5**, 113–133.

Batchelor, G. K., Howells, I. K., and A., Townsend A. 1959. Small-scale variations of convected quantities like temperature in turbulent fluid, Part 2: The case of large conductivity. *J. Fluid Mech.*, **5**, 134–139.

Biferale, L., and Procaccia, I. 2005. Anisotropy in turbulent flows and in turbulent transport. *Phys. Rep.*, **414**, 43–164.

Biferale, L., Calzavarini, L., Lanotte, A. S., and Toschi, F. 2004. Universality of anisotropic turbulence. *Physica A*, **338**, 194–200.

Bilger, R. W. 1989. Turbulent diffusion flames. *Ann. Rev. Fluid Mech.*, **21**, 101–135.

Bogucki, D., Domaradzki, J. A., and Yeung, P. K. 1997. Direct numerical simulations of passive scalars with $Pr > 1$ advected by turbulent flow. *J. Fluid Mech.*, **343**, 111–130.

Boratav, O. N., and Pelz, R. B. 1998. Coupling between anomalous velocity and passive scalar increments in turbulence. *Phys. Fluids*, **10**, 2122–2124.

Borgas, M. S., Sawford, B. L., Xu, S, Donzis, D. A., and Yeung, P. K. 2004. High Schmidt number scalars in turbulence: Structure functions and Lagrangian theory. *Phys. Fluids*, **16**, 3888–3899.

Bos, W. J. T., Touil, H., and Bertoglio, J.-P. 2005. Reynolds number dependency of the scalar flux spectrum in isotropic turbulence with a uniform scalar gradient. *Phys. Fluids*, **17**, 125108.

Brethouwer, G., Hunt, J. C. R., and Nieuwstadt, F. T. M. 2003. Micro-structure and Lagrangian statistics of the scalar field with a mean gradient in isotropic turbulence. *J. Fluid Mech.*, **474**, 193–225.

Castaing, B., Gunaratne, G., Heslot, F., Kadanoff, L., Libchaber, A. Thomae, S., Wu, X-Z., Zaleski, S., and Zanetti, G. 1989. Scaling of hard thermal turbulence in Rayleigh–Benard convection. *J. Fluid Mech.*, **204**, 1–30.

Celani, A., Lanotte, A., Mazzino, A., and Vergassola, M. 2000. Universality and saturation of intermittency in passive scalar turbulence. *Phys. Rev. Lett.*, **84**, 2385–2388.

Celani, A., Lanotte, A., Mazzino, A., and Vergassola, M. 2001. Fronts in passive scalar turbulence. *Phys. Fluids*, **13**, 1768–1783.

Chambers, A.J., and Antonia, R.A. 1984. Atmospheric estimate of power-law exponents μ and μ_θ. *Boundary-Layer Met.*, **28**, 343–352.

Chasnov, J. R., Canuto, V. M., and Rogallo, R. S. 1988. Turbulence spectrum of a passive temperature field: Results of a numerical simulation. *Phys. Fluids*, **31**, 2065–2067.

Chen, S., and Cao, N. 1997. Anomalous scaling and structure instability in three-dimensional passive scalar turbulence. *Phys. Rev. Lett.*, **78**, 3459–3462.

Chertkov, M., Falkovich, G., and Lebedev, V. 1996. Nonuniversaality of the scaling exponents of a passive scalar convected by a random flow. *Phys. Rev. Lett.*, **76**, 3707–3710.

Ching, E. S. 1993. Probability densities of turbulent temperature fluctuations. *Phys. Rev. Lett.*, **70**, 283–286.

Colucci, P. J., Jaberi, F. A., Givi, P., and Pope, S. B. 1998. Filtered density function for large-eddy simulation of turbulent reacting flows. *Phys. Fluids*, **10**, 499–515.

Corrsin, S. 1951. On the spectrum of isotropic temperature fluctuations in an isotropic turbulence. *J. Appl. Phys.*, **22**, 469–473.

Danaila, L, and Antonia, R. A. 2009. Spectrum of a passive scalar in moderate Reynolds number homogeneous isotropic turbulence. *Phys. Fluids*, **21**, 111702.

Danaila, L., and Mydlarski, L. 2001. Effect of gradient production on scalar fluctuations in decaying grid turbulence. *Phys. Rev. E*, **64**, 016316.

Danaila, L., Anselmet, F., Zhou, T., and Antonia, R. A. 1999. A generalization of Yaglom's equation which accounts for the large-scale forcing in heated decaying turbulence. *J. Fluid Mech.*, **391**, 359–372.

Dimotakis, P. E. 2005. Turbulent mixing. *Ann. Rev. Fluid Mech.*, **37**, 329–356.

Doering, C. R. 2009. The 3D Navier–Stokes problem. *Ann. Rev. Fluid Mech.*, **41**, 109–128.

Donzis, D. A., and Sreenivasan, K. R. 2010. The bottleneck effect and the Kolmogorov constant in isotropic turbulence. *J. Fluid Mech.*, **657**, 171–188.

Donzis, D. A., and Yeung, P. K. 2010. Resolution effects and scaling in numerical simulations of passive scalar mixing in turbulence. *Physica D*, **239**, 1278–1287.

Donzis, D. A., Sreenivasan, K. R., and Yeung, P. K. 2005. Scalar dissipation rate and Dissipative anomaly in isotropic turbulence. *J. Fluid Mech.*, **532**, 199–216.

Donzis, D. A., Yeung, P. K., and Sreenivasan, K. R. 2008a. Energy dissipation rate and enstrophy in isotropic turbulence: resolution effects and scaling in direct numerical simulations. *Phys. Fluids*, **20**, 045108.

Donzis, D. A., Yeung, P. K., and Pekurovsky, D. 2008b. Turbulence simulations on $O(10^4)$ processors. In: *TeraGrid'08 Conference, Las Vegas, NV, June 2008*.

Donzis, D. A., Sreenivasan, K. R., and Yeung, P. K. 2010. The Batchelor spectrum for mixing of passive scalars in isotropic turbulence. *Flow, Turb. & Combust.*, **85**, 549–566.

Dopazo, C, Martin, J., and J., Hierro. 2007. Local geometry of isoscalar surfaces. *Phys. Rev. E.*, **76**, 056316.

Eswaran, V., and Pope, S. B. 1988. Direct numerical simulations of the turbulent mixing of a passive scalar. *Phys. Fluids*, **31**, 506–520.

Eyink, G. L. 1996. Intermittency and anomalous scaling of passive scalars in any space dimension. *Phys. Rev. E.*, **54**, 1497–1503.

Falkovich, G. 1994. Bottleneck phenomenon in developed turbulence. *Phys. Fluids*, **6**, 1411–1413.

Ferchichi, M., and Tavoularis, S. 2002. Scalar probability density function and and fine structure in uniformly sheared turbulence. *J. Fluid Mech.*, **461**, 155–182.

Fox, R. O. 1999. The Lagrangian spectral relaxation model for differential diffusion in homogeneous turbulence. *Phys. Fluids*, **11**, 1550–1571.

Fox, R. O. 2003. *Computational Models for Turbulent Reacting Flows*. Cambridge University Press.

Fox, R. O., and Yeung, P. K. 2003. Improved Lagrangian mixing models for passive scalars in isotropic turbulence. *Phys. Fluids*, **15**, 961–985.

Frisch, U. 1995. *Turbulence*. Cambridge University Press.

Gibson, C. H. 1968. Fine structure of scalar fields mixed by turbulence. 2. Spectral theory. *Phys. Fluids*, **11**, 2316–2327.

Gollub, J. P., Clarke, J., Gharib, M., Lane, B., and Mesquita, O. N. 1991. Fluctuations and transport in a stirred fluid with a mean gradient. *Phys. Rev. Lett.*, **67**, 3507–3510.

Gotoh, T. 1989. Passive scalar diffusion in two-dimensional turbulence in the Lagrangian renormalized approximation. *J. Phys. Soc. Jpn.*, **58**, 2365.

Gotoh, T., and Watanabe, T. 2012. Scalar flux in a uniform mean scalar gradient in homogeneous isotropic steady turbulence. *Physica D*, **241**, 141–148.

Gotoh, T., Nagaki, J., and Kaneda, Y. 2000. Passive scalar spectrum in the viscous-convective range in two-dimensional steady turbulence. *Phys. Fluids*, **12**, 155–168.

Gotoh, T., Fukayama, D., and Nakano, T. 2002. Velocity field statistics in homogeneous steady turbulence obtained using a high-resolution direct numerical simulation. *Phys. Fluids*, **14**, 1065–1081.

Grant, H. L., A., Hughes B., Vogel, W. M., and Moilliet, A. 1968. The spectrum of temperature fluctuations in turbulent flow. *J. Fluid Mech.*, **34**, 423–442.

Gylfason, A., and Warhaft, Z. 2004. On higher order passive scalar structure functions in grid turbulence. *Phys. Fluids*, **16**, 4012–4019.

Herring, H. R., and Kraichnan, R. H. 1979. A numerical comparison of velocity-based and strain-based Lagrangian-history turbulence approximations. *J. Fluid Mech.*, **91**, 581–597.

Higgins, K., Ooi, A., and Chong, M. S. 2006. Batchelor's spectrum from an axisymmetric strained scalar field. *Phys. Fluids*, **18**, 065111.

Hill, R. J. 2002. Structure-function equations for scalars. *Phys. Fluids*, **14**, 1745–1756.

Holzer, M., and Siggia, E. D. 1994. Turbulent mixing of a passive scalar. *Phys. Fluids*, **6**, 1820–1837.

Ishihara, T., and Kaneda, Y. 1992. Stretching and distortion of material line elements in two-dimensional turbulence. *J. Phys. Soc. Jpn.*, **61**, 3547–3558.

Ishihara, T., Gotoh, T., and Kaneda, Y. 2009. Study of high Reynolds number isotropic turbulence by direct numerical simulations. *Ann. Rev. Fluid Mech.*, **41**, 165–180.

Jaberi, F. A., Miller, R. S., Mandia, C. K., and Givi, P. 1996. Non-Gaussian scalar statistics in homogeneous turbulence. *J. Fluid Mech.*, **313**, 241–282.

Jayesh, and Warhaft, Z. 1992. Probability distribution, conditional dissipation, and transport of passive temperature fluctuations in grid-generated turbulence. *Phys. Fluids*, **A 4**, 2292–2306.

Jukkien, M. C., Castiglione, P., and Tabeling, P. 2000. Experimental observation of Batchelor dispersion of passive tracers. *Phys. Rev. Lett.*, **85**, 3636.

Kaneda, Y. 1986. Inertial range structure of turbulent velocity and scalar fields in a Lagrangian renormalized approximation. *Phys. Fluids*, **29**, 701–708.

Kaneda, Y., and Ishihara, T. 2006. High resolution direct numerical simulation of turbulence. *J. Turb.*, **7** (20), 1–17.

Kaneda, Y., Ishihara, T., Yokokawa, M., Itakura, T., and Uno, A. 2003. Energy dissipation rate and energy spectrum in high resolution direct numerical simulations of turbulence in a periodic box. *Phys. Fluids*, **15**, L21–L24.

Kerstein, A. R. 1991. Linear-eddy modelling of turbulent transport. Part 6. Microstructure of diffusive scalar mixing fields. *J. Fluid Mech.*, **231**, 361–394.

Kimura, Y., and Kraichnan, R. H. 1993. Statistics of an advected passive scalar. *Phys. Fluids*, **A5**, 2264–2277.

Kraichnan, R. H. 1966. Isotropic turbulence and inertial-range structure. *Phys. Fluids*, **9**, 1728–1752.

Kraichnan, R. H. 1968. Small-scale structure of scalar field convected by turbulence. *Phys. Fluids*, **11**, 945–953.

Kraichnan, R. H. 1974. Convection of a passive scalar by a quasi-uniform random straining field. *J. Fluid Mech.*, **64**, 737–762.

Kraichnan, R. H. 1994. Anomalous scaling of a randomly advected passive scalar. *Phys. Rev. Lett.*, **72**, 1016–1019.

Kurien, S., and Sreenivasan, K. R. 2001. Measures of anisotropy and universal properties of turbulence. Pages 53–111 of: *New Trends in Turbulence, NATO Advanced Study Institute, Les Houches, France.* Springer and EDP-Sciences.

Kurien, S., Aivalis, K., and Sreenivasan, K. R. 2001. Anisotropy of small-scale scalar turbulence. *J. Fluid Mech.*, **448**, 279–288.

Lévêque, E., Ruiz-Chavarria, G., Baudet, C., and Ciliberto, S. 1999. Scaling laws for the turbulent mixing of a passive scalar in the wake of a cylinder. *Phys. Fluids*, **11**, 1869–1879.

Li, Y., and Meneveau, C. 2006. Intermittency trends and Lagrangian evolution of non-Gaussian statistics in turbulent flow and scalar transport. *J. Fluid Mech.*, **558**, 133–142.

Lumley, J. L. 1964. The spectrum of nearly inertial turbulence in a stably stratified fluid. *J. Atmos. Sci.*, **21**, 99–102.

Lumley, J. L. 1967. Similarity and the turbulence spectrum. *Phys. Fluids*, **10**, 855–858.

Mills, R. R., Kistler, A. L., O'Brien, V., and Corrsin, S. 1958. *Turbulence and temperature fluctuations behind a heated grid.* NASA TN 4288.

Moin, P., and Mahesh, K. 1998. Direct numerical simulation: a tool in turbulence research. *Ann. Rev. Fluid Mech.*, **30**, 539–578.

Moisy, F., Willaime, H., Anderson, J. S., and Tabeling, P. 2001. Passive scalar intermittency in low temperature helium flows. *Phys. Rev. Lett.*, **86**, 4827–4830.

Mydlarski, L. 2003. Mixed velocity-passive scalar statistics in high-Reynolds number turbulence. *J. Fluid Mech.*, **475**, 173–203.

Mydlarski, L., and Warhaft, Z. 1998. Passive scalar statistics in high-Péclet-number grid turbulence. *J. Fluid Mech.*, **358**, 135–175.

Newman, G. R., and Herring, J. R. 1979. A test field model study of a passive scalar in isotropic turbulence. *J. Fluid Mech.*, **94**, 163–194.

Oakey, A. N. 1982. Determination of the rate of dissipation of turbulent energy from simultaneous temperature and velocity shear microstructure measurements. *J. Phys. Oceanogr.*, **12**, 256–271.

Obukhov, A. M. 1949. Structure of the temperature field in a turbulent flow. *IZv. Akad. Nauk SSSR. Ser. Geogr. Geofiz.*, **13**, 58–69.

O'Gorman, P. A., and Pullin, D. I. 2005. Effect of Schmidt number on the velocity-scalar cospectrum in isotropic turbulence with a mean scalar gradient. *J. Fluid Mech.*, **532**, 111–140.

Orlandi, P., and Antonia, R. A. 2002. Dependence of the nonstationary form of Yaglom's equation on the Schmidt number. *J. Fluid Mech.*, **451**, 99–108.

Overholt, M. R., and Pope, S. B. 1996. Direct numerical simulation of a passive scalar with imposed mean gradient in isotropic turbulence. *Phys. Fluids*, **8**, 3128–3148.

Piller, M., Nobile, E., and Hanratty, T. 2002. DNS study of turbulent transport at low Prandtl numbers in a channel flow. *J. Fluid Mech.*, **458**, 419–441.

Pope, S. B. 1985. PDF methods for turbulent reactive flows. *Progr. Energy & Combust. Sci.*, **11**, 119–192.

Pope, S. B. 2000. *Turbulent Flows.* Cambridge University Press.

Pumir, A. 1994. A numerical study of the mixing of a passive scalar in three dimensions in the presence of a mean gradient. *Phys. Fluids*, **6**, 2118–2132.

Pumir, A., Shraiman, B., and Siggia, E. D. 1991. Exponential tails and random advection. *Phys. Rev. Lett.*, **66**, 2984–2987.

Qian, J. 1986. Turbulent passive scalar field of a small Prandtl number. *Phys. Fluids*, **29**, 3586–3589.

Qian, J. 1995. Viscous range of turbulent scalar of large Prandtl number. *Fluid Dyn. Res.*, **15**, 103–112.

Rogallo, R. S. 1981. *Numerical experiments in homogeneous turbulence.* NASA TM 81315. NASA Ames Research Center, Moffett Field, CA.

Rogers, M. M., Mansour, N. N., and Reynolds, W. C. 1989. An algebraic model for the turbulent flux of a passive scalar. *J. Fluid Mech.*, **203**, 77–101.

Ruiz-Chavarria, G., Baudet, C., and Ciliberto, S. 1996. Scaling laws and dissipation scale of a passive scalar in fully developed turbulence. *Physica D*, **99**, 369–380.

Schumacher, J., and Sreenivasan, K. R. 2005. Statistics and geometry of passive scalars in turbulence. *Phys. Fluids*, **17**, 125107.

Schumacher, J., Sreenivasan, K. R., and Yeung, P. K. 2003. Schmidt number dependence of derivative moments for quasi-static straining motion. *J. Fluid Mech.*, **479**, 221–230.

Schumacher, J., Sreenivasan, K. R., and Yeung, P. K. 2005. Very fine structures in scalar mixing. *J. Fluid Mech.*, **531**, 113–122.

Shraiman, B. I., and Siggia, E. D. 1994. Lagrangian path integrals and fluctuations in random flows. *Phys. Rev. E*, **49**, 2912–2927.

Shraiman, B. I., and Siggia, E. D. 1998. Anomalous scaling for a passive scalar near the Batchelor limit. *Phys. Rev. E*, **57**, 2965–2977.

Shraiman, B. I., and Siggia, E. D. 2000. Scalar turbulence. *Nature*, **405**, 639–646.

Sirivat, A., and Warhaft, Z. 1983. The effect of a passive cross-stream temperature gradient on the evolution of temperature variance and heat flux in grid turbulence. *J. Fluid Mech.*, **128**, 323–346.

Sreenivasan, K. R. 1991. On local isotropy of passive scalars in turbulent shear flows. *Proc. Roy. Soc. Lond., Ser. A*, **434**, 165–182.

Sreenivasan, K. R. 1998. An update on the energy dissipation rate in isotropic turbulence. *Phys. Fluids*, **10**, 528–529.

Sreenivasan, K. R. 2004. Possible effects of small-scale intermittency in turbulent reacting flows. *Flow, Turb. & Combust.*, **72**, 115–131.

Sreenivasan, K. R., and Antonia, R. A. 1997. The phenomenology of small-scale turbulence. *Ann. Rev. Fluid Mech.*, **29**, 435–472.

Sreenivasan, K. R., and Kailasnath, P. 1993. An update on the intermittency exponent in turbulence. *Phys. Fluids A*, **5**, 512–513.

Sreenivasan, K. R., and Prasad, R. P. 1989. New results on the fractal and multifractal structure of the large Schmidt number passive scalars in fully turbulent flows. *Physica D*, **38**, 322–329.

Sreenivasan, K. R., Antonia, R. A., and Britz, D. 1979. Local isotropy and large structures in a heated turbulent jet. *J. Fluid Mech.*, **94**, 745–775.

Sreenivasan, K. R., Tavoularis, S., Henry, R., and Corrsin, S. 1980. Temperature fluctuations and scales in grid-generated turbulence. *J. Fluid Mech.*, **100**, 597–621.

Sreenivasan, K.R. 1996. The passive scalar spectrum and the Obukhov-Corrsin constant. *Phys. Fluids*, **8**, 189–196.

Stolovitzky, G., Kailasnath, P., and Sreenivasan, K. R. 1995. Refined similarity hypotheses for passive scalars mixed by turbulence. *J. Fluid Mech.*, **297**, 275–291.

Tavoularis, S., and Corrsin, S. 1981. Experiments in nearly homogeneous shear flow with a uniform mean temperature gradient. *J. Fluid Mech.*, **104**, 311–347.

Thoroddsen, S. T., and Van Atta, C. W. 1992. Exponential tails and skewness of density-gradient probability density functions in stably stratified turbulence. *J. Fluid Mech.*, **244**, 547–566.

Vedula, P., Yeung, P. K., and Fox, R. O. 2001. Dynamics of scalar dissipation in isotropic turbulence: a numerical and modelling study. *J. Fluid Mech.*, **433**, 29–60.

Wang, L. P., Chen, S., and Brasseur, J. G. 1999. Examination of hypotheses in the Kolmogorov refined turbulence theory through high-resolution simulations. Part 2. Passive scalar field. *J. Fluid Mech.*, **400**, 163–197.

Warhaft, Z. 2000. Passive scalars in turbulent flows. *Ann. Rev. Fluid Mech.*, **32**, 203–240.

Warhaft, Z., and Lumley, J. L. 1978. An experimental study of the decay of temperature fluctuations in grid-generated turbulence. *J. Fluid Mech.*, **88**, 659–6840.

Watanabe, T., and Gotoh, T. 2004. Statistics of a passive scalar in homogeneous turbulence. *New J. Phys.*, **6**, 40.

Watanabe, T., and Gotoh, T. 2006. Intermittency in passive scalar turbulence under the uniform mean scalar gradient. *Phys. Fluids*, **18**, 058105.

Watanabe, T., and Gotoh, T. 2007a. Inertial-range intermittency and accuracy of direct numerical simulation for turbulence and passive scalar turbulence. *J. Fluid Mech.*, **590**, 117–146.

Watanabe, T., and Gotoh, T. 2007b. Scalar flux spectrum in isotropic turbulence with a uniform mean scalar gradient. *Phys. Fluids*, **19**, 121701.

Watanabe, T., and Gotoh, T. 2010. Coil-stretch transition in ensemble of polymers in isotropic turbulence. *Phys. Rev. E.*, **81**, 066301.

Yaglom, A. M. 1949. On the local structure of a temperature field in a turbulent flow. *Dokl. Akad. Nauk SSSR*, **69**, 743–746.

Yakhot, V., and Sreenivasan, K. R. 2005. Anomalous scaling of structure functions and dynamic constraints on turbulence simulations. *J. Stat. Phys.*, **121**, 823–841.

Yeh, T. T., and Atta, C. W. Van. 1973. Spectral transfer of scalar and velocity fields in heated-grid turbulence. *J. Fluid Mech.*, **58**, 233–261.

Yeung, P. K. 1996. Multi-scalar triadic interactions in differential diffusion with and without mean scalar gradients. *J. Fluid Mech.*, **321**, 235–278.

Yeung, P. K. 1998. Correlation and conditional statistics in differential diffusion: scalars with uniform mean gradients. *Phys. Fluids*, **10**, 2621–2635.

Yeung, P. K. 2001. Lagrangian characteristics of turbulence and scalar transport in direct numerical simulations. *J. Fluid Mech.*, **427**, 241–274.

Yeung, P. K., Xu, S., and Sreenivasan, K. R. 2002. Schmidt number effects on turbulent transport with uniform mean scalar gradient. *Phys. Fluids*, **14**, 4178–4191.

Yeung, P. K., Xu, S., Donzis, D. A., and Sreenivasan, K. R. 2004. Simulations of three-dimensional turbulent mixing for Schmidt numbers of the order 1000. *Flow, Turb. & Combust.*, **72**, 333–347.

Yeung, P. K., Donzis, D. A., and Sreenivasan, K. R. 2005. High-Reynolds-number simulation of turbulent mixing. *Phys. Fluids*, **17**, 081703.

Yeung, P. K., Pope, S. B., and Sawford, B. L. 2006b. Reynolds number dependence of Lagrangian statistics in large numerical simulations of isotropic turbulence. *J. Turb.*, **7(58)**, 1–12.

Yokokawa, M., Itakura, T., Uno, A., Ishihara, T., and Kaneda, Y. 2002 (November). 16.4-Tflops direct numerical simulation of turbulence by a Fourier spectral method on the Earth Simulator. In: *Proceedings of the Supercomputing Conference, Baltmore, MD. Nov. 2002*.

Yu, H., and Meneveau, C. 2010. Lagrangian refined Kolmogorov similarity hypothesis for gradient time evolution and correlation in turbulent flows. *Phys. Rev. Lett.*, **104**, 084502.

4
A Lagrangian View of Turbulent Dispersion and Mixing

Brian L. Sawford and Jean-François Pinton

4.1 Introduction

For good practical reasons, most experimental observations of turbulent flow are made at fixed points x in space at time t and most numerical calculations are performed on a fixed spatial grid and at fixed times. On the other hand, it is possible to describe the flow in terms of the velocity and concentration (and other quantities of interest) at a point moving with the flow. This is known as a Lagrangian *description* of the flow ((Monin and Yaglom, 1971)). The position of this point $x^+(t; x_0, t_0)$ is a function of time and of some initial point x_0 and time t_0 at which it was identified or "labelled". Its velocity is the velocity of the fluid where it happens to be at time t, $u^+(t; x_0, t_0) = u(x^+(t), t)$. We will use the superscript $(+)$ to denote Lagrangian quantities, and quantities after the semi-colon are independent parameters. We refer to a point moving in this way as a fluid particle.

Flow statistics obtained at fixed points and times are known as Eulerian statistics. On the other hand, statistics obtained at specific times by sampling over trajectories, which at some reference times passed through fixed points, are known as Lagrangian statistics. For example, the mean displacement at time t of those particles that passed through the point x_0 at time t_0 is just $\langle x^+(t; x_0, t_0) - x_0 \rangle$. In both cases, the measurement time t can be earlier or later than the reference time. That is, the trajectories can be followed both backwards and forwards in time. The essential feature is that for Eulerian statistics the measurement locations are fixed, while for Lagrangian statistics the reference locations are fixed.

A Lagrangian view of turbulent diffusion goes back as least as far as G.I. Taylor's famous work (Taylor, 1921), which provided the foundation for a kinematic approach to turbulent transport (Batchelor, 1952) and, over the last 20 years or so, to wide-ranging applications of Lagrangian stochastic models. These developments have been documented in reviews covering environmental problems in the atmosphere (Sawford, 1985, 1993, 2001; Wilson and Sawford, 1996; Rodean, 1996), the oceans (Griffa, 1996) and engineering flows (Pope, 1985, 1994, 2000). More recently increased computational power and vastly improved flow visualisation techniques and data-capture capabilities has led to numerical and experimental access to Lagrangian flow information at Reynolds numbers large enough to exhibit an approach to asymptotic behaviour, at least for some quantities of interest (see Yeung (2002); Salazar and Collins (2008); Toschi and Bodenschatz (2009) for recent reviews). Also recently there has been a surge of interest, and significant progress, in probing fundamental aspects of turbulent dynamics of flow and scalar fields using Lagrangian methods (Falkovich *et al.*, 2001). In the present review, we aim for a broad coverage of all these Lagrangian aspects of fluid flow, with particular emphasis on the links with practical models.

Lagrangian statistics are of interest in their own right as an alternative representation of the flow, as inputs to practical models and as a means of testing the predictions of these models. However, perhaps the most important aspect of a Lagrangian description of the flow is that it is a natural way to treat the turbulent transport of scalar materials, and that is the motivation for many practical models, and hence also a major focus of our review.

Consider, for example, a particle which moves under the combined influence of a prescribed turbulent velocity field (for example a realisation of a direct numerical solution of the Navier–Stokes equations from a single initial flow) and molecular diffusion, so that its position changes according to the stochastic differential equation (Pope, 1998; Frisch *et al.*, 1999; Shraiman and Siggia, 2000)

$$d\boldsymbol{x}^+ = \boldsymbol{u}\left(\boldsymbol{x}^+(t), t\right) + \sqrt{2\kappa}d\boldsymbol{\xi}(t) \tag{4.1}$$

where $d\boldsymbol{\xi}(t)$ is a 3-dimensional incremental Wiener process (Gardiner, 1983). We refer to a particle moving in this way as a Brownian particle and for simplicity we consider constant property flows; i.e. we take the fluid density and the viscosity and molecular diffusivity to be constant.

In general, the nth-order moments of the scalar concentration field $c(\boldsymbol{x}, t)$ due to a source $S(\mathbf{x}, t)$ can be written in terms of the n-particle displacement PDF by averaging over realisations of both the velocity field and the Brownian motion to give

$$\begin{aligned}&\langle c\left(\boldsymbol{x}_1, t_1\right) \cdots c\left(\boldsymbol{x}_n, t_n\right)\rangle \\ &= \int_{-\infty}^{t_1} \cdots \int_{-\infty}^{t_n} \int_V \cdots \int_V P\left(\boldsymbol{x}_1, t_1, \ldots, \boldsymbol{x}_n, t_n | \boldsymbol{x}_1', t_1', \ldots, \boldsymbol{x}_n', t_n'\right) \\ &\quad S\left(\boldsymbol{x}_1', t_1'\right) \cdots S\left(\boldsymbol{x}_n', t_n'\right) d\boldsymbol{x}_1' \cdots d\boldsymbol{x}_n' dt_1' \cdots dt_n'.\end{aligned} \tag{4.2}$$

From a practical point of view, the one-point moments $\langle c^n(\boldsymbol{x},t)\rangle$ are of most interest, while at a more fundamental level there is also great interest in the structure functions $\langle (c(\boldsymbol{x},t) - c(\boldsymbol{x}+\boldsymbol{r},t))^n\rangle$.

The result, Eq. (4.2), is exact and is the fundamental connection between the displacement statistics of molecular trajectories and statistics of the scalar field. In principle it provides a powerful formalism for computing statistics of the scalar field, and indeed great theoretical advances have been made through the underlying Lagrangian methods (Falkovich et al., 2001).

In general of course, Brownian trajectories and those of fluid particles diverge, and the statistical properties of the fluid motion along molecular trajectories differ from those along fluid trajectories. Similarly, trajectories of physical particles with inertia, or of bubbles, deviate from those of fluid particles.

Turbulence itself appears in different forms, ranging from fully three-dimensional boundary layer flows, jets and plumes encountered in the environment and engineering, to quasi-two-dimensional flow on global scales in the atmosphere (and ocean). The different dynamics of 2D and 3D turbulence (vortex stretching is absent in 2D) is of fundamental interest and is reflected in different cascade regimes. Sometimes density stratification is important, particularly in the atmosphere and ocean, and in engineering the interaction between chemistry and turbulent mixing is important in combustion and other chemical reactors. All of these aspects of turbulence are amenable to Lagrangian treatment and modelling, but of necessity we must adopt a more restricted overview here. In particular, we note that Lagrangian aspects of 2D turbulence are covered in recent reviews (Kellay and Goldburg, 2002; Tabeling, 2002), and so do not elaborate here.

This review covers measurement, numerical calculation and modelling of Lagrangian statistics of molecular, fluid and inertial particle trajectories in 3D turbulence, and covers the Lagrangian description and modelling of the statistics of a scalar concentration field. We attempt to establish linkages between fundamental theory, measurement and calculation, and practical modelling for environmental and engineering flows. The material in the following sections is based on the hierarchy embodied in Eq. (4.2). In Section 4.2, we deal with single particle trajectories and statistics, which are relevant to dispersion in a fixed reference frame, also known as absolute dispersion, and to the mean concentration field. Section 4.3 deals with the joint trajectories and statistics of two particles. These describe the process of turbulent relative dispersion, which is fundamentally connected with turbulent mixing, concentration fluctuations about the mean, scalar dissipation and the dissipation anomaly. In Section 4.4 we review work on the trajectories of three and four particles, which determine area, volume and shape statistics and higher moments of the scalar concentration field, and are related to anomalous scaling of the scalar structure functions. Finally in Section 4.5 we attempt an overview of the benefits of a Lagrangian approach and look at some of the gaps in our knowledge and possible future directions.

4.2 Single particle motion and absolute dispersion

4.2.1 Fluid particles in isotropic turbulence

We begin this section by considering the statistics along the trajectory of a single fluid particle in each realisation of the flow. Such trajectories are influenced by all scales of motion. We restrict attention for the moment to stationary isotropic turbulence in incompressible flow. Such flow can be characterised by the velocity component variance $\sigma_u^2 = \frac{2}{3}k$, where k is the turbulence kinetic energy, the mean rate of dissipation of turbulence kinetic energy $\langle \epsilon \rangle$ and the kinematic viscosity ν. These parameters define length and time scales representative of the energy-containing eddies, $L = \sigma_u^3 / \langle \epsilon \rangle$ and $T_E = \sigma_u^2 / \langle \epsilon \rangle$ and corresponding velocity, length and time scales representative of the smallest eddies where viscosity eliminates the turbulence, $v_\eta = (\langle \epsilon \rangle \nu)^{1/4}$, $\eta = \left(\nu^3 / \langle \epsilon \rangle\right)^{1/4}$ and $t_\eta = (\nu / \langle \epsilon \rangle)^{1/2}$. The subscript E on the large eddy time scale emphasises that this is an Eulerian scale. The turbulence Reynolds number is given by $Re = \sigma_u L / \nu$, but is often characterised by the Taylor scale Reynolds number $R_\lambda = (15 Re)^{1/2}$.

Under the special circumstances outlined above one-time Lagrangian statistics and one-point-one-time Eulerian statistics are equivalent. For example, the mean velocities vanish $\langle \boldsymbol{u}^+(t; \boldsymbol{x}_0, t_0) \rangle = \langle \boldsymbol{u}(\boldsymbol{x}, t) \rangle = 0$ and the Lagrangian and Eulerian velocity covariances are given by $\left\langle u_i^+(t) u_j^+(t) \right\rangle = \langle u_i(\boldsymbol{x}, t) u_j(\boldsymbol{x}, t) \rangle = \sigma_u^2 \delta_{ij}$. Our notation has been simplified here to reflect the lack of dependence on the reference time and location.

Lagrangian trajectories, as a collection of three dimensional curves, can be characterized by the statistics of their curvature and torsion. It has been observed in numerical simulations (Braun et al., 2006) and in experimental data from 3D-PTV (Xu et al., 2007) that the PDF of the curvature of Lagrangian trajectories has strong power-law tails. These can essentially be understood on the basis of purely Gaussian statistics, although the evolution of curvature along single trajectories reveals the influence of intermittency (Xu et al., 2007).

Correlation functions

The first non-trivial (and perhaps the most important) Lagrangian information occurs at the level of two-time statistics. For example, the Lagrangian velocity correlation function $R_L(\tau)$, defined by $\langle u_i^+(t) u_j^+(t+\tau) \rangle = \sigma_u^2 R_L \delta_{ij}$, encapsulates the effect of the spatial and temporal structure on the rate of decorrelation of the velocity along a trajectory and depends only on the lag τ. The decorrelation rate is quantified by the Lagrangian integral time scale $T_L = \int_0^\infty R_L(\tau) d\tau$. The ratio of the Lagrangian and Eulerian time scales is an important characteristic of the turbulence and an essential parameter in practical models.

Formally, one can rewrite the Lagrangian velocity correlation function as $R_L(\tau = t - t') \sim \int R_L(t - t'|\boldsymbol{x}; \boldsymbol{x}_0) G(\boldsymbol{x} - \boldsymbol{x}', t - t') d\boldsymbol{x}$, where the single particle Green function is $G(\boldsymbol{x} - \boldsymbol{x}', t - t') \equiv P(\boldsymbol{x}, t|\boldsymbol{x}', t')$. Corrsin (Corrsin, 1959; Shlien and Corrsin, 1974) have conjectured that one may replace $R_L(t - t'|\boldsymbol{x}; \boldsymbol{x}_0)$ in the above expression by $R_E(\boldsymbol{x} - \boldsymbol{x}', t - t')$ in order to link the Eulerian and Lagrangian correlations. This idea has been tested experimentally using particle tracking (Ott and Mann, 2005) and it was concluded that the simpler relation $R_L(\tau) \sim R_E(\boldsymbol{0}, \tau)$ yields a much better estimate.

One of the advantages of the Lagrangian approach is that kinematic relationships connect the velocity of a fluid particle to its displacement,

$$\boldsymbol{x}^+(t) - \boldsymbol{x}_0 = \int_{t_0}^t \boldsymbol{u}^+(t') dt', \tag{4.3}$$

and to its acceleration,

$$\boldsymbol{A}^+(t) = d\boldsymbol{u}^+/dt. \tag{4.4}$$

As a result of the first of these kinematic relations, the mean-square displacement of fluid particles (also known as the dispersion) is determined by the Lagrangian velocity correlation function:

$$\left\langle x_i^+(t) x_j^+(t) \right\rangle = \sigma_x^2 \delta_{ij} = 2\sigma_u^2 \delta_{ij} \int_0^t \int_0^{t'} R_L(\tau) d\tau dt' \tag{4.5}$$

where without loss of generality we have set $\boldsymbol{x}_0 = 0$ and $t_0 = 0$. Two limits are important,

$$\sigma_x^2 = \begin{cases} \sigma_u^2 t^2 & (t \ll T_L) \\ 2\sigma_u^2 T_L t & (t \gg T_L). \end{cases} \tag{4.6}$$

The small-time limit is known as the ballistic limit and represents the regime where particles remember their initial velocity perfectly and travel in straight lines. The large time limit is a diffusion limit with turbulent diffusivity $\sigma_u^2 T_L$. The dispersion of fluid particles is a very important quantity since, ignoring the effects of molecular diffusion or inertia, both of which we discuss below, it

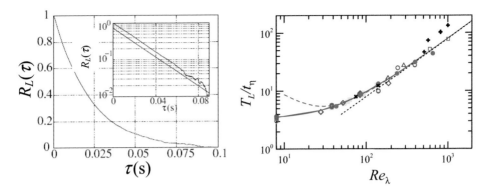

Figure 4.1 (a) Lagrangian velocity autocorrelation function, measurement from (Mordant et al., 2001) at $R_\lambda = 740$ – inset: plot in semi-logarithmic scale. (b) Data for Lagrangian integral time scale. Symbols: ■, (Yeung and Pope, 1989; Yeung, 2001); •, (Yeung et al., 2006a); □, Yeung (Pers. comm.); △, (Biferale et al., 2005a) (adjusted after pers. comm.); ◇, (Overholt and Pope, 1996) indirect); ○, (Borgas et al., 2004) indirect; +, (Mordant et al., 2004a) lab, ×, (Mordant et al., 2004a) DNS. Lines: – – – , second order stochastic theory Eq.(4.8); ——, empirical fit Eq. (4.9); - - -, large Re limit Eq. (4.7). Indirect estimates are based on Eq. (4.21).

defines the spread of a tracer cloud about its point of release in an absolute reference frame and figures prominently, for example, in simple practical models for the mean concentration of pollutant in dispersing clouds and plumes (Pasquill and Smith, 1993). A knowledge of the Lagrangian velocity correlation function, and in particular the Lagrangian integral time scale, is thus crucial to many practical problems of this sort and the development of models incorporating memory effects has been an important step forward.

DNS and experimental results for the Lagrangian velocity correlation function are very well fitted by a decaying exponential function for lags much larger than the Kolmogorov time scale (Yeung and Pope, 1989; Yeung, 2002; Mordant et al., 2004a). Figure 4.1(a) shows an experimental measurement of the velocity correlation function from acoustic tracking (Mordant et al., 2001). Figure 4.1(b) shows DNS and laboratory results for the ratio T_L/t_η as a function of Reynolds number. The dotted line shows the asymptotic form at large Reynolds number

$$\frac{T_L}{t_\eta} = \frac{T_L}{T_E}Re^{1/2} = \frac{T_L}{T_E}\frac{R_\lambda}{\sqrt{15}} \qquad (4.7)$$

which with the Lagrangian to Eulerian time scale ratio $T_L/T_E = 0.31$ is a good representation of the DNS results. If we assume an exponential decay for the correlation function, then matching to the structure function in the inertial sub-range requires $T_L/T_E = 2/C_0$, where C_0 is the Lagrangian velocity structure function constant, so the asymptotic form corresponds to $C_0 = 6.5$. The dashed line is given by

$$\frac{T_\mathrm{L}}{t_\eta} = \frac{2}{C_0}\frac{R_\lambda}{\sqrt{15}}\left[1 + \frac{15^{1/2}C_0^2}{20}R_\lambda^{-1}\left(1 + 110R_\lambda^{-1}\right)\right] \qquad (4.8)$$

with $C_0 = 6.5$, and is based on a stochastic model for the acceleration (Sawford, 1991) described in more detail below (see Eq. (4.13)). The solid line is the empirical form

$$\frac{T_\mathrm{L}}{t_\eta} = \left[4.77 + (R_\lambda/12.6)^{4/3}\right]^{3/4} \qquad (4.9)$$

which has been constructed to match the large Reynolds number limit Eq. (4.7). The laboratory values generally lie above the DNS, corresponding to a lower asymptotic value of $C_0 \approx 4$ (Mordant et al., 2004a). Perhaps this difference reflects differences in forcing and boundary conditions between the DNS and laboratory flows.

Second order structure function

The Lagrangian velocity structure function D_2^L is defined through the velocity difference variance $\left\langle \left(u_i^+(t) - u_j^+(t+\tau)\right)^2 \right\rangle = D_2^L(\tau)\delta_{ij}$. It is closely related to the correlation function since $D_2^L(\tau) = 2\sigma_u^2\left(1 - R_\mathrm{L}(\tau)\right)$, and carries essentially the same information, but provides a different view of the Lagrangian structure of the turbulence. Because it is defined in terms of velocity differences, the structure function highlights the small-time structure of the velocity field. In particular, according to Kolmogorov similarity theory (Monin and Yaglom, 1975), for $\tau \ll T_\mathrm{L}$ the structure function is a universal function of the lag τ, $\langle\epsilon\rangle$ and ν. Performing a Taylor expansion for small time $\tau \ll t_\eta$ and using dimensional arguments in the inertial sub-range where viscosity is no longer important, we have

$$D_2^L(\tau) = \begin{cases} \frac{1}{3}\langle A_i^2\rangle \tau^2 & \tau \ll t_\eta \\ C_0 \langle\epsilon\rangle \tau & t_\eta \ll \tau \ll T_\mathrm{L}. \end{cases} \qquad (4.10)$$

The time for which $\tau \ll t_\eta$ is known is the dissipation sub-range, while that for which $t_\eta \ll \tau \ll T_\mathrm{L}$ is known as the inertial sub-range (Monin and Yaglom, 1975). The acceleration variance can be represented in terms of the dissipation scales as $\langle A_i^2\rangle = 3a_0\langle\epsilon\rangle/t_\eta = 3a_0\langle\epsilon\rangle^{3/2}/\nu^{1/2}$. Because the second relation in Eq. (4.10) is linear in the dissipation rate, it is expected to be independent of intermittency corrections and therefore universal. Consequently the Lagrangian structure function inertial sub-range constant C_0 is

4: A Lagrangian View of Turbulent Dispersion and Mixing 139

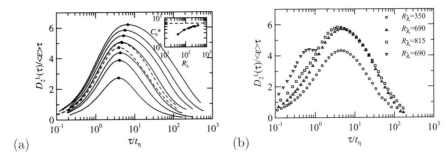

Figure 4.2 (a) Second-order Lagrangian structure function scaled by Kolmogorov variables at different Reynolds numbers based on (Yeung et al., 2006a) with new data at $R_\lambda = 1000$ [Yeung, pers. comm.] (solid lines) and (Biferale et al., 2008) (dashed lines). Open circles mark the location of the peak in each curve. The inset shows these peak values (C_0^*) versus the time-averaged Reynolds number in each simulation, compared with a dashed line at the value 6.5. (b) Experimental data are from optical Particle Tracking Velocimetry, in a von Kármán flow, from (Biferale et al., 2008).

a very important universal constant of turbulence. Through its connection with T_L (for an exponential correlation) C_0 is also an important parameter in practical dispersion models.

Direct numerical simulations (DNS) in forced isotropic turbulence show that the universal inertial sub-range form of the Lagrangian velocity structure function shown in Eq. (4.10) is approached only at large Reynolds numbers (Yeung et al., 2006a). Figure 4.2(a) shows the structure function in inertial sub-range compensated form from DNS for a range of Reynolds numbers. Similar behaviour, although with a less well-ordered dependence on Reynolds number, has been observed in laboratory experiments (Ouellette et al., 2006; Biferale et al., 2008; Mordant et al., 2001, 2004a) – see Fig. 4.2(b) after Biferale et al. (2008). According to Eq. (4.10), the inertial sub-range should be manifested as a plateau of height C_0 on scales intermediate between t_η and T_L. We should remark though that even the simplest relationship such as equation (4.7) implies a limited separation T_L/t_η: less than one order of magnitude at $R_\lambda \sim 100$ and less than 2 orders of magnitude at $R_\lambda \sim 1000$ with most laboratory flows confined between these two limits. As a result the inertial sub-range is virtually nonexistent at $R_\lambda \sim 100$ and at best limited for $R_\lambda \sim 1000$. Thus we see a peak rather than a plateau and this peak increases with Reynolds number. All numerical or experimental estimates of C_0 are derived from this peak value, which we denote by C_0^*, rather than from a true inertial sub-range plateau. At least for the DNS, C_0^* appears to converge to a value $C_0 \approx 7$ for large Reynolds number, as indicated in the insert to 4.2(a). The differences between numerical and experimental investigations may reflect differences in forcing and boundary

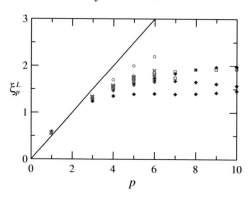

Figure 4.3 Lagrangian structure function scaling exponents from: laboratory data, (Mordant et al., 2001) ▲, $R_\lambda = 740$; (Mordant et al., 2004a), ×, $R_\lambda = 510 - 1000$; (Xu et al., 2006), +, $R_\lambda = 200 - 815$; and DNS; (Biferale et al., 2005a), ○, $R_\lambda = 284$; (Mordant et al., 2004a), □, $R_\lambda = 75 - 140$. Line is the Kolmogorov prediction $\xi_p^L = p/2$

conditions between the DNS and laboratory flows. In von Kármán flows it has been noted that the anisotropy on injection persists to the smallest scales (Ouellette et al., 2006), at Reynolds numbers R_λ approaching 1000.

Higher order structure functions, intermittency

Both DNS and laboratory data for higher order Lagrangian structure functions $D_p^L(\tau) = \langle |\boldsymbol{u}^+(t) - \boldsymbol{u}^+(t+\tau)|^p \rangle$ show intermittency corrections, $D_p^L(\tau) \propto \tau^{\xi_p^L}$ with $\xi_p^L \neq p/2$. Note here that absolute values are used in the definition of the structure functions, because the (signed) Lagrangian velocity increments are observed to be skew symmetric. Unlike the Eulerian increments in space which bear the signature of irreversibility, the one-point, two-time Lagrangian increments are invariant under time reversal.

Although at the Reynolds numbers for which data are available none of the structure functions shows a scaling range directly (see for example Fig. 4.2 for the second-order structure function), a scaling range is observed when the extended self-similarity (ESS) ansatz (Benzi et al., 1993) is used. Indeed a scaling behavior can be measured when $D_p^L(\tau)$ is plotted against $D_2^L(\tau)$, as for example in Fig. 5 in (Mordant et al., 2001). Scaling exponents evaluated in this way from both laboratory and DNS data are shown in Fig. 4.3. The results show a clear intermittency. While there is a lot of scatter mainly due to the experimental and numerical uncertainties in the tracking of Lagrangian tracers, the trend seems to be for the exponents to saturate to a value around $\xi_\infty^L \approx 1.7 \pm 0.3$ for $p > 6$, with no significant dependence on Reynolds number. This scaling behavior has also been observed in several models of Lagrangian motion in a field of random vorticies (Wilczek et al., 2008; Rast and Pinton, 2009).

Given the lack of a true inertial range for the Reynolds numbers at which numerical or experimental studies are being done, it is helpful to get a unified description of the scaling properties from dissipative to integral scale. This is provided by the multifractal analysis, introduced for inertial range scaling by Frisch (1985) and later extended over the entire range of Eulerian scales (Meneveau, 1996). When applied to the Lagrangian domain (Chevillard et al., 2003), this formalism shows that the statistics of the Lagrangian structure functions can be described using a single singularity spectrum $D(h)$, from the smallest time increments (*i.e.* the statistics of fluid particle accelerations, as measured by (Voth et al., 1998, 2002)) to the inertial range behavior (Mordant et al., 2001). A recent extensive compilation of experimental and numerical studies (Arneodo et al., 2008; Biferale et al., 2008) focuses on the behavior of the fourth order structure function, since it converges and shows deviation from Kolmogorov predictions. There is very good agreement with the multifractal theory despite the experiments being done with a variety of techniques and the numerics covering a range of turbulent Reynolds numbers. For $\tau/t_\eta \to 0$, $\xi_4^L \to 2$ the non-intermittent value, for $1 \leq \tau/t_\eta \leq 10$ there is a dip to a minimum at $\xi_4^L \approx 4$, and for $\tau/t_\eta \geq 10$ there is a plateau at $\xi_4^L \approx 1.6$. The plateau is identified as the inertial subrange scaling regime, displaying the effects of intermittency. The dip in the local scaling exponent has sometimes been ascribed to the effects of vortex trapping of particles (Biferale et al., 2008). This may be misleading however because such a dip is fully accounted for in the multifractal formalism which attributes no particular role to vortex or straining regions. Recent laboratory data (Berg et al., 2009) show a similar structure also for the behavior of higher order exponents ξ_6^L and ξ_8^L.

Intermittency is also reflected in the PDF for velocity increments, which shows increasingly stretched exponential tails as the time lag decreases until the acceleration PDF is recovered at zero lag (Mordant et al., 2001, 2004a). As mentioned, the PDFs at all time lags between t_η and T_L can be recovered once the singularity spectrum $D(h)$ is given. In addition, the multifractal formalism allows Eulerian and Lagrangian intermittencies to be connected. Using a Kolmogorov-Richardson argument (Borgas, 1993) to relate fluctuations in time of the dissipation rate encountered by a Lagrangian particle to the fluctuations in space of this quantity, one obtains a relationship between singularity spectra (Chevillard et al., 2003): $D^L(h) = -h + (1+h)D^E(h/(1+h))$. Recent theoretical and numerical investigations using mixed Eulerian-Lagrangian statistics (Kamps et al., 2009; Homann et al., 2009) show that there may be corrections to the above Kolmogorov-Richardson argument, at least at finite Reynolds numbers.

Acceleration statistics

It follows from the kinematic relation Eq. (4.4) that the acceleration covariance can also be written in terms of the velocity correlation function $R_A(\tau) = \frac{1}{3}\langle A_i^+(t) A_i^+(t+\tau)\rangle = -\sigma_u^2 d^2 \langle R_L(\tau)\rangle / d\tau^2$ (Tennekes and Lumley, 1972; Borgas and Sawford, 1991), and then from Eq. (4.10) that

$$\frac{1}{3}\langle A_i^+(t) A_i^+(t+\tau)\rangle = \begin{cases} \frac{1}{3}\langle A_i^2\rangle & \tau \ll t_\eta \\ 0 & t_\eta \ll \tau \ll T_L. \end{cases} \quad (4.11)$$

The vanishing covariance in the inertial sub-range simply means that the acceleration covariance remains much smaller than the naïve dimensional form $\langle \epsilon \rangle / \tau$ there (Borgas and Sawford, 1991). The empirically observed exponential decay of the velocity correlation function for $\tau \gg t_\eta$, implies an exponentially-decaying negative contribution of the form

$$-\left(\sigma_u^2/T_L^2\right)\exp(-\tau/T_L)$$

to the acceleration covariance over that range. This can be clearly seen in 4.4(a), although the time at which the correlation crosses zero is somewhat larger than in DNS for example due to the effects of inertia on the laboratory tracer particles. The structure that emerges for the acceleration covariance is thus a small-scale positive peak of height $\frac{1}{3}\langle A_i^2\rangle$ and width t_η and a weak negative exponentially-decaying large-scale component. This structure ensures that the kinematic constraint $\int_0^\infty R_A(\tau)\,d\tau = 0$ is satisfied. If we scale the covariance in terms of the large scale parameters σ_u and T_L, we see a small scale peak of height $\frac{1}{3}\langle A_i^2\rangle T_L^2/\sigma_u^2 \propto Re^{1/2}$ and width $t_\eta/T_L \propto Re^{-1/2}$ and a large scale component $-\exp(-\tau/T_L)$. As $Re \to \infty$, the small scale peak becomes higher and narrower, while conserving its area and thus has the character of a δ-function.

The δ-function component of the acceleration covariance corresponds to a white-noise component in the acceleration and has motivated the development of Markov models that represent the Lagrangian velocity by a stochastic differential equation of the form

$$du_i^+ = -\frac{C_0 \langle \epsilon \rangle}{2\sigma_u^2} u_i^+ dt + \sqrt{C_0 \langle \epsilon \rangle}\, d\xi_i(t) \quad (4.12)$$

This equation models the velocity correlation as an exponential with integral time scale $T_L = 2\sigma_u^2/C_0 \langle \epsilon \rangle$ and the coefficient of the random term ensures

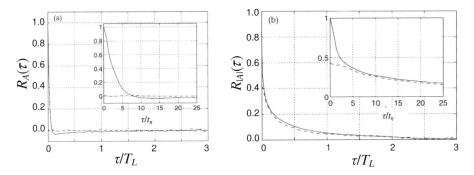

Figure 4.4 Particle acceleration autocorrelation functions, at $R_\lambda = 810$, from Mordant et al. (2004a). (a) Autocorrelation of one component (solid line) and cross-correlation of two components of acceleration (dashed line). (b) Autocorrelation function of the absolute value of one component (solid line) and cross-correlation of the absolute values of two components of acceleration (dashed line). In each case, the inset shows the curves as a function of τ/t_η, instead of τ/T_L.

consistency with the structure function Eq. (4.10) in the inertial sub-range. It models the acceleration covariance as a positive δ-function component and a weak negative exponential component. It is thus a good representation of the true acceleration covariance in the limit $t_\eta/T_\mathrm{L} \to 0$; i.e. $Re \to \infty$. The form of the drift term, and its relationship to the diffusion term, ensures stationarity of the Lagrangian velocity. Eq. (4.12) is the prototype for an enormous range of applications to many types of turbulent flows in environmental and engineering applications.

The SDE (4.12) has been modified to allow for finite Reynolds numbers by treating the acceleration as a continuous Markov process (Sawford, 1991)

$$dA_i^+ = -\frac{C_0 \langle \epsilon \rangle}{2\sigma_u^2}(1+Re_*)^{1/2} A_i^+ dt - \left(\frac{C_0 \langle \epsilon \rangle}{2\sigma_u^2}\right)^2 Re_*^{1/2} u_i^+ dt$$

$$+ \frac{C_0 \langle \epsilon \rangle}{2\sigma_u^2}\sqrt{C_0 \langle \epsilon \rangle Re_* \left(1+Re_*^{-1/2}\right)} d\xi_i(t)$$

$$du_i^+ = A_i^+ dt \qquad (4.13)$$

where $Re_* = \left(16a_0^2/C_0^2\right) Re$. The model given by Eq. (4.13) is less well-founded than that for the velocity, since it predicts Gaussian statistics for the acceleration in contradiction with observations (Yeung and Pope, 1989; Voth et al., 1998, 2002; La Porta et al., 2001; Mordant et al., 2001, 2002, 2004a,b). Nevertheless it gives a good representation of low-order moments such as the acceleration correlation and the velocity structure function, and it reduces to Eq. (4.12) in the limit $Re \to \infty$. The acceleration model gives

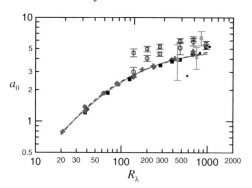

Figure 4.5 DNS and laboratory results for the non-dimensional acceleration $\frac{1}{3}\langle A_i^2\rangle/(\langle\epsilon\rangle/t_\eta) = a_0$. Symbols: ●, (Yeung and Pope, 1989); ♦, (Yeung et al., 2006a) ; ■, (Gotoh and Fukayama, 2001); +, (Vedula and Yeung, 1999); ×, (Volk et al., 2008b); ○,□ (Sawford et al., 2003); ●, (Volk et al., 2011). Lines: ———, Eq. (4.15); - - - , Eq. (4.16)

the Lagrangian integral time scale

$$T_{\rm L} = \frac{2\sigma_u^2}{C_0\langle\epsilon\rangle}\left(1 + Re_*^{-1/2}\right). \qquad (4.14)$$

For many practical purposes, the effect of finite Reynolds number can be incorporated in the velocity SDE (4.12) by adjusting the integral time scale according to Eq. (4.14); i.e. by adjusting the value of C_0 to $\tilde{C}_0 = C_0\bigl(1 + Re_*^{-1/2}\bigr)^{-1}$.

Modelling at this level requires a knowledge of the acceleration variance. Figure 4.5 shows DNS and laboratory results for the non-dimensional acceleration variance $a_0 = \frac{1}{3}\langle A_i^2\rangle t_\eta/\langle\epsilon\rangle$ as a function of Reynolds number. According to Kolmogorov similarity theory, a_0 should approach a constant value at sufficiently large Reynolds number and the solid line is the function

$$a_0 = 5/\left(1 + 110 R_\lambda^{-1}\right) \qquad (4.15)$$

which has been constructed to fit the DNS results with a constant asymptotic value.

On the other hand, allowing for intermittency corrections using the so-called p-model of multifractal turbulence, gives $a_0 \propto Re^{0.135}$ at large Reynolds numbers (Sawford et al., 2003). The multifractal prediction (Chevillard et al., 2006) assuming a lognormal model for the Lagrangian statistics, leads to $a_0 \sim Re^{0.16}$. The function

$$a_0 = 1.9 R_\lambda^{0.135}/\left(1 + 85 R_\lambda^{-0.135}\right) \qquad (4.16)$$

shown as the dashed line fitting the DNS results in Figure 4.5 is consistent with these intermittent behaviours. While the intermittent form Eq. (4.16) is favoured on theoretical grounds, the observations do not extend to sufficiently high Reynolds numbers to confirm that, and for many practical purposes the interpolation based on Kolmogorov theory is likely to be adequate. The Cornell laboratory values reported in Sawford *et al.* (2003) lie above the DNS results while those of Volk *et al.* (2008b) are broadly consistent with the DNS, but with relatively large error bars. Some of these differences may be due to particle inertia effects and to uncertainty in estimating the dissipation rate in the laboratory flows.

While the asymptotic behavior at large Reynolds numbers is at present not resolved, we remark here that the evolution of the acceleration variance, given by the Heisenberg–Yaglom scaling $\frac{1}{3}\langle A_i^{+2} \rangle = a_0 \epsilon^{3/2} \nu^{-1/2}$, can be re-expressed as $\frac{1}{3}\langle A_i^{+2} \rangle = a_0 \sigma_u^{9/2} L^{-3/2} \nu^{-1/2}$. It suggests the relationship $\langle A_i^{+2} | u \rangle \propto u^{9/2}$ for the conditional acceleration variance, which is confirmed experimentally (Crawford *et al.*, 2005).

On more fundamental grounds the neglect of intermittency in Eqs. (4.12) and (4.13) is a serious limitation. Higher-order moments of Lagrangian velocity increments show departures from Kolmogorov scaling, and consequently as we have already noted the PDFs for velocity increments and the acceleration are strongly non-Gaussian. Different approaches have been used to account for these effects of intermittency. The dissipation rate itself can be treated as a stochastic variable and modelled explicitly as an additional SDE (Pope and Chen, 1990; Reynolds, 2003; Lamorgese *et al.*, 2007) or as a multiplicative cascade process (Borgas and Sawford, 1994a). This has motivated interest in Lagrangian statistics conditioned on the dissipation (Yeung *et al.*, 2007). Alternatively, the stochastic term in Eq. (4.12) can be modelled directly as a multifractal random walk (Mordant *et al.*, 2002, 2004a). Both approaches account for the nonlinearity of the dependence of the scaling exponents of the velocity increments on the order of the moment and the highly non-Gaussian nature of the velocity increments and the acceleration. These models use (Langevin-type) equations of evolutions driven by forces with fluctuating intensity in their multiplicative and/or additive terms. In order to reproduce the long time correlations found in the amplitude of the acceleration of the Lagrangian tracers (see Fig. 4.4(b)), these models have a slowly varying intensity of the noise (with time scales of the order of $T_L \gg t_\eta$. In these models (Aringazin, 2004; Aringazin and Mazhintove, 2004) it was found that the best agreement with experimental data is obtained when one correlates the additive and multiplicative noise terms (Laval *et al.*, 2001) in the generalized Langevin processes.

4.2.2 Brownian particles in isotropic turbulence

Our discussion so far has been in terms of fluid particles, but the kinematic analysis above was also applied to the dispersion of molecular trajectories by Saffman (1960). Corresponding to Eq. (4.5), for the dispersion of molecular trajectories we have

$$\left\langle x_i^{+(B)}(t) x_j^{+(B)}(t) \right\rangle = 2\sigma_u^2 \delta_{ij} \int_0^t \int_0^{t'} R_L^{(B)}(\tau) \, d\tau dt' + 2\kappa t \delta_{ij} \qquad (4.17)$$

where the superscript (B) denotes quantities along a Brownian trajectory. This differs from the corresponding result for fluid particles in two respects. First there is the direct effect of the molecular dispersion given by the term $2\kappa t \delta_{ij}$. Secondly, the interaction between the molecular motion and the turbulent flow is reflected in the fact that the velocity correlation along the molecular trajectories differs from that along fluid trajectories. Saffman showed that for small times the interaction reduces the dispersion compared with that for fluid particles. His exact result is

$$\left\langle x_i^{+(B)} x_j^{+(B)} \right\rangle = \left(2\kappa t + \sigma_u^2 t^2 - \frac{1}{9} \frac{\kappa \langle \epsilon \rangle}{\nu} t^3 \right) \delta_{ij} \qquad (t \ll Sct_\eta, t_\eta) \qquad (4.18)$$

Using more heuristic arguments, for large times he obtained the result

$$\left\langle x_i^{+(B)} x_j^{+(B)} \right\rangle = \left(1 - \frac{a\kappa}{\sigma_u^2 t_\eta} \right) \left\langle x_i^+(t) x_j^+(t) \right\rangle + 2\kappa t \delta_{ij} \qquad (t \gg T_L) \qquad (4.19)$$

and estimated $a \approx 0.23$ by comparison with experimental data. Using the diffusive limit given by the second line of Eq. (4.6), we can write Eq. (4.19) in terms of the integral time scales

$$1 - T_L^{(B)}/T_L = a\sqrt{15} Sc^{-1} R_\lambda^{-1} \qquad (t \gg T_L) \qquad (4.20)$$

The only direct estimates of which we are aware for the Lagrangian time scale $T_L^{(B)}$ along molecular trajectories are those of Yeung and Borgas (1997). Figure 4.6(a) shows that Saffman's theory is in good agreement with these results.

It can be shown that in the scalar field produced by isotropic turbulence acting on a uniform scalar gradient G there is a stationary state in which the production P and mean rate of dissipation of scalar variance $\langle \epsilon_c \rangle$ balance and are given by

$$\langle \epsilon_c \rangle = P = 2\sigma_u^2 T_L^{(B)} G^2 \qquad (4.21)$$

Thus it is possible to obtain estimates of $T_L^{(B)}$ from DNS of scalar fields in a uniform scalar gradient (Borgas et al., 2004). The result of these direct and indirect estimates is shown in Fig. 4.6(b).

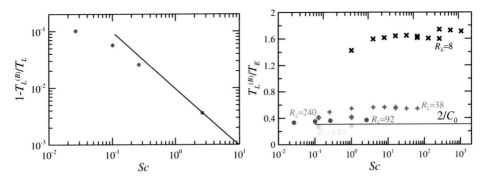

Figure 4.6 (a) Effect of molecular diffusion on the Lagrangian integral time scale. Symbols (Yeung and Borgas, 1997); line, Eq. (4.20) (Saffman, 1960) (b) Lagrangian integral time scale for fluid velocity along Brownian trajectories. Symbols: (Borgas et al., 2004), ■, ♦, + and, ×; (Yeung and Borgas, 1997), •; line is the large Reynolds number limit for fluid particles $T_L/T_E = 0.31$ corresponding to the dotted line in Fig. 4.1(b)

We see that unless Sc is very small (i.e. unless the turbulent diffusivity is much larger than the viscosity), $T_L^{(B)}$ converges to the fluid particle time scale very quickly with increasing Reynolds number. In practical terms the small-time correction Eq. (4.18) is negligible and so the effects of molecular diffusion at finite Re can be accommodated by simply adjusting the integral time scale, i.e. by using $T_L^{(B)}$ in Eq. (4.12) rather than T_L, but even this correction will be small except at very low Re and low Sc.

4.2.3 Inertial particles in isotropic turbulence

Another case of interest concerns the dynamics of inertial particles. The equation for the motion of physical particles with finite mass and size in turbulence is not fully known. Some of the physical effects include the inertia of the particle, the added volume effect, gravitational settling (or rise in the case of bubbles), drag etc., some of which depend on the particle Reynolds number. Often in computational work it is assumed that the particle Reynolds number is low enough that Stokes drag can be used and the equation of motion for the particle is

$$\frac{d\boldsymbol{v}_p^+}{dt} = \beta \frac{d\boldsymbol{u}_p^+}{dt} + \frac{1}{\tau_p}\left(\boldsymbol{u}_p^+ - \boldsymbol{v}_p^+\right) + \left(\frac{\rho_p - \rho_f}{\rho_p + (1/2)\rho_f}\right)\boldsymbol{g} \qquad (4.22)$$

where $\boldsymbol{v}_p^+$ is the velocity of the particle, $\boldsymbol{u}_p^+$ is the fluid velocity at the location of the particle, $\beta = 3\rho_f/(\rho_f + 2\rho_p)$ represents the added mass, ρ_f and ρ_p are the densities of the fluid and the particle respectively, $\tau_p = d^2/(3\beta\nu)$ is the response time for a particle of diameter d and $\boldsymbol{g}$ is the gravitational acceleration. Neglecting the gravitational term, the particle motion can be characterised by β and the Stokes number $St = \tau_p/t_\eta$. Note that since one invokes the velocity $\boldsymbol{u}_p^+$ of the fluid at the location of the particle, one assumes in Eq. (4.22) that the flow field is uniform over the particle size. This restricts the approach to $d \leq \eta$ and the diameter d is then introduced essentially on dimensional grounds, but none of the physical effects associated with size are incorporated. As we see below, this is fine in the limit of a very small particle having a much larger density than that of the fluid. However, in the case of a particle with a density of the order of that of the fluid, more complex effects arise because the momentum exchanged with the surrounding fluid can be of the order of the particle net momentum. The interaction of the particle with its own wake can become quite strong, and corrections to Eq. (4.22) are needed.

Early work on inertial particles focussed on the effects of inertia and gravity on the dispersion of heavy particles; i.e. for $\beta = 0$ (Wells and Stock, 1983; Snyder and Lumley, 1971; Csanady, 1963; Squires and Eaton, 1991a). One of the important aspects of these studies is the so-called "crossing trajectories" effect, in which the trajectories of fluid and inertial particles separate. As a consequence, the statistics of the flow along the trajectories of inertial particles differ from those along fluid trajectories. This is reflected at the simplest level by the dependence of the integral time scale of velocity fluctuations along inertial particle trajectories $T_{\rm L}^{(p)}$ on the still-fluid settling velocity, $\boldsymbol{V}_s = \boldsymbol{g}\tau_p$. Using simple theoretical ideas, Csanady (1963) derived corrections for the integral time scale of the form

$$T_{\rm L}^{(p)} = T_{\rm L} \left(1 + \alpha^2 V_s^2/\sigma_u^2\right)^{-1/2} \qquad (4.23)$$

for velocity fluctuations in the direction of the settling velocity (i.e. here in the direction of gravity). For velocity fluctuations perpendicular to the settling velocity the correction term is a factor of four times larger as a consequence of continuity (as $V_s/\sigma_u \to \infty$ the particles move in straight lines through the fluid and so the velocity correlations along the trajectory reduce to the longitudinal and lateral Eulerian correlations). Squires and Eaton (1991a) found good agreement with Csanady's predictions in DNS of decaying grid turbulence. The overall effect of settling is to reduce the dispersion relative to that of fluid particles as a result of the smaller Lagrangian timescale given by Eq. (4.23) and to make the dispersion anisotropic, so that in the limit $V_s/\sigma_u \to \infty$ dispersion perpendicular to the settling velocity is reduced compared with that parallel to it.

Another aspect studied in early numerical work was the effect of turbulence in enhancing the settling velocity of heavy particles (Maxey, 1987), and its connection with the clustering of these particles in non-vortical regions of the flow (Wang and Maxey, 1993). Wang and Maxey (1993) show that these effects are controlled by the small scales of the flow. For $V_s/v_\eta = 1$ the enhancement of the settling velocity as a function of the particle inertia shows a peak of up to 50% for $0.5 \leq St \leq 1$. The relative enhancement also peaks at $V_s/v_\eta = 1$ for fixed inertia $St = 1$. Clustering is strongest for $St = 1$ but depends only weakly on the still-fluid settling velocity for $V_s/v_\eta < 3$, showing that clustering itself is primarily an inertial effect. This clustering effect has become a strong focus of numerical and experimental work (Squires and Eaton, 1991b; Sundaram and Collins, 1997; Reade and Collins, 2000; Salazar et al., 2008; Saw et al., 2008; Aliseda et al., 2002) and theoretical (Balkovsky et al., 2001; Falkovich and Pumir, 2004) work. Sundaram and Collins (1997) and Reade and Collins (2000) show that particle collisions are enhanced by up to two orders of magnitude, peaking at around $St = 4$.

With improved experimental (Voth et al., 1998, 2002; Volk et al., 2008a,b; Ayyalasomayajula et al., 2006) and numerical techniques (Bec et al., 2006) modern studies have also shifted the emphasis to acceleration statistics. The goal is to investigate the statistics of the forces acting on inertial particles, and studies have focused on three main cases: (i) small and heavy particles ($\beta \to 0$, $St \ll 1$) where the equation of motion is expected to be dominated by the Stokes drag term; (ii) the case of light small particles ($\beta \to 3$, $St \ll 1$), where the added mass term comes into play because the momentum of the displaced fluid dominates the particle's own inertia. In both cases, some dynamical behavior can be explained by the heuristic argument according to which lighter elements tend to travel to low pressure regions, while heavier elements are centrifuged into the strain-dominated regions. This view has however been recently challenged by a sweep-stick mechanism for the clustering of inertial particles at zero-acceleration points (Chen et al., 2006; Coleman and Vassilicos, 2009). (iii) neutrally buoyant particles ($\beta = 1$) of increasing size progressively reveal that the flow at their surface is non uniform, and leads to a more complex situation.

Point particles

We mean particles with a size smaller that the dissipation scale η, and a density much larger (or smaller) than that of the fluid. Bec et al. (2006) show that the acceleration variance for inertial particles decreases monotonically with Stokes number. On the other hand, the acceleration variance of the fluid along the particle trajectory initially decreases strongly, tracking the particle acceleration, but then reaches a minimum at $St \approx 0.5$ before increasing at large Stokes number to the variance for fluid elements. For small (but non-zero) values of the Stokes number, the reduction in the variance for both the inertial particles themselves and for the fluid along the particle trajectory is due to clustering of particles in non-vortical regions where the fluid accelerations are smaller. For large Stokes numbers, the particles tend to move in straight lines, unaffected by the fluid motion, and so there is no bias in sampling fluid accelerations along the particle trajectories. Data from experiments and numerical simulation are in broad agreement showing that accelerations for bubbles (heavy particles) decorrelate faster (slower) than for fluid particles. However there are some discrepancies between numerical and experimental estimates for the acceleration variance and PDF, probably as a consequence of the approximations made in the equation of motion for inertial particles in the numerical work. Some experiments (Ayyalasomayajula et al., 2006) and DNS (Bec et al., 2006) show that the tails of the normalised acceleration PDF (i.e. the PDF for A_i^+/σ_A, where σ_A is the rms component acceleration) are narrower than those for fluid particles. Gerashchenko et al. (2008) measured acceleration statistics for inertial particles in a turbulent boundary layer and found that as the wall is approached the mean acceleration and the acceleration variance increase due to the interaction between the strong shear and the increased Stokes number.

Some attempts have been made to extend stochastic models like Eq. (4.12) to inertial particle trajectories (Zhuang et al., 1989; Sawford and Guest, 1991). Sawford and Guest (1991) used Eq. (4.22) for heavy particles, $\beta = 0$, and a stochastic equation for the fluid velocity along the particle trajectory modified to allow for the effects of crossing trajectories on the integral time scale ((Csanady, 1963) – see Eq. (4.23)). Using a value $\alpha = 1.5$ they obtained good agreement with the data of Snyder and Lumley (1971) for both the dispersion and velocity variance of a range of heavy particles. A slightly different approach was used by Zhuang et al. (1989) who accounted explicitly for the crossing trajectories effect by keeping track of the separation between a fluid particle trajectory and a heavy particle trajectory and adjusting the de-correlation of the fluid velocity along the particle trajectory by a factor depending on this separation.

Neutrally buoyant, finite size, particles

Experiments have considered the effect of increasing size for particles of density equal to that of the fluid ($\rho_p/\rho_f = 1$, i.e. $\beta = 1$). As the diameter d of spherical particles in increased, experimental studies (Voth et al., 2002; Qureshi et al., 2007, 2008; Brown et al., 2009; Volk et al., 2011) have shown

4: A Lagrangian View of Turbulent Dispersion and Mixing

that the variance of the particle acceleration $\langle A_d^{+2}\rangle$ varies as $\langle A^{+2}\rangle(d/\eta)^{-2/3}$, where $\langle A^{+2}\rangle$ is the variance of Lagrangian tracer particles as measured using particles with $d \leq \eta$ (effectively fluid particles). This scaling behavior can be derived using standard Kolmogorov phenomenology (Monin and Yaglom, 1971) by assuming that the particle's acceleration results from pressure gradients (Qureshi et al., 2007): the pressure second order structure function at scale d varies as $d^{4/3}$ so that pressure gradients at scale d scale as $d^{4/3-2} = d^{-2/3}$ – cf. Fig. 4.7(a). The 'point-particle' models using equation (4.22) cannot account for this feature: as the particle size increases, the drag becomes small compared to the other terms, with the unphysical result that the acceleration variance remains equal to that of the fluid as d increases. A first correction to Eq. (4.22), neglecting gravity effects, is (Calzavarini et al., 2009):

$$\frac{d\boldsymbol{v}_p^+}{dt} = \beta \left\langle \frac{d\boldsymbol{u}_p^+}{dt} \right\rangle_{V_P} + \frac{1}{\tau_p}\left(\langle \boldsymbol{u}_p^+\rangle_{S_p} - \boldsymbol{v}_p\right), \qquad (4.24)$$

where V_P and S_p are the volume of the particle and its area, respectively. It corresponds to the inclusion of Faxén corrections (Faxén, 1922; Gatignol, 1983; Auton et al., 1988) which account for the non-uniformity of velocity gradients at the particle scale. In this way, a decrease of the particle acceleration variance with increasing particle size is observed in numerical simulations (Calzavarini et al., 2009), but there are still discrepancies with experimental data. Refined models and numerical schemes attempt to resolve the dynamics of the boundary layer around the moving particle (Naso and Prosperetti, 2010; Homann and Bec, 2010); they are at present limited to low Reynolds number flows.

Another observation from experiments, is that the probability density functions of the particle acceleration for increasing diameter d are quite similar when normalized by the variance. There is a slight decrease of the wings of the distributions (Volk et al., 2011), which, however do not approach a Gaussian, even for particles much larger than the Kolmogorov scale (Gasteuil (2009) – see Fig. 4.7(b)).

4.2.4 Fluid particles in other flows

We begin our discussion of the modelling of single particle dispersion for more general flows by considering extension of the SDE Eq. (4.12). The most general form for a continuous stochastic differential equation representing the Lagrangian velocity which is consistent with the locally isotropic inertial

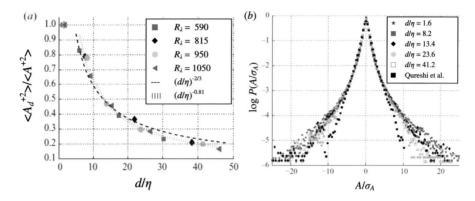

Figure 4.7 (a) Variance of one component of acceleration of the particles versus particle size (Volk et al., 2011). Dashed line: Kolmogorov scaling $\langle A_d^2 \rangle / \langle A^2 \rangle \propto (d/\eta)^{-2/3}$; (b) PDFs of particle acceleration at $R_\lambda = 950$, normalized by their variance, compared with wind tunnel data of Qureshi et al. (2007; 2008) at $R_\lambda = 160$.

sub-range structure given by the second line of Eq. (4.10) is

$$du_i^+ = a_i\left(\bm{u}^+, \bm{x}^+, t\right) dt + \sqrt{C_0 \langle \epsilon \rangle} d\xi_i \quad (4.25)$$

Thus the full specification of the SDE for a given flow (defined by its one-point-one-time Eulerian velocity statistics; i.e. in general its Eulerian velocity PDF $P_{\mathrm{E}}(\bm{u}; \bm{x}, t)$) reduces to specifying the drift term $\bm{a}(\bm{u}^+, \bm{x}^+, t)$. This has been shown to consist of two components, (Pope, 1985)

$$a_i\left(\bm{u}, \bm{x}, t\right) = \langle A_i | \bm{u}; \bm{x}, t \rangle + C_0 \langle \epsilon \rangle \frac{\partial P_{\mathrm{E}}}{\partial u_i} \quad (4.26)$$

The first term on the RHS is a physical term, the Eulerian conditional mean acceleration $\langle A_i | \bm{u}; \bm{x}, t \rangle$, which is evaluated at the fluid particle in Eq. (4.25), and the second term arises as a consequence of the Markov assumption and is a response to the white noise forcing. The connection between this Markov drift term and the forcing is a consequence of the fluctuation-dissipation theorem which is assumed to hold in a first order approximation. The conditional mean acceleration is related to the velocity PDF through the exact PDF evolution equation (Pope, 1985)

$$\frac{\partial P_{\mathrm{E}}}{\partial t} + \frac{\partial u_i P_{\mathrm{E}}}{\partial x_i} + \frac{\partial \langle A_i | \bm{u}; \bm{x}, t \rangle P_{\mathrm{E}}}{\partial u_i} = 0 \quad (4.27)$$

In practice the physical part of the drift term is usually deduced from Eq. (4.27) using an assumed form for the Eulerian velocity PDF, but then is not uniquely defined since Eq. (4.27) is satisfied by $\langle A_i | \bm{u}; \bm{x}, t \rangle + \bm{\varphi}\left(\bm{u}, \bm{x}, t\right)$ where

$\varphi(\boldsymbol{u}, \boldsymbol{x}, t)$ is any rotational function in velocity space, i.e. any function such that $\partial \varphi_i / \partial u_i = 0$ (Thomson, 1987). Alternatively, this term can be modelled directly to be consistent with appropriate physical constraints (Haworth and Pope, 1986).

There seem to be little DNS or experimental data for Lagrangian statistics in more general flows. Shen and Yeung (1997) and Sawford and Yeung (2001) present DNS results for the dispersion of fluid particles in uniform shear turbulence. At the low Reynolds numbers of these DNS there is some anisotropy present in the structure functions, but the component values of C_0^* are broadly consistent with isotropic values and seem to becoming less anisotropic with increasing Re. Lagrangian statistics have also been obtained by DNS in channel flow by Choi et al. (2004). Experimentally, Lagrangian data are difficult to obtain from open flows with a non-zero mean velocity and many confined flows are not homogeneous and isotropic. Only recently devices have been proposed with the potential of producing flows with a controlled anisotropy in the large scales (Variano *et al.*, 2004; Hwang and Eaton, 2004; Zimmermann *et al.*, 2010).

4.3 Two particle motion and relative dispersion

4.3.1 Fluid particles

The dispersion of a cloud of particles relative to their centre of mass is known as relative dispersion. In practical terms it determines the size and concentration distribution in dispersing puffs, and the distribution of material around the instantaneous centreline of dispersing plumes, but as we shall see relative dispersion is also of great importance fundamentally through its connection with scalar mixing, dissipation and concentration fluctuations. The most commonly used measure of relative dispersion is the variance of the particle displacements relative to the centre of mass, which can be simply related to the mean-square separation of a pair of particles. Thus, the separation of pairs of particles captures much of the physics of relative dispersion and is our focus here. We return to consider multi-particle clusters in the next Section.

The relative dispersion of a pair of particles is one of the classical problems of turbulence dating back to Richardson (1926). Richardson identified the accelerating nature of relative dispersion and explained it in terms of the increasing diffusivity of the increasingly large eddies influencing the separation as the separation itself increased. He derived empirically his famous 4/3-law describing the dependence of the turbulent diffusivity on the separation, $K \propto r^{+4/3}$. By treating the separation as a diffusion process with this scale-dependent diffusivity he also derived his t^3-law for the mean-square separation $\left\langle r^{+2}(t) \right\rangle$. Obukhov (1941) put Richardson's theory on a more fundamental theoretical basis by invoking Kolmogorov's similarity theory on inertial sub-range scales, so that for $\eta \ll r_0 \ll \left\langle r^{+2}(t) \right\rangle^{1/2} \ll L$, where r_0 is the initial separation, the statistics of relative dispersion depend solely on the mean dissipation rate and travel time. Then on dimensional grounds Richardson's t^3-law can be written

$$\left\langle r^{+2}(t) \right\rangle = g \left\langle \epsilon \right\rangle t^3 \qquad (4.28)$$

where g is known as Richardson's constant. The condition $r_0 \ll \left\langle r^{+2}(t) \right\rangle^{1/2}$ restricts Eq. (4.28) to times $t \gg t_0$, where $t_0 = r_0^{2/3} \left\langle \epsilon \right\rangle^{-1/3}$ is the time for which the initial separation is important.

Batchelor (1950)) independently developed the similarity theory for relative dispersion, and in addition to Richardson's t^3-law also derived the small-time result

$$\left\langle r^{+2}(t) \right\rangle - r_0^2 = \left\langle (\Delta u(r_0))^2 \right\rangle t^2$$
$$= \tfrac{11}{3} C \left\langle \epsilon \right\rangle^{2/3} r_0^{2/3} t^2 \qquad (\eta \ll r_0; t \ll t_0) \qquad (4.29)$$

where $\left\langle (\Delta u(r_0))^2 \right\rangle$ is the Eulerian velocity structure function, the second line follows from Kolmogorov similarity theory and C is known as the Eulerian structure function constant. Equation (4.29) corresponds to particles separating with straight-line trajectories according to their initial velocity, and so the regime $t \ll t_0$ is known as the ballistic limit or sometimes the Batchelor limit and the time scale t_0 is known as the Batchelor time scale.

For initial separations within the dissipation sub-range, $r_0 \ll \eta$, the separation is governed by a random, but locally uniform in space, strain field and so in the ballistic regime

$$\left\langle r^{+2}(t) \right\rangle - r_0^2 = \left\langle (\Delta u(r_0))^2 \right\rangle t^2$$
$$= \tfrac{1}{3} r_0^2 (t/t_\eta)^2 \qquad (r_0 \ll \eta; t \ll t_\eta) \qquad (4.30)$$

For $t \gg t_\eta$, but $r^+(t) \ll \eta$ (which requires r_0 to be sufficiently small), the separation grows exponentially

$$\left\langle r^{+2}(t) \right\rangle = r_0^2 \exp\left(\gamma t/t_\eta\right) \qquad (r^+ \ll \eta; t \gg t_\eta) \qquad (4.31)$$

4: A Lagrangian View of Turbulent Dispersion and Mixing   155

As a consequence, the time taken for the rms separation to grow to the Kolmogorov scale is $t/t_\eta = \ln\left(\eta^2/r_0^2\right)/\gamma$ and in the limit $r_0 \to 0$, the particles never separate.

For completeness we note that at large enough times $t \gg T_L$ the particles separate beyond the scale of the largest eddies and so move independently, and their mean square separation is given by twice the single particle dispersion

$$\left\langle r^{+2}(t) \right\rangle = 2\sigma_x^2 \delta_{ii} = 12\sigma_u^2 T_L t \qquad (t \gg T_L) \qquad (4.32)$$

This is known as the diffusive range.

We now know that there are corrections to Kolmogorov's similarity theory due to spatial and temporal variations in the dissipation rate, a phenomenon known as small-scale intermittency. However, since Richardson's t^3-law is linear in the dissipation rate we expect it to be unaffected by these corrections, and therefore to be a universal law of turbulence. It, and the value of the Richardson constant, are therefore of fundamental importance. We expect the Batchelor range on the other hand to be subject to intermittency corrections, although to our knowledge this has not been observed

Despite this fundamental importance, unambiguous Richardson scaling has proven difficult to demonstrate and only recently have reliable estimates for Richardson's constant been obtained. The difficulty has been in obtaining sufficient separation in the time scales t_0 and T_L in order to observe a significant Richardson range uncontaminated by the Batchelor and diffusive ranges, especially if we seek to observe Richardson scaling over a range of values of the initial separation and Reynolds number. For instance, measurements in 3D in a von Kármán flow (Bourgoin et al., 2006) show a Batchelor regime rather than a Richardson one.

Plots of DNS results for $\left\langle r^{+2}(t) \right\rangle - r_0^2$, or in compensated form $\left(\left\langle r^{+2}(t) \right\rangle - r_0^2\right)/\left(\langle\epsilon\rangle t^3\right)$, against t/t_η typically show growth close to t^3 (or a plateau in the compensated plot) for $t/t_\eta > 10$ for $r_0/\eta \approx 4$, with faster growth for smaller initial separations and slower growth for larger initial separations (see Fig. 5 in (Yeung and Borgas, 2004) at $R_\lambda = 230$ or Fig 4(a) in (Sawford et al., 2008) for R_λ up to 650). The t^3 regions in these plots imply a Richardson constant $g \approx 0.6$ but there is only a rough indication even at $R_\lambda = 650$ that the curves for larger and smaller values of r_0 are converging towards the same Richardson regime.

A clearer picture emerges if the time is scaled by t_0, rather than t_η, as shown in Fig. 4.8 adapted from Fig. 4(b) in (Sawford et al., 2008). In this scaling, both the Batchelor range (ignoring the effects of intermittency) and

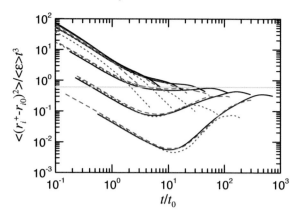

Figure 4.8 Relative dispersion plots in inertial sub-range scaling using the length and time scales r_0 and t_0. Initial separations are nominally, from bottom to top, $r_0/\eta = 1/4, 1, 4, 16, 64$, and 256. Lines: - - -, $R_\lambda = 38$; – – –, $R_\lambda = 240$; ———, $R_\lambda = 650$. The horizontal line is at a value of 0.6

the Richardson range collapse to a single curve over inertial-sub-range values of the initial separation, $\eta \ll r_0 \ll L$, as is apparent in Figure 8 for $r_0 > 16$. This curve is truncated by the diffusive behaviour Eq. (4.32) at large times, but with decreasing values of r_0 (within the inertial sub-range), and for increasing Reynolds number, it shows an increasing tendency to approach a plateau, corresponding to Richardson's t^3-law with $g \approx 0.6$ as indicted by the horizontal line, before finally entering the diffusive regime.

For small values of the initial separation, viscous effects are significant for small times and in t_0 scaling the compensated dispersion curves depend strongly on the initial separation and the dispersion is increasingly retarded as the initial separation decreases. However, for sufficiently large times viscous effects become unimportant and if the Reynolds number is large enough we expect the dispersion to exhibit a Richardson regime before the onset of diffusive behaviour. We see in Fig. 4.8 that in the compensated curves for $r_0/\eta \leq 4$ the peak at large times increases with increasing Reynolds number towards the line indicating $g \approx 0.6$.

In analysing laboratory data at $R_\lambda = 90$, Ott and Mann (2000) showed that taking the cube-root of the mean-square separation reduces the influence of the Batchelor regime and shows a clearer Richardson regime, manifested as a linear growth with time

$$\left\langle r^{+2}(t) \right\rangle^{1/3} = A_0 + (g\langle \epsilon \rangle)^{1/3} t \qquad (t \gg t_0) \qquad (4.33)$$

and estimated $g = 0.5 \pm 0.2$ for $r_0/\eta \approx 10$. Ishihara and Kaneda (2002) applied this approach to DNS results at $R_\lambda = 283$. They estimated g for a range of initial separations, and showed a convergence to $g = 0.7$ for $30 \leq r_0/\eta \leq 45$. Berg et al. (2006) also used this technique to analyse laboratory data at $R_\lambda = 172$ and DNS results at $R_\lambda = 284$. They found values $g = 0.55 \pm 0.05$ for $2 \leq r_0/\eta \leq 26$ for the laboratory data. For the DNS data their estimates range from $g = 0.45$ at $r_0/\eta = 3.1$ to $g = 0.6$ at $r_0/\eta = 22$.

Sawford et al. (2008) noted that there is some uncertainty in the range in t/t_0 over which the linear growth in $\left\langle r^{+2}(t) \right\rangle^{1/3}$ should occur, and proposed a local slope method to eliminate this uncertainty. They analysed DNS results over the range $140 \leq R_\lambda \leq 650$, including the data of Ishihara and Kaneda and those used by Berg et al., and for $R_\lambda = 390$ showed a convincing collapse to a Richardson regime over $3 \leq t/t_0 \leq 10$ for initial separations $8 \leq r_0/\eta \leq 64$. They found only a weak decrease in g with Reynolds number and estimated the large-Reynolds number asymptotic value to be $g = 0.55 - 0.57$, depending on the functional form assumed to make the extrapolation.

Berg et al. (2006) also estimated Richardson's constant for dispersion backwards in time; i.e. for the earlier dispersion of particles which at some later time have a specified separation r_0. For their laboratory data at $R_\lambda = 172$, they estimated the backwards Richardson constant $g_b = 1.15 \pm 0.05$ and for DNS at $R_\lambda = 284$, their estimates range from $g_b = 1.09$ at $r_0/\eta = 3.1$ to $g_b = 1.31$ at $r_0/\eta = 22$.

4.3.2 Stochastic modelling of relative dispersion for fluid particles

Stochastic modelling of relative dispersion is in principle a straightforward generalisation of that for single particle motion in Section 4.2.4 and has been reviewed comprehensively (Sawford, 2001). In its fully 3D form, the generalisation involves a SDE like Eq. (4.25) written in the phase space consisting of the velocities and positions of the two particles ($\boldsymbol{u}^{(1)}, \boldsymbol{u}^{(2)}, \boldsymbol{x}^{(1)}, \boldsymbol{x}^{(2)}$). The drift term is now related to the two-point Eulerian velocity PDF through a generalisation of Eqs. (4.26) and (4.27) and typically is determined from some specified form for the Eulerian PDF. Whereas in the one-particle case the physical part of the drift term vanishes in isotropic turbulence and is therefore given uniquely, for two-particles this term is non-trivial and non-unique. Thomson (1990) derived a solution for the drift term for a Gaussian velocity PDF and Borgas and Sawford (1994b) explored the non-uniqueness of the drift term for a Gaussian PDF, showing for example, the strong dependence of Richardson's constant on the specification of the drift term. It is well-known that the Eulerian two-point PDF is not Gaussian, but it is very difficult to represent non-Gaussian behaviour in the full two-particle phase space.

Kurbanmuradov (1997) derived non-Gaussian models in the reduced phase space consisting of the separation r and the velocity difference parallel to the separation $u_\| = u_i r_i/r$. These models are known as quasi-one-dimensional (Q1D) models because they attempt to capture the 3D physics of relative dispersion in terms of one component of the relative velocity. A non-Gaussian form for $P_E(u_\|; r)$ can be constructed analytically and so the drift term can also be derived analytically. Non-Gaussianity is an essential feature of turbulence and is intimately related to the dissipation of turbulence kinetic energy, time irreversibility and the cascade of energy from large to small scales. In the context of relative dispersion, non-Gaussianity in the Eulerian PDF results in an asymmetry in relative dispersion forwards and backwards in time and in particular in different Richardson constants for backwards and forwards relative dispersion (Sawford et al., 2005). Recently Pagnini (2008) extended this approach to include the velocity component perpendicular to the separation. The results of these models, for example in predicting Richardson's constant, are strongly dependent on the details of the formulation of the Eulerian PDF (Sawford et al., 2005).

As for single particle models, the constant C_0 determines the ratio of Lagrangian and Eulerian time scales. In the limit $C_0 \to \infty$, which corresponds to a vanishing Lagrangian time scale, all these stochastic models reduce to a diffusion limit

$$dr^+ = \left(\frac{dK}{dr} + \frac{2K}{r}\right) dt + \sqrt{2K}\, d\xi \qquad (4.34)$$

where, in the inertial sub-range $K(r) = k_0 \epsilon^{1/3} r^{4/3}$, Richardson's famous 4/3-law (Richardson, 1926). Richardson's diffusion equation in 3D is just the Fokker–Planck equation corresponding to (4.34).

4.3.3 Relative dispersion of Brownian particles

In his analysis of Brownian trajectories at small scales Saffman (1960) also presented results for relative dispersion, which in the present context can be written as

$$\left\langle r^{+^2}(t) \right\rangle - r_0^2 = 12\kappa t + \frac{4}{3}\kappa (\langle \epsilon \rangle)/\nu t^3 \qquad \left(\left\langle r^{+^2}\right\rangle \ll \eta^2; t \ll t_\eta\right) \qquad (4.35)$$

This result only holds so long as $\left\langle r^{+^2} \right\rangle \ll \eta^2$, since it relies on a Taylor expansion in space of the velocity field, and for $t \ll t_\eta$, since it also relies on a Taylor expansion in time. Thus the range of the validity of Eq. (4.35) depends on the Schmidt number. The first of these constraints also requires $r_0 \ll \eta$.

For strong molecular diffusion ($Sc \ll 1$), the constraint on the magnitude of the separation requires $t \ll Sc t_\eta$ which is a stronger condition on Eq. (4.35) than $t \ll t_\eta$. For $t \gg Sc t_\eta$, $\left\langle r^{+^2} \right\rangle \gg \eta^2$ and Saffman's result is no longer applicable and the relative dispersion is determined by the interaction of the molecular diffusion with the inertial sub-range eddies. This corresponds to the so-called inertial-diffusive sub-range of the scalar variance spectrum, for which there is no accepted theory. Within this regime the relevant parameters are κ and $\langle \epsilon \rangle$, which determine the time scale $(\kappa/\langle \epsilon \rangle)^{1/2} = Sc^{-1/2} t_\eta$. For $t \gg Sc^{-1/2} t_\eta$ only the dissipation remains as parameter and we expect Richardson scaling to apply in the range $Sc^{-1/2} t_\eta \ll t \ll T_{\mathrm{L}}$, although to our knowledge this has never been observed.

For weak molecular diffusion ($Sc \gg 1$), there will be a period of growth where the relative dispersion remains within the dissipation sub-range for times much larger than the Kolmogorov scale and Saffman's expansion breaks down. Within this regime there is a period of exponential growth (see e.g. Borgas et al. (2004))

$$\left\langle r^{+^2}(t) \right\rangle = r_0^2 \exp(2\tilde{s} t) + \frac{6\kappa}{\tilde{s}} \exp(2\tilde{s} t - 1) \quad (t_\eta \ll t \ll t_\eta \ln Sc; r_0 \ll \eta) \tag{4.36}$$

where $\tilde{s} = C_{\mathrm{B}} t_\eta$ and C_{B} is the Batchelor constant. This is the viscous-convective regime (Batchelor, 1959; Kraichnan, 1974). Again, for large enough times $t_\eta \ln Sc \ll t \ll T_{\mathrm{L}}$ and for large enough Reynolds number when the relative dispersion is under the influence of inertial sub-range eddies, we expect Richardson scaling to apply.

To summarise, molecular diffusion is responsible for the initial stages of the relative dispersion of initially coincident Brownian particles and the details of the early time behaviour depends on the Schmidt number. Provided the Reynolds number is large enough (depending on the Schmidt number) we expect that at large enough times Brownian particles will behave like fluid particles and exhibit a Richardson scaling regime followed by diffusive behaviour for $t \gg T_{\mathrm{L}}$.

Connection with scalar mixing and dissipation

Scalar dissipation is associated with the mixing that occurs when two parcels of fluid with different concentrations come into close proximity. The statistics of the prior locations of a pair of particles are determined by the process of backwards relative dispersion, so in general we expect that mixing and scalar dissipation are closely connected with this process.

We can make this connection explicit for the simple case of a scalar field with a constant uniform mean gradient, which we can construct using a source of the form

$$S(\boldsymbol{x}, t) = Gz\delta(t) \tag{4.37}$$

Then from the backwards form of Eq. (4.2) which connects the displacement statistics of Brownian particles to scalar statistics, we have for stationary isotropic turbulence

$$\langle c(\boldsymbol{x}, t) \rangle = \int_{-\infty}^{t_1} P(\boldsymbol{x}', t'|\boldsymbol{x}, t) Gz'\delta(t') d\boldsymbol{x}'dt' = G\langle z^+(0)|\boldsymbol{x}, t\rangle = Gz \tag{4.38}$$

since the mean displacement vanishes. Thus, by construction the mean scalar gradient is uniform and stationary.

From Eq. (4.2) the mean square concentration is determined by two-particle displacement statistics. For the source given by Eq. (4.37) Sawford et al. (2005) show that the scalar dissipation is related to the backwards relative dispersion by

$$\langle \epsilon_c \rangle = 2\kappa \left\langle (\partial c'/\partial x_i)^2 \right\rangle = \tfrac{1}{2}G^2 d\left\langle r_z^{+^2}(0)^2 | \boldsymbol{x}, t\right\rangle / dt - 2\kappa G^2 \tag{4.39}$$

That is, the scalar dissipation is given by the rate of change of the backwards relative dispersion of initially coincident Brownian particles corrected for the direct molecular effect.

It is interesting to consider the consequences of Eq. (4.39) in the limit $\kappa \to 0$. If we set $\kappa = 0$ while keeping the viscosity finite (i.e. we have the limit $Sc = \nu/\kappa \to \infty$, then the particles become fluid particles and we see from Eq. (4.31) that they never separate in the limit $\eta \gg r_0 \to 0$ and so the scalar dissipation vanishes. On the other hand, if we keep the Schmidt number fixed while taking the limit $\kappa, \nu \to 0$, which corresponds to the limit $Re \to \infty$, then even for $r_0 = 0$ we see from Eqs. (4.35) and (4.36) that for a finite Reynolds number, no matter how large, the particles initially separate due to molecular diffusion - that is the scalar dissipation is non-zero even though the diffusivity vanishes in the limit. This is known as the dissipation anomaly. If we take this limit while letting the initial separation vanish like

4: A Lagrangian View of Turbulent Dispersion and Mixing

$r_0 \gg \eta \to 0$, then the effect of molecular diffusion can be neglected and the particles are effectively fluid particles which separate because in the limit the initial separation always remains under the influence of inertial sub-range eddies. So in this limit, the dissipation is non-zero and fluid particles can be used to calculate scalar statistics.

The scalar dissipation anomaly has been demonstrated explicitly in DNS simulations for a scalar field with a uniform mean gradient (Borgas *et al.*, 2004; Donzis *et al.*, 2005) in stationary isotropic turbulence, for which the scalar field reaches a stationary state with constant scalar variance and scalar dissipation. For $Sc \gg 1$ Borgas et al. (2004) show that the normalised scalar variance rate is reasonably well represented by the expression

$$\langle c'^2 \rangle / (\langle \epsilon_c \rangle T_E) = 0.61 + 10 R_\lambda^{-1} + \tfrac{1}{2} C_B \sqrt{15} R_\lambda^{-1} \ln Sc \qquad (4.40)$$

where C_B is the Batchelor constant (Batchelor, 1959). Donzis et al. (2005) derive a similar result and also present a corresponding expression for $Sc \ll 1$. According to Eq. (4.40) the normalised stationary scalar variance converges to a constant value with increasing Reynolds number at fixed finite Schmidt number, but diverges with increasing Schmidt number at fixed finite Reynolds number. Both Borgas et al. and Donzis et al. estimate a value $C_B \approx 5$ for the Batchelor constant. Earlier estimates ranged from 0.9 ((Saffman, 1963) to 9.5 (Smyth, 1999).

4.3.4 Micromixing models for the scalar dissipation

Fully satisfactory stochastic models for two-particle dispersion have yet to be realised, even for isotropic turbulence, and two-particle models capable of predicting the scalar variance and the scalar dissipation in general flows are a long way off. A different Lagrangian approach has been developed to deal with these problems, and indeed that of predicting higher moments of the scalar field and the scalar PDF (Pope, 1985). In this approach, the SDEs for single-particle trajectories (e.g. (4.25)) are augmented by an equation describing the evolution of the scalar concentration along the trajectory

$$dc^+/dt = \theta(\boldsymbol{u}, \boldsymbol{x}, t) \qquad (4.41)$$

The specification of the function θ constitutes what is known as a mixing model. Pope (2000) lists several properties that should be satisfied by a mixing model. For example, the mean concentration should be unchanged by mixing. This approach is known as PDF modelling, and as with Lagrangian methods in general requires no modelling of the advection terms.

In the simplest mixing model the concentration along the trajectory relaxes back to the local mean concentration with a time scale t_m known as the mixing time scale, so $\theta = -(c - \langle c(\boldsymbol{x}, t) \rangle)/t_m$. This is known as the interaction by exchange with the mean (IEM) model (Villermaux and Devillon, 1972; Pope, 1985; Dopazo, 1975). However, it has been shown that the IEM model does not conserve the mean concentration, but that a simple intension in which the mean concentration is replaced by the mean conditioned on the local velocity $\theta = -(c - \langle c | \boldsymbol{u}; \boldsymbol{x}, t \rangle)/t_m$ does (Pope, 1994, 1998). This is known as the interaction by exchange with the conditional mean (IECM) model. Many other mixing models have been developed in an attempt to improve the representation of the scalar field, but the IEM model is widely used in modelling mixing, usually with chemistry which can be treated exactly in this framework, in engineering flows (see e.g. Pope (2000)).

In most engineering applications the mixing time scale is assumed to be proportional to the turbulence time scale $t_m = k/(C_m \langle \epsilon \rangle)$, particularly where the scalar field is distributed throughout the flow volume. However, in many flows, particularly environmental flows, the scalar is initially confined to a region much smaller than the flow domain of interest, as in for example pollutant released from a chimney into the atmosphere or from a pipe into a river or ocean. In this case the mixing time scale of the scalar, as deduced from a relation such as $t_m = \langle c'^2 \rangle / \langle \epsilon_c \rangle$ which holds for the IEM model, is typically initially much smaller than the turbulence time scale and grows with time as the width of the plume of scalar material grows with time. With a modification to allow for this time-dependent mixing time scale, the mixing model approach has been applied succesfully to a wide range of laboratory and environmental flows (Sawford, 2004; Luhar and Sawford, 2005; Cassiani et al., 2007; Viswanathan and Pope, 2008).

4.4 *n*-particle statistics

Multi-particle Lagrangian statistics are of interest for two reasons. Firstly, from Eq. (4.2) we see that nth-order moments of the scalar field are determined by the joint displacement statistics of n particles. Secondly, clusters of more than two particles carry information about the evolution of the shape of structures moving with the turbulent flow and so provide a new way at looking at the flow dynamics (Pumir et al., 2001).

4.4.1 Triads and tetrads

We commence by considering the joint motion in isotropic turbulence of four particles which define a tetrahedron that evolves with time in both size and shape. It is useful to introduce the separations

$$\rho^{+(1)} = \tfrac{1}{\sqrt{2}} \left(x^{+(2)} - x^{+(1)} \right)$$
$$\rho^{+(2)} = \tfrac{1}{\sqrt{6}} \left(2x^{+(3)} - x^{+(2)} - x^{+(1)} \right)$$
$$\rho^{+(3)} = \tfrac{1}{\sqrt{12}} \left(3x^{+(4)} - x^{+(3)} - x^{+(2)} - x^{+(1)} \right) \quad (4.42)$$

Since the turbulence is homogeneous, the centre-of-mass of the cluster is not important.

Now, the moment of inertia tensor can be written as $M_{ij} = \tilde{\rho}_{ij}\tilde{\rho}_{ji}$, where $\tilde{\rho}_{ij}$ is the matrix whose columns are the separation vectors $\rho^{(i)}$, with $i = 1, 2, 3$. Denoting the eigenvalues of $\boldsymbol{M}$ by $g_1 \geq g_2 \geq g_3$, then we can define a measure of the scale (size) of the tetrahedron by (Pumir et al., 2000; Biferale et al., 2005a)

$$R^2 = \tfrac{1}{8} \sum_{i,j=1}^{4} \left(x^{+(i)} - x^{+(j)} \right)^2 = g_1 + g_2 + g_3 \quad (4.43)$$

The dimensionless eigenvalues $I_i = g_i/R^2$ are determined by the aspect ratio of the tetrahedron and so are known as shape factors. For example, $I_1 \approx I_2 \gg I_3$ corresponds to pancake-like objects and $I_1 \gg I_2 \approx I_3$ corresponds to needle-shape objects. The volume is

$$V = \tfrac{1}{3} \det \tilde{\rho} = \tfrac{1}{3} \sqrt{g_1 g_2 g_3} = \tfrac{1}{3} R^3 \sqrt{I_1 I_2 I_3} \quad (4.44)$$

so the non-dimensional volume V/R^3 is another measure of the shape of a tetrahedron.

The same approach can be applied to three particles forming a triangle with

$$R^2 = \tfrac{1}{6} \sum_{i,j=1}^{3} \left(x^{+(i)} - x^{+(j)} \right)^2 = g_1 + g_2 \quad (4.45)$$

which corresponds to a degenerate tetrahedron with $\rho^{+(3)} = 0$, $g_3 = 0$ and $V = 0$. The area of a triangle is given by

$$\mathscr{A} = \tfrac{\sqrt{3}}{2} |\rho^{+(1)} \times \rho^{+(2)}| = \tfrac{\sqrt{3}}{2} \sqrt{g_1 g_2} = \tfrac{\sqrt{3}}{2} R^2 \sqrt{I_1 I_2} \quad (4.46)$$

and $\mathscr{A}/R^2$ is another measure of the shape of a triangle. Note that the definitions of volume and area given by Eqs. (4.44) and (4.46) differ by numerical factors from those of Pumir et al. (2000).

Since all the quantities we are concerned with involve particle separations Kolmogorov similarity arguments can be used to predict the time evolution of their statistics in the ballistic, Richardson and diffusion regimes identified in Section 4.3. Essentially the scale of a triangle or tetrahedron carries the same information as the separation of a pair of particles and the new information in three and four particle statistics is embodied in the shape factors.

In the inertial sub-range $T_L \gg t \gg t_0$ Kolmogorov scaling gives

$$\langle V \rangle \approx \langle \epsilon \rangle^{3/2} t^{9/2} \qquad (4.47)$$

although because the dissipation appears in a non-linear form, in this case we would expect intermittency corrections. The area of a triangle on the other hand scales like the mean-square separation, and so should be independent of intermittency

$$\langle \mathscr{A} \rangle \approx \langle \epsilon \rangle t^3 \qquad (4.48)$$

For large times, $t \gg T_L$ the particles move independently with Gaussian statistics. For triangles an analytical treatment is possible, giving (Borgas, 1998; Pumir et al., 2000)

$$\langle \mathscr{A} \rangle / \langle r^{+2} \rangle = \sqrt{3}/6$$
$$\langle I_1 \rangle = 5/6; \quad \langle I_2 \rangle = 1/6 \qquad (4.49)$$

while for tetrahedra, Monte Carlo estimation gives

$$\langle V \rangle / \langle r^{+2} \rangle^{3/2} = 0.03618$$
$$\langle I_1 \rangle = 0.7480; \quad \langle I_2 \rangle = 0.2222; \quad \langle I_3 \rangle = 0.02974 \qquad (4.50)$$

Both numerical ((Pumir et al., 2000; Yeung, 2002; Borgas, 1998; Biferale et al., 2005b)) and experimental ((Xu et al., 2008; Lüthi et al., 2007) simulations of various three and four particle statistics have been reported at Reynolds numbers ranging from 82 to 400 for the numerical results and 172 to 815 for laboratory results. No convincing Richardson scaling corresponding to Eqs. (4.47) and (4.48) has been reported for the mean tetrahedron volume or the mean triangle area, but the large time limits Eqs. (4.49) and (4.50) have been observed in numerical simulations ((Pumir et al., 2000; Hackl et al., 2011)).

Since the shape factors are dimensionless quantities, within the inertial sub-range we expect their statistics to be stationary; i.e. all the moments should be constant and the PDF should be invariant with time. Direct estimates for the mean shape factors in the inertial sub-range are not completely consistent. Experimental results of Xu et al. (2008) at $R_\lambda = 690$ for approximately regular tetrads with initial sides $r_0/\eta = 330 - 660$ show a tendency to plateau at $\langle I_1 \rangle \approx 0.75, \langle I_2 \rangle \approx 0.22$ and $\langle I_3 \rangle \approx 0.03$ for $t/t_0 \approx 1$. These

4: A Lagrangian View of Turbulent Dispersion and Mixing

are close to the Gaussian values. Using numerical simulations at $R_\lambda = 284$ Biferale et al. (2005b) found that shape factor PDFs, conditionally sampled to exclude tetrahedra with eigenvalues outside the inertial sub-range, are invariant in time for times in the inertial sub-range. From these conditional sampled PDFs they estimated $\langle I_1 \rangle \approx 0.85, \langle I_2 \rangle \approx 0.135$ and $\langle I_3 \rangle \approx 0.011$. These values represent a greater tendency for flat and elongated structures than the Gaussian case. This tendency is reinforced by a strong peak at small values of I_2 in the PDF. Biferale et al. also presented results for the mean doubling time of the eigenvalues; i.e. the mean time for a tetrahedron to increase its value of g_i by a constant factor. These fixed-scale statistics tend to show clearer inertial sub-range scaling than do fixed-time statistics. Biferale et al. showed that this scaling range can be characterised by values $I_1 = 0.82, I_2 = 0.16$ and $I_3 = 0.02$. It is not clear however, how these values relate to the mean values for example, although they show the same tendency to represent flatter, more elongated structures than the Gaussian case.

4.4.2 Scalar structure functions and intermittency

Anomalous scaling of the scalar structure functions in the inertial sub-range has been related to the breaking of scale invariance by conserved Lagrangian statistical integrals, i.e. by functions $\phi(\underline{x})$, where $\underline{x} = (x^{(1)}, \ldots, x^{(n)})$, such that

$$\langle \phi(\underline{x}^+(t;\underline{x},0)) \rangle = \int \phi(\underline{x}')P_n(\underline{x}',t|\underline{x},0)d\underline{x}' = \phi(\underline{x}) \qquad (4.51)$$

For the Kraichan flow these statistical integrals of motion have been identified as zero modes of the n-particle Fokker–Planck operator (Chertkov et al., 1995; Gawędzki and Kupiainen, 1995; Shraiman and Siggia, 1995). Thus anomalous scaling and scalar intermittency is intimately connection with the evolution of both the scale and shape of clusters of n particles. A technical overview of these important developments can be found in Falkovich et al. (2001), while a more accessible summary for the non-expert can be found in Falkovich and Sreenivasan (2006).

4.5 Conclusions

Lagrangian methods provide a natural way to treat the transport and dispersion and mixing in turbulence. In some ways modelling applications successfully used in environmental and engineering flows have preceded theoretical and empirical understanding. More recently though, through increasingly powerful computers and new experimental techniques there has been more fundamental interest in Lagrangian aspects of turbulence. This has been accompanied by significant theoretical advances in the understanding of scalar fields in turbulence using Lagrangian methods. In this review we have tried to give an overview of these more recent developments and to link them back to practical modelling work. There remain many open questions, some of which we highlight here.

On the experimental side:

- Fluid particles: Lagrangian tracking methods may in the near future give access to a resolved measurement of the velocity gradient tensor (and dissipation) in high Reynolds number flows. Several recent theoretical approaches (Chevillard *et al.*, 2006; Naso *et al.*, 2007; Li *et al.*, 2009) could be tested.
- Inertial particles: models need to be elaborated for the dynamics of particles that are not in the limits $d \ll \eta$ or $(d \sim \eta, \rho_p \gg \rho_f)$ which are begining to be understood. Experimental data are gradually becoming available for the study of finite density effects and finite size effects.
- For 2-particle dispersion, the Richardson behavior remains elusive. Measurements are needed to connect with the problem of scalar mixing and dispersion.

In these problems, the advent of optical methods using fast cameras is a major improvement. In the coming years, 3D PIV and 3D PTV will be bridged and one will access a fully resolved measurement of both the Eulerian flow and the dynamics of individual tracers, at least in the inertial range of scales. This will enable a better description of the interactions between structure/geometry and statistical behavior.

On the numerical and theoretical side:

- Direct numerical simulations are approaching sufficiently high Reynolds numbers to manifest scaling behaviour in Lagrangian statistics and to see convergence towards large Reynolds number limits. However, the influence of the large scale forcing, and the effect of periodic boundary conditions on Lagrangian statistics are not well-understood. The question of the influence of the resolution of the smallest scales also needs to be addressed for Lagrangian statistics, particularly higher-order statistics.
- Lagrangian intermittency is well established, but there remain open questions about the connection between Eulerian and Lagrangian intermit-

tency. Little or nothing is known about intermittency in multiparticle statistics.
- Modelling of single particle trajectories is well-established at a practical level for fluid particles, but developments in modelling multi-particle trajectories and inertial particles, particularly finite size particles, are needed.

Both numerically and experimentally there is a lack of information about Lagrangian statistics in complex flows, such as wall-bounded turbulence, jets, shear turbulence stratified or convective turbulence, although such kmowledge is very important practically. More work in these areas is badly needed.

We conclude that interest and new developments in Lagrangian treatments of fluid transport will continue for some time to come.

References

Aliseda, A. Cartellier, A., Hainaux, F. and Lasheras, J.C. (2002). Effect of preferential concentration on the settling velocity of heavy particles in homogeneous isotropic turbulence. *J. Fluid Mech.*, **468**, 77–105.

Aringazin, A.K. (2004). Conditional Lagrangian acceleration statistics in turbulent flows with Gaussian-distributed velocities. *Phys. Rev. E*, **70**, 036301-1–8.

Aringazin, A.K. and Mazhitov. M.I. (2004). One-dimensional Langevin models of fluid particle acceleration in developed turbulence. *Phys. Rev. E*, **69**, 026305-1–17.

Arneodo, A., Benzi, R., Berg, J., Biferale, L., Bodenschatz, E., Busse, A., Calzavarini, E., Castaing, B., Cencini, M., Chevillard, L., Fisher, R. T.,, Grauer, R.,, Homann, H., Lamb, D. Lanotte, A. S., Leveque E., Luthi, B., Mann, J., Mordant, N., Muller, W.-C., Ott, S., Ouellette, N. T., Pinton, J.-F., Pope, S. B., Roux, S. G., Toschi, F., Xu, H. and Yeung, P. K. (2008). Universal intermittent properties of particle trajectories in highly turbulent flows. *Phys. Rev. Lett.*, **100**, 254504-1–4.

Auton, T., Hunt, J.C.R. and Prud'homme, M. (1988). The force exerted on a body in inviscid unsteady non-uniform rotational flow. *J. Fluid Mech.*, **197**, 241–257.

Ayyalasomayajula, S., Gylfason, A., Collins, L.R., Bodenschatz, E. and Warhaft, Z. (2006). Lagrangian measurements of inertial particle accelerations in grid generated wind tunnel turbulence. *Phys. Rev. Lett.*, **97**, 144507-1–4.

Balkovsky, E., Falkovich, G. and Fouxon, A. (2001). Intermittent distribution of inertial particles in turbulent flows. *Phys. Rev. Lett.*, **86**, 2790–93.

Batchelor, G.K. (1950) The application of the similarity theory of turbulence to atmospheric diffusion. *Q.J.R. Meteorol. Soc.*, **76**, 133–146.

Batchelor, G.K. (1952). Diffusion in a field of homogeneous turbulence. II. The relative motion of particles. *Proc. Cambridge Philos. Soc.*, **48**, 345–62.

Batchelor, G. K. (1959). Small-scale variation of convected quantities like temperature in a turbulent fluid. Part 1. General discussion and the case of small conductivity. *J. Fluid Mech.*, **5**, 113–133.

Bec, J., Biferale, L., Boffetta, G., Celani, A., Cencini, C., Lanotte, A., Musacchio, S. and Toschi, F. (2006). Acceleration statistics of heavy particles in turbulence. *J. Fluid Mech.*, **550**, 349–358.

Benzi, R., Ciliberto, S., Tripiccione, R., Baudet. C., Massaioli, F. and Succi, S. (1993). Extended self-similarity in turbulent flows. *Phys. Rev. E*, **48**, R29–R32.

Berg, J., Lüthi, B., Mann, J. and Ott, S. (2006). Backwards and forwards relative dispersion in turbulent flow: An experimental investigation. *Phys. Rev. E*, **74**, 016304.

Berg, J., Ott, S., Mann, J. and Lüthi, B. (2009). Experimental investigation of Lagrangian structure functions in turbulence, *Phys. Rev. E*, **80**, 026316-1–11.

Biferale, L., Boffetta, G., Celani, A., Devenish, B. J., Lanotte, A. and Toschi, F. (2005a). Lagrangian statistics of particle pairs in homogeneous isotropic turbulence. *Phys. Fluids*, **17**, 115101-1–9.

Biferale, L., Boffetta, G., Celani, A., Devenish, B. J., Lanotte, A. and Toschi, F. (2005b). Multiparticle dispersion in fully developed turbulence. *Phys. Fluids*, **17**, 111701-1–4.

Biferale, L., Bodenschatz, E., Cencini, M., Lanotte, A. S., Ouellette, N. T., Toschi, F. and Xu H. (2008). Lagrangian structure functions in turbulence: A quantitative comparison between experiment and direct numerical simulation. *Phys. Fluids*, **20**, 065103-1–12.

Borgas, M.S. (1993). The multifractal Lagrangian nature of turbulence. *Phil. Trans. R. Soc. A*, **342**, 379–411.

Borgas, M.S. (1998). Meandering plume models in turbulent flows. In *Proc. 13th Australasian Fluid Mechanics Conference*, Melbourne Australia: Monash University, 139–142.

Borgas, M.S. and Sawford, B.L. (1991). The small-scale structure of acceleration correlations and its role in the statistical theory of turbulent dispersion. *J. Fluid Mech.*, **228**, 295–320.

Borgas, M.S. and Sawford, B.L. (1994a). Stochastic equations with multifractal random increments for modelling turbulent dispersion, *Phys. Fluids*, **6**, 618–633.

Borgas, M.S. and Sawford, B.L. (1994b). A family of stochastic models for two-particle dispersion in isotropic homogeneous stationary turbulence, *J. Fluid Mech.*, **279**, 69–99.

Borgas, M.S., Sawford, B.L., Xu, S., Donzis D. and Yeung, P.K. (2004). High Schmidt number scalars in turbulence: structure functions and Lagrangian theory. *Phys. Fluids*, **16**, 3888–3899.

Bourgoin, M., Ouellette, N.T., Xu, H., Berg, J. and Bodenschatz, E. (2006). The role of pair dispersion in turbulent flow. *Science*, **311**, 935–838.

Braun, W., De Lillo, F. and Eckhardt, B. (2006). Geometry of particle paths in turbulent flows, *J. Turbul.* **7**, N62, 1–10.

Brown, R., Warhaft, Z., Voth, G. (2009). Acceleration statistics of neutrally buoyant spherical particles in intense turbulence. *Phys. Rev. Lett.*, **103**, 194501-1–4.

Calzavarini, E., Volk, R., Lévêque, E., Bourgoin, B., Toschi, F. and Pinton, J.-F. (2009). Acceleration statistics of finite-size particles in turbulent flow: the role of Faxén corrections. *J. Fluid Mech.*, **630**, 179–189.

Cassiani M., Radicchi, A., Albertson, J. D. and Giostra, U. (2007). An efficient algorithm for scalar PDF modeling in incompressible turbulent flows; numerical analysis with evaluation of IEM and IECM micro-mixing models. *J. Computat. Phys.*, **223**, 519–550.

Castiglione, P. and Pumir, A. (2001). Evolution of triangles in a two-dimensional turbulent flow. *Phys. Rev. E*, **64**, 056303-1–11.

Celani, A. and Vergassola, M. (2001). Statistical geometry in scalar turbulence. *Phys. Rev. Lett.*, **86**, 424-1–4.

Chen, L., Goto, S. and Vassilicos, J.C. (2006). Turbulent clustering of stagnation points and inertial particles, *J. Fluid Mech.*, **553**, 143–154.

Chertkov, M., Falkovich, G., Kolokoloov, I. and Lebedev, V. (1995). Normal and anomalous scaling of the fourth-order structure function of a randomly advected passive scalar. *Phys. Rev. E*, **52**, 4924–4941.

Chevillard, L., Roux, S.G., Lévêque, E. Mordant, N. Pinton, J.-F. and Arnèodo, A. (2003). Lagrangian velocity statistics in turbulent flows: Effects of dissipation. *Phys. Rev. Lett.*, **91**, 214501-1–4.

Chevillard, L., Castaing, B., Lévêque, E. and Arnéodo, A. (2006). Unified multifractal description of velocity increments statistics in turbulence: Intermittency and skewness. *Physica D*, **218**, 77–82.

Choi, Jung-Il, Yeo, K. and Lee, C. (2004). Lagrangian statistics in turbulent channel flow. *Phys. Fluids*, **16**, 779–793.

Coleman, S.W. and Vassilicos, J.C., (2009). A unified sweep-stick mechanism to explain particle clustering in two- and three-dimensional homogeneous, isotropic turbulence. *Phys. Fluids*, **21**, 113301-1–10.

Corrsin, S. (1959). Progress report on some turbulent diffusion research. *Adv. Geophys.*, **6**, 161–164.

Crawford, A.M., Mordant, N. and Bodenschatz, E. (2005). Joint statistics of the Lagrangian acceleration and velocity in fully developed turbulence, *Phys. Rev. Lett.*, **94**, 024501-1–4.

Csanady, G.T. (1963). Turbulent diffusion of heavy particles in the atmosphere. *J. Atmos. Sci.*, **20**, 201–208.

Donzis, D.A., Sreenivasan, K.R., Yeung, P.K. (2005). Scalar dissipation rate and dissipative anomaly in isotropic turbulence. *J. Fluid Mech.*, **532**, 199–216.

Dopazo, C. (1975). Probability density function approach for a turbulent axisymmetric heated jet. Centreline evolution. *Phys. Fluids*, **18**, 397–410.

Falkovich, G., Gawędzki, K. and Vergassola, M. (2001). Particles and fields in fluid turbulence. *Rev. Mod. Phys* **73**, 913.

Falkovich, G. and Pumir, A. (2004). Intermittent distribution of heavy particles in a turbulent flow. *Phys. Fluids*, **16**, L47–50.

Falkovich, G. and Sreenivasan, K.R. (2006). Lessons from hydrodynamic turbulence. *Physics Today*, (April), 43–49.

Faxén, H. (1922). Der widerstand gegen die bewegung einer starren kugel in einer zähen Flüssigkeit, die zwischen zwei parallelen ebenen Wänden eingeschlossen ist. *Annalen der Physik*, **373**, 99–119.

Frisch, U. (1985). Fully developed turbulence and intermittency. In *Proceedings of the International School on Turbulence and Predictability in Geophysical Fluid Dynamics and Climate Dynamics*. M. Ghil, R. Benzi and G. Parisi (eds.) Amsterdam: North-Holland.

Frisch, U., Mazzino, A., Noullez, A. and Vergassola, M. (1999). Lagrangian method for multiple correlations in passive scalar advection. *Phys. Fluids*, **11**, 2178–2186.

Gardiner, C.W., (1983). *Handbook of Stochastic Methods for Physics, Chemistry and the Natural Sciences*. Berlin: Springer.

Gasteuil, Y. (2009). PhD Thesis, Instrumentation Lagrangienne en Turbulence: Mise en Öuvre et Analyse, École Normale Supérieure de Lyon.

Gatignol, R. (1983). The Faxén formulae for a rigid particle in an unsteady non-uniform stokes flow. *J. Mec. Theor. Appl.*, **1**, 143–160.

Gawędzki, K. and Kupiainen, A. (1995). Anomalous scaling of the passive scalar. *Phys. Rev. Lett.*, **75**, 3834–3837.

Gerashchenko, S., Sharp, N., Neuscamman, S. and Warhaft, Z. (2008). Lagrangian measurements of inertial particle accelerations in a turbulent boundary layer. *J. Fluid Mech.*, **617**, 255–281.

Gotoh, T. and Fukayama, D. (2001). Pressure spectrum in homogeneous turbulence. *Phys. Rev. Lett.*, **86**, 3775–3778.

Griffa, A. (1996) Applications of stochastic particle models to oceanographic problems. In *Stochastic Modeling in Physical Oceanography*, R. Adler, P. Muller, and B. Rozovskii (eds.), Birkhäuser Verlag, 114–140.

Hackl, J.F., Yeung, P.K. and Sawford, B.L. (2011). Multi-particle and tetrad statistics in numerical simulations of turbulent relative dispersion. *Phys. Fluids*, **23**, 063103-1-20.

Haworth, D.C. and Pope, S.B. (1986). A generalized Langevin model for turbulent flows. *Phys. Fluids*, **29**, 387–405.

Homann, H., Kamps, O., Friedrich, R. and Grauer, R. (2009). Bridging from Eulerian to Lagrangian statistics in 3D hydro- and magnetohydrodynamic turbulent flows. *New J. Phys.*, **11**, 073020-1-15.

Homann, H. and Bec, J. (2010). Finite-size effects in the dynamics of neutrally buoyant particles in turbulent flow. *J. Fluid Mech.*, **651**, 91–91.

Hwang, W. and Eaton, J.K. (2004). Creating homogeneous and isotropic turbulence without a mean flow. *Exp. Fluids*, **36**, 444–454.

Iliopoulos, I. and Hanratty, T.J. (1995) Turbulent dispersion in a non-homogeneous field. *J. Fluid Mech.*, **392**, 45–71.

Ishihara, T. and Kaneda, Y. (2002). Relative diffusion of a pair of fluid particles in the inertial subrange of turbulence. *Phys. Fluids*, **14**, L69–L72.

Kamps, O., Friedrich, R. and Grauer, R. (2009). Exact relation between Eulerian and Lagrangian velocity increment statistics. *Phys. Rev. E*, **79**, 066301-1-5.

Kellay, H. and Goldburg, W.I. (2002). Two-dimensional turbulence: a review of some recent experiments. *Rep. Prog. Phys.*, **65**, 945–894.

Kraichnan, R.H. (1974). Convection of a passive scalar by a quasi-uniform random straining field. *J. Fluid Mech.*, **64**, 737–762.

Kurbanmuradov, O.A. (1997). Stochastic Lagrangian models for two-particle relative dispersion in high-Reynolds number turbulence. *Monte Carlo Methods and Appl.*, **3**, 37–52.

La Porta, A., Voth, G.A., Crawford, A. M., Alexander, J. and Bodenschatz, E. (2001). Fluid particle accelerations in fully developed turbulence. *Nature*, **409**, 1017–1019.

Lamorgese A.G., Pope, S.B., Yeung, P.K. and Sawford, B.L. (2007). A conditionally cubic-Gaussian stochastic Lagrangian model for acceleration in isotropic turbulence, *J. Fluid Mech.*, **582**, 423–448.

Laval, J.-P., Dubrulle, B. and Nazarenko, S. (2001). Nonlocality and intermittency in three-dimensional turbulence. *Phys. Fluids*, **13**, 1995–2012.

Li, Y., Chevillard, L., Eyink, G. and Meneveau, C. (2009). Matrix exponential-based closures for the turbulent subgrid-scale stress tensor. *Phys. Rev. E*, **79**, 016305-1–9.

Lüthi, B., Ott, S., Berg, J., Mann, J. (2007). Lagrangian multi-particle statistics. *J. Turbul.*, **8**, N45, 1–17.

Luhar, A.K. and Sawford, B.L. (2005). Micromixing modelling of concentration fluctuations in inhomogeneous turbulence in the convective boundary layer. *Bound. Layer Meteorol.*, **114**, 1–30.

Maxey, M.R. (1987). The gravitational settling of aerosol particles in homogeneous turbulence and random flow fields. *J. Fluid Mech.*, **174**, 441–465.

Meneveau, C. (1996). On the cross-over between viscous and inertial-range scaling of turbulence structure functions. *Phys. Rev. E*, **54**, 3657–3663.

Monin, A.S. and Yaglom, A.M. (1971). *Statistical Fluid Mechanics: Mechanics of Turbulence Vol. 1*. Cambridge MA: MIT Press.

Monin, A.S. and Yaglom, A.M. (1975). *Statistical Fluid Mechanics: Mechanics of Turbulence Vol. 2*. Cambridge MA: MIT Press.

Mordant, N., Metz, P., Michel, O. and Pinton, J.-F. (2001). Measurement of Lagrangian velocity in fully developed turbulence. *Phys. Rev. Lett.*, **87**, 214501-1–4.

Mordant, N., Delour, J., Lévêque, E., Arnéodo, A. and Pinton, J.-F. (2002). Long time correlations in Lagrangian dynamics: A key to intermittency in turbulence. *Phys. Rev. Lett.*, **89**, 254502-1–4.

Mordant, N., Lévêque, E. and Pinton, J.-F. (2004a). Experimental and numerical study of the Lagrangian dynamics of high Reynolds turbulence. *New J. Phys.*, **6**, 116-1–44.

Mordant, N., Crawford, A.M. and Bodenschatz, E. (2004b). Experimental Lagrangian acceleration probability density function measurement. *Physica D*, **193**, 245–251.

Naso, A., Pumir, A. and Chertkov, M. (2007). Statistical geometry in homogeneous and isotropic turbulence., *J. Turbul.* **8**, N39, 1–13.

Naso, A. and Prosperetti, A., (2010). The interaction between a solid particle and a turbulent flow. *New J. Phys.*, **12**, 033040-1–20.

Nelkin, M. (1990). Multifractal scaling of velocity derivatives in turbulence. *Phys. Rev. A*, **42**, 7226–9.

Obukhov, A.M. (1941). On the distribution of energy in the spectrum of turbulent flow. *Izv. Akad. Nauk SSSR, Ser. Geogr. Geofiz.* **5**, 453–468.

Ott, S. and Mann, J. (2000). An experimental investigation of the relative diffusion of particle pairs in three-dimensional turbulent flow. *J. Fluid Mech.*, **422**, 207–223.

Ott, S. and Mann, J. (2005). An experimental test of Corrsin's conjecture and some related ideas. *New J. Phys.*, **7, N142**, 1–24.

Ouellette, N.T., Xu, H., Bourgoin, M. and Bodenschatz, E. (2006). Small-scale anisotropy in Lagrangian turbulence. *New J. Phys.*, **8**, 102–1–10.

Overholt, M.R. and Pope, S.B. (1996). Direct numerical simulation of a passive scalar with imposed mean gradient in isotropic turbulence. *Phys. Fluids*, **8**, 3128–3148.

Pagnini, G. (2008). Lagrangian stochastic models for turbulent relative dispersion based on particle pair rotation. *J. Fluid Mech.*, **616**, 357–395.

Pasquill, F. and Smith, F.B. (1993). *Atmospheric Diffusion*. Chichester: Ellis Horwood.

Pope, S.B. and Chen, Y.L. (1990). The velocity-dissipation probability density function model for turbulent flows. *Phys. Fluids A*, **2**, 1437–1449.

Pope, S.B. (1985). PDF methods for turbulent reactive flows. *Prog. Energy Combust. Sci.*, **11**, 119–192.

Pope, S.B. (1994). Lagrangian PDF methods for turbulent flows. *Annu. Rev. Fluid Mech.*, **26**, 23–63.

Pope, S. B. (1998). The vanishing effect of molecular diffusivity on turbulent dispersion: implications for turbulent mixing and the scalar flux. *J. Fluid Mech.*, **359**, 299–312.

Pope, S.B. (2000). *Turbulent Flows*. Cambridge: Cambridge University Press.

Pumir A., Shraiman, B.I. and Chertkov M. (2000). Geometry of Lagrangian dispersion in turbulence. *Phys. Rev. Lett.*, **85**, 5324–5327.

Pumir, A., Shraiman, B.I. and Chertkov M. (2001). The Lagrangian view of energy transfer in turbulent flow. *Europhys. Lett.*, **56**, 379–385.

Qureshi, N.M., Bourgoin, M., Baudet, C., Cartellier, A. and Gagne, Y. (2007). Turbulent transport of material particles: An experimental study of finite size effects. *Phys. Rev. Lett.*, **99**, 184502–1–4.

Qureshi, N.M., Arrieta, U., Baudet, C., Cartellier, A., Gagne, Y. and Bourgoin, M. (2008). Acceleration statistics of inertial particles in turbulent flow. *Eur. Phys. J. B*, **66**, 531–536.

Rast, M. and Pinton, J.-F. (2009). Point-vortex model for Lagrangian intermittency in turbulence. *Phys. Rev. E*, **79**, 046314–1–4.

Reade, W.C. and Collins, L.R. (2000). Effect of preferential concentration on turbulent collision rates. *Phys. Fluids*, **12**, 2530–2540.

Reynolds, A.M. (2003). Superstatistical mechanics of tracer-particle motions in turbulence. *Phys. Rev. Lett.*, **91**, 084503–1–4.

Richardson, L.F. (1926). Atmospheric diffusion shown on a distance neighbour graph. *Proc. Roy. Soc. Lond. A*, **110**, 709–737.

Rodean, H.C. (1996). *Stochastic Lagrangian Models of Turbulent Diffusion*. Meteorological Monographs, Vol. 26, No. 48, Boston: Am. Meteor. Soc.

Salazar, J.P.L.C., de Jong, J., Cao, L., Woodward, S.H., Meng, H. and Collins, L.R. (2008). Experimental and numerical investigation of inertial particle clustering in isotropic turbulence. *J. Fluid Mech.*, **600**, 245–256.

Salazar, J.P.L.C. and Collins, L.R. (2008). Two-particle dispersion in isotropic turbulent flows. *Annu. Rev. Fluid Mech.*, **41**, 405–432.

Saffman, P.G. (1960). On the effect of the molecular diffusivity in turbulent diffusion. *J. Fluid Mech.*, **8**, 273–283.

Saffman, P.G. (1963). On the fine-scale structure of vector fields convected by a turbulent fluid. *J. Fluid Mech.*, **16**, 545–572.

Saw, E.W., Shaw, R.A., Ayyalasomayajula, A., Chuang, P. and Gylfarson, A. (2008). Inertial clustering of particles in high-Reynolds-number turbulence. *Phys. Rev. Lett.*, **100**, 214501-1–4.

Sawford, B.L. (1985). Lagrangian statistical simulation of concentration mean and fluctuation fields. *J. Climat. Appl. Meteorol.*, **24**, 1152–1166.

Sawford, B.L. (1991). Reynolds number effects in Lagrangian stochastic models of dispersion. *Phys. Fluids*, **A3**, 1577–1586.

Sawford, B.L. (1993). Recent developments in the Lagrangian stochastic theory of turbulent dispersion. *Boundary-Layer Meteorol.*, **62**, 197–215.

Sawford, B.L. (2001). Turbulent relative dispersion. *Annu. Rev. Fluid Mech.*, **33**, 289–317.

Sawford, B.L. (2004). Micro-mixing modelling of scalar fluctuations for plumes in homogeneous turbulence, *Flow, Turb. Combust.*, **72**, 133–160.

Sawford, B.L. (2006). A study of the connection between exit-time statistics and relative dispersion using a simple Lagrangian stochastic model. *J. Turbul.*, **7**, (13), 1–10.

Sawford B.L. and Yeung, P.K. (2001). Lagrangian statistics in uniform shear flow: Direct numerical simulation and Lagrangian stochastic models. *Phys. Fluids*, **13**, 2626–2634.

Sawford, B.L. and Guest, F.M. (1991). Lagrangian statistical simulation of the turbulent motion of heavy particles. *Boundary-Layer Meteorol.*, **54**, 147–166.

Sawford, B.L., Yeung, P.K., Borgas, M.S., Vedula, P., La Porta, A., Crawford, A.M. and Bodenschatz, E. (2003). Conditional and unconditional acceleration statistics in turbulence. *Phys. Fluids*, **15**, 3478–3489.

Sawford, B.L., Yeung, P.K. and Borgas, M.S. (2005). Comparison of backwards and forwards relative dispersion in turbulence. *Phys. Fluids*, **17**, 095109-1–9.

Sawford, B.L., Yeung, P.K. and Hackl J. F. (2008). Reynolds number dependence of relative dispersion statistics in isotropic turbulence. *Phys. Fluids*, **20**, 065111-1–13.

Shen P. and Yeung, P.K. (1997). Fluid particle dispersion in homogeneous turbulent shear flow. *Phys. Fluids*, **9**, 3472.

Shlien, D.J. and Corrsin, S., (1974). A measurement of Lagrangian velocity autocorrelation in approximately isotropic turbulence. *J. Fluid Mech.*, **62**, 255–271.

Shraiman, B. and Siggia, E. (1995). Anomalous scaling of a passive scalar in turbulent flow. *C. R. Acad. Sci. Ser. II*, **321**, 279–284.

Shraiman, B.I. and Siggia, E.D. (2000). Scalar turbulence. *Nature*, **405**, 639–646.

Smyth, W.D. (1999). Dissipation-range geometry and scalar spectra in sheared stratified turbulence. *J. Fluid Mech.*, **401**, 209–242.

Snyder, W.H. and Lumley, J.L. (1971). Some measurements of particle velocity autocorrelation functions in a turbulent flow. *J. Fluid Mech.*, **48**, 41–71.

Squires, K.D. and Eaton, J.K. (1991a). Measurements of particle dispersion obtained from direct numerical simulations of isotropic turbulence. *J. Fluid Mech.*, **226**, 1–35.

Squires, K.D. and Eaton, J.K. (1991b). Preferential concentration of particles by turbulence. *Phys. Fluids A*, **3**, 1169–1178.

Sundaram, S. and Collins, L.R. (1997). Collision statistics in an isotropic particle-laden turbulent suspension. Part 1. Direct numerical simulations. *J. Fluid Mech.*, **335**, 75–109.

Tabeling, P. (2002). Two-dimensional turbulence: a physicist approach. *Phys. Rep.*, **362**, 1–62.

Taylor, G.I. (1921). Diffusion by continuous movements. *Proc. London Math. Soc. Ser. 2*, **20**, 196–211.

Tennekes, H. and Lumley, J.L. (1972). *A First Course in Turbulence*. Cambridge MA: MIT.

Thomson, D.J. (1987). Criteria for the selection of stochastic models of particle trajectories in turbulent flows. *J. Fluid Mech.*, **180**, 529–556.

Thomson, D.J. (1990). A stochastic model for the motion of particle pairs in isotropic high-Reynolds-number turbulence, and its application to the problem of concentration variance. *J. Fluid Mech.*, **210**, 113–153.

Toschi, F. and Bodenschatz, E. (2009). Lagrangian properties of particles in turbulence. *Annu. Rev. Fluid Mech.*, **41**, 375–404.

Variano, E.A., Bodenschatz, E. and Cowen, E.A. (2004). A random synthetic jet array driven turbulence tank. *Exp. Fluids*, **37**, 613–615.

Vedula, P. and Yeung, P.K. (1999). Similarity scaling of acceleration and pressure statistics in numerical simulations of isotropic turbulence. *Phys. Fluids*, **11**, 1208–1220.

Villermaux, J. and Devillon, J.C. (1972). In *Proceedings of the Second International Symposium on Chemical Reaction Engineering*, New York: Elsevier.

Viswanathan, S. and Pope, S.B. (2008). Turbulent dispersion from line sources in grid turbulence. *Phys. Fluids*, **20**, 101514–1–25.

Volk, R., Calzavarini, E., Verhillea, G., Lohse, D., Mordant, N., Pinton, J.-F. and Toschi, F. (2008a). Acceleration of heavy and light particles in turbulence: Comparison between experiments and direct numerical simulations. *Physica D*, **237**, 2084–2089.

Volk, R., Mordant, N., Verhille, G. and Pinton, J.-F. (2008b). Measurement of particle and bubble acceleration in turbulence. *Europhys. Lett.*, **81**, 34002.

Volk, R., Calzavarini, E., Lévêque, E. and Pinton, J.-F. (2011). Dynamics of inertial range particles in a turbulent flow, *J. Fluid Mech.*, **668**, 223–235.

Voth, G.A., Satyanarayan, K. and Bodenschatz, E. (1998). Lagrangian acceleration measurements at large Reynolds numbers. *Phys. Fluids*, **10**, 2268–2280.

Voth, G.A., La Porta, A., Crawford, A.M., Alexander J. and Bodenschatz, E. (2002). Measurement of particle accelerations in fully developed turbulence. *J. Fluid Mech.*, **469**, 121–160.

Wang, L-P. and Maxey, M.R. (1993). Settling velocity and concentration distribution of heavy particles in homogeneous isotropic turbulence. *J. Fluid Mech.*, **256**, 21–68.

Wells, M.R. and Stock, D.E. (1983). The effects of crossing trajectories on the dispersion of particles in a turbulent flow. *J. Fluid Mech.*, **136**, 31–62.

Wilczek, M., Jenko, M. and Friedrich, R. (2008). Lagrangian particle statistics in turbulent flows from a simple vortex model. *Phys. Rev. E*, **77**, 056301–1–6.

Wilson, J.D. and Sawford, B.L. (1996). Review of Lagrangian stochastic models for trajectories in the turbulent atmosphere. *Boundary-Layer Meteorol.*, **78**, 191–210.

Xu, H., Bourgoin, M., Ouellette, N.T. and Bodenschatz, E. (2006). High order Lagrangian velocity statistics in turbulence. *Phys. Rev. Lett.*, **96**, 024503–1–4.

Xu, H., Ouellette, N. T. and Bodenschatz, E. (2007). Curvature of Lagrangian trajectories in turbulence. *Phys. Rev. Lett.*, **98**, 050201–1–4.

Xu, H., Ouellette, N.T. and Bodenschatz, E. (2008). Evolution of geometric structures in intense turbulence. *New J. Phys.*, **10**, 013012–1–9.

Yeung, P.K. (2001). Lagrangian characteristics of turbulence and scalar transport in direct numerical simulations, *J. Fluid Mech.*, **427**, 241–274.

Yeung, P.K. (2002). Lagrangian investigations of turbulence, *Annu. Rev. Fluid Mech.*, **34**, 115–42.

Yeung, P.K. and Borgas, M.S. (1997). Molecular path statistics in turbulence: simulation and modeling. Proceedings of the First AFOSR International Conference on Direct Numerical Simulation and Large Eddy Simulation (DNS/LES), Ruston, LA: Louisiana Tech University.

Yeung, P.K. and Borgas, M.S. (2004). Relative dispersion in isotropic turbulence: Part 1. Direct numerical simulations and Reynolds number dependence. *J. Fluid Mech.*, **503**, 93–124.

Yeung, P.K. and Pope, S.B. (1989). Lagrangian statistics from direct numerical simulations of isotopic turbulence. *J. Fluid Mech.*, **207**, 531.

Yeung P.K., Pope, S.B. and Sawford, B.L. (2006a). Reynolds number dependence of Lagrangian statistics in large numerical simulations of isotropic turbulence. *J. Turbul.*, **7**, N58, 1–12.

Yeung, P.K., Pope, S.B., Lamorgese, A.G. and Donzis, D.A. (2006b). Acceleration and dissipation statistics of numerically simulated isotropic turbulence. *Phys. Fluids*, **18**, 065103–1–14.

Yeung, P.K., Pope, S.B., Kurth, E.A. and Lamorgese, A.G. (2007). Lagrangian conditional statistics, acceleration and local relative motion in numerically simulated isotropic turbulence. *J. Fluid Mech.*, **582**, 399–422.

Zimmermann, R., Xu, H., Gasteuil, Y., Bourgoin, M., Volk, R., Pinton, J.-F. and Bodenschatz, E. (2011). The Lagrangian exploration module: An apparatus for the study of statistically homogeneous and isotropic turbulence. *Rev. Sci. Instr.*, **81**, 055112–1–8.

Zhuang, Y., Wilson, J.D. and Lozowski, E.P. (1989). A trajectory-simulation model for heavy particle motion in turbulent flow. *J. Fluids Eng.*, **111**, 492–494.

5
The Eddies and Scales of Wall Turbulence

Ivan Marusic and Ronald J. Adrian

The wide ranges of length and velocity scales that occur in turbulent flows make them both interesting and difficult to understand. The length scale range is widest in high Reynolds number turbulent wall flows where the dominant contribution of small scales to the stress and energy very close to the wall gives way to dominance of larger scales with increasing distance away from the wall. Intrinsic length scales are defined statistically by the two-point spatial correlation function, the power spectral density, conditional averages, proper orthogonal decomposition, wavelet analysis and the like. Before two- and three-dimensional turbulence data became available from PIV and DNS, it was necessary to attempt to infer the structure of the eddies that constitute the flow from these statistical quantities or from qualitative flow visualization. Currently, PIV and DNS enable quantitative, direct observation of structure and measurement of the scales associated with instantaneous flow patterns. The purpose of this chapter is to review the behavior of statistically-based scales in wall turbulence, to summarize our current understanding of the geometry and scales of various coherent structures and to discuss the relationship between instantaneous structures of various scales and statistical measures.

5.1 Introduction

Wall-bounded turbulent flows are pertinent in a great number of fields, including geophysics, biology, and most importantly engineering and energy technologies where skin friction, heat and mass transfer, flow-generated noise, boundary layer development and turbulence structure are often critical for system performance and environmental impact. Wall turbulence is also perhaps the most intriguing of the generic inhomogeneous turbulent flows, as it is here that the panoply of scales is richest, owing to the progression from large to small that occurs from the edge or center of such flows to the wall, and the fact that the ratio of large-scale to small-scale lengths increases without bound with increasing Reynolds number.

Among the reviews of wall turbulence in the literature (Townsend 1976, Sreenivasan 1989, Robinson 1991, Gad-el-Hak & Bandyopadhyay 1994, Fernholz & Finley 1996, Smits & Dussauge 1996, Adrian 2007, Marusic et al. 2010, Klewicki 2011, Smits et al. 2011, and Jimenez & Kawahara 2012), one can find many different approaches and viewpoints, emphasizing the, as yet, unsettled nature of this subject. Often in science, differing views are neither entirely right nor entirely wrong, the correct answer being found in a unification that is only possible after the conflicts are resolved. In the context of turbulence this means bringing together approaches based on scaling theories, statistical analysis, stability analysis and coherent structure models. This chapter attempts to contribute to this process by describing what we know about the organized (coherent) motions and eddies of wall turbulence and their scales of length and velocity in terms of well known statistical quantities. This process rests on the kinematic properties of the motions, and it does not address matters such as the origins and interactions of coherent structures, which we shall refrain from discussing in detail.

Before proceeding further it is appropriate to give some introductory comments on the concepts of scales and eddies.

5.1.1 Scales

We distinguish two types of scales: extrinsic and intrinsic. Extrinsic scales are variables that come from the boundary conditions, initial conditions, applied forces and fluid properties of the flow. Examples are the pipe radius, channel half-height, free-stream velocity and viscosity. They can, in principle, be varied at the will of the experimentalist or computational simulator. Intrinsic scales are determined by the flow's response to extrinsic conditions. For example, the friction velocity or the momentum thickness of the boundary layer are intrinsic scales.

Both extrinsic and intrinsic scales are obviously important, but the issues remain as to what the relevant scales are for a meaningful description of the eddying motions, and what the important scales are for *scaling* of the turbulence statistics. In turbulence, one aim is to obtain scaling laws that will have a predicative capability. For wall turbulence, as will be discussed in the next section, there remains considerable debate over the correct scaling of even the mean velocity, let alone higher-order moment statistics.

5.1.2 Eddies

One approach towards understanding turbulence is to attempt to break the complex, random field of turbulent motions into elementary recurring motions, which can then be considered as the key building-blocks or components of the turbulent flow. These motions are somewhat loosely and variously referred to as *eddies, organized motions* or *coherent structures*, and they have been the focus of many studies for over four decades (Townsend 1976; Cantwell 1981; Hussain 1986; Adrian 2007). This resembles the atomistic approach, which has enjoyed enormous success in chemistry and physics. But, it is clear that it cannot be so simple when applied to continuous, random fluid motions, especially when they range over many scales. The difficulty presents itself in the very first step of defining what exactly constitutes an eddy or a coherent structure, and it must be confessed that here there are no unequivocal definitions. Worse, inconsistent usages are rife in the literature, so that commonly agreed upon meanings do not exist. We will not attempt to solve this problem here, but for the purposes of clarity within this chapter we will discuss some of the properties associated with the terms used.

A turbulent structure is characterized by the scales and pattern of its motion, as manifested in the velocity field, the vorticity field or, less desirably, the pattern of a dye or marker. The terms 'structure' and 'motion' imply very little about the flow other than that it has some pattern capable of description. Thus, 'motion' will be used when very little is known about the flow element, as for example in 'very large scale motions' whose structure was quite unknown for many years after their discovery.

We take our meaning for 'eddies' from the usage of Townsend (1956, 1976) who, more than any other, pursued a structural description of wall turbulence. Townsend (1976) defined eddies to be "flow patterns with spatially limited distributions of vorticity and comparatively simple forms". Eddies, like the 'whorls' in the poem by L. F. Richardson, possess rotation. In its most common usage an eddy is a tube of vorticity bundle having a dimension along the axis of rotation that is long compared to its diameter, but we shall not make this restriction unless specifically stated. (It is not clear if Townsend's intent in requiring vorticity of limited spatial extent is to make the eddy have finite, measurable size, or if it is to make the law of Biot-Savart more applicable.) We append two additional properties to Townsend's definition. First, eddies should be 'typical' in the sense that they occur in regions of space-time large enough to make significant contributions to statistical measures such as the mean, root mean square or spectrum. In other words, they should be worth identifying and analyzing because they are important to the turbulence. Second, they should be recognizable for a time that is comparable to or greater than the eddy turn-over time defined by the ratio of the eddy diameter divided by its rotational velocity. Put another way, the pattern should remain temporally coherent for at least one rotation of the eddy. If the pattern destroys itself in less than one rotation, the flow is not usefully described as an eddy.

'Vortices' and 'eddies' are often used as interchangeable terms, but it seems useful to use vortex more precisely. Following Chakraborty *et al.* (2005) we shall use 'vortex filament' to mean a long, thin bundle of vortex lines, in distinction with other common rotational flows such as a two-dimensional vortex sheet or a uniform shear flow. Following common usage, we shall take vortex to mean vortex filament. This is not universal, but it is very common in turbulence where, for example, no one would refer to a vortex sheet or a shear flow as a vortex. Locally, the flow around a vortex filament is a spiraling pattern in planes perpendicular to the axis of the filament. Such patterns can be identified by a variety of methods, most of them based on the velocity gradient tensor.

'Coherent structures' and 'organized motions' are essentially synonymous terms. The dictionary defines 'coherent' as 'being marked by an orderly or logical relation of parts that afford comprehension or recognition'. This is exactly what we want the concept of coherent structures to do for us. Comprehension or recognition can be afforded by a structure having properties that correspond to elementary flow fields that we understand very well, such as saddle points or vortices. Coherent structures need not have vorticity. For example early studies of coherent structures in wall turbulence focused on the so-called VITA events, which were later shown to be associated with irrotational saddle points that are now known to rest between the hairpin vortices that occur in packets. Hence, a coherent structure may contain regions of rotational and irrotational flow. And like eddies, coherent structures must make significant contributions to the statistics of the flow and be coherent in space and time.

With these definitions, eddies, vortices, coherent structures and incoherent structures are all motions; vortices and eddies are subsets of coherent structures; and vortices are a subset of eddies.

5.2 Background

5.2.1 Canonical wall-bounded turbulent flows

Wall-bounded turbulent flows manifest themselves in a variety of configurations depending on the nature of the imposed external pressure gradients and strain-rates, the curvature and complexity of the wall geometry, etc. Here we consider what are often referred to as the *canonical* flows, where the term canonical refers to the simplicity of the geometry and not necessarily the complexity of the associated physics. The three canonical wall-bounded turbulent flows are the flat-plate zero-pressure-gradient boundary layer flow, fully-developed channel flow and fully-developed pipe flow. A schematic showing cross-sectional (spanwise/wall-normal) views of the instantaneous fluctuating streamwise velocity for each of the three flows is shown in Figure 5.1. The pipe and channel flows are also often called 'internal' flows, since the turbulent flow is confined within a closed geometry, while the boundary layer flow is unconfined, always developing and not quite unidirectional. Figure 1 highlights the essential differences in the outer-flow region (away from the walls) between the flows. The pipe and channel flows contain turbulent flow throughout the closed geometry, while the boundary layer has a significant region of intermittent turbulence, i.e. nominally irrotational, free-stream fluid interspersed with rotational turbulent flow.

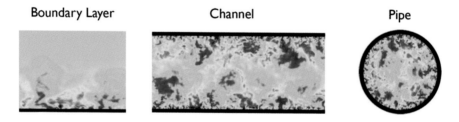

Figure 5.1 Comparison of representative cross-sections for the canonical wall-bounded flows. The bulk flow is out of the page. Here, $\delta^+ = 1010$, $h^+ = 934$ and $R^+ = 1180$, respectively. Figure adapted from Monty et al. (2009).

Here we only consider flows that are incompressible, and the reader is referred to Smits & Dussauge (1996) and Elsinga et al. (2010) for reviews of compressibility effects. Further specifics for each of the canonical flows are given next.

Pipe and Channel Flows

Channel and pipe flows have two homogenous directions (streamwise-spanwise for channels, and streamwise-azimuthal for pipes) and a fixed outer-length scale (h, the channel half-height, and R, the pipe radius). These flows are regarded as *fully developed* when the mean flow quantities (that is, velocity field and pressure gradient) and all turbulence quantities (i.e., Reynolds stresses, spectra, skewness, flatness, etc.) become independent of streamwise location. Invariably, the lower the order of the statistic, the shorter the required development length. For this reason, there have been significant differences in the lengths of experimental facilities used in the past. For example, Nikuradse (1932) used a pipe of length $80R$, while Perry, Henbest & Chong (1986) used a pipe of length $796R$. Based on a survey of recent experiments by Lien et al. (2005), Doherty et al. (2007) and others, Marusic et al. (2010) conclude that for statistics up to fourth order (e.g. flatness) development length should exceed $160R$ for pipes and $260h$ for channels. An additional parameter for channel flows is the width-to-height aspect ratio of the cross-section. It should be large enough to ensure that side-wall influences can be ignored (typically greater than 10).

For pipes and channels the extrinsic scales are U_B, the bulk velocity, R (or h), the outer-length scale, and ν, the kinematic viscosity of the fluid. When the pipe and channel flows are fully developed the skin-friction is balanced by a favorable pressure gradient:

$$\tau_w = -h\frac{dP_w}{dx}; \qquad 2\tau_w = -R\frac{dP_w}{dx}, \qquad (5.1)$$

and the streamwise momentum equation is

$$0 = -\frac{1}{\rho}\frac{dP_w}{dx} + \nu\frac{d^2U}{dy^2} - \frac{d\overline{uv}}{dy} \qquad (5.2)$$

for channel flow, and

$$0 = -\frac{1}{2\rho}\frac{dP_w}{dx} + \nu\frac{d^2U}{dy^2} - \frac{d\overline{uv}}{dy} \qquad (5.3)$$

for pipe flow. Here τ_w is mean wall-shear stress, P_w is the mean pressure at the wall, ρ is fluid density, and throughout this chapter u, v and w are the fluctuating velocities in the streamwise (x), wall-normal (y) and spanwise (z) directions respectively. U and V are the mean (ensemble averaged) streamwise and wall-normal velocities respectively, and overbars indicate ensemble averaged quantities.

Flat plate boundary layers

The flat plate boundary layer consists of a semi-infinite plate that is aligned with a flow of constant free-stream velocity, resulting in a mean zero pressure gradient (ZPG). Experimentally, the free-stream needs to have a sufficiently low turbulence intensity that it does not influence the statistics of the boundary layer. Coles (1962) showed that free-stream turbulence intensities exceeding 0.5% have a noticeable influence in the outer mean velocity profile, and therefore typically an upper limit of 0.3% is used when validating wind tunnel or water channel facilities for boundary layer investigations.

For ZPG boundary layers the extrinsic scales are U_∞, the free-stream velocity, x and ν. The streamwise momentum equation is given by

$$U\frac{\partial U}{\partial x} + V\frac{\partial U}{\partial y} = \nu\frac{\partial^2 U}{\partial y^2} - \frac{\partial \overline{uv}}{\partial y}, \qquad (5.4)$$

where the pressure gradient term has been set identically to zero. This is impossible to achieve in experiments and typically boundary layers are regarded as ZPG when the coefficient of pressure varies less than ± 0.01 (maximum of 1% variation in U_∞) over the length of the plate with no discernible pressure gradient trend within this region. Recently, Chauhan et al. (2008) have suggested even more restrictive requirements with the free-stream velocity varying less than 0.5% over the length of the plate.

5.2.2 Classical layers based on the mean velocity

Clauser (1956), Coles & Hirst (1969), Sreenivasan (1989), Gad-el-Hak & Bandyopadhyay (1994) and others present comprehensive reviews of what is now commonly referred to as "classical" scaling. In this view, which is largely related to the mean velocity behaviour, the boundary layer is held to be composed of two principal regions that follow distinct scalings: a near-wall region where viscosity is important, and an outer region where it is not. The inner scales are taken to be $y^* = \nu/U_\tau$ and $U_\tau = \sqrt{\tau_w/\rho}$, where τ_w is the wall shear stress. Therefore, with inner scaling

$$y^+ = \frac{y}{y^*}, \quad U^+ = \frac{U}{U_\tau}, \quad t^+ = \frac{tU_\tau}{y^*} \quad \text{etc.}$$

In the outer region, it is assumed that the appropriate length scale is the boundary layer thickness δ (taken to be h or R in the cases of channels and pipes respectively), or a scale related to δ. The classical velocity scale continues to be U_τ, since U_τ sets up the inner boundary condition for the outer flow. The von Karman number, or friction Reynolds number, is the ratio of the inner and outer length scales. That is, $Re_\tau = \delta/y^*$. The other important length scales from the mean velocity are δ^*, the displacement thickness, and θ, the momentum thickness.

In the classical description the inner and outer regions are divided into four layers,

- viscous sublayer: $y^+ < 5$; $U^+ = y^+$
- buffer layer: $5 < y^+ < 30$; $U^+ = f_1(y^+)$
- log. layer: $30 < y^+ < 0.15 Re_\tau$; $U^+ = \kappa^{-1}\ln y^+ + A$;
  $(U_\infty - U)^+ = -\kappa^{-1}\ln y/\delta + B$

- wake layer: $\qquad\qquad y^+ > 0.15 Re_\tau; \qquad\qquad (U_\infty - U)^+ = f_2(y/\delta)$

and the locations where they appear across a turbulent boundary layer are indicated for a given experiment in Figure 5.2. The viscous sublayer is the region where a linear profile of velocity with distance from the wall exists to a leading order, while the buffer region is the viscous dominated region coincident with the peak production of turbulence and consequently also the peak streamwise turbulence intensity (as seen in the figure). The log layer is the common part of the inner and outer regions, and the wake layer is the outer most part of the boundary layer where viscosity is assumed not to play any role for the mean relative and turbulent energy containing motions (Townsend 1956, 1976; Rotta 1962).

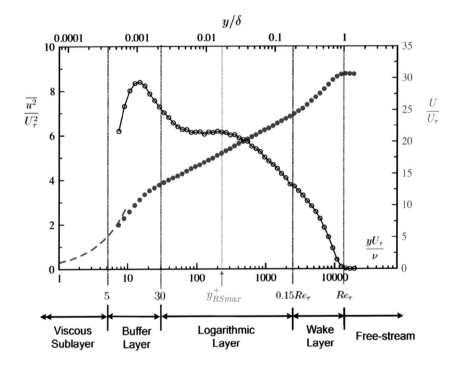

Figure 5.2 The four classical layers of wall turbulence, shown labelled with the mean velocity and streamwise turbulence intensity profiles for a ZPG turbulent boundary layer at $Re_\tau = 13600$. (Data from Hutchins et al. 2009). The red line indicates the location of the peak Reynolds shear stress for this case.

The values given above, and on Figure 5.2, for the start and end of the four layers are nominal, and there are considerable differences in estimates from various investigators. For example, estimates for the start of the logarithmic region and the end of the buffer region range from the classically accepted value of 30 (e.g. Spalart 1988, and many others) to 200–600, based on more stringent modern assessments (Nagib *et al.* 2007; McKeon *et al.* 2004). (In addition, there are a number of theories based on the existence of an additional meso-layer or meso-layers, which are reviewed by Klewicki *et al.* 2009 and Marusic *et al.* 2010). The lower estimate of $y^+ = 30$ for the start of the log layer is mainly from curve-fits of the log law to the mean velocity profiles as shown in Figure 5.2, while the upper estimates of 200–600 are based on studies of the mean velocity wall-normal derivative and from analysis of the Princeton *superpipe* (Zagarola & Smits 1998) data respectively. However, it is noted that the $y^+ \approx 200$–600 estimates appear to be consistent with the location where $\overline{u^2}^+$ begins to also exhibit a logarithmic profile as predicted from Townsend's (1976) attached eddy hypothesis (see Perry *et al.* 1986, Marusic & Perry 1995) for sufficiently high Reynolds numbers, as seen in Figure 5.2.

Scaling

Here, we only briefly mention the types of scaling that have been proposed for wall turbulence. A more comprehensive survey can be found in a recent review by Marusic *et al.* (2010) and Klewicki (2010).

The classical scaling is as described above with the mean velocity following a law of the wall, law of the wake description (Coles 1956)

$$U^+ = f(y^+) + \Pi \, g(y/\delta). \tag{5.5}$$

The parameter Π is the Coles wake factor, and the classical arguments lead to a logarithmic variation in the mean velocity profile. Several extensions to this classical approach have been proposed. Wei *et al.* (2005) argue, based on the ratio of viscous and Reynolds stress gradients, that the classical inner and outer layers need to be supplemented by intermediate 'meso-layers' (see also Klewicki *et al.* 2007); this is an elaboration of the same concept formulated earlier by Long & Chen (1981), Afzal (1982), and Sreenivasan (1987). In particular, Wei *et al.* propose a layer centered on the location of maximum Reynolds stress y^+_{RSmax} for which they derive the scaling $y^+_{RSmax} \sim (Re_\tau)^{1/2}$, in accord with Sreenivasan & Sahay (1997). A second type of extension has been proposed by Wosnik, Castillo & George (2000) for pipe and channel flows. Using near-asymptotics with the friction velocity as the outer velocity scale, they derived a log law, albeit with an extra additive constant for y^+ and a weak Reynolds number dependence for κ, which corresponds, at least qualitatively, to the variation seen in some experiments at low Reynolds number. Other studies have also noted the possibility of a log law with a shifted origin, including Oberlack (2000), Lindgren *et al.* (2004) and Spalart, Coleman & Johnstone (2008).

Apart from extensions to the classical description, there are also alternative formulations to the problem, the main one being a power-law representation. Barenblatt (1993) showed that if Reynolds number effects persist at all Reynolds numbers the similarity is incomplete and, as a consequence, the mean velocity follows power-laws with Reynolds-number-dependent exponents for pipes and channels, and (with further attributes) also in boundary layers. George & Castillo (1997) also obtain power laws for boundary layers but this is based on asymptotic arguments and the requirement that U_∞, and not U_τ, is the appropriate velocity scale in the outer region. The debate over power-law versus log law may continue until very clear differences can be shown in high-fidelity experimental data at high Reynolds numbers. The real difficulty is that experiments are unlikely to ever reveal asymptotic conditions for a boundary layer. Indeed, one can evaluate the viability of an asymptotic theory only in the context of finite-Reynolds-number corrections, which no theory has satisfactorily produced thus far. Recent efforts by Monkewitz *et al.* (2007, 2008) have tackled this problem and compared a large number of high Reynolds number experimental data to the classical scaling and the two main power law theories, and conclude that the log law is empirically superior. Apart from U_∞ and U_τ as candidate velocity scales for the outer region, another is the "Zagarola & Smits" outer velocity scale $(U_\infty - U)$, however for sufficiently high Reynolds number this becomes proportional to U_τ.

It is clear that while the log law finds support, there are still a number of unanswered questions, including the robustness of the log law, its range of application and the universality of its constants (Nagib & Chauhan 2008; Bourassa & Thomas 2009), and these are the subject of ongoing studies.

Apart from the mean flow, high Reynolds number studies over the past few years have revealed that the scaling behaviour of the Reynolds stresses and other second-order statistics are considerably more complicated than traditionally believed. For instance, wall-scaling is a standard approach in RANS-based CFD where all components of the Reynolds stress tensor are assumed to scale with inner variables in the near-wall region. That is, $\overline{u_i u_j}^+ = f_{ij}(y^+)$ for the logarithmic region and below. However, many studies (Metzger & Klewicki 2001, Marusic & Kunkel 2003, Hutchins & Marusic 2007, Jimenez & Hoyas 2008 and others) in boundary layers and channels now suggest that in the inner region

$$\overline{u^2}^+ = f_{11}(y^+, Re_\tau), \ \overline{w^2}^+ = f_{33}(y^+, Re_\tau), \ \overline{v^2}^+ = f_{22}(y^+), \ -\overline{uv}^+ = f_{12}(y^+). \tag{5.6}$$

(Recent experiments by Hultmark et al. 2010 suggest that this may not hold for pipes, so it, too, is subject to further study.)

The peak value of $\overline{u^2}^+$, as seen at $y^+ \approx 15$ in Figure 5.2, and the observation that it increases (weakly) with increasing Reynolds number are of particular interest. Degraaff & Eaton (2000) were able to collapse their experimental data with a "mixed" scaling (see also, Johansson & Alfredsson 1982). That is, $\overline{u^2}/(U_\infty U_\tau)$ instead of $\overline{u^2}^+$. This may be interpreted as a U_∞ influence near the wall, or as argued by Marusic & Kunkel (2003), this is equivalent to an Re_τ logarithmic dependence for $\overline{u^2}^+$, which is consistent with equation (5.6) and the Townsend (1976) attached eddy hypothesis.

5.2.3 Scales from turbulence statistics

In addition to the length and velocity scales obtained from the mean velocity, there are a large number of other notable scales that have been proposed from the analysis of second and higher-order moment statistics. We list some of the main ones in Table 5.1.

Reynolds stresses $\overline{u_i u_j}$:	Root-mean-squares of normal stresses: $(\overline{u_i^2})^{1/2}$
	Friction velocity: U_τ
	y^+_{RSmax}: location of max. Reynolds shear stress
	y^+ location of max. production $\mathcal{P} = -\overline{uv}(\partial U/\partial y)$
	y^+ location of max. $\overline{u^2}^+$
Spectra/Correlations:	Integral length scales L_{11}, L_{22}, L_{33}
	Taylor microscales λ_{T11}, λ_{T22}
	Kolmogorov scales: η, υ
	u-spectra: Λ_{VLSM}, Λ_{SS}, Λ_{LSM}, Λ_{diss}
	Structure angle θ_S

Table 5.1 *Scales from second-moment statistics and correlations.*

Here, the Kolmogorov and Taylor microscales are defined by

$$\eta = \left(\frac{\nu^3}{\epsilon}\right)^{1/4}, \quad \upsilon = (\nu\epsilon)^{1/4}, \quad \frac{\overline{u^2}}{\lambda_{Tij}^2} = \overline{\left(\frac{\partial u_i}{\partial x_j}\right)^2}. \tag{5.7}$$

(There is no sum on i,j in (5.7).) Here, ϵ is the turbulent dissipation rate, which here is usually, perforce, *estimated* from experiments using the isotropic assumptions, namely

$$\epsilon = 15\nu \overline{\left(\frac{\partial u}{\partial x}\right)^2} = 15\nu \int_0^\infty k_1^2 \phi_{11} dk_1 = 15\nu \int_0^\infty k_1^3 \phi_{11} d(\ln k_1), \quad (5.8)$$

where k_1 is streamwise wavenumber and ϕ_{11} is the u-spectrum.

The integral length scales are defined by the area under the autocorrelation curves

$$L_{ij}(\hat{\mathbf{s}}) = \frac{1}{R_{ij}(\mathbf{x},\mathbf{x})} \int_0^\infty R_{ij}(\mathbf{x},\mathbf{x}+s\hat{\mathbf{s}}) ds; \quad R_{ij} = \overline{u_i(\mathbf{x}) u_j(\mathbf{x},\mathbf{x}+s\hat{\mathbf{s}})} \quad (5.9)$$

where here we are concerned with $ij = 11$, 22 or 33, and $(\hat{\mathbf{s}})_i = s\hat{e}_i$ where $i = 1$, 2 or 3 (streamwise, wall-normal, spanwise).

The length scales Λ_{VLSM}, Λ_{SS}, Λ_{LSM} and Λ_{diss} correspond to wavelengths obtained from the fluctuating streamwise velocity spectra as illustrated in Figure 5.3. Following Kim & Adrian (1999) and Hutchins & Marusic (2007), Λ_{VLSM} and Λ_{SS} correspond to the lowest wavenumbers in the flows that carry a significant kinetic energy. "VLSM" indicates *very-large scale motions* in pipes and channels, and "SS" indicates *superstructures* in boundary layers. (The distinction between the superstructures of the turbulent boundary layer and the VLSMs of pipe and channel flow is based somewhat on the shorter length of the superstructures. As more is learned about the properties and origins of these structures, these terms may blur.) "LSM" indicates large-scale motions (Adrian 2007). Λ_{diss} corresponds to the streamwise wavelength that coincides with the peak contribution to dissipation, using the definition given in equation (5.8).

Following Brown & Thomas (1977) and Marusic & Heuer (2007), the structure angle is defined as

$$\theta_S = \arctan(y/\Delta x_m), \quad (5.10)$$

where the terms in the equation are shown schematically in Figure 5.4. Here y is the wall-normal position where velocity is measured, and Δx_m is the spatial delay corresponding to a peak in cross-correlation between the fluctuating velocity and wall shear stress. Most commonly, Δx is estimated by using Taylor's hypothesis of frozen turbulence where a time series is converted to streamwise spatial distance (x) using $U(y)$ as the convection velocity.

5.2.4 Scales of filamentary vortices

As discussed in the Introduction, filamentary vortices have in general a three-dimensional shape characterized by at least three scales - the vortex core

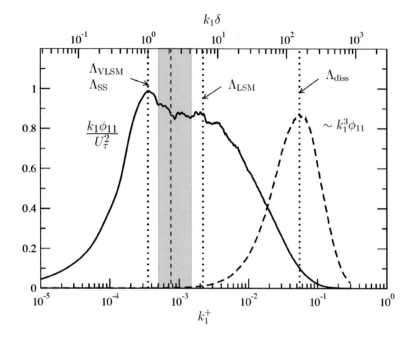

Figure 5.3 Data from ZPG boundary layer at $Re_\tau = 2800$; $y/\delta = 0.055$ (Hutchins et al. 2009). Wavelength scales as indicated for very-large scale motions (VSLM) or superstructures (SS) and large-scale motions (LSM) on the premultiplied u-spectra. Λ_{diss} corresponds to the wavelength that coincides with the peak contribution to dissipation. The red line indicates the demarcation between LSM and VSLM used by Balakumar & Adrian (2007), with shaded area indicating a 50% variation in this wavelength.

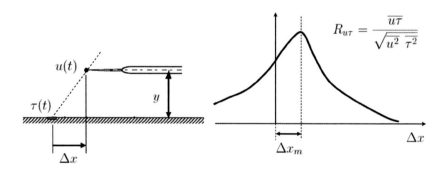

Figure 5.4 Schematic showing the quantities used to define structure angle.

diameter d_c, circulation Γ, and filament length l_f. Determining these scales relies heavily on the criteria one adopts to define a vortex, and the reader is referred to Chakraborty et al. (2005) for a full review and detailed discussion and comparison of various schemes. Many schemes are based on local

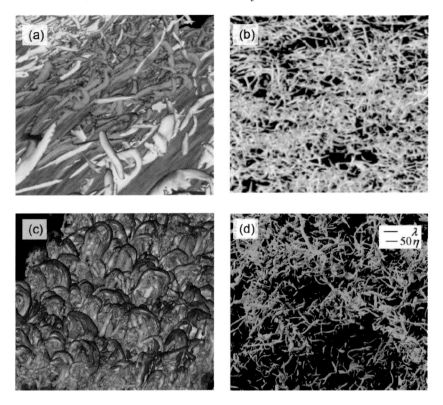

Figure 5.5 Visualization results of filamentary vortices from DNS of turbulent flows. The first three (a-c) are from wall turbulence, while (d) is a simulation of isotropic turbulence in a periodic box. (a) From Ganapathisubramani et al. (2006) - isosurface plot of λ_{ci} in channel flow at $Re_\tau = 950$. (b) From Kang et al. (2007) - isosurfaces of Q in channel flow at $Re_\tau = 1270$. (c) From Wu & Moin (2009) - isosufaces of Q for developing turbulent boundary layer at $R_\theta \approx 800$. (d) From Okamoto et al. (2007) - isosufaces of vorticity in isotropic turbulence at $Re_\lambda = 732$.

or point-wise methods for vortex identification and rely on having planar or volumetric data of the full velocity vector, and often the velocity gradient tensor. Examples include criteria based on Q the second invariant of velocity gradient tensor (Hunt et al. 1988), Δ the discriminant of the velocity gradient tensor (Chong et al. 1990), λ_2 (Jeong & Hussain 1995), λ_{ci} the swirl strength (Zhou et al. 1999), vorticity, enstrophy and additional local and non-local criteria (Chakraborty et al. 2005; Cucitore et al. 1999). Figure 5.5 shows a collage of visualization results from DNS data showing typical filamentary vortices for various criteria relying on components of the velocity gradient tensor.

Other approaches have also been proposed for extracting the scales of filamentary vortices, most being based on templates of vortex signatures and applying these to planar velocity fields. This was done by Carlier & Stanislas (2005) who used an Oseen vortex signature to extract vortex core scales from PIV data in a turbulent boundary layer.

5.3 Scales of coherent structures in wall turbulence

In this section we present a summary of the scales of the main eddies or coherent structures that we feel are most prominently observed across the different regions of wall turbulence. Different viewpoints to ours exist, but it is not feasible to present a thorough review of the differing viewpoints or to summarize the extensive literature on this topic here. For this the reader is referred to the review material presented by Adrian (2007), Robinson (1991), Panton (2001), Schoppa & Hussain (2006), and Wu & Moin (2009).

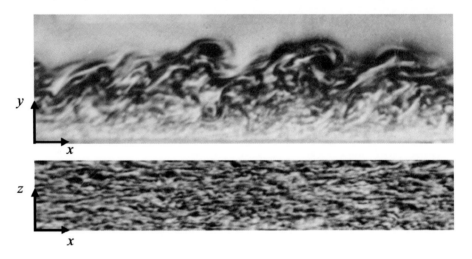

Figure 5.6 Flow visualization in a turbulent boundary layer showing simultaneous views in a streamwise-wall-normal plane and a streamwise-spanwise plane, as viewed through the underside of the transparent wall. Flow is from left to right and the visualization details are as described in Cantwell *et al.* (1978). Photos courtesy of Don Coles.

5.3.1 Near-wall quasi-streamwise vortices (QSV)

In the near-wall region (say, $0 < y^+ < 40$) the mean velocity gradient and the associated stretching have their largest values. The dominant motions are long in the streamwise direction, and thin in the spanwise and wall-normal directions. Their orientation is not perfectly streamwise, so they are normally described as being quasi-streamwise. This goes back to the flow visualization studies of Hama in the 1950s and was extensively studied by

Kline et al. (1967). The lower image in Figure 5.6 is a flow visualization by Cantwell et al. (1978) where the view is through the underside of a transparent wall. This clearly shows the streaky structure of the near-wall region. The streakiness was first shown to be associated with low-speed streaks of fluid being lifted up from the wall, and later it was demonstrated that the uplift was motion induced by quasi-streamwise oriented, counter-rotating vortex filaments on either side of a low speed streak. The mean spanwise spacing between low-speed streaks scales with the viscous wall unit, and has a nominal value of 100^+. On closer examination, the mean spacing varies with y, as shown in Figure 5.7. The mean vortex diameter of the QS vortex velocity field is found to be 40 in viscous wall units or about $40/2.4 \approx 17\eta^+$, somewhat thicker than the accepted value for the worm diameter in isotropic turbulence (Herpin et al. 2009). The similarity between the diameter of the QS vortex and the thickness of the buffer layer is no coincidence. The QS vortices dominate the buffer layer and determine its mean structure. The location of the maximum streamwise kinetic energy at $y^+ \approx 12$ is likely associated with the maximum induced velocity between two QS vortices being located near the height of their centers.

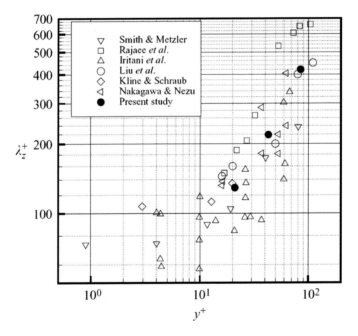

Figure 5.7 Mean spanwise spacing of low-speed streaks as a function of wall-normal location plotted in inner variables. Figure from Tomkins & Adrian (2005).

Measurements of the streamwise length of the QS vortices range from O(100) to O(10,000), a much greater spread than the scatter in the spanwise spacing. This suggests that the streak length does not scale with inner units, but it may also be that the streak length is difficult to measure. The u-spectra in the buffer region does show a clear peak at a streamwise wavelength of 1000^+, and this coincides with the reported streak lengths found by Praturi & Brodkey (1978) and Blackwelder & Eckelmann (1979).

5.3.2 QSV to HPV

In the region nominally $40 < y^+ < 100$, the QS vortices lift up from the wall and begin to form hairpin vortices. A hairpin typically assumes the fully formed shape of an omega by the time that its head is approximately 100^+ above the wall (Zhou et al. 1999). Thus, this region may be regarded as a transition layer in which the first hairpin vortices (HPVs) appear. The concept of a hairpin vortex dates back to Theodorsen (1952), who visualized vortex filaments oriented spanwise to the mean flow and perturbed by a small upward motion. The consequent kinematic response leads to a hairpin shaped vortex, which is stretched and intensified and inclined downstream at nominally 45 degrees. Hairpin vortices were generally ignored for decades, possibly because the notion of coherent structures in turbulence was not recognized until the 1970s (eg. Brown & Roshko 1974). Renewed interest in HPVs followed from work at Stanford (eg. Offen & Kline 1975) where hairpin structures were identified as a likely explanation for transport mechanisms near the wall, but most prominently from the experimental flow visualization studies of Head & Bandyopadhyay (1981). At that time, and since then, there has been considerable controversy as to the existence of HPVs. Most recently the controversy has extended to visualizations from direct numerical simulation (DNS) data, (see Marusic 2009) with Wu & Moin (2009) reporting a predominance of HPVs, as shown in Figure 5.5(c), for a spatially evolving boundary layer, while other numerical studies such as Schlatter et al. (2009, 2010), Ferrante & Elghobashi (2005) and others, with different initial conditions and simulation schemes, report mainly inclined vortices unconnected by the head or 'arch' of a hairpin. Further study of these databases is needed to determine if heads can be revealed by more refined visualization methods, as reported in Wallace et al. (2010).

We subscribe to the school in favour of HPVs, particularly in the sense that this structure is best representative of the statistical average, and most significantly, has the features required to explain localized transport mechanisms in wall turbulence. It has consequently been used to model wall turbulence using the attached eddy hypothesis of Townsend (1976). In this

5: The Eddies and Scales of Wall Turbulence

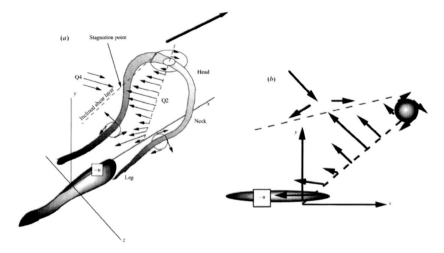

Figure 5.8 (a) Schematic of hairpin eddy attached to the wall; (b) signature of the hairpin eddy in the streamwise/wall-normal plane (from Adrian et al. 2000).

model the HPVs are the representative attached eddies distributed across a range of scales and population densities for statistics above the buffer region (Perry & Chong 1982; Perry & Marusic 1995).

Figure 5.8 shows a schematic of a first generation near-wall HPV growing up out of a pair of QS vortices. The spacing between the QS vortices is about 40–50^+, and the maximum width of the omega shaped hairpin is slightly larger, initially, and grows with time. The lifted portion of the hairpin lies in a plane inclined at $\sim 45°$ to the wall. The DNS on which this sketch is based started from an initial field equal to the conditional average given a second quadrant ejection at $y^+ = 50$. This initial disturbance already showed inclined vortices lifting from counter-rotating vortices, so some of the connection was 'built-into' the simulation. However, similar forms are found in fully turbulent simulations, c.f. Figure 5.5a and in PIV measurements, so the connection between QS vortices and hairpins is well established. The average diameter of the transverse vortex core of the head of a near-wall hairpin is about $16\eta^+$ according to Herpin et al. (2009), close to the average diameter of the QS vortex.

5.3.3 HPV packets and packet hierarchy

Once in the classical logarithmic region and above ($30 < y^+ < \delta^+$) there is strong support for a range of scales of HPVs, and thus this region may be reasonably referred to as the HPV layer. The size of the individual HPV is seen to scale approximately linearly with distance from the wall. In the attached eddy model (Perry & Marusic 1995) this behaviour leads to the logarithmic law of the wall and, for sufficiently high Reynolds numbers in the log region, the k_1^{-1} law for the u-spectra (Nickels et al. 2005).

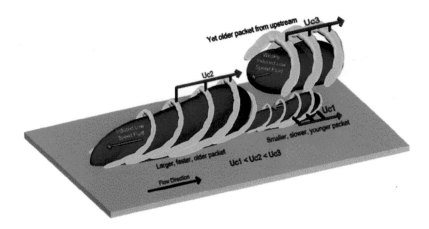

Figure 5.9 Schematic from Adrian et al. (2000) showing conceptual scenario of HPVs attached to the wall and growing in an environment of overlying larger hairpin packets.

Adrian et al. (2000) were the first to extract planar quantitative information about HPVs from experimental data. For this, they considered particle image velocimetry (PIV) results in the streamwise-wall-normal plane for a range of Reynolds numbers in ZPG boundary layers. Strong support was found for HPV signatures consistent with the schematic in Figure 5.8, together with a range of scale of these eddies as deduced by Head & Bandyopadhyay (1981) and as presumed in the modeling work of Perry & Chong (1982) and Perry & Marusic (1995). However, what was also revealed by Adrian et al. (2000) was that the HPVs most often occurred in spatially coherent packets that were aligned in the streamwise direction. Groups of 5-10 hairpins, which extended over a length of 2δ, were found to convect at a uniform streamwise velocity, and it was also noted that there was a range or hierarchy of scales of packets. Figure 5.9 shows a schematic from Adrian et al. (2000) explaining the concept of the organization, and how different scales of packets may coexist and possibly interact. Bandyopadhyay (1980) and Head & Bandyopadhyay (1981), from their flow visualization studies, had previously proposed that hairpin vortices travel in groups, and theoretical and DNS studies by Smith et al. (1991) and Zhou et al. (1999) at low Reynolds number had demonstrated that a single HPV with sufficient circulation could spawn a trailing group of hairpins that were convected at the same speed as the leading structure.

In a kinematic sense, the streamwise alignment of the HPVs into packets is able to explain many interesting (and previously puzzling) observations in wall turbulence. For example, the signature from a packet of HPVs convecting past a fixed probe would produce multiple ejections associated with individual bursts, as observed by Bogard & Tiederman (1986) and Tardu (1995). The induced flow between the legs and head of an organized streamwise packet of HPVs produces an elongated and spatially coherent region of low speed fluid, as highlighted by the dark blue-coloured regions in Figure 5.9. These elongated zones of uniform streamwise velocity explain the long tails in two-point correlations of streamwise velocity found in Kovasznay et al. (1970), Townsend (1976) and Brown & Thomas (1977).

Further evidence for packets came from streamwise-spanwise PIV measurements and inclined plane PIV measurements (Ganapathisubramani et al. 2003; Tomkins & Adrian 2003, Ganapathisubramani et al. 2005, Hutchins et al. 2005). Tomkins & Adrian (2003) took conditional averages of the u–w velocity field on $(x$–$z)$ planes, given a local minimum of the velocity in a low-speed streak, showing the average location of the vortex legs on each side of the streak. The mean spacing between the legs was found to be a remarkably linear function of y (as shown in Figure 5.10). Ganapathisubramani et al. (2003) used a feature-detection algorithm to identify and examine vortical structures and packets occurring in individual velocity fields using stereoscopic PIV. They identified many elongated packets within the log region but few in the outer region. The packets typically had vortices with two legs (counter-rotating) on either side of the low-speed region. Sometimes legs were offset in the streamwise direction. Not all vortices had two legs, implying the existence of asymmetric (cane-type) structures as well. Studies of DNS fields (Guezennec et al. 1989) and experimental flows (Stanislas et al. 2008) both conclude that the most probable form of a hairpin vortex is a one legged cane. Careful examination of canes in DNS fields reveals that apparently one-legged canes often have a weak second leg, revealed by reducing the contour level used for visualization or by tracing vorticity lines (Wallace et al. 2010). Thus, it is probably best to refer to them as 'weak-legged' rather than 'one-legged'.

Although the packets identified in the log layer occupied nominally 5% of the total streamwise-spanwise area examined, they were found to contribute 25–28% of the total Reynolds shear stress $-\overline{uv}$. This estimate is conservative as it considers the contribution from the identified low-speed regions only and does not account for the associated 'sweeping'/Q4 ($u > 0$; $v < 0$)

contributions from either side of this region. Later studies by Balakumar & Adrian (2007), considering the partitioning of energy in premultiplied spectra as in Figure 5.3 for streamwise wavelengths $\lambda_x/\delta > 3$, conclude that 40–65% of the kinetic energy and 30–50% of the Reynolds shear stress is accounted for in the long modes, which are probably related to the hairpin packets. What is clear is that the hairpin packets are a key contributor to the transport and energy-containing phenomena in wall turbulence.

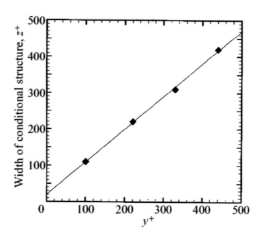

Figure 5.10 Spanwise size of conditional eddy in the $(x-z)$ plane as a function of wall-normal location. Figure from Tomkins & Adrian (2003).

Evidence for packets at high Reynolds number is less detailed, but fully consistent with low Reynolds number DNS and laboratory experiments. At high Reynolds numbers it is difficult to observe the vorticity, so evidence for hairpin packets rests on observations of low speed streaks of size and shape consistent with the low momentum zones induced by HPVPs, and the characteristic 'ramp' shape of the envelope of an HPVP. The shape is remarkably straight, with evidence of heads of hairpins along the envelop and a fairly constant angle with respect to the wall. Brown & Thomas (1978) inferred this angle to be 18° at moderate Reynolds number and Zhou et al. (1999) and Adrian et al. (2000) found 13–18°. Observations in the very high Reynolds number atmospheric boundary layer (Hommema & Adrian 2003, Morris et al. 2007) give clear evidence for the existence of ramp-shaped low momentum zone on scales up to 1–2 m that are consistent with hairpin packets. (Larger regions could not be observed with their apparatus.) Such organized coherence is also consistent with the findings of Phong-Anant et al. (1980) and Antonia et al. (1982) who reported temperature-ramp signatures, and their accompanying statistics, in the atmospheric surface layer. Support for vortex packets also comes from moderate Reynolds number studies in a water channel boundary layer by Dennis & Nickels (2011a,2011b), and Tomographic PIV observation of a Mach 2 boundary layer also give clear three-dimensional visualizations of hairpins wrapped around low momentum streaks (Elsinga et al. 2010). Thus the existence of packets beyond low Reynolds number seems probable, and the ramp angle between 13-18° degrees is not very sensitive to Reynolds number. This is consistent with measurements of structure angle by Marusic & Heuer (2007), who reported $\theta_S \approx 14°$ for experiments covering three orders of magnitude change in Reynolds number.

Figure 5.9 also indicates that a hierarchy of packets exists (although only three are shown at that instance). Here, the weak backward induction of the larger packets causes them to propagate upstream with respect to their surrounding environment more slowly than a smaller packet, meaning that their forward velocity with respect to the wall must be faster. The sketch in Figure 5.9 should not be interpreted to imply that the packets always align in the streamwise or the wall-normal directions, although it seems that there is a tendency for streamwise alignment (Adrian 2007). Also, in DNS simulations the hairpin shapes appear to be more distorted than Figure 5.9 indicates, with many missing heads. Another implication of this schematic is that at a given streamwise (x) location, the structure changes from new packets created in the vicinity of x to progressively older and larger packets created farther upstream as one traverses the thickness of the boundary layer from the wall to its outer edge. Therefore, the uniform momentum zones are like geographic strata, and the variations in wall-normal distance depend on the upstream history of the flow, which is important in flows that are developing and/or are in a nonequilibrium state. Of particular interest is the possibility that large hairpin packets create smaller hairpins so that smaller, nearer-wall packets have a phase relationship to larger packets. However, as yet there is no definitive determination of the processes by which packets might trigger new packets to form a pattern.

Statistical evidence

As mentioned above, statistical evidence for long streamwise motions in the logarithmic region dates back to two-point correlation studies based on hot-wire data using Taylor's hypothesis. The earliest studies are as documented by Townsend (1976) going back to Favre et al. (1957). Based on numerical simulation data, Moin & Kim (1985) concluded that the two-point correlations strongly support a flow model with hairpin vortices inclined at 45° to the wall in a channel flow, and that the size of vortices increased with wall normal location. Willmarth & Lu (1972) and Blackwelder & Kaplan (1976) performed detailed studies using conditional and phase averaging techniques to obtain average structure. They conditionally averaged on sweep and ejection events and concluded that sweeps and ejections of certain sizes make an appreciable contribution to the Reynolds shear stress and therefore to the drag on the surface. Kovasznay et al. (1970) constructed contour plots of space-time correlation of the three velocity components from fixed-probe data at a wall-normal location of $y/\delta = 0.5$, and concluded that the individual 'bulges' or 'bursts' are strongly three-dimensional. They also found that these events are elongated in the streamwise direction (2-3δ mean length) and comparatively compact in the spanwise direction (1–1.5δ width). Similar findings were also found by Krogstad & Antonia (1994). More recent two-point correlation studies by Ganapathisubramani et al. (2005) and Hutchins et al. (2005) using multi-plane PIV data in a ZPG boundary layer confirm the earlier findings and present a clear three-dimensional view, an example of which is shown in Figure 5.11. Here, it is clear that the two-point u correlation results indicate that when the reference point is in the log region, an elongated streamwise structure appears that is inclined in the downstream direction, and extends down towards the buffer region. When the reference point is in the outer wake region, the conditionally averaged structure is seen to be more compact with a limited three-dimensional extent. This outer region behaviour is expected for boundary layers where the turbulence is intermittently dispersed within essentially irrotational free-stream fluid, as is clearly seen in the visualization shown in Figure 5.6. Such behaviour is not so clear in pipes and channels where the bulk flow is always fully turbulent.

All of the above mentioned observations for two-point velocity correlations are consistent with a hairpin packet model as discussed by Hutchins et al. (2005). Other statistics also endorse this model. Christensen & Adrian (2001) used linear stochastic estimation (LSE) on PIV data to compute the conditionally averaged velocity field associated with swirling motion (taken to be the signature of a HPV head). The PIV data were in $(x-y)$ planes for turbulent channel flow. They found that the conditional structure consisted of a series of spatially organized hairpin vortex heads along a line inclined at 12°–13° with the wall. Hambleton et al. (2006) also performed a linear stochastic estimation analysis on PIV planar data in a ZPG boundary layer and obtained similar results to those of Christensen & Adrian (2001) for the $(x-y)$ plane. Interestingly, Hambleton et al. (2006) also obtained a simultaneous result for an $(x-z)$ plane as their study involved simultaneous dual-orthogonal-plane stereo-PIV. A typical result from the LSE analysis

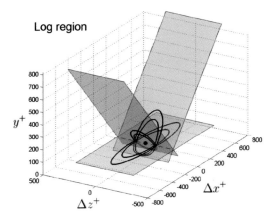

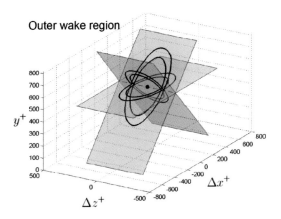

Figure 5.11 Planar R_{uu} contours at 0.3 and 0.4, shown for three available datasets ((x, y), 45° and 135° planes) for ZPG boundary layer at $Re_\tau = 1100$, as reported by Ganapathisubramani *et al.* (2005). Top: Reference point nominally in log region. Bottom: in outer wake region ($y/\delta = 0.5$).

from Hambleton *et al.* (2006) is shown in Figure 5.12. Here the velocity-direction field obtained from an LSE conditioned point nominally in the log layer reveals a flow pattern strikingly similar to that expected from an averaged HPV packet signature.

Interpreting the relationship between averaged statistics and observations from instantaneous realizations can be difficult. For example, while the above mentioned two-point correlations were found to be consistent with hairpin packets this does not exclude other possible scenarios or form of coherent structures. A more incisive approach follows that of Townsend (1976) and Perry & Chong (1982) where the aim is to calculate averaged statistical

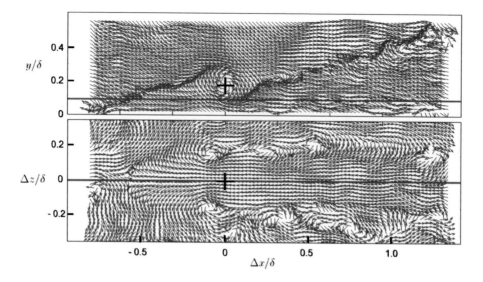

Figure 5.12 Direction field from linear stochastic estimation of velocity based on positive signed swirl (swirl consistent with mean vorticity) at $y_{ref}/\delta = 0.19$ (condition point is marked with a cross). Results are as described by Hambleton et al. (2006). The solid red line indicates the location of the orthogonal PIV plane.

quantities (such as Reynolds stresses, spectra and two-point correlations) by assuming a representative eddy shape and then distributing these eddies with a range of scales with varying population density for each of the scales. This way, the assumption of the representative eddy can potentially be tested against experimental data, and one is not left guessing how the range of the scales that are present may be influencing the intrepretation. For example, Marusic (2001) calculated the structure angle (equation 5.10) using the attached eddy model and found that simple eddies inclined at 45° return θ_S values that vary across the layer, considerably different to 45° – consistent with experimental data. Townsend (1976) and Perry & Chong (1982) found that asymptotically an inverse-power law population density distribution (i.e. double the length scale has half the population) was needed to recover a log flow for the mean velocity (and consequently a k_1^{-1} law for the u and w spectra, together with equivalent logarithmic behaviour for $\overline{u^2}^+$ and $\overline{w^2}^+$), but their results were restricted to a qualitative description. The attached eddy model was later refined by Perry & Marusic (1995) and Marusic & Perry (1995) and they were able to make quantitative predictions of the turbulence statistics given only the mean-flow information. This was found to work very well for the Reynolds stresses, although differences were found in the spectra. An essential feature of the attached eddy model is that the representative eddies are assumed to be uncorrelated with each other, and this is in contrast to the findings of Adrian et al. (2000) of spatially correlated packets of HPVs. Marusic (2001) revisited the attached eddy calculations and found that if one used a train or packet of HPVs as the representative eddy (rather than individual HPVs) then the calculations were able to reproduce the correlation statistics in the log layer very well, as is seen in Figure 5.13. This further supports the notion that hairpin packets are important in a statistical sense and emphasizes that a hierarchy of packets exists across the log layer.

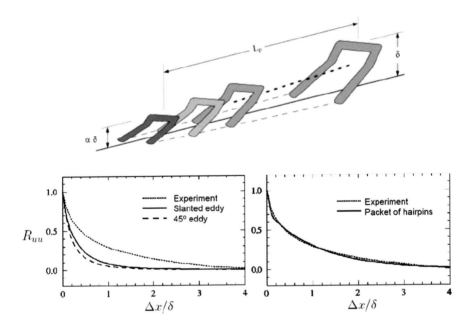

Figure 5.13 Attached eddy calculations of Marusic (2001) for the streamwise velocity autocorrelation compared to experimental data at $y/\delta = 0.1$ in a ZPG boundary layer at $Re_\tau = 4700$. (Bottom left) Models based on randomly scattered individual eddies, slanted or $45°$, do not agree well with the experimental correlation for large separations. (Bottom right) Model based on packets of hairpins (as shown in top panel) explains the long correlation tail.

5.3.4 LSM, VLSM and Superstructures

Kim & Adrian (1999) investigated the nature of u-spectra in turbulent pipe flow and concluded that the spectra in the log region may be interpreted as bimodal, with very long scale motions (VLSM) being distinct from large scale motions (LSM). This conjecture is yet to be confirmed, as the interpretation of spectra is clouded by the nature of any nonlinear interactions across scales and by possible low-wavenumber distortion resulting from using Taylor's hypothesis, as was recently discussed by del Alamo & Jimenez (2009). However, a delineation of the trends can be made, as was done by Balakumar & Adrian (2007) as shown in Figure 5.14. Here, the variations in

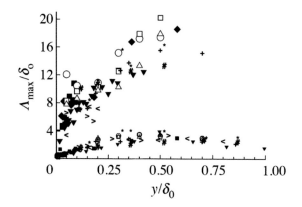

Figure 5.14 Wavelengths of the spectral peaks in pipe, channel and boundary layer flow as functions of wall-normal distance. Λ_{max} presents either Λ_{LSM} or $\Lambda_{VLSM} = \Lambda_{SS}$. From Balakumar & Adrian (2007).

the wavelengths of the peaks are plotted versus the distance from the wall for various wall-bounded flows. The shorter wavelength peak Λ_{LSM}, which represents the LSM wavelength carrying the highest energy density increases until $y/\delta \approx 0.5$, and then reverses the trend to decrease towards the channel or pipe centreline or the boundary-layer edge. This behaviour points to the presence of a similar underlying mechanism in the channel, boundary layer and pipe flows that acts to create the 2–3δ motions. The long-wavelength peak, which represents the VLSM wavelength carrying the highest energy density, increases to approximately $20R$ for the pipe flow while the values of Λ_{max} (wavelength of the spectral peak) are shorter for the channel flow reaching a wavelength of approximately 13–15h near $y/h \approx 0.5$. Generally, the behaviour and trends observed for the pipe flow are also found for the channel flow. However, the boundary layer spectra make an early transition from the presumed bimodal distribution to a unimodal distribution near $y/\delta \approx 0.2$ beyond the edge of the log layer. In ZPG boundary layers Λ_{VLSM} is generally restricted to 6–8δ, which is considerably shorter than in the pipe or channel. On the basis of the shapes of the streamwise power spectra and the uv co-spectra, Balakumar & Adrian (2007) nominally placed the dividing line between LSM and VLSM at 3δ. From Figure 5.14, such a di-

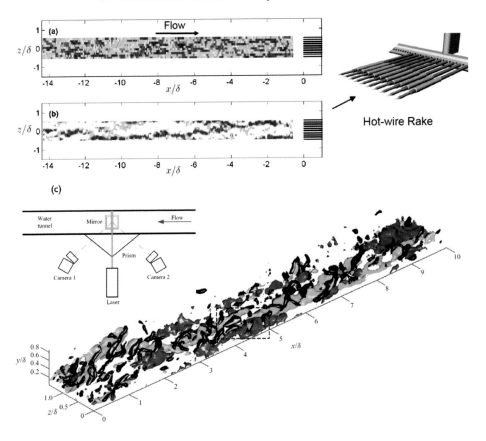

Figure 5.15 Superstructure signatures: (a) From rake of hot wire traces from Hutchins & Marusic (2007); u signal at $y/\delta = 0.15$ for $Re_\tau = 14400$ (b) Same with only low speed regions highlighted. (c). High-frame rate stereo-PIV measurements from Dennis & Nickels (2011a,b) in a turbulent boundary layer at $Re_\theta = 4700$, showing similar features to the hot-wire rake measurements. Here the black isocontours show swirl strength, indicating the corresponding location of vortical structures with the low-speed (blue) and high-speed (red) regions.

vision should be clear above $y/\delta \sim 0.05$, at least for the Reynolds numbers considered by Balakumar & Adrian (2007).

Hutchins & Marusic (2007) investigated the length of the large scale motions in boundary layers by using a spanwise array of hot-wires (and sonic anemometers in the atmospheric surface layer) and found evidence of very long organized motions in the log layer, often in excess of 10δ, a sample realization of which is shown in Figs. 5.15(a,b). Hutchins & Marusic refer to these motions as "superstructures" because of their very large extent (both in the wall-normal and streamwise direction), their clear signature as an outer peak in the u-spectrogram across the layer, and because they account for a major proportion of the Reynolds shear stress (Ganapathisubramani et al. 2003; Balakumar & Adrian 2007). The shorter lengths noted in the spectra (6δ) were believed to be caused by the spanwise meandering nature of the motions. Figure 5.15(c) shows a similar image from recent work by Dennis & Nickels (2011), who used high-frame rate PIV with Taylor's hypothesis to obtain a volumetric image of these events. Monty et al. (2008) performed similar measurements to Hutchins & Marusic in pipe flow, and found organized motions up to $30R$ in length, consistent with the VLSM of Kim & Adrian (1999) in pipes, where it was conjectured the VLSM to be a concatenation of packets. One of the uncertainties when interpreting images such as in Figure 5.15 is the effect of using Taylor's hypothesis to convert a time series to a spatial signature. Dennis & Nickels (2008) investigated this using time-resolved PIV with a field of view of 6δ in the streamwise direction. They concluded that Taylor's hypothesis could reasonably be used up to these lengths for the largest organized motions.

The issue of whether the superstructures found in boundary layers by Hutchins & Marusic (2007) were the same as the VLSM found in pipes by Kim & Adrian (1999) and Guala et al. (2006) was investigated by Monty et al. (2009) who made detailed comparisons of the spectra in a pipe, channel and ZPG boundary layer with matched Re_τ and measurement probe resolution. Figure 5.16 is taken from their paper and indicates that the largest energetic scales in pipe and channels are distinctly different to those found in a boundary layer. Although the large-scale phenomena have been shown to be qualitatively similar (Hutchins & Marusic 2007; Monty et al. 2007), their contributions to the energy continues to move to longer wavelengths with distance from the wall in internal flows. The opposite occurs in boundary layers, where outer-flow structures shorten very rapidly beyond the log region. That is, VLSMs in internal flows are likely different from superstructures in boundary layers; qualitatively the structures are similar, however, the VLSM energy in internal flows resides in larger wavelengths and at greater distances from the wall than superstructures in boundary layers. Interestingly, for $y/\delta < 0.5$ the different energy distributions in pipes and channels and ZPG boundary layers occur in regions where the streamwise turbulence intensity is equal. Monty et al. (2009) concluded that this result suggests that all three flows might be of a similar type structure, with energy simply redistributed from shorter to longer scales for the pipe and channel flow cases. Whether the quantitative differences are due to the interaction of the opposite wall in internal flows, or the intermittency of the outer region in boundary layers remains uncertain.

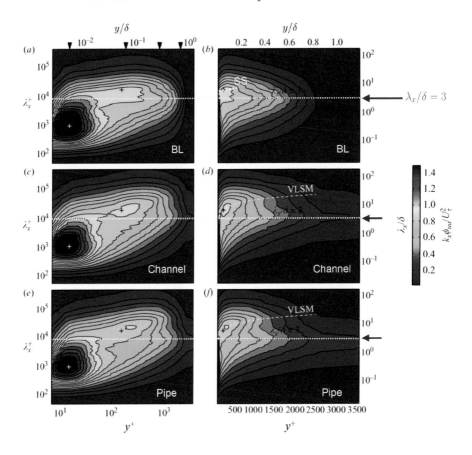

Figure 5.16 Maps of premultiplied u spectra as a function of streamwise wavelength (λ_x) and distance from the wall from Monty et al. (2009). Figures show (from top to bottom) turbulent boundary layer (a, b), channel (c, d) and pipe (e, f), respectively. Right-hand figures have a linear abscissa, to emphasize the differences in the outer region between VLSM in pipes and channel and the superstructures in boundary layers. Red arrows indicate the wavelength $\lambda_x/\delta = 3$.

Another possible explanation for the apparent differences between superstructures in boundary layers and VLSMs in pipes and channels could be that they are both produced by a concatenation of packets/LSM, but that this organization is distinctly different between the two types of flows. This may be due to differences in the lateral organization, or in boundary layers, the interruption of wall-normal extent due to the entrained irrotational free-stream flow, which will limit the extent in the streamwise direction – akin to crystal growth limitations in materials due to grain boundaries.

5.3.5 Inner-outer interactions

An important aspect of LSM, VLSM and superstructures, which are primarily associated with the outer region, is their role in interacting with the inner near-wall region, including their influence on the fluctuating wall-shear stress. There has been debate over many decades as to whether the inner and outer regions do interact, or whether they can be considered as independent, as assumed in all classical scaling approaches. Considerable evidence now exists that outer scales are important for characterising near-wall events (Rao *et al.* 1971, Blackwelder & Kovasznay 1972, Ginvald & Nikora 1988, Wark & Nagib 1991, Hunt & Morrison 2000, DeGraaff & Eaton 2000, Metzger & Klewicki 2001, Abe *et al.* 2004, Hoyas & Jimenez 2006, Hutchins & Marusic 2007, Orlu & Schlatter 2011 and others). The clearest manifestation of outer layer influence on the near-wall layer is the Reynolds number dependence of some of the turbulence moments when plotted in terms of wall variables. There is also strong evidence to suggest that these outer-region motions extend down to the wall, and modulate the flow in the inner layer, including the buffer layer. This interaction was quantified by Mathis *et al.* (2009), and formed the basis of an algebraic model by Marusic *et al.* (2010b), wherein the statistics of the streamwise fluctuating velocity in the near-wall region could be predicted given only the large-scale velocity signature in the outer logarithmic region of a given flow.

While questions still remain as to the origin of these dynamical interactions, and alternative theories based on instability mechanisms continue to be debated, it is clear that future scaling arguments need to account for inner and outer regions that are far from independent.

5.4 Relationship between statistical fine-scales and eddy scales

There has been considerable interest in quantifying the size of the fine-scale eddies found in wall turbulence (Das *et al.* 2006, del Alamo & Jimenez 2006; Stanislas *et al.* 2008 and many others), and the question arises as to their relationship to the coherent structures discussed in the preceding. Their size is normally stated in terms of the Kolmogorov scale (η) implying that the smallest eddies are in some way universal across all turbulent flows, because the Kolmogorov scaled high wave number spectrum is thought to be universal. From a survey of the literature it appears that the nominal accepted core diameter of filamentary vortices is in the range 10–12η. This applies to several turbulent flows, as is noted for the visualization examples shown in Figure 5.5, which includes flows from channels, boundary layers and

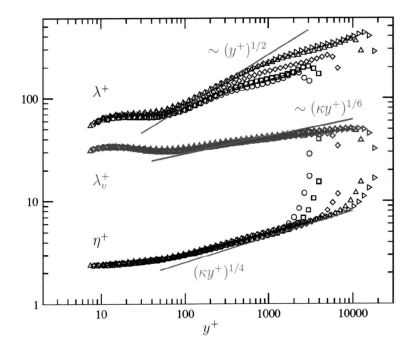

Figure 5.17 Taylor and Kolmogorov length scales with inner scaling across turbulent boundary layers. The blue coloured symbols are profiles of λ_w^+, a modified Taylor microscale. Data as in Hutchins *et al.* (2009). Symbols indicated Re_τ values: ○ 2800, □ 3900, ◇ 7300, △ 13600, ▷ 19000.

homogeneous isotropic turbulence. An alternative scaling uses the Taylor microscale, (λ), possibly because the numerical coefficient is then closer to unity (diameter $\sim 0.3\lambda$), suggesting the scale λ more closely represents the physical diameter. The real issue, of course, is to use the length scale that scales the core diameter with varying Reynolds number.

The inhomogeneity of wall turbulence complicates the scaling further, because, both η and λ vary with wall-normal distance, as seen in Figure 5.17 for results from a series of ZPG boundary flows across a range of Reynolds number. Here η and λ are estimated using the approximation for dissipation rate given in equation (5.8), and the red lines on Figure 2.40 indicate the expected slopes in the logarithmic layer if one assumes that the production of turbulent kinetic energy equals the dissipation rate and $-\overline{uv} \approx U_\tau^2$. (For the λ^+ line, an additional assumption is made that $\overline{u^2}^+ \approx$ constant in equation (5.7), and here the agreement of the data with the 1/4 power law is seen to not be as convincing.) It is noted that $\eta^+ \approx 2.4$ in the viscous buffer region and follows inner-scaling reasonably well across the logarithmic region. In contrast λ^+ does not scale with inner variables across most of the layer. These trends are in agreement with those reported by Stanislas et al. (2008).

Stanislas et al. (2008) also investigated experimentally the strength and size of the filamentary vortices across the logarithmic layers of two turbulent boundary layers and found that the vortex core diameter scaled with η, with a mean core vorticity of $1.6v/\eta$. Similar findings have been reported in DNS studies (Tanahashi et al. 2004, Kang et al. 2005, Das et al. 2006), and as stated above the nominal accepted core size of filamentary vortices in the logarithmic layer is in the range 10–12η.

Questions remain as to whether this scaling holds at very high Reynolds number, as many studies, particularly DNS, have limited Reynolds number ranges. The problem with Kolmogorov scaling is that the length scale varies with Reynolds number in the outer region (Figure 5.17). The Taylor microscale behaves even more poorly, being sensitive to Reynolds number even in the log layer. Thus, if either scale is used to represent the diameters of the small eddies, the coefficient must vary with Reynolds number across the outer layer. But, we expect the dissipation range to be Reynolds number independent, so fine scale eddies lying in the range should also have diameters that are Reynolds number independent. Combined with the large/small values of the numerical coefficient, the evidence suggests that neither η nor λ is the best scale for the small eddy core diameter.

To resolve this problem, we propose a new length scale which, for want of a better term, we shall call the 'vortex micro-scale'. It is derived from the following principle: To represent the eddies in the dissipation range, their scales of length and velocity can depend only on the power spectrum of velocity in dissipation range. This criterion rules out the Taylor microscale, which mixes the energy from the energy containing range with the dissipation rate, and, perhaps surprisingly, it rules out the Kolmogorov scale which mixes viscosity with the spectral energy transfer rate established by the large scales. Figure 5.3 shows what wavelengths contribute to the dissipation (area under $k_1^3 \phi_{11}$ on this log-linear plot), and the energy-containing scales (area under $k_1 \phi_{11}$). As Reynolds number increases at a fixed y^+ location (in the log layer, say) it is known that the broadband energy under the $k_1 \phi_{11}/U_\tau^2$ curve will increase (with increasing contribution at lower k_1^+) while the contribution to Λ_{diss}^+ is approximately unchanged for $k_1^+ \gtrsim 0.01$. Therefore, at sufficiently high Reynolds number there is a separation of scales between the main energy-containing motions and dissipative motions that scale with η. Thus, it is possible to define a wavenumber k_d^+ bounding the low wavenumber end of the dissipation spectrum such that the spectral energies at all wavenumbers greater than k_d^+ are independent of Reynolds number.

The principle stated above combined with dimensional analysis of the dissipation range spectrum yields a vortex microscale of length defined by

$$(\lambda_v^+)^2 = \frac{\int_{k_d^+}^{\infty} \phi_{11} dk_1^+}{\int_{k_d^+}^{\infty} (k_1^+)^2 \phi_{11} dk_1^+}. \tag{5.11}$$

Numerically we take the integrals from a fixed inner-scaled wavenumber $k_d^+ = 0.01$. In the limit $k_d^+ \to 0$ and under the assumption of isotropic fine scale turbulence, the vortex microscale approaches the Taylor microscale, which constitutes an upper bound. Using this definition, the λ_v^+ profiles, shown in blue in Figure 2.40, are seen to collapse across the boundary layer and are independent of Reynolds number. Numerically, $\lambda_v^+ = 30\text{--}50$, with a nearly constant value ~ 35 across the log layer. Thus, the mean diameter of the small eddies is approximately $0.7\text{--}0.8$ λ_v^+ in the log layer, and as a rule of thumb it is approximately one λ_v^+ throughout the boundary layer for all Reynolds numbers studied thus far.

5.5 Summary and conclusions

A review has been presented of the scales of the coherent turbulent eddies that are currently understood to occur in wall turbulence. A summary of these main scales is tabulated in the Appendix, together with the values of more traditional statistical scales. Throughout this chapter, the numbers quoted are always means, and when a range is indicated, it is the range of variation of the mean value. Individual realizations have values fluctuating about the means, sometimes substantially. Before measuring a scale, one must, necessarily, define a class of eddies and have a means of separating them from the total flow field. The numerical values depend on both the definition and the means of identification, and both may change as our understanding of turbulence progresses. Hence, in the future, numerical values presented here for the various scales may change, or perhaps even be re-defined.

On the basis of spectral analysis, the average length scales of turbulent motion are taken conventionally to span the range from the Kolmogorov scale to the thickness of the shear flow, δ, h or R. The length scales of coherent eddies are generally longer, ranging from $\sim 16\eta$, the mean diameter of a filamentary vortex ('worm') to $6\text{--}30\delta$, or more, where the latter scale of the very large-scale motions is associated with streamwise length. (It is conceivable that the filamentary eddy cores are made up of rolled up vortex sheets, whose thickness will likely scale with η, as in Lundgren's 1982 model, but thus far they are not observable in the data or the computational simulations.)

The long, quasi-streamwise vortex filaments in the buffer layer and the inclined, cane or hairpin eddies in the lower logarithmic layer are the smallest coherent eddies in the near-wall flow. Their mean spanwise spacing, $\lambda^+ = 100$ at $y^+ = 20$ is one of the most reliable empirical numbers in turbulence. The diameter of the QS vortex velocity field in the spanwise wall-normal plane is about 40^+, but the core diameter is smaller, about 17^+, corresponding to a worm. The symmetric hairpin is an idealization that represents a class of eddies that includes highly distorted hairpins, parts of hairpins, canes and vortex filaments inclined at $\sim 45°$ to the wall. (While it is not common to employ Latin names in fluid mechanics it may be useful in the future to group all of these eddies into the class of *turbo propensus*, meaning 'inclined eddy'. Hopefully this would de-emphasize the connotations of shape that are intrinsic to the term 'hairpin'.)

The first fully formed hairpins are about 100^+ tall (wall to head). Thereafter, the hairpin typically expands in all three directions at approximately equal rates, suggesting self-similar growth (cf. del Alamo *et al.* 2006), and perhaps explaining the linear variation of length scales in the logarithmic layer. The parent hairpin spawns similar hairpins upstream and downstream, and these offspring can produce more hairpins, leading to the formation of a packet of hairpins travelling together. Within the packets formed in the near-wall region, individual hairpins are spaced roughly 200^+ apart in the streamwise direction, and packets contain up to 5–10 hairpins. Each hairpin creates a low speed streak about 100–200^+ long in the buffer layer, so the packets and the sequence of low-speed streaks within them are about 500–2000^+ long. This corresponds well with measurements of the mean low speed streak length in the buffer layer of order 1000^+.

The linear increase of the spacing between hairpin necks as a function of y and the relatively constant angle of the envelope of the hairpin packets, $\sim 15°$, suggest that the hairpins in the packets are self-similar in the logarithmic layer and that they are associated with the logarithmic variation of the mean velocity, which is implied by linear variation of the mixing length.

A hierarchy of hairpin and packet scales exists. Figure 5.18 shows a cartoon to illustrate this, and compares the wall-normal length-scales with the traditional scales based on the mean velocity. It is noted that the diagram does not purport to show organization of the pattern, only the scale hierarchy.

If the packets start with a spanwise spacing $\sim 100^+$, linear growth also implies that spanwise adjacent packets must interact. Vortex cut and connect processes may destroy or distort the hairpins and packets may merge into

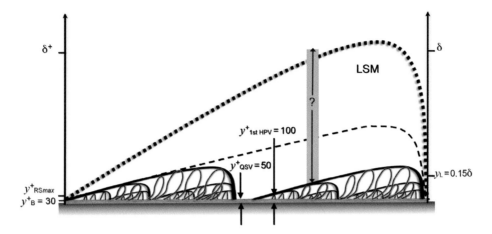

Figure 5.18 Summary of eddy scales in a wall turbulent flow and comparison to traditional scales based on the mean velocity. The first two scales are the height of the QS vortices and the height of the first fully formed hairpin. After that first generation packets grow. The extent to which they grow is not established, as indicated by the gray bar. The rest of the sketch shows what might happen if the packets were to grow in a self-similar fashion by successive scale doublings and concomitant reduction in the number of packets until the last bulge is reached. For this case, $\delta^+ \approx 2000$. The question mark indicates that it is not known if a bulge is similar to the packets or a different mechanism.

larger packets. Whatever the process, the details are bound to be chaotic, and if there is progressive growth into larger scales, it would only be described in the broadest terms. Even so, there is reason to pursue the packet model because there is evidence that packets exist well up into the logarithmic layer at very high Reynolds numbers. Laboratory data and visualization in the logarithmic layer of the atmospheric boundary layer indicate low speed regions within linear ramps with angles ~15–18°, suggesting that some form of hairpin packet persists, despite numerous vortex interactions during the growth process. Thus, while we do not yet understand the behavior of vortex packets much beyond the first, near-wall generation, the vortex packet paradigm may still be useful in explaining the scale variation observed statistically throughout the logarithmic layer. Further research at high Reynolds numbers is needed to conclusively determine these formation processes in the upper logarithmic layer and whether alternative mechanisms originating from the outer region may become prominent.

The logarithmic layer accounts for no more than 15–20% of the layer thickness. Above the logarithmic layer, i.e. in the wake region, there is a change in the characters of both the mean statistics and the eddy structure, but we have no clear picture of how this occurs. It is noted that the attached-eddy modeling study of Perry & Marusic (1995) relied on having distinctly different representative eddies for the outer region. This is clearly an area for future research, both to understand what form the eddies take in this region and to understand the mechanisms forming them.

The large-scale motions – LSMs – have previously been identified as the turbulent bulges in boundary layers. Their mean streamwise length of 2-4δ fits well with the shorter wavelength peak in the streamwise power spectrum above the logarithmic layer (Corrsin & Kistler 1955). The visualizations by Falco (1977) and Head and Bandyopadhyay (1981) in boundary layers give the impression that the bulges may also be hairpin vortex packets. Similar spectral behavior in channel flow suggests that it is reasonable to expect bulge-like motions in the internal flows as well, although they will be obscured by interactions with bulges from other parts of the interior surface.

The super-structures in boundary layer flow and VLSMs in pipe and channel flow have been defined, nominally, to be the motions longer than 3δ, the mean bulge length. Since individual bulges can be longer than the mean, there is clearly a gray zone in which it is difficult to distinguish bulges from VLSMs. In boundary layers, the mean super-structure streamwise wavelength reaches $\sim$6-8δ at the edge of the logarithmic layer. In the wake region the energy of the VLSM peak is so low that it cannot be discerned from the LSM. In contrast, in internal flows the wavelength that is attributed to the VLSM, corresponding to the location of the low-wavenumber peak in pre-multiplied u-spectra (see Figure 5.3), continues to increase up to about $0.6R$ (or h) where it achieves values of 20–$30R$ (or h). Closer to the centerline the VLSM spectral peak become indiscernible. The superstructure terminology has been adopted to distinguish the differences between the boundary layer and the internal flows. The spanwise width of VLSM low speed streaks is also greater than the width of the super-structures. It is unclear how much of this difference is due to the different geometries or due to the problem of defining equivalent outer length scales for the differing geometries.

Despite the differences noted above, the combination of LSM and VLSM or superstructure share a common burden in carrying over half of the streamwise turbulent kinetic energy and perhaps even more of the Reynolds shear stress in the outer region.

These large scale motions also extend from the outer flow down to the wall, and modulate the flow in the inner layer, including the buffer layer. This coupling may be part of a cycle in which high speed sweeps from the outer large scales intensify the near-wall shear, leading to increased rate of production of near-wall eddies. The intensified eddies increase the rate of low momentum transfer upwards, thereby slowing down the large-scale sweep. Regardless of details, any coupling is likely to link the scales of the eddies so that the inner scaling law has some dependence on outer scales, and perhaps vice versa for the outer law. Modelling efforts along these lines show promise, and it is hoped that further understanding of eddy structure, scaling and dynamics will lead to improved scaling laws for all turbulence statistics.

Acknowledgements The authors gratefully acknowledge the support of the Miegunyah Fellowship from the Roger and Mab Grimswade Foundation (RA), NSF Grant CBET-0933848 (RA), and the Australian Research Council (IM).

Appendix

Below is a tabulated summary of various scales for coherent structures (here "transition layer" refers to QSV to HPV transition region (typically $40 < y^+ < 100$)); those obtained from a DNS dataset (del Alamo et al. 2004); and from a range of hot-wire time-series data from Hutchins et al. (2009). y^+_{LOG} indicates the nominal centre of the logarithmic region, which here is estimated to be at $y^+ \approx 3.9 Re_\tau^{1/2}$. It is noted that these numbers are provisional.

Coherent structure organization

	Angle deg.°	Core diameter d_0^+	Width l^+_{width}	Length l^+_x	Spanwise spacing l^+_z
Near-wall region	≈ 0	20	40	1000	100
Transition layer	≈ 0	20	40	1000	100
Log layer	45	25	$O(y^+)$	$O(y^+)$	–
LSM	14	30	$0.4\delta^+$	$3\delta^+$	–
VLSM/SS	≈ 0	30	$0.4\delta^+$	$O10(\delta^+)$	–

Channel flow DNS, $Re_\tau = 934$ (del Alamo et al. 2004)

	y^+			
	15	100	y^+_{LOG}	$0.5\delta^+$
η^+	1.7	2.6	2.7	4.1
λ^+_{T11}	58	57	61	104
λ^+_{T22}	27	26	28	46
L_{11}/δ	0.85	1.58	1.64	2.08
L_{22}/δ	0.05	0.16	0.18	0.34
L_{33}/δ	0.06	0.13	0.16	0.34

ZPG TBL experiments (Hutchins et al. 2009)

x (m)	U_1 (m/s)	U_τ (m/s)	y^* (μm)	θ (mm)	δ (mm)	Re_θ	Re_τ
5.0	12.0	0.442	35	8.7	98	6760	2800
11.0	11.9	0.426	36	12.7	140	9850	3900
21.0	9.8	0.330	44	30.6	326	19600	7300
21.0	20.6	0.671	23	27.1	315	36000	13600
21.0	30.2	0.960	16	24.2	303	47200	19000

ZPG TBL experiment

Taylor microscale: λ_{T11}^+

Re_τ	\multicolumn{4}{c}{y^+}	
	15	100	y_{LOG}^+	$0.5\delta^+$
2800	66	79	100	156
3900	65	80	110	181
7300	69	84	134	242
13600	70	93	171	336
19000	69	90	182	383

Kolmogorov scale: η^+

2800	2.5	3.0	3.5	5.3
3900	2.4	3.0	3.5	5.7
7300	2.4	3.0	3.9	6.6
13600	2.5	3.1	4.3	7.9
19000	2.5	3.1	4.4	8.5

Integral length scale: L_{11}/δ

2800	0.35	0.95	1.03	0.91
3900	0.29	0.82	1.02	0.90
7300	0.22	0.62	0.90	0.85
13600	0.16	0.44	0.80	0.80
19000	0.14	0.38	0.71	0.81

References

Adrian, R. J. 2007. Hairpin vortex organization in wall turbulence. *Phys. Fluids*, **19**, 041301.

Adrian, R. J., Meinhart, C. D., and Tomkins, C. D. 2000. Vortex organization in the outer region of the turbulent boundary layer. *J. Fluid Mech.*, **422**, 1–53.

Afzal, N. 1982. Fully turbulent flow in a pipe - an intermediate layer. *Ingenieur Arch.*, **52**, 355–377.

Antonia, R. A., Chambers, A. J., and Bradley, E. F. 1982. Relationships between structure functions and temperature ramps in the atmospheric surface layer. *Boundary Layer Met.*, **23**(4), 395–403.

Balakumar, B. J., and Adrian, R. J. 2007. Large- and very-large scale motions in channel and boundary-layer flows. *Proc. R. Soc. Lond. A*, **365**, 665–681.

Bandyopadhyay, P. 1980. Large structure with a characteristic upstream interface in turbulent boundary layers. *Phys. Fluids*, **23**, 2326–2327.

Barenblatt, G. I. 1993. Scaling laws for fully developed turbulent shear flows. Part 1. Basic hypotheses and analysis. *J. Fluid Mech.*, **248**, 513–520.

Blackwelder, R. F., and Eckelmann, H. 1979. Streamwise vortices associated with the bursting phenomenon. *J. Fluid Mech.*, **94**, 577–594.

Blackwelder, R. F., and Kaplan, R. E. 1976. On the wall structure of the turbulent boundary layer. *J. Fluid Mech.*, **76**, 89–112.

Bourassa, C., and Thomas, F. O. 2009. An experimental investigation of a highly accelerated turbulent boundary layer. *J. Fluid Mech.*, **634**, 359–404.

Brown, G. L., and Roshko, A. 1974. On density effects and large structure in turbulent mixing layers. *J. Fluid Mech.*, **64**, 775–816.

Brown, G. R., and Thomas, A. S. W. 1977. Large structure in a turbulent boundary layer. *Phys. Fluids*, **20**, S243–S251.

Cantwell, B., Coles, D., and Dimotakis, P. 1978. Structure and entrainment in the plane of symmetry of a turbulent spot. *J. Fluid Mech.*, **87**, 641–672.

Cantwell, B.J. 1981. Organised motion in turbulent flow. *Annu. Rev. Fluid Mech.*, **13**, 457–515.

Carlier, J., and Stanislas, M. 2005. Experimental study of eddy structures in a turbulent boundary layer using particle image velocimetry. *J. Fluid Mech.*, **535**, 143–188.

Chakraborty, P., Balachandar, S., and Adrian, R. J. 2005. On the relationships between local vortex identification schemes. *J. Fluid Mech.*, **535**, 189–2004.

Chauhan, K. A., Monkewitz, P. A., and Nagib, H. M. 2009. Criteria for assessing experiments in zero pressure gradient boundary layers. *Fluid Dynamics Research*, **41**(2), 021404.

Chong, M. S., Perry, A. E., and Cantwell, B. J. 1990. A General Classification of Three-Dimensional Flow Fields. *Phys. Fluids A*, **2**(5), 765–777.

Christensen, K. T., and Adrian, R. J. 2001. Statistical evidence of hairpin vortex packets in wall turbulence. *J. Fluid Mech.*, **431**, 433–443.

Coles, D. E. 1956. The law of the wake in the turbulent boundary layer. *J. Fluid Mech.*, **1**, 191–226.

Coles, D. E. 1962. *The turbulent boundary layer in a compressible field*. Tech. rept. USAF The Rand Cooperation. Rep. R-403-PR, Appendix A.

Coles, D. E., and Hirst, E. A. 1969. Compiled data. In: *Proc. Computation of Turbulent Boundary Layers Vol. II AFOSR-IFP-Stanford Conference 1968*.

Corrsin, S., and Kistler, A. L. 1955. *Free stream boundaries of turbulent flows*. Tech. rept. 1244. NACA Tech. Report.

Cucitore, R., Quadrio, M., and Baron, A. 1999. On the effectiveness and limitations of local criteria for the identification of a vortex. *Eur. J. Mech. B Fluids*, **18**, 261–282.

Das, S. K., Tanahashi, M., Shoji, K., and Miyauchi, T. 2006. Statistical properties of coherent fine eddies in wall-bounded turbulent flows by direct numerical simulation. *Theor. Comput. Fluid Dyn.*, **20**(55-71).

DeGraaff, D. B., and Eaton, J. K. 2000. Reynolds-number scaling of the flat-plate turbulent boundary layer. *J. Fluid Mech.*, **422**, 319–346.

del Álamo, J. C., and Jiménez, J. 2006. Linear energy amplification in turbulent channels. *J. Fluid Mech.*, **559**, 205–213.

del Álamo, J. C., and Jiménez, J. 2009. Estimation of turbulent convection velocities and corrections to Taylor's approximation. *J. Fluid Mech.*, **640**, 5–26.

del Álamo, J. C., Jiménez, J., Zandonade, P., and Moser, R. D. 2006. Self-similar vortex clusters in the turbulent logarithmic region. *J. Fluid Mech.*, **561**, 329–358.

Dennis, D. J. C., and Nickels, T. B. 2011a. Experimental measurement of large-scale three-dimensional structures in a turbulent boundary layer. Part 1: Vortex packets. *J. Fluid Mech.*, **673**, 180–217.

Dennis, D. J. C., and Nickels, T. B. 2011b. Experimental measurement of large-scale three-dimensional structures in a turbulent boundary layer. Part 2: Long structures. *J. Fluid Mech.*, **673**, 218–244.

Doherty, J., Ngan, P., Monty, J. P., and Chong, M. S. 2007. The development of turbulent pipe flow. Pages 266–270 of: et al, P. Jacobs (ed), *Proc. 16th Australasian Fluid Mech. Conf.* University of Queensland.

Elsinga, G. E., Adrian, R. J., van Oudheusden, B. W., and Scarano, F. 2010. Three-dimensional vortex organization in a high-Reynolds-number supersonic turbulent boundary layer. *J. Fluid Mech.*, **644**, 35–60.

Falco, R. E. 1977. Coherent motions in the outer region of turbulent boundary layers. *Phys. Fluids*, **20**(10), S124–S132.

Fernholz, H. H., and Finley, P. J. 1996. The incompressible zero-pressure-gradient turbulent boundary layer: an assessment of the data. *Prog. Aerospace Sci.*, **32**, 245–311.

Ferrante, A., and Elghobashi, S. 2005. Reynolds number effect on drag reduction in a microbubble-laden spatially developing turbulent boundary layer. *J. Fluid Mech.*, **543**, 93–106.

Gad-el-Hak, M., and Bandyopadhyay, P. 1994. Reynolds number effects in wall-bounded turbulent flows. *Appl. Mech. Rev.*, **47**, 307–365.

Ganapathisubramani, B., Longmire, E. K., and Marusic, I. 2003. Characteristics of vortex packets in turbulent boundary layers. *J. Fluid Mech.*, **478**, 35–46.

Ganapathisubramani, B., Hutchins, N., Hambleton, W. T., Longmire, E. K., and Marusic, I. 2005. Investigation of large-scale coherence in a turbulent boundary layer using two-point correlations. *J. Fluid Mech.*, **524**, 57–80.

Ganapathisubramani, B., Longmire, E. K., and Marusic, I. 2006. Experimental investigation of vortex properties in a turbulent boundary layer. *Phys. Fluids*, **18**, 055105.

George, W. K., and Castillo, L. 1997. Zero-pressure-gradient turbulent boundary layer. *Applied Mechanics Reviews*, **50**, 689–729.

Guala, M., Hommema, S. E., and Adrian, R. J. 2006. Large-scale and very-large-scale motions in turbulent pipe flow. *J. Fluid Mech.*, **554**, 521–542.

Guezennec, Y., Piomelli, U., and Kim, J. 1989. On the shape and dynamics of wall structure in turbulent channel flow. *Phys. Fluids A*, **1**, 764–766.

Hambleton, W. T., Hutchins, N., and Marusic, I. 2006. Simultaneous orthogonal-plane particle image velocimetry measurements in a turbulent boundary layer. *J. Fluid Mech.*, **560**, 53–64.

Head, M. R., and Bandyopadhyay, P. R. 1981. New aspects of turbulent structure. *J. Fluid Mech.*, **107**, 297–337.

Herpin, S., Coudert, S., Foucaut, J. M., Soria, J., and Stanislas, M. 2009. Study of vortical structures in turbulent near wall flows. In: *Proc. 8th Intl. Symp. on Particle Image Velocimetry*. Melbourne, Aug. 25-28.

Hommema, S. E., and Adrian, R. J. 2003. Packet structure of surface eddies in the atmospheric boundary layer. *Boundary-Layer Meteorol.*, **106**(1), 147–170.

Hunt, J. C. R., Wray, A. A., and Moin, P. 1988. *Eddies, Stream and Convergence Zones in Turbulent Flows*. Tech. rept. CTR-S88. Center Turbulence Research.

Hutchins, N., and Marusic, I. 2007. Evidence of very long meandering streamwise structures in the logarithmic region of turbulent boundary layers. *J. Fluid Mech.*, **579**(1-28).

Hutchins, N., Hambleton, W. T., and Marusic, I. 2005. Inclined cross-stream stereo PIV measurements in turbulent boundary layers. *J. Fluid Mech.*, **541**, 21–54.

Hutchins, N., Nickels, T. B., Marusic, I., and Chong, M. S. 2009. Hot-wire spatial resolution issues in wall-bounded turbulence. *J. Fluid Mech.*, **635**, 103–136.

Jeong, J., and Hussain, F. 1995. On the Identification of a Vortex. *J. Fluid Mech.*, **258**, 69–94.

Jimenez, J., and Kawahara, G. 2012. Dynamics of wall-bounded turbulence. (This Volume) *The Nature of Turbulence*. Cambridge Uni. Press.

Johansson, A. V., and Alfredsson, P. H. 1983. On the structure of turbulent channel flow. *J. Fluid Mech.*, **122**, 295–314.

Kang, S. H., Tanahashi, M., and Miyauchi, T. 2007. Dynamics of fine scale eddy clusters in turbulent channel flows. *J. Turbulence*, **8**(52), 1–20.

Kim, K. C., and Adrian, R. J. 1999. Very large-scale motion in the outer layer. *Phys. Fluids*, **11**(2), 417–422.

Klewicki, J. C. 2011. Reynolds Number Dependence, Scaling, and Dynamics of Turbulent Boundary Layers. *J. Fluids Eng.*, **132**, 094001, 1–48.

Klewicki, J. C., Fife, P., Wei, T., and McMurtry, P. 2007. A physical model of the turbulent boundary layer consonant with mean momentum balance structure. *Phil. Trans. R. Soc. Lond. A*, **365**, 823–839.

Klewicki, J. C., Fife, P., and Wei, T. 2009. On the logarithmic mean profile. *J. Fluid Mech.*, **638**, 73–93.

Kline, S. J., Reynolds, W. C., Schraub, F. A., and Rundstadler, P. W. 1967. The structure of turbulent boundary layers. *J. Fluid Mech.*, **30**, 741–773.

Kovasznay, L. S. G., Kibens, V., and Blackwelder, R. F. 1970. Large-scale motion in the intermittent region of a turbulent boundary layer. *J. Fluid Mech.*, **41**, 283–326.

Krogstad, P.-A., and Antonia, R. A. 1994. Structure of turbulent boundary layers on smooth and rough walls. *J. Fluid Mech.*, **277**, 1–21.

Lien, K., Monty, J. P., S., Chong M., and Ooi, A. 2004. The entrance length for fully developed turbulent channcel flow. In: *Proc. 15th Australasian Fluid Mech. Conf.*

Lindgren, B., Osterlund, J. M., and Johansson, A. V. 2004. Evaluation of scaling laws derived from Lie group symmetry methods in zero-pressure-gradient turbulent boundary layers. *J. Fluid Mech.*, **502**, 127–152.

Long, R. R., and Chen, T. C. 1981. Experimental evidence for the existence of the 'mesolayer' in turbulent systems. *J. Fluid Mech.*, **105**, 19–59.

Lundgren, T. S. 1982. Strained spiral vortex model for turbulent fine structure. *Phys. Fluids*, **25**(12), 2193–2203.

Marusic, I. 2001. On the role of large-scale structures in wall turbulence. *Phys. Fluids*, **13**(3), 735–743.

Marusic, I. 2009. Unravelling turbulence near walls. *J. Fluid Mech.*, **632**, 431–442.

Marusic, I., and Heuer, W. D. C. 2007. Reynolds Number Invariance of the Structure Inclination Angle in Wall Turbulence. *Phys. Rev. Lett.*, **99**, 114504.

Marusic, I., and Kunkel, G. J. 2003. Streamwise turbulence intensity formulation for flat-plate boundary layers. *Phys. Fluids*, **15**, 2461–2464.

Marusic, I., and Perry, A. E. 1995. A wall wake model for the turbulent structure of boundary layers. Part 2. Further experimental support. *J. Fluid Mech.*, **298**, 389–407.

Marusic, I., McKeon, B. J., Monkewitz, P. A., Nagib, H. M., Smits, A. J., and Sreenivasan, K. R. 2010. Wall-bounded turbulent flows at high Reynolds numbers: Recent advances and key issues. *Phys. Fluids*, **22**, 065103.

McKeon, B. J., Li, J., Jiang, W., Morrison, J. F., and Smits, A. J. 2004. Further observations on the mean velocity distribution in fully developed pipe flow. *J. Fluid Mech.*, **501**, 135–147.

Metzger, M. M., and Klewicki, J. C. 2001. A comparative study of near-wall turbulence in high and low Reynolds number boundary layers. *Phys. Fluids*, **13**(3), 692–701.

Monkewitz, P. A., Chauhan, K. A., and Nagib, H. M. 2007. Self-contained high-Reynolds-number asymptotics for zero-pressure-gradient turbulent boundary layers. *Phys. Fluids*, **19**, 115101.

Monkewitz, P. A., Chauhan, K. A., and Nagib, H. M. 2008. Comparison of mean flow similarity laws in zero pressure gradient turbulent boundary layers. *Phys. Fluids*, **20**, 105102.

Monty, J. P., Hutchins, N., Ng, H.C.H., Marusic, I., and Chong, M. S. 2009. A comparison of turbulent pipe, channel and boundary layer flows. *J. Fluid Mech.*, **632**, 431–442.

Morris, S. C., Stolpa, S. R., Slaboch, P. E., and Klewicki, J. C. 2007. Near surface particle image velocimetry measurements in a transitionally rough-wall atmospheric surface layer. *J. Fluid Mech.*, **580**, 319–338.

Nagib, H. M., and Chauhan, K. A. 2008. Variations of von Kármán coefficient in canonical flows. *Phys. Fluids*, **20**, 101518.

Nagib, H. M., Chauhan, K. A., and Monkewitz, P. A. 2007. Approach to an asymptotic state for zero pressure gradient turbulent boundary layers. *Phil. Trans. R. Soc. Lond. A*, **365**, 755.

Nickels, T. B., Marusic, I., Hafez, S. M., and Chong, M. S. 2005. Evidence of the k^{-1} Law in a High-Reynolds-Number Turbulent Boundary Layer. *Phys. Rev. Letters*, **95**, 074501.

Nikuradse, J. 1932. *VDI Forschungsheft Arb. Ing.-Wes.*, **356**(Also NACA TT F-10 359).

Oberlack, M. 2001. A unified approach for symmetries in plane parallel turbulent shear flows. *J. Fluid Mech.*, **427**, 299–328.

Offen, G. R., and Kline, S. J. 1975. A proposed model of the bursting process in turbulent boundary layers. *J. Fluid Mech.*, **70–2**, 209–228.

Okamoto, N., Yoshimatsu, K., Schneider, K., Farge, M., and Kaneda, Y. 2007. Coherent vortices in high resolution direct numerical simulation of homogeneous isotropic turbulence: A wavelet viewpoint. *Phys. Fluids*, **19**, 115109.

Panton, R. L. 2001. Overview of the self-sustaining mechanisms of wall turbulence. *Prog. Aero. Sci.*, **37**, 341–383.

Perry, A. E., and Chong, M. S. 1982. On the mechanism of wall turbulence. *J. Fluid Mech.*, **119**, 173–217.

Perry, A. E., and Marusic, I. 1995. A wall-wake model for the turbulence structure of boundary layers. Part 1. Extension of the attached eddy hypothesis. *J. Fluid Mech.*, **298**, 361–388.

Perry, A. E., Henbest, S. M., and Chong, M. S. 1986. A theoretical and experimental study of wall turbulence. *J. Fluid Mech.*, **165**, 163–199.

Phong-Anant, D., Antonia, R. A., Chambers, A. J., and Rajagopalan, S. 1980. Features of the organized motion in the atmospheric surface layer. *J. Geophys. Res.*, **85**, 424–432.

Praturi, A. K., and Brodkey, R. S. 1978. A stereoscopic visual study of coherent structures in turbulent shear flow. *J. Fluid Mech.*, **89**, 251–272.

Robinson, S. K. 1991. Coherent motions in turbulent boundary layers. *Annu. Rev. Fluid Mech.*, **23**, 601–639.

Rotta, J. C. 1962. Turbulent boundary layers in incompressible flow. *Prog. Aero. Sci.*, **2**, 1–219.

Schlatter, P., Orlu, R., Li, Q., Brethouwer, G., Fransson, J. H. M., Johansson, A. V., Alfredsson, P. H., and Hennningson, D. S. 2009. Turbulent boundary layers up to $Re_\theta = 2500$ studied through simulation and experiment. *Phys. Fluids*, **21**, 051702.

Schlatter., P., Li, Q., Brethouwer, G., Johansson, A. V., and Hennningson, D. S. 2010. Simulation of spatially evolving turbulent boundary layers up to $Re_\theta = 4300$. *Int. J. Heat and Fluid Flow*, **31**, 251–261.

Schoppa, W., and Hussain, F. 2002. Coherent structure generation in near-wall turbulence. *J. Fluid Mech.*, **453**, 57–108.

Smith, C. R., Walker, J. D. A., Haidari, A. H., and Soburn, U. 1991. On the dynamics of near-wall turbulence. *Phil. Trans. R. Soc. Lond. A*, **336**, 131–175.

Smits, A. J., and Dussauge, J. P. 1996. *Turbulent Shear Layers in Supersonic Flow*. AIP Press.

Smits, A. J., McKeon, B. J., and Marusic, I. 2011. High Reynolds number wall turbulence. *Annu. Rev. Fluid Mech.*, **43**, 353–375.

Spalart, P. R. 1988. Direct simulation of turbulent boundary layer up to $R_\theta = 1410$. *J. Fluid Mech.*, **187**, 61–98.

Spalart, P. R., Coleman, G. R., and Johnstone, R. 2008. Direct numerical simulation of the Ekman layer: A step in Reynolds number, and cautious support for a log law with a shifted origin. *Phys. Fluids*, **20**, 101507.

Sreenivasan, K. R. 1987. A unified view of the origin and morphology of the turbulent boundary layer structure. In: Liepmann, H.W., and Narasimha, R. (eds), *Turbulence Management and Relaminarization*. Springer-Verlag.

Sreenivasan, K. R. 1989. The turbulent boundary layer. In: Gad-el-Hak, M. (ed), *Frontiers in Experimental Fluid Mechanics*. Springer-Verlag.

Sreenivasan, K. R., and Sahay, A. 1997. The persistence of viscous effects in the overlap region and the mean velocity in turbulent pipe and channel flows. Pages 253–272 of: Panton, R. (ed), *Self-Sustaining Mechanisms of Wall Turbulence*. Southampton U.K.: Comp. Mech. Publ.

Stanislas, M., Perret, L., and Foucaut, J.-M. 2008. Vortical structures in the turbulent boundary layer: a possible route to a universal representation. *J. Fluid Mech.*, **602**, 327–382.

Tanahashi, M., Kang, S.-J., Miyamoto, T., Shiokawa, S., and Miyauchi, T. 2004. Scaling law of fine scale eddies in turbulent channel flows up to $Re_\tau = 800$. *Int. J. Heat Fluid Fl.*, **25**, 331–340.

Tardu, S. 1995. Characteristics of single and clusters of bursting events in the inner layer, Part 1: Vita events. *Exp. Fluids*, **20**, 112–124.

Tomkins, C. D., and Adrian, R. J. 2003. Spanwise structure and scale growth in turbulent boundary layers. *J. Fluid Mech.*, **490**, 37–74.

Tomkins, C. D., and Adrian, R. J. 2005. Energetic spanwise modes in the logarithmic layer of a turbulent boundary layer. *J. Fluid Mech.*, **545**, 141–162.

Townsend, A. A. 1956. *The Structure of Turbulent Shear Flow*. Cambridge University Press.

Townsend, A. A. 1976. *The Structure of Turbulent Shear Flow*. 2nd edition, Cambridge University Press.

Wallace, J. M., Park, G. I., Wu, X., and Moin, P. 2010. Boundary layer turbulence in transitional and developed states. In: *Proc. Summer Program 2010, Center Turb. Research*. Stanford Univ., Stanford CA.

Wei, T., Fife, P., Klewicki, J. C., and McMurtry, P. 2005. Properties of the mean momentum balance in turbulent boundary layer, pipe and chanel flows. *J. Fluid Mech.*, **522**, 303–327.

Willmarth, W. W., and Lu, S. S. 1972. Structure of the Reynolds stress near the wall. *J. Fluid Mech.*, **55**, 65–92.

Wosnik, M., Castillo, L., and George, W. K. 2000. A theory for turbulent pipe and channel flows. *J. Fluid Mech.*, **421**, 115–145.

Wu, X., and Moin, P. 2009. Direct numerical simulation of turbulence in a nominally-zero-pressure-gradient flat-plate boundary layer. *J. Fluid Mech.*, **630**, 5–41.

Zagarola, M. V., and Smits, A. J. 1998. Mean-flow scaling of turbulent pipe flow. *J. Fluid Mech.*, **373**, 33–79.

Zhou, J., Adrian, R. J., Balachandar, S., and Kendall, T. M. 1999. Mechanisms for generating coherent packets of hairpin vortices in channel flow. *J. Fluid Mech.*, **387**, 353–396.

6
Dynamics of Wall-Bounded Turbulence

Javier Jiménez and Genta Kawahara

6.1 Introduction

This chapter deals with the dynamics of wall-bounded turbulent flows, with a decided emphasis on the results of numerical simulations. As we will see, part of the reason for that emphasis is that much of the recent work on dynamics has been computational, but also that the companion chapter by Marusic and Adrian (2012) in this volume reviews the results of experiments over the same period.

The first direct numerical simulations (DNS) of wall-bounded turbulence (Kim et al., 1987) began to appear soon after computers became powerful enough to allow the simulation of turbulence in general (Siggia, 1981; Rogallo, 1981). Large-eddy simulations (LES) of wall-bounded flows had been published before (Deardorff, 1970; Moin and Kim, 1982) but, after DNS became current, they were de-emphasized as means of clarifying the flow physics, in part because doubts emerged about the effect of the poorly resolved near-wall region on the rest of the flow. Some of the work summarized below has eased those misgivings, and there are probably few reasons to distrust the information provided by LES on the largest flow structures, but any review of the physical results of numerics in the recent past necessarily has to deal mostly with DNS. Only atmospheric scientists, for whom the prospect of direct simulation remains remote, have continued to use LES to study the mechanics of the atmospheric surface layer (see, for example, Deardorff, 1973; Siebesma et al., 2003). A summary of the early years of numerical turbulence research can be found in Rogallo and Moin (1984) and Moin and Mahesh (1998). Here, we review the advances that have taken place in its application to wall-bounded flows since the first of those two papers, and, in particular, what we have learned during that time about the physics of the flow. The interactions with experimental results will be discussed where appropriate.

First of all, it is important to emphasize that numerical simulations and experiments are different, although related, approximations to reality, and that they should not be expected to agree exactly with each other. For example, computing a channel in a doubly periodic domain is an approximation, in the same way as an experimental channel with a finite aspect ratio. In both cases, the parameters have to be chosen so that the resulting system approaches 'closely enough' the ideal channel of real interest, which is infinitely long and wide. Simulations only need to reproduce experiments, and vice versa, in the sense that they are both intended to approximate the same real object, and neither of the two is automatically to be preferred to the other.

Luckily, the strengths and weaknesses of the two techniques tend to be complementary. Simulations give precise answers to precisely posed and controlled problems, with precisely known initial and boundary conditions. No simulation can be run without fully stipulating all those factors, and the only limitation to our knowledge in that respect is how many data we choose to store. The situation is typically different in experiments, for which there are often unknown factors that are difficult to document, such as inflow conditions and wind tunnel imperfections. Those imperfections are different from the ones in numerics, but they are in no sense more 'natural'. For example, it is tempting to consider that the random perturbations in the free stream of a wind tunnel are a 'more natural' trigger for transition than the deterministic perturbations typically introduced in simulations, but there is nothing random about the natural world, and the perturbations present in different experiments are as different from each other, and as different from the ones presumably present in our intended applications, as any deterministic perturbation imposed in a simulation. In most cases, randomness is not a virtue, but a measure of our ignorance.

Another difference that is often mentioned is that experiments are imperfect measurements of a 'true' system, while simulations are perfect measurements of something that is at best only an approximation. Both things are true, but only in the weak sense that it is not completely clear whether the Navier–Stokes equations represent fluid flow, or whether they have continuous unique solutions. The first concern is probably not worse than worrying about whether the air in a wind tunnel is a perfect gas, but the second is slightly more substantive. Numerical methods typically assume some continuity in the solutions that they represent, in which case their error bounds are extremely well understood and kept under tight control in most simulations. The error associated with the numerical approximation is usually negligible compared with the statistical uncertainty. On the other hand, most simulation codes are designed to be 'robust', and to behave gracefully when an unexpected discontinuity is found. For that reason, they would probably miss a hypothetical isolated finite-time singularity weak enough to have only a local effect. Note that if those singular, or perhaps exceptionally sharp, events were anything but rare, they would be caught and avoided by most numerical techniques, and that identifying rare events experimentally is also hard; but it is probably true that experiments have a better chance of doing so than simulations.

Another important difference between simulations and experiments is that the latter are typically cheaper to run, at least once the initial investment is discounted, and that it is easier to rerun an experiment than to repeat a large DNS. As a consequence, experiments can scan a wider class of flows for a given effort, and they remain the method of choice when the behaviour of some simple quantity is required over a wide range of parameters. They are also able to collect more cheaply long statistical series, although, for reasons unclear to the present authors, it is often true that experimental data are noisier than numerical ones.

A characteristic of simulations is that everything is measured, and that everything can be stored and reprocessed in the future. Recent experimental techniques, such as PIV, approach the full-field capability of simulations (see the chapter by Marusic and Adrian in this volume), but the ability of DNS to create time-resolved three-dimensional fields of the three velocity components, gradients, and pressure, is still unmatched. That is an obvious advantage of simulations, but it is also responsible for their higher cost. It is no cheaper to compute the mean velocity profile of a boundary layer than to compute the whole flow field, but it is infinitely cheaper experimentally. That is one of the reasons why experiments, at least apparently, cover a wider range of Reynolds numbers than simulations. It is relatively easy to create a high-Reynolds-number flow in the laboratory, even if it is hard to measure it in detail beyond low order magnitudes such as the mean velocity, or some fluctuation intensities. Simulations, on the other hand, need to compute everything to estimate even those quantities, and the cost of increasing the Reynolds number has to be paid 'in full'.

Even so, the Reynolds numbers of simulations have increased steadily in the past quarter century, and it is important to realize that this process is not open-ended. The goal of turbulence theory, and of the supporting simulations and experiments, should not be to reach ever increasing Reynolds numbers, but to describe turbulence well enough to be able to make useful predictions under any circumstance. It has been understood since Kolmogorov (1941) that the key complication of turbulence is its multiscale character, and it is probably true that, if we could compile a detailed data base of the space-time evolution of enough flows with a reasonably wide range of scales, such a data base would contain all the information required to formulate a theory of turbulence. Of course, such a data set would not be a theory, but it is doubtful whether further increasing the Reynolds number of the simulations, or of the experiments, would provide much additional help in formulating one (Jiménez, 2012).

It is difficult to say a-priori when that stage will be reached, but a rough estimation is possible. The large, energy-containing, structures of wall-bounded flows are of the order of the boundary layer thickness δ (which we will also use for the channel half-width or for the pipe radius), and the smallest near-wall viscous structures have sizes of the order of $100\nu/u_\tau$, where ν is the fluid viscosity, and $u_\tau = \tau_w^{1/2}$ is the friction velocity, defined in term of the wall shear stress τ_w. Variables scaled with u_τ and ν are said to be in wall units, and will be denoted by a '+' superscript. Note that, since we restrict ourselves in this chapter to incompressible flows, we will always assume the fluid density to be $\rho = 1$, and drop it from our equations. Capitals are used for instantaneous values, lower-case symbols for fluctuations with respect to the mean, and primes for the root-mean-squared intensities of those fluctuations. The average $\langle \cdot \rangle$ is conceptually defined over many equivalent independent experiments, unless otherwise noted. We denote by x, y and z the streamwise, wall-normal and spanwise coordinates, respectively, and the corresponding velocity components by U, V and W. With this notation, the ratio between the largest and smallest scales in the flow is $\delta^+/100$, where $\delta^+ = u_\tau \delta/\nu$ is the friction Reynolds number.

The channel simulations of Kim et al. (1987) had $\delta^+ = 180$, and therefore had essentially no scale range, but the more recent numerical channels by del Álamo et al. (2004), Abe et al. (2004) or Hoyas and Jiménez (2006), and the boundary layers by Lee and Sung (2007, 2011), Simens et al. (2009), Wu and Moin (2010), Schlatter et al. (2009); Schlatter and Örlü (2010), and Sillero et al. (2010), with $\delta^+ \approx 1000 - 2000$, are comparable to most well-resolved experiments, and have a full decade of scale disparity. It will be shown in §6.4 that those simulations are already providing a lot of information on the multiscale dynamics of wall-bounded flows. Moreover, because the Reynolds numbers of simulations and experiments are beginning to be comparable, it is now possible to validate, for example, the structural models derived from experimental observations (e.g. Adrian, 2007) with the time-resolved three-dimensional flow fields of simulations, and vice versa.

It is probably true that a further factor of 5–10 in δ^+, which would give us a range of scales close to 100, would provide us with all the information required to understand many of the dynamical aspects of wall-bounded turbulence. Using the usual estimate of δ^{+3} for the cost of simulations, and the present rate of increase in computer speed of 10^3 per decade, it should be possible to compile such a data base within the next decade (Jiménez, 2003).

This brings us to discuss the question of structure versus statistics, which has been a recurrent theme in turbulence theory from its beginning. Thus, while Richardson (1920) framed the multiscale nature of turbulence in terms of 'little and big whorls', the older decomposition paper of Reynolds (1894) had centred solely on the statistics of the fluctuations, and proved to be more fruitful for the practical problem of turbulence modelling. Even the classical paper of Kolmogorov (1941), which is usually credited with introducing the concept of a turbulent cascade, is a statistical description of the fluctuation intensity versus scale, and it was only the slightly less famous companion paper by Obukhov (1941) that put the cascade concept in terms of interactions among eddies. It can be argued that it was not until the visualizations of large coherent structures in free-shear layers by Brown and Roshko (1974), and of sublayer streaks in boundary layers by Kline et al. (1967), that the structural view of turbulence gained modern, although still far from universal, acceptance.

This is not the place to discuss the relative merits of the two points of view, which, in any case, should be judged in relation to each particular application, but simulations have come down decisively on the side of structure. This is in part because, as we have seen, simulations are an expensive way of compiling statistics, in the same way that experiments are not very good at extracting structure, but it also points to one of the characteristic advantages of numerics, which is their ability to simulate unphysical systems.

Structural models often take a deterministic view of the flow, and deal with how its different parts interact with each other. Some of the most powerful tools for analysing interactions in physics have long been conceptual experiments, which often involve systems that cannot be physically realized, such as, for example, point masses. In simple dynamical situations, the outcome of such experiments can often be guessed correctly, and used to judge the soundness of a particular model, but in complex phenomena, such as turbulence, the guessing almost always has to be substituted by the numerical simulation of the modified system. We will see in §6.3 examples of how the dynamics of near-wall turbulence was clarified in part through experiments of this kind, in which some of the inherent limitations of numerics, such as spatial periodicity, were put to good use in isolating what was essential, and what accidental. Similar techniques have been used for isotropic turbulent flows, which are outside the scope of the present article, but some examples can be found in Kida (1985) or She (1993).

It should be noted that conceptual experiments are not completely beyond the reach of the laboratory. For example, rough walls can be used as conceptual tests for the importance of near-wall processes in wall turbulence, since roughness destroys the detailed interactions that dominate the flow over smooth walls (Flores and Jiménez, 2006). In the same way, the use by Bradshaw (1967) of adverse pressure gradients to explore the importance of 'inactive' modes in boundary layers, remains one of the most beautiful examples of the experimental use of those techniques. However, the freedom afforded by numerical simulations to create artificial systems is difficult to match experimentally.

There are finally two approaches to the study of turbulence in which simulation techniques are predominant. The first one is the study of equilibrium, or otherwise simple, solutions of the Navier–Stokes equations that may be important in turbulent flows. It is generally understood that turbulence is chaotic, and it would be a surprise if a steady structure, or even a steady wave, were to be found in a natural turbulent flow. But it is also a common experience that such flows contain 'coherent' structures with long lifetimes. Examples range from the already cited large-scale coherent eddies of jets and shear layers, or from the sublayer streaks in wall-bounded turbulence, to the small-scale long-lived vortices in the dissipative range of many turbulent flows (Vincent and Meneguzzi, 1991; Jiménez et al., 1993). Not only are those structures believed to contribute substantially to the overall statistical properties of their respective flows but, once they are properly understood, they point to efficient control strategies (Ho and Huerre, 1984).

The prevalence of such structures raises the question of whether they can be identified with underlying solutions of the equations of motion, which are almost certain to be unstable, and therefore experimentally unobservable, but which can be extracted numerically as properties of the averaged velocity profile of the flow under study. For wall-bounded turbulence, the first solutions of this kind were obtained by Nagata (1990), and many more have been found since then. They are not only conceptually important, but their signature can be identified in full-scale turbulence (Jiménez et al., 2005). Moreover, not only the equilibrium structures, but their connections in phase space have been examined more recently, and appear to be related to the temporal modulation of the near-wall velocity fluctuations (Halcrow et al., 2009). This approach is reviewed in §6.5.

The last interesting result of numerical experiments is the study of the possible relation between linear dynamics and the large-scale structures of turbulent flows. Turbulence is nonlinear, but a consideration of the time scales of the different processes shows that the dominant effect in the creation of the largest scales of shear flows is the energy transfer from the mean shear to the fluctuations, which is a linear process. It was understood from the beginning that the coherent eddies of free-shear layers were reflections of the Kelvin-Helmholtz instability of the mean velocity profile (Brown and Roshko, 1974; Gaster et al., 1985), but it was thought for a long time that wall-bounded flows, whose mean velocity profiles are linearly stable, could not be explained in the same way. That changed when it was realized in the early 1990's that even stable linear perturbations can grow substantially by extracting energy from the mean flow, and that it is possible to relate such 'transient' growth to the observed coherent structures in wall-bounded turbulence (Butler and Farrell, 1993; del Álamo and Jiménez, 2006). Space considerations prevent us from discussing that question here in detail, but occasional references to it will be made where appropriate.

6.2 The classical theory of wall-bounded turbulence

Wall-bounded turbulence includes pipes, channels and boundary layers. We will restrict ourselves to cases with little or no longitudinal pressure gradients, since otherwise the flow tends to relaminarize or to separate. In the first case it stops being turbulent, and in the second one it loses many of its wall-bounded characteristics, and tends to resemble free-shear flows. Wall-bounded turbulence is of huge technological importance. About half of the energy spent worldwide in moving fluids along pipes and canals, or vehicles through air or water, is dissipated by turbulence in the immediate vicinity of walls. Turbulence was first studied scientifically in attached wall-bounded flows (Hagen, 1839; Darcy, 1854), but those flows have remained to this day worse understood than their homogeneous or free-shear counterparts. That is in part because what is sought in both cases is different. In the classical conceptual model for isotropic turbulence, energy resides in the largest eddies, and cannot be dissipated until it is transferred by a self-similar cascade of 'inertial' eddies to the smaller scales of the order of the Kolmogorov viscous length η, where viscosity can act (Richardson, 1920; Kolmogorov, 1941). The resulting energy spectrum, although now recognized as only an approximation, describes well the experimental observations, not only for

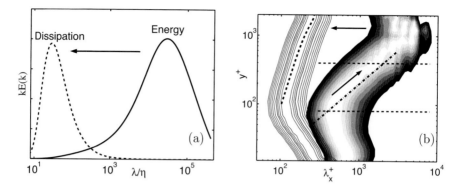

Figure 6.1 Spectral energy density, $kE(k)$. (a) In isotropic turbulence, as a function of the isotropic wavelength $\lambda = 2\pi/k$. (b) In a numerical turbulent channel with half-width $\delta^+ = 2003$ (Hoyas and Jiménez, 2006), plotted as a function of the streamwise wavelength λ_x, and of the wall distance y. The shaded contours are the density of the kinetic energy of the velocity fluctuations, $k_x E_{\mathbf{uu}}(k_x)$. The lines are the spectral density of the surrogate dissipation, $\nu k_x E_{\omega\omega}(k_x)$, where $\boldsymbol{\omega}$ are the vorticity fluctuations. At each y the lowest contour is 0.86 times the local maximum. The horizontal lines are $y^+ = 80$ and $y/\delta = 0.2$, and represent conventional limits for the logarithmic layer. The diagonal through the energy spectrum is $\lambda_x = 5y$; the one through the dissipation spectrum is 40η. The arrows indicate the implied cascades.

isotropic turbulence, but also for small-scale turbulence in general. A sketch can be found in Figure 6.1(a).

However, isotropic theory gives no indication of how energy is fed into the turbulent cascade. In shear flows, the energy source is the gradient of the mean velocity, and the mechanism is the interaction between that gradient and the average momentum fluxes carried by the velocity fluctuations (Tennekes and Lumley, 1972). We have already mentioned that, in free-shear flows such as jets or mixing layers, this leads to large-scale instabilities of the mean velocity profile (Brown and Roshko, 1974), with sizes of the order of the flow thickness. The resulting 'integral' eddies contain most of the energy, and the subsequent transfer to the smaller scales is thought to be essentially similar to the isotropic case.

In contrast, wall-bounded flows force us to face squarely the role of inhomogeneity. That can be seen in Figure 6.1(b), in which each horizontal section is the equivalent to the spectra in Figure 6.1(a) for a given wall distance. The energy is again at large scales, while the dissipative eddies are smaller, but the sizes of the energy-containing eddies change with the distance to the wall, and so does the range of scales over which the energy has to cascade. It turns out that, except very near or very far from the wall, where there are small imbalances between the production and the dissipation of turbulent kinetic energy, most of the energy generated at a given distance from the wall is dissipated locally (Hoyas and Jiménez, 2008), but the eddy sizes containing most of the energy at one wall distance are in the midst of the inertial cascade when they are observed farther away. The Reynolds number, defined as the scale disparity between energy and dissipation at some given location, also changes with wall distance, and the main emphasis in wall turbulence is not on the local inertial energy cascade, but on the interplay between different scales at different distances from the wall.

Models for wall-bounded turbulence also have to deal with spatial fluxes that are absent from the homogeneous case. The most important one is that of momentum. Consider a turbulent channel, driven by a pressure gradient between infinite parallel planes, and decompose the flow quantities into mean values and fluctuations with respect to those means. Using streamwise and spanwise homogeneity, and assuming that the averaged velocities are stationary, the mean streamwise momentum equation is

$$\partial_x \langle P \rangle = -\partial_y \langle uv \rangle + \nu \partial_{yy} \langle U \rangle, \qquad (6.1)$$

Streamwise momentum is fed across the channel by the mean pressure gradient, $\partial_x \langle P \rangle$, which acts over the whole cross section, and is removed by viscous friction at the wall, to where it is carried by the averaged momentum flux of the fluctuations $-\langle uv \rangle$. This tangential Reynolds stress resides in eddies of roughly the same scales as the energy, and it is clear from Figure 6.1(b) that the sizes of the stress-carrying eddies change as a function of the wall distance by as much as the scale of the energy across the inertial cascade. This implies that momentum is transferred in wall-bounded turbulence by an extra spatial cascade. Momentum transport is present in all shear flows, but the multiscale spatial cascade is characteristic of very inhomogeneous situations, such as wall turbulence, and complicates the problem considerably.

The wall-normal variation of the range of scales across the energy cascade divides the flow into several distinct regions. Wall-bounded turbulence over smooth walls can be described by two sets of scaling parameters (Tennekes and Lumley, 1972). Viscosity is important near the wall, and length and velocity in that region scale in wall units. There is no scale disparity in this region, as seen in Figure 6.1(b). Most large eddies are excluded by the presence of the impermeable wall, and the energy and dissipation are at similar sizes. If y is the distance to the wall, y^+ is a Reynolds number for the size of the structures, and it is never large within this layer, which is typically defined at most as $y^+ \lesssim 150$ (Österlund et al., 2000). It is conventionally divided into a viscous sublayer, $y^+ \lesssim 10$, where viscosity is dominant, and a 'buffer' layer in which both viscosity and inertial effects have to be taken into account.

The velocities also scale with u_τ away from the wall, because the momentum equation 6.1 requires that the Reynolds stress, $-\langle uv \rangle$, can only change slowly with y to compensate for the pressure gradient. This uniform velocity scale is the extra constraint introduced in wall-bounded flows by the momentum transfer. The length scale in the region far from the wall is the flow thickness δ, but, between the inner and the outer regions, there is an intermediate layer where the only available length scale is the distance y to the wall .

Both the constant velocity scale across the intermediate region, and the absence of a length scale other than y, are only approximations. It will be seen below that large-scale eddies of size $O(\delta)$ penetrate to the wall, and that the velocity does not scale strictly with u_τ even in the viscous sublayer. On the other hand, Figure 6.1(b) shows that, for $y/\delta \lesssim 0.2$, the length scale of the energy-containing eddies is approximately proportional to y, and, if both approximations are accepted, it follows from relatively general arguments that the mean velocity in this 'logarithmic' layer is (Townsend, 1976)

$$\langle U \rangle^+ = \kappa^{-1} \log y^+ + A. \qquad (6.2)$$

This form agrees well with experimental evidence, with an approximately universal Kármán constant, $\kappa \approx 0.4$, but the intercept A depends on the details of the near-wall region, because 6.2 does not extend to the wall. For smooth walls, $A \approx 5$. In spite of its simplicity and good experimental agreement, the theoretical argument leading to 6.2 has been challenged (Barenblatt et al., 2000) and extended (Afzal and Yajnik, 1973). For example, it is theoretically possible to include a coordinate offset inside the logarithm in 6.2, which is expected to scale roughly in wall units (Wosnik et al., 2000; Oberlack, 2001; Spalart et al., 2008). Since that question is essentially unrelated to the simulation results, it will not be pursued here, but a short critical discussion, including a reanalysis of the DNS results, can be found in Jiménez and Moser (2007).

The viscous, buffer, and logarithmic layers are the most characteristic features of wall-bounded flows, and constitute the main difference between them and other types of turbulence. Even if they are geometrically thin with respect to the layer as a whole, they are extremely important. We saw in the introduction that the ratio between the inner and the outer length scales is $10^{-2}\delta^+$, where the friction Reynolds number δ^+ ranges from 200 for barely turbulent flows, to 10^6 for large water pipes. In the latter, the near-wall layer is only about 10^{-4} times the pipe radius, but it follows from 6.2 that, even in that case, 35% of the velocity drop takes place below $y^+ = 50$. Because there is relatively little energy transfer among layers, except in the viscous and buffer regions, those percentages also apply to where the energy is dissipated. Turbulence is characterized by the expulsion towards the small scales of the energy dissipation, away from the large energy-containing eddies. In the limit of infinite Reynolds number, this is believed to lead to non-differentiable velocity fields. In wall-bounded flows that separation occurs not only in the scale space for the velocity fluctuations, but also in the shape of the mean velocity profile for the momentum transfer. The singularities are expelled both from the large scales, and from the centre of the flow towards the logarithmic and viscous layers near the walls.

The near-wall viscous layer is relatively easy to simulate numerically because the local Reynolds numbers are low, and difficult to study experimentally because it is usually very thin in laboratory flows. Its modern study began experimentally in the 1970's (Kline et al., 1967; Morrison et al., 1971), but it got its strongest impulse with the advent of high-quality direct numerical simulations in the late 1980's and in the 1990's (Kim et al., 1987). We will see in the next section that it is one of the turbulent systems about which most is known.

Most of the velocity difference that does not reside in the near-wall viscous region is concentrated just above it, in the logarithmic layer, which is also unique to wall turbulence. It follows from 6.2 that the velocity difference above the logarithmic layer is about 20% of the total when $\delta^+ = 200$, and that it decreases logarithmically as the Reynolds number increases. In the limit of very large Reynolds numbers, all the velocity drop is in the logarithmic layer.

The logarithmic layer is intrinsically a high-Reynolds number phenomenon. Its existence requires at least that its upper limit should be above the lower one, so that $0.2\delta^+ \gtrsim 150$, and $\delta^+ \gtrsim 750$. The local Reynolds numbers y^+ of the eddies are also never too low. The logarithmic layer has been studied experimentally for a long time, but numerical simulations with even an incipient logarithmic region have only recently become available. It is worse understood than the viscous layers, and will be reviewed in §6.4.

6.3 The dynamics of the near-wall region

The region below $y^+ \approx 100$ is dominated by coherent streaks of the streamwise velocity and by quasi-streamwise vortices. The former are an irregular

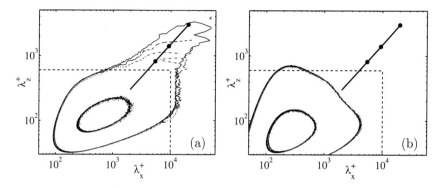

Figure 6.2 Two-dimensional spectral energy density $k_x k_z E(k_x, k_z)$ in the near-wall region ($y^+ = 15$), in terms of the streamwise and spanwise wavelengths. Numerical channels, —·—, $\delta^+ = 547$ (del Álamo and Jiménez, 2003); - - - -, 934 (del Álamo et al., 2004); ———, 2003 (Hoyas and Jiménez, 2006). Spectra are normalized in wall units, and the two contours for each spectrum are 0.125 and 0.625 times the maximum of the spectrum for the highest Reynolds number. The heavy straight line is $\lambda_z = 0.15\lambda_x$, and the heavy dots are $\lambda_x = 10\delta$ for the three cases. The dashed rectangle is the spectral region used in Figure 6.3(a) to isolate the near-wall structures. (a) Kinetic energy. (b) Enstrophy, $|\omega|^2$.

array of long ($x^+ \approx 10^3$–10^4) sinuous alternating streamwise jets superimposed on the mean shear, with an average spanwise separation of the order of $z^+ \approx 100$. The quasi-streamwise vortices are slightly tilted away from the wall, and stay in the near-wall region only for $x^+ \approx 100$ (Moin and Moser, 1989). Several vortices are associated with each streak (Jiménez et al., 2004), with a longitudinal spacing of the order of $x^+ \approx 300$.

The streaks and the vortices are easily separated in the two-dimensional spectral densities in Figure 6.2, which are taken at the kinetic energy peak near $y^+ = 15$. The streaks are represented by the spectra of the kinetic energy in Figure 6.2(a), which are dominated by the streamwise velocity. The vortices are represented by the enstrophy spectra in Figure 6.2(b), which are very similar to those of the wall-normal velocity at this distance to the wall. The three spectra in each figure correspond to turbulent channels at different Reynolds numbers, and differ from one another almost exclusively in the long and wide structures in the upper-right corner of the kinetic energy spectra, with sizes of the order of $\lambda_x \times \lambda_z = 10\delta \times 1.5\delta$. A rectangle with these dimensions has been added to Figure 6.4, where the large-scale modulation of the flow can be easily seen.

In this section, we deal mostly with the rotational structures in the spectral region in which $\lambda_x^+ \lesssim 10^4$ and $\lambda_z^+ \lesssim 600$. Figure 6.3(a) shows that, when the statistics are computed within that window, they are essentially indepen-

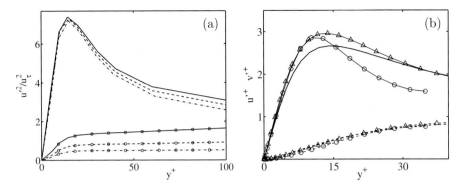

Figure 6.3 Profiles of the velocity fluctuations. (a) Squared intensities of the streamwise velocity. Simple lines are computed within the spectral window $\lambda_x^+ \times \lambda_z^+ = 10^4 \times 600$; those with symbols are outside that window. Lines and conditions as in Figure 6.2. (b) Root-mean-squared intensities. Simple lines are a full channel with $\delta^+ = 180$ (Kim et al., 1987); —△—, a minimal channel with $\delta^+ = 180$ (Jiménez and Moin, 1991); —○—, a permanent-wave autonomous solution (Jiménez and Simens, 2001). ———, streamwise velocity; ----, wall-normal velocity.

dent of the Reynolds number, especially below $y^+ \approx 100$. The larger-scale structures outside the window become stronger as the Reynolds number increases, and contain no enstrophy or tangential Reynolds stresses at this wall distance. They extend into the logarithmic layer, where they are both rotational and carry stresses (Hoyas and Jiménez, 2006), and correspond to the 'inactive eddies' of Townsend (1961). They will be discussed in §6.4. In this part of the flow, they are responsible for the growth of the turbulent energy with the Reynolds number (deGraaff and Eaton, 2000).

Note that, even within the rotational region, the longer end of the spectrum is also wider. It was shown by Jiménez et al. (2004) that this is mostly due to meandering of the structures, and that even the longest near-wall low-velocity streaks are seldom wider than $\Delta z^+ = 50$. We will see below that this is also the order of magnitude of the height of those streaks, which are therefore roughly equilateral wavy cylinders. Meandering, as well as a certain amount of branching, is easily seen in the top panel of Figure 6.4 and in Figure 6.6(e), and has been documented in the logarithmic layer by Hutchins and Marusic (2007).

Soon after they were discovered by Kline et al. (1967), it was proposed that the streaks and the vortices were involved in a regeneration cycle in which the vortices are the results of an instability of the streaks (Swearingen and Blackwelder, 1987), while the streaks are caused by the advection of the mean velocity gradient by the vortices (Bakewell and Lumley, 1967). Both processes have been documented and sharpened by numerical experi-

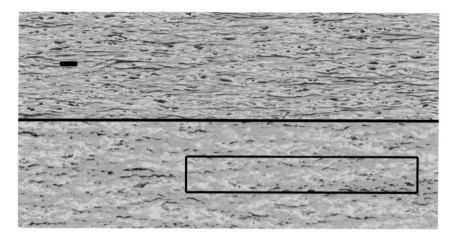

Figure 6.4 Streamwise velocity in wall-parallel planes. The top panel is at $y^+ = 12$, and the bottom one at $y^+ = 50$. Channel at $\delta^+ = 547$ (del Álamo and Jiménez, 2003). The flow is from left to right, and each panel is approximately 10^4 by 5×10^3 wall units. The velocity range in the top panel is $U^+ = 0.6$–17 (blue to red), and in the bottom one, 10–26. The small black rectangle in the top panel is a minimal box, of size 400×100 wall units. The one in the lower panel is $10\delta \times 1.5\delta$.

ments. For example, Jiménez and Pinelli (1999) showed that disturbing the streaks inhibits the formation of the vortices, but only if it is done between $y^+ \approx 10$ and $y^+ \approx 60$, suggesting that it is predominantly between those limits where the regeneration cycle works. There is a substantial body of numerical (Hamilton et al., 1995; Waleffe, 1997; Schoppa and Hussain, 2002) and analytic (Reddy et al., 1998; Kawahara et al., 2003) work on the linear instability of model streaks. The inflection points of their distorted velocity profiles are unstable to sinuous perturbations, and the eigenfunctions closely resemble the shape and location of the observed vortices. The model implied by these instabilities is a cycle in which streaks and vortices are created, grow, generate each other, and eventually decay (see Jiménez and Pinelli, 1999, for additional references).

Although the flow in the buffer layer is disorganized and seemingly chaotic, the spatial chaos is not required to reproduce the turbulence statistics. Jiménez and Moin (1991) presented simulations in which the near-wall region was substituted by an ordered 'crystal' of surprisingly small identical 'minimal' units, of size 400×100 wall units in the streamwise and spanwise directions. A box of that size is superimposed on the top panel of Figure 6.4. Each unit contains a single streak that crosses the box (and is therefore essentially infinite), and, on average, a single pair of staggered quasi-streamwise vortices, in spite of which it reproduces fairly well the statistics of

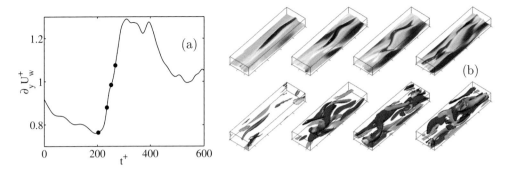

Figure 6.5 Evolution of a minimal Poiseuille flow during a bursting event. $L_x^+ \times \delta^+ \times L_z^+ = 460 \times 180 \times 125$. (a) Evolution of the mean velocity gradient at the wall. The dots correspond to the snapshots on the right. (b) The top frames are isosurfaces $U^+ = 8$, coloured by the distance to the wall, up to $y^+ = 30$ (blue to red). The bottom ones are $\omega_x^+ = 0.2$ (purple), and -0.2 (cyan). The flow is from bottom-left to top-right, and time marches towards the right. The top of the boxes is $y^+ = 60$, and the axes move with velocity $U_a^+ = 7.6$.

the full flow (Figure 6.3b). Moreover, it was shown by Jiménez and Pinelli (1999), that the dynamics of the layer below $y^+ \approx 60$ is autonomous, in the sense that the streaks and the vortices continue regenerating themselves even when the flow above them is artificially removed, and that those local interactions still result in approximately correct statistics.

Even further removed from the real flow are the three-dimensional nonlinear equilibrium solutions of the Navier–Stokes equations discussed in §6.5, which were first obtained numerically at about the same time as the minimal units just mentioned. Their fluctuation intensity profiles, included in Figure 6.3(b), are also strongly reminiscent of experimental turbulence.

On the other hand, although its statistics are well approximated by those steady solutions, the real flow is not steady. The time evolution of the near-wall structures is most easily studied in minimal boxes, because each of them contains only a few structures whose evolution can be traced by integral measures over the whole box. All those measures 'burst' quasi-periodically on times of the order of $T^+ = Tu_\tau^2/\nu \approx 400$ (Jiménez and Moin, 1991; Hamilton et al., 1995; Jiménez et al., 2005). An example is given in Figure 6.5, which depicts the rising phase of a burst. The streak becomes wavy, friction increases sharply, and the streamwise vorticity grows. It is harder to follow the time evolution of individual structures in real flows, but when their statistics are compiled over randomly chosen sub-boxes of 'minimal' size, their distributions agree well with those of the temporal variability of the minimal flows (Jiménez et al., 2005), suggesting that full flows also burst.

The period given above is longer than the survival time of individual vortices, which decay viscously with a characteristic time $T^+ \approx 60$ if their production is artificially inhibited by damping the streaks (Jiménez and Pinelli, 1999). A similar time scale, $T^+ \approx 50$, can be extracted from the space-time spectra of the wall-normal velocity in natural flows (del Álamo et al., 2006). The discrepancy between the vortex lifetimes and the bursting period suggests that the regeneration cycle consists of a relatively quiescent phase followed by shorter eruptions. Jiménez et al. (2005) analysed several minimal flows, and concluded that the bursting phase takes about one-third of the cycle, or $T^+ \approx 100$, half of which is taken by the growth of the instability, and the other half by its decay.

It is likely that the quasi-periodic bursting of the minimal boxes is an artefact of the spatial periodicity, and that the streaks in the real flow move away from the location where they are created before bursting again. In a minimal periodic flow, the spatial periodicity brings the streak back into the simulation box but, in a real one, bursts would be more or less independent of each other. All that probably remains is the average time elapsed from the moment in which a burst creates a vortex and begins to form a streak, until the streak becomes unstable and erupts into a new burst. That time is presumably related to the growth rate of the instabilities of the streak itself, which is proportional to its internal velocity gradient, of the order of $u_\tau/\Delta z$. The implied growth times, $T^+ = O(\Delta z^+)$, are about a hundred wall units, compatible with the bursting times mentioned above. Note that the approximate agreement of those orders of magnitude is unlikely to be a coincidence, and should rather be seen as determining the size of the streaks. Schoppa and Hussain (2002) noted that the streaks found in real flows are typically stable, or at most marginally unstable, and the likeliest interpretation of that observation is that the streaks grow until they become unstable, after which they burst, and are quickly destroyed. On the other hand, the viscous decay time of the streak is $T^+ = O(\Delta z^{+2}) = O(3000)$, much longer than the bursting period, suggesting again that the length of the streaks is determined by their instabilities, rather than by their viscous decay.

The streaks are wakes created in the mean velocity profile as the vortices are advected and sheared. For example, Jiménez et al. (2004) studied the relation between streaks and vortices using as surrogates connected regions in which u or v were more than one standard deviation away from their means in planes parallel to the wall. Two such objects were considered related if their rectangular bounding boxes intersected. Figure 6.6(a) shows the probability density functions (PDFs) of the position of the $v > v'$ ejections with

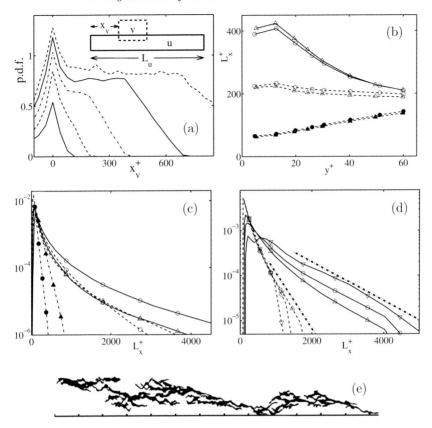

Figure 6.6 Numerical channels as in Figure 6.2. (a) Compensated PDF, $L_u p(x_v|L_u)$, of the positions of the front ends of the bounding boxes of $v > v'$ ejections with respect to the front-end of the low-velocity streaks they intersect, for several bands of streak lengths, limited by $L_u^+ = 50, 100, 200, 400, 700, 1100$. Channel with $\delta^+ = 2003$ and $y^+ = 13$. Longer PDFs correspond to longer streaks. (b) Average lengths of the bounding boxes of strong velocity fluctuations, as functions of y. Lines with open symbols are $|u| \geq u'$. ———, Low-velocity; – – – –, high-velocity. Lines with closed symbols are $|v| \geq v'$. ○, $\delta^+ = 934$; △, 2003. (c) PDFs of the lengths of the velocity perturbations. Lines as in (b), but ○, $y^+ = 13$; △, $y^+ = 50$. $\delta^+ = 2003$. (d) PDFs of the lengths of the u–boxes, restricted to the band $13 \leq y^+ \leq 50$. Lines as in (b), but ▽, $\delta^+ = 547$; ○, 934; △, 2003. The heavy chaindotted lines are exponentials with $L_c^+ = 350$ and 10^3. (e) Long sublayer streak, defined as a connected region in the plane $y^+ = 13$ for which $u < -u'$, in a channel at $\delta^+ = 2003$. The tick marks in the streamwise (x) axis are 1000 wall units apart.

respect to the head of the streak to which they are related, as functions of the streak length. The ejections are uniformly distributed along the streaks, except at the streak head, which is always more likely to have an ejection.

If, in this wake model, we assume an advection velocity $U_{ad}^+ \approx 8$ near the wall (Kim and Hussain, 1993), the vortices would create streaks of length

$L_x^+ \approx 800$ before decaying, which is consistent with the spectral maximum in Figure 6.2(a). Moreover, the ratio (one to three) between the active lifetimes of the bursts and the full regeneration period suggests that there should be a similar ratio between active and total lengths along the streaks, which is also roughly correct. The mean length of the vortices, as measured by the regions of high wall-normal velocity in the buffer layer, is approximately 80 wall units (Figure 6.6b), while we have already seen that their mean spacing along the streaks is of the order of $\lambda_x^+ = 300$.

Figure 6.6(b) shows that the character of the streaks changes around $y^+ \approx 30$–40, which is also where viscosity stops being an important dynamical factor for the energy-containing eddies (Jiménez et al., 2004). Figure 6.6(c) shows the PDFs of the lengths of the streaks (u) and vortices (v) in two planes at different distances from the wall. In the viscous region, $y^+ = 12$, the average length of low-speed streaks, $L_x^+ \approx 500$, is about twice that of the high-speed ones, but they shorten away from the wall, and settle to $L_x^+ \approx 250$ beyond $y^+ = 50$. Above that height, the lengths of the two types of streaks are essentially the same and, at least at the Reynolds number of Figure 6.6(c), the longest streaks are the low-velocity ones in the viscous layer. The simplest explanation is that the high-velocity regions have a higher velocity gradient at the wall, and, being associated with downwashes, are not as tall as the low-velocity ones. The viscous dissipation is proportional to the square of the velocity derivatives, and, at $y^+ \approx 10$, it is predominantly due to the wall-normal gradients. Thus, although there is very little asymmetry between the streaks of both signs regarding the wall-parallel velocity gradients, the dissipation due to the wall-normal ones is much higher in the thinner fast streaks than in the taller low-speed ones. That asymmetry weakens as the effect of viscosity decreases away from the wall, and the two velocity signs then behave similarly. The disappearance of the long streaks above the viscous region can also be seen in the spectral densities shown in Figure 6.8(a), in the next section, which correspond to the logarithmic layer, but which includes the kinetic-energy spectrum at $y^+ = 15$, for comparison. Even at the relatively high Reynolds number of this figure, which would tend to reinforce the larger structures of the outer region, the energy in $\lambda_x^+ \times \lambda_z^+ \approx 10^4 \times 100$ has disappeared at $y^+ = 100$, and the remaining long structures seem to belong to the logarithmic-layer family. The same distinction is clear in Figure 6.4. The streaks in the upper panel, at $y^+ = 12$, are mostly absent from the lower one, at $y^+ = 50$, and the longest structures at the higher level are wider and more disorganized. On the other hand, the positive and negative v-structures, corresponding to the vortices, have lengths similar to each other, and both grow longer away from the wall.

The viscous low-velocity streaks are not only long in the mean, but the PDF of their lengths has a very long superexponential tail, which can only be explained by the interaction of several bursts. A plausible model for that interaction is suggested by Figure 6.6(d). It was shown by del Álamo et al. (2006) that the largest vortical structures at each wall distance are objects that remain attached to the wall even when they extend far into the flow (Townsend, 1961), and that they are associated with long low-velocity features. Consequently, part of the reason why the PDFs in Figure 6.6(c) are long is that they include the roots of taller objects that do not really belong to their wall distance. In an attempt to separate those roots from the PDFs of objects which are local to the viscous region, we display in Figure 6.6(d) the difference between the histograms at the planes $y^+ = 12$ and $y^+ = 50$. The long low-speed streaks of the viscous layer still appear, but their PDF is now a fairly good exponential, with a characteristic length $L_c^+ \approx 1000$, which is the same for the three Reynolds numbers in the figure. The distributions of the high-velocity streaks are also exponential, but with a shorter length scale, $L_c^+ \approx 250$. Exponential probability distributions suggest a Poisson process (Feller, 1971), which can be incorporated into several plausible models for the viscous region.

The simplest one is that streaks grow by the aggregation of smaller units. Consider elementary streaks created by individual bursts with average length L_0, and assume that each such unit has a probability $q < 1$ of connecting with another one, either by chance, or by creating a new burst in its wake. The probability of a composite streak of $n \geq 1$ units is $p(n) = (1-q)q^{n-1}$, which can be written it terms of the streak length, $L = nL_0$, as $p(L) \sim \exp(-L/L_c)$ for $L \geq L_0$, where $L_c = -L_0/\log(q)$. The two factors entering L_c can be separated if we take into account that the PDF peaks at $L = L_0$, in which case the best fit to Figure 6.6(d) is approximately $L_0^+ \approx 500$ and $q \approx 0.6$. Those are sensible numbers, given the previous discussion on the effects of a single burst, and they are given some credence by the appearance of the long streaks, such as the one in Figure 6.6(e), but they should be used with care because the subtraction method used to generate Figure 6.6(d) is hard to justify for the very short events near the mode of the PDF. For example, note that the shortest streaks around L_0 become more frequent as the Reynolds number increases, causing a systematic lowering of the offset of the exponential tails. Moreover, an equally valid model would be that streaks tend to form infinitely long networks, but are cut randomly by some external influence, presumably from the logarithmic or outer layers. The

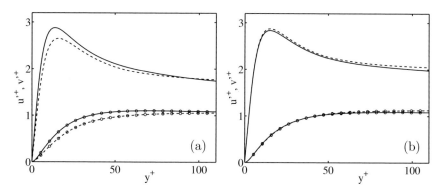

Figure 6.7 Fluctuation intensities computed over sub-boxes with dimensions $\lambda_x \times \lambda_z = 1.5\delta \times 0.75\delta$. Numerical channel at $\delta^+ = 2003$ (Hoyas and Jiménez, 2006). Simple lines are u', and those with symbols are v', conditioned to boxes in which the friction velocity is more than one standard deviation above (———) or below (- - - -) average. (a) Wall distance and velocities normalized with the global friction velocity. (b) Normalized with the friction velocity local to each sub-box.

two models give identical length distributions, and can probably only be distinguished by a full space-time analysis of this region of the flow.

It is interesting that the PDFs of the lengths of the v-events, which represent the vortices, are also exponential, although with a shorter length scale $L_c^+ \approx 200$ near the wall, suggesting that they also form packets. Although this is reminiscent of the experimental observations of vortex packets in the logarithmic region (Adrian, 2007), the difference in length scales makes it unlikely that the two phenomena are directly related.

There is an obvious large-scale modulation in the near-wall panel of Figure 6.4, on scales that we have associated with the outer layer. Mathis et al. (2009) discussed that modulation, and concluded that it is stronger than a simple linear superposition of the velocity of the different spectral modes. It turns out that most of it can be absorbed into the variability of the local friction velocity, which changes from place to place because of the effect of large-scale down- and up-drafts. That can be seen in Figure 6.7, which displays velocity statistics compiled over sub-boxes of sizes comparable to the outer structures. The standard deviation for each sub-box is computed with respect to the average velocity over that sub-box, and the boxes are classified according to their mean friction velocity, which varies with respect to the global mean by about $\pm 5\%$. Not surprisingly, boxes with high friction velocities correspond to mean downdrafts, and those with low friction velocities to updrafts, with maximum differences of the wall-normal velocity averaged over the sub-box, $\langle v \rangle^+ \approx 0.1$, peaking at $y/\delta \approx 0.2$. Figure 6.7(a) compares statistics over boxes whose friction velocities are more than one standard deviation above or below the global mean. The former are more intense than the latter, but most of the difference can be compensated by normalizing both the velocities and the wall distance with the local mean friction velocity averaged over the box, as in Figure 6.7(b). That makes sense, because the boxes, which at the Reynolds number of the simulation are 3000×1500 wall units long and wide, are much larger than the elementary structures of the viscous layer, which therefore 'live' in an environment defined by the mean box properties (Jiménez, 2012).

In summary, our best understanding of the viscous and buffer layers is, at present, that the peak of the energy spectrum, at $\lambda_x^+ \times \lambda_z^+ = 1000 \times 100$, is associated with structures that can be approximated by the steady travelling-wave solutions of the Navier-Stokes equations discussed in §6.5, but which in reality undergo strong quasi-periodic bursting. The bursts create vortices whose wakes are elemental streaks with lengths of the order of 1000 wall units, and which aggregate into the longer observed composite objects, either by chance or by generating new bursts in their tails. Those structures are restricted to $y^+ \lesssim 50$, which is the only part of the flow in which high- and low-velocity streaks are substantially different from each other. They give way above that level to more symmetric velocity fluctuations, in which viscosity is not the determining factor.

6.4 The logarithmic and outer layers

Immediately above the viscous region we find the logarithmic and outer layers. They are expensive to compute, and the first simulations with even an incipient logarithmic region have only recently began to appear. For example, the logarithmic layer in the numerical channel in Figure 6.1(b) is defined as the range, $y^+ = 80$ to 400, in which the wavelength of the spectral energy peak grows linearly with y, and only spans a factor of five. Even so, those simulations, as well as the simultaneous advances in experimental methods reviewed by Marusic and Adrian (2012) in this volume, have greatly improved our knowledge of the kinematics of the outer-layer structures, and are beginning to give some indications about their dynamics.

Before considering those results, it should be stressed that structural models mean something different for the outer and buffer layers. Near the wall, the local Reynolds numbers are low, and the structures are smooth, and it is possible to speak of 'objects', and to write equations for them. Both things are harder above the buffer layer. The integral scales are $O(y)$, the velocity fluctuations are $O(u_\tau)$, and the turbulent Reynolds number is

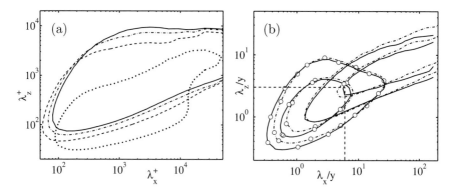

Figure 6.8 (a) Spectral densities of the kinetic energy in the logarithmic region of a numerical channel at $\delta^+ = 2003$, versus wavelengths scaled in wall units. Isolines are 0.125 times the maximum of each spectrum. ········, $y^+ = 15$; ----, 100; —·—, 200; ———, 300. (b) Two-dimensional spectral densities of u (without symbols) and v (with symbols), versus wavelengths scaled with the wall distance. Lines as in (a). The dashed rectangle is $(\lambda_x, \lambda_z) = (6, 3)y$. Isolines are 0.3 and 0.6 times the maximum of each spectrum.

$Re = O(u_\tau y/\nu \equiv y^+)$. The definition of the outer layers, $y^+ \gg 1$, implies that most of their structures have large internal Reynolds numbers, and are themselves turbulent. The energy-containing eddies have cascades connecting them to the dissipative scales, and algebraic spectra that correspond to non-smooth geometries. They are 'eddies', rather than 'vortices', because turbulent vorticity is always at the viscous Kolmogorov length scale, separated from the energy-containing scales by a ratio $Re^{3/4}$. Some examples are given in Figure 6.9. We can only expect dynamical descriptions of those structures in a statistical sense, perhaps coupled to stochastic models for the turbulent cascade 'underneath'.

Perhaps the first new information provided by the numerics about the logarithmic layer was spectral, although some of the early analyses of large-eddy simulations of channels contributed significantly to the understanding of the bursting event (Kim, 1985). Very large scales had been found experimentally in the outer layers of turbulent wall flows (Jiménez, 1998; Kim and Adrian, 1999), but DNS provided the first data about their two-dimensional spectra, and about their wall-normal correlations (del Álamo and Jiménez, 2003; del Álamo et al., 2004). Figure 6.8(a) shows that the spectral densities of the kinetic energy in the logarithmic region have elongated shapes along lines $\lambda_z^2 \sim y\lambda_x$, which del Álamo and Jiménez (2003) interpreted as the signature of statistically conical structures similar to the 'attached' eddies proposed by Townsend (1976). Note that this description only holds above $y^+ \approx 100$.

As mentioned in the previous section, the buffer-layer spectrum in Figure 6.8(a) is dominated by streaks with $\lambda_z^+ \approx 100$, but they have essentially disappeared at $y^+ = 100$. The vorticity is isotropic above that height, with the three components centred around 40 Kolmogorov viscous units (Figure 6.1b), but the large-scale velocities are very anisotropic, with the long structures mostly associated with the streamwise component (Figure 6.8b).

Figures 6.8(a) and 6.8(b) show that the cores of the velocity spectra scale relatively well with the wall distance, although, as in all turbulent flows, their short wavelengths scale with the Kolmogorov viscous length, and the long- and wide-wavelength limits of the streamwise velocity scale with the channel height. The velocities below $\lambda_x^+ \approx 500$ are essentially isotropic (not shown), with similar intensities for the three components. Note that the scaling with y of the wavelengths in Figure 6.8(b) can be read as meaning that the structures that reach from the wall to height y are not much longer than $6y$ in the logarithmic layer. Longer structures are also taller. We will see later that the dashed rectangle in Figure 6.8(b) represents the minimal box that can sustain turbulence up to a given y, presumably because it is able to contain at least one complete structure with that height.

Indeed, when three-dimensional flow fields eventually became available from simulations, it was found that there is a self-similar hierarchy of compact ejections extending from the buffer layer into the outer flow, within which the coarse-grained dissipation is more intense than elsewhere (del Álamo et al., 2006). An example is shown in Figure 6.9(a). They correspond to the ejections represented by the v-spectra in Figure 6.8(b), and, when the flow is conditionally averaged around them, as in Figure 6.9(b), they are associated with extremely long, conical, low-velocity regions whose intersection with a fixed y is a parabola that explains the quadratic behaviour of the spectrum of u. These are probably the same objects variously described as VLSM or "superstructures" in the paper by Marusic and Adrian (2012) in this volume, and are not just statistical constructs. Individual cones are observed as low-momentum 'ramps' in streamwise sections of instantaneous flow fields (Meinhart and Adrian, 1995), and two examples can be seen in the instantaneous streamwise velocity isosurface in Figure 6.9(c). As in the buffer layer, the longest low-velocity structures at each height appear to be composite objects, formed by the concatenation of smaller subunits of dimensions of the order of the v-ejections mentioned above (Figure 6.9d). We have seen that their near-wall footprints are seen in the spectra of the buffer layer as the 'tails' in Figure 6.2(a).

When the cones reach heights of the order of the flow thickness, they stop growing, and become long cylindrical 'streaks', similar to those of the

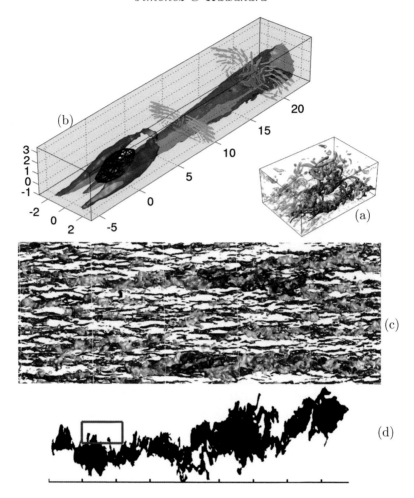

Figure 6.9 (a) The coloured object is a cluster of connected vortices in a $\delta^+ = 550$ channel (del Álamo and Jiménez, 2003), from red near the wall, to yellow near $y = \delta$. Grey objects are unconnected vortices. (b) Averaged velocity field conditioned to a vortex cluster. The black mesh is an isosurface of the PDF of the vortex positions. The blue volume surrounding the cluster is the isosurface $u^+ = 0.3$, and the red one downstream is the isosurface $u^+ = -0.1$. The vector plots represent (v, w) in the planes $r_x/y_c = 10, 20$ (from del Álamo et al., 2006). (c) Isosurface of the streamwise fluctuation velocity, $u^+ = -2$, in a computational channel with $\delta^+ = 550$. The flow is from left to right, in a partial domain $(15.5 \times 5.5)\,\delta$ in the streamwise and spanwise directions. Colours are distance from the wall, from blue at the wall, to red at the central plane. Figures (a) and (c), courtesy of O. Flores. (d) Long logarithmic-layer streak, defined as a connected region in the plane $y^+ = 200$ for which $u < -u'$, in a channel at $\delta^+ = 2003$. The tick marks in the streamwise (x) axis are 1000 wall units apart, and the red box is $(6, 3)y$.

sublayer, but with spanwise scales of about 1–2δ. They are fully turbulent objects, and neither simulations nor experiments have provided hard

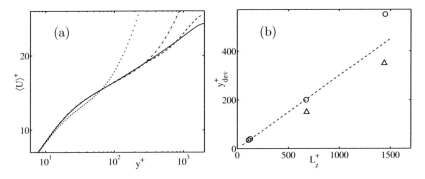

Figure 6.10 (a) Mean profiles of small-box simulations, compared with a full-box one. $\delta^+ \approx 1800$. ...., $L_x^+ \times L_z^+ = 440 \times 110$; —·—, 1350×670; ----, 2900×1450; ———, full. (b) Height of the logarithmic profile in small-box simulations, as a function of the spanwise box dimension. The open circles are the cases in (a). The two symbols at $L_z^+ \approx 100$ have $\delta^+ = 180$ and 1800. The triangles are as in (a), but with shorter or longer boxes, 770×1440 and 2700×680. The dashed line is $y_{dev} = 0.3 L_z$.

estimates of their maximum length, although the longest available spectra suggests lengths of about 25δ (Jiménez and Hoyas, 2008). The wall-normal dimension of these 'global modes' is of the order of the flow thickness, and they extend from the central plane to the wall (del Álamo and Jiménez, 2003; del Álamo et al., 2004). The probability density functions of the velocity strongly suggest that these large scales account for the variation of the turbulence intensities with the Reynolds number, both near and far from the wall (Jiménez and Hoyas, 2008), both of which disappear when wavelengths longer than about 6δ are filtered from the flow field (Jiménez, 2007).

It is interesting that the logarithmic layer can also be simulated in relatively small numerical boxes, periodic in the two wall-parallel directions (Flores and Jiménez, 2010). Those boxes are not minimal in the sense of those discussed in §6.3 for the buffer layer. When the simulation box is made smaller, turbulence does not decay, but becomes restricted to a thinner layer near the wall. In that sense, the minimal boxes of Jiménez and Moin (1991) are the innermost members of a hierarchy in which progressively smaller wall-attached structures are isolated in progressively smaller numerical boxes. The especial feature of the minimal boxes of the buffer layer is that they cannot be restricted any further, while larger boxes isolate more complicated structures, fully multiscale, that reach from the wall farther into the core flow.

That can be seen in Figure 6.10(a), which compares the mean velocity profiles of three numerical channels with different periodic box sizes to the profile of a full channel, all at the same nominal Reynolds number. The

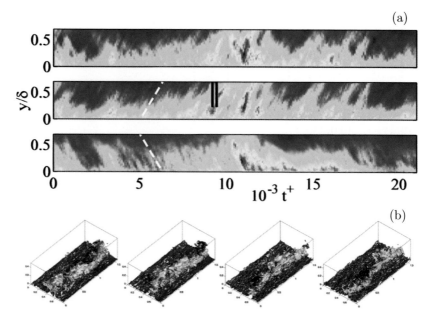

Figure 6.11 (a) Temporal evolution of the Reynolds stress, $-\langle uv \rangle$ for a small-box simulation, $\delta^+ = 1800$, $L_x^+ = 2900$, $L_z^+ = 1450$. In the top panel, $\langle uv \rangle$, is averaged over each wall-parallel plane; in the central one, it is averaged only over events for which $(u < 0, v > 0)$, and in the bottom one, for $(u > 0, v < 0)$. The heavy dashed lines are $dy/dt = \pm u_\tau$. (b) Time evolution of the $U^+ = 15$ isosurface during the burst marked by the two vertical lines in the middle panel of (a). Colours are distance from the wall, with a maximum (red) $y/\delta \approx 0.4$. The flow is from bottom-left to top-right, and time increases towards the right (from Flores and Jiménez, 2010).

smallest box corresponds to the minimal dimensions in §6.3, and its profile diverges just above the buffer layer. The other two, with larger boxes, reproduce well the logarithmic profile over taller regions. If we arbitrarily define the limit of healthy turbulence as the point where the mean profiles begin to diverge from that of a full-sized flow, it turns out to be roughly proportional to the box size. The critical dimension appears to be the spanwise periodicity, $y_{dev} \approx 0.3 L_z$, as seen in Figure 6.10(b), which includes the three cases in Figure 6.10(a), plus other combinations in which one of the box dimensions was doubled or halved independently of the other one. This result agrees with the early experiments of Toh and Itano (2005), who were able to simulate a healthy logarithmic profile in a numerical channel with a fairly wide box, but whose streamwise dimension was of the order of the buffer-layer minimal boxes ($L_z^+ = 380 \approx \delta^+$).

As in the case of the buffer layer, small-box simulations can be used to study the dynamics of the elementary energy-containing structures of the logarithmic layer. They also contain a single large-scale streamwise-velocity streak, and they also burst intermittently. An example is given in Figure 6.11(a), which displays the time evolution of the Reynolds stress, $-\langle uv \rangle$, averaged over wall-parallel planes, as a function of the wall distance. The top panel of the Figure reveals that the stress-producing events are temporally intermittent. The averaging can be separated into points in which the velocity perturbations are in the different quadrants of the (u, v) plane. The two lower panels of Figure 6.11(a) display the time history of the average stress for points in the two main stress-producing quadrants. It is not surprising that ejections, $(u < 0, v > 0)$, move away from the wall, while sweeps, $(u > 0, v < 0)$, move towards it, but it is interesting that the vertical velocities are similar in both cases (of the order of u_τ), and remarkably uniform over the region depicted, suggesting that the two events are parts of some common larger structure.

The evolution of the velocity field during the burst is shown in Figure 6.11(b), and looks remarkably similar to the events in the buffer layer. The single streak in the simulation box becomes increasingly wavy, and is eventually destroyed by the instability, but it should be emphasized that, in spite of the obvious similarities between Figures 6.11(b) and 6.5(b), the boxes in Figure 6.11(b) are about fifteen times wider than those in the buffer layer, and that the streak is now a fully turbulent multiscale object. The bursting period can be estimated from the temporal evolution of different integrated quantities, and depends on the distance to the wall ($u_\tau T \approx 6y$), rather than on the size of box. Wider boxes simply continue the linear trend farther from the wall. As in the buffer layer, we know less about the temporal evolution of the structures in full-size flows, but the temporal variability of the statistics of the minimal boxes is essentially identical to the variability among randomly chosen sub-boxes of the same size in full channels, suggesting that the processes are similar in the full and restricted systems (for details, see Flores and Jiménez, 2010).

On the other hand, it is also unlikely in this case that the minimal boxes represent the complete behaviour of real flows. In a hypothetical cycle in which the instabilities of the large-scale low-velocity cones create the intermittent ejections associated with the vortex tangles, which in turn create new streaks, the minimal boxes can probably only represent correctly the part connected with a single instability event. For example, the formation of the composite streak in Figure 6.9(d) clearly requires more space than the minimal dimensions, and the processes responsible for the concatenation of these subunits into larger wholes, especially regarding their apparent alignment into streamwise streaks, are not understood.

This qualitative model does not include interactions among different wall distances. Each streak instability creates an ejection of comparable size, which in turn recreates a commensurate streak. The fact that small boxes sustain healthy turbulence up to a certain wall distance without the corresponding larger scales suggests that the overlying structures are unnecessary for the smaller ones. Numerical experiments in which the viscous wall cycle is artificially removed, but whose outer-flow ejections and streaks remain essentially identical to those above smooth walls (Flores and Jiménez, 2006; Flores et al., 2007), suggest that the small-scale structures below a given ejection are also essentially accidental. Experimentally, this is equivalent to the classical observation that the outer parts of turbulent boundary layers are independent of wall roughness (Townsend, 1976; Jiménez, 2004). Note, however, that this independence among scales cannot be complete, because the central constraint of wall-bounded turbulence is that the mean momentum transfer has to be balanced among the different wall distances. That is what fixes the relative intensity of the momentum-carrying structures of different sizes.

It is not difficult to construct conceptual feed-back models in which locally weak structures, with too little Reynolds stress, result in the local acceleration of the mean velocity profile, which in turn leads to local enhancements of the velocity gradient and to the strengthening of the local fluctuations. But one should beware of two-scale models of turbulence, which in this case would be one scale of fluctuations and the mean flow. Everything that we know about turbulence points to its multiscale character, without privileged spectral gaps. It is more likely that any interaction leading to the adjustment of the intensities of the structures at different wall distances takes place between structures of roughly similar sizes, without necessarily passing through the mean flow. Elucidating such a mechanism will require either fairly large Reynolds numbers, or clever postprocessing techniques.

The models outlined in this section have been mostly derived from computational experiments. The experimental observations of the past decade have suggested a different scenario in which the basic object is a hairpin vortex growing from the wall, whose induced velocity creates the low momentum ramps mentioned above (Adrian et al., 2000). The hairpins regenerate each other, creating packets that are responsible for the long observed streaks (Christensen and Adrian, 2001).

Some of the differences between the two models are probably notational. For example, individual hairpins can be made to correspond to the instabilities of the shear layers around the streaks, especially if hairpins are allowed to be irregular or incomplete. The formation of vortex packets would correspond to the lengthening of the streaks in the wake of the bursts. Similarly, the respective emphases on vortices and on larger eddies might be influenced by the relatively coarse resolution of most experiments, which are unable to resolve individual vortices. Relying on conditional averages, as in Figure 6.9(b), or on limited statistics based on selected 'recognizable' objects, might give an impression of symmetry that does not apply to more typical individual structures. But it is harder to reconcile the respective treatments of the importance of the wall. The 'numerical' model emphasizes the effect of the local velocity shear, while the 'experimental' one appears to require the formation of hairpins at, or near, the buffer region. We have argued above that the evidence supports the former, but the experimental observations are also plausible. The theoretical results are ambiguous. The original simulations supporting the growth of hairpin packets were carried out by Zhou et al. (1999) on a laminar flow with a mean turbulent velocity profile, using molecular viscosity. They show hairpins growing away from the wall. But Flores and Jiménez (2005) have argued that large structures feel the effect of small-scale turbulent dissipation, and should be studied using some kind of subgrid eddy viscosity. When that is done, vorticity rises very little before it is dissipated.

Only recently has it become possible to compile flow animations fully resolved in both space and time (Lozano-Durán and Jiménez, 2010), and their analysis has only just began. The resolution of the controversy about the flow of causality in the logarithmic layer will have to wait for those results.

6.5 Coherent structures and dynamical systems

We have seen in the previous sections several instances in which the behaviour of the flow can be qualitatively explained in terms of deterministic interactions among structures. In fact, although apparently infinite-dimensional, the turbulent motion of a viscous fluid in a finite domain, governed by the Navier–Stokes equations, i.e., by partial differential evolution equations is a finite-dimensional dynamical system. Physically, the reason is that small-scale motions are smoothed by viscosity.

There is a mathematical proof of the existence of approximate inertial manifolds for the Navier–Stokes equations and this proof leads to an approximate relation between the dominant modes and the higher-order ones that can be truncated (Foias et al., 2001). Within such a manifold, the Navier–Stokes equations can be approximated by a finite-dimensional system of ordinary differential evolution equations. For example, Keefe et al.

(1992) demonstrated numerically that a spatially periodic turbulent channel flow with $\delta^+ = 80$, simulated with a spatial resolution of $16 \times 33 \times 16$, was really confined to a finite-dimensional strange attractor whose Kaplan–Yorke dimension was estimated to be 780, about 10% of the total number of numerical degrees of freedom. That measure of the dimension of the attractor is based on the Lyapunov exponents of the system, and approximately measures the dimension of the subspace whose volume neither increases nor decreases during the evolution of the flow (Frederickson et al., 1983).

Dynamical-system theory tells us that coherent structures in turbulent flows may be thought of as low-dimensional invariant sets in phase space, in the neighbourhood of which the system spends a substantial fraction of time, as suggested in Jiménez (1987). Spatio-temporally organized structures appear when a turbulent state approaches such an invariant set. Possible invariant solutions are simple saddles in phase space, such as steady travelling waves, or orbits that are periodic in time in some moving frame of reference. It is known that the statistical properties of chaotic low-dimensional dynamical systems can be estimated from the coherent states represented by unstable periodic orbits (Cvitanović, 1987; Artuso et al., 1990a,b; Christiansen et al., 1997). In this section, we will restrict ourselves to such relatively simple solutions, even if coherent structures can also correspond to more complex sets in much lower dimensional systems, such as the strange attractors of some highly reduced dynamical systems obtained from the Navier–Stokes equations by proper orthogonal decomposition (see Holmes et al., 1996).

The computation of such solutions in ideal, infinitely large, plane channels, or in experimental rectangular ducts with inflow and outflow boundary conditions, is at the moment beyond our computational capabilities, but we have argued above that the purpose of theory is not necessarily to reproduce experiments, but to give information on how the cases of interest could eventually be predicted. Numerical simulations in doubly periodic domains are known to reproduce well the statistics of experiments, and we have seen that even the minimal flow units invented by Jiménez and Moin (1991) for plane Poiseuille turbulence, and later applied by Hamilton et al. (1995) to plane Couette turbulence, represent well at least part of the buffer layer. Such reduced systems are small enough for their coherent structures to be described in terms of simple invariant solutions, and we will see below that several have been found for doubly- or singly-periodic 'minimal' domains in plane Couette, Poiseuille, Hagen–Poiseuille, and square-duct flows.

The first solutions of this kind were obtained by Nagata (1990) in plane Couette flow. That flow is stable to infinitesimal disturbances at all Reynolds numbers, so that nonlinear solutions cannot simply be found by continuation from the laminar state. Nagata (1990) found his solution by first imposing a spanwise rotation on the system, which led to a sequential series of bifurcations of two- and three-dimensional steady solutions from the laminar state. He then extended one of those three-dimensional nonlinear solutions to the non-rotating case. Clever and Busse (1992) obtained the same three-dimensional solution by initially imposing a temperature difference between the two horizontal walls of a Couette flow, and the same solution was again found by Waleffe (1998, 2003) who, based on the physical insight previously gained by Hamilton et al. (1995) into the self-sustaining process of turbulent coherent structures, initiated his continuation from an artificial flow with streamwise-independent longitudinal vortices induced by an imposed body force. The three-dimensional equilibrium solution found by those three groups arises from a saddle-node bifurcation at a finite value of the Reynolds number, above which it splits into two solution branches. As we will see later, the solutions of the upper branch contain a wavy low-velocity streak and a pair of staggered counter-rotating streamwise vortices, and are remarkably similar to the coherent structures of the turbulent buffer layer. The solutions in the lower branch are closer to the laminar state. Other equilibrium solutions have been found more recently for Couette flow by Nagata (1997); Schmiegel (1999); Gibson et al. (2008, 2009); Itano and Generalis (2009). They are not necessarily related to the one originally identified by Nagata (1990), but they have been used to discuss coherent structures and the subcritical transition to turbulence.

In contrast to plane Couette flow, laminar plane Poiseuille flow in a channel is unstable to infinitesimal disturbances beyond a certain Reynolds number, from where a two-dimensional equilibrium travelling-wave solution bifurcates subcritically. A three-dimensional solution originating from the two-dimensional wave was found by Ehrenstein and Koch (1991), while Waleffe (1998, 2001, 2003) and Itano and Toh (2001) found families of three-dimensional steady travelling waves that are not known to be connected to the laminar state. In Waleffe (1998, 2001, 2003), he used the artificial-force approach to construct nonlinear steady travelling waves with a reflectional symmetry with respect to the channel central plane, which also contain streaks and vortices. His Poiseuille solution can be continuously connected to the Nagata (1990) solution for plane Couette flow (Waleffe, 2003). Itano and Toh (2001) solution also includes streaks and vortices, but they are localized near one of the two walls. A three-dimensional steady travelling with a very similar structure was found by Jiménez and Simens (2001) in a so-called 'autonomous' flow that is confined to the vicinity of one wall under the action of a damping filter (Jiménez and Pinelli, 1999). That is the case used in Figure 6.3(b) to compare the fluctuation profiles of simple solutions with those of real turbulence.

The laminar Hagen–Poiseuille flow in a circular pipe is also linearly stable at all Reynolds numbers, but Faisst and Eckhardt (2003) and Wedin and Kerswell (2004) have recently discovered three-dimensional steady travelling-wave solutions by using the self-sustaining approach proposed by Waleffe (1998, 2003). Both groups obtain the same solution, which possesses discrete rotational symmetry with respect to the pipe axis, with wavy low-velocity streaks flanked by staggered streamwise vortices. The solution with three-fold rotational symmetry arises from a saddle-node bifurcation at the lowest Reynolds number, although it was later found that travelling waves without any discrete rotational symmetry exist at much lower Reynolds numbers (Pringle and Kerswell, 2007).

As in circular pipes, the laminar flow in a square duct is linearly stable, and no travelling-wave solutions were known until three three-dimensional steady travelling waves were found using the artificial-force approach or internal heating. The two found by Wedin et al. (2009) and Okino et al. (2010) have low- or high-velocity streaks and streamwise vortices on only two opposite walls, while the other two walls of the duct are empty. The one found by Uhlmann et al. (2010), on the other hand, has streaks and vortices on the four walls, so that its streamwise-averaged cross-flow velocity exhibits an eight-vortex pattern closely resembling the mean secondary flow of experimental turbulent ducts (Figure 6.12). This correspondence gives some support to the numerical evidence that the mean secondary motion of low-Reynolds-number square-ducts is a direct consequence of coherent structures (Pinelli et al., 2010).

We already mentioned in §6.3 and §6.4 that turbulent flow is really unsteady, even in small boxes, and that something like a bursting cycle is necessary for regenerating the fluctuations. However, even if we have just seen that three-dimensional equilibrium solutions reproducing buffer-layer coherent structures are available for a variety of wall-bounded flows, only a few periodic solutions have recently been discovered.

In plane Couette flow, Clever and Busse (1997) analysed the linear stability of Nagata (1990) three-dimensional steady solution, and identified a Hopf bifurcation from where a temporally-periodic three-dimensional nonlinear solution arises. They continued that solution within its stable parameter range by forward time integration, but found that its properties did not differ too much from those of the steady solution due to the small amplitude of the oscillations. Using an iterative method to minimize the recurrence

6: Dynamics of Wall-Bounded Turbulence

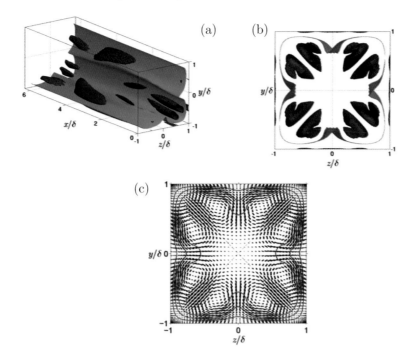

Figure 6.12 Shape of the eight-vortex steady upper-branch travelling wave in a duct, at bulk Reynolds number $Re = U_b \delta / \nu = 1404$, and streamwise wavelength $L_x/\delta = 2\pi$ (Uhlmann et al., 2010). (a, b) The green sheet near the wall is an isosurface of the streamwise velocity, 0.55 times its maximum. The blue and red tubular structures are the streamwise vorticity, ± 0.65 times its maximum ($\omega_x^+ = \pm 0.0655$). Figure (a) only shows surfaces to one side of a duct diagonal. (c) The contour lines are isolines of the streamwise-averaged streamwise velocity, (0.25, 0.5, 0.75 times its maximum). The arrows are the streamwise-averaged secondary velocity.

error, Kawahara and Kida (2001) found a more unsteady three-dimensional unstable periodic solution for plane Couette flow, which reproduces much better the full regeneration cycle of near-wall coherent structures, i.e., the formation and development of low-velocity streaks by decaying streamwise vortices, the bending of the streaks followed by the regeneration of the vortices, the breakdown of the streaks, and the rapid development of the vortices. They also obtained a gentler periodic solution that represents a weak spanwise standing-wave motion of the low-velocity streak.

More recently Viswanath (2007) has developed a new method based on Newton–Krylov iteration and on a locally-constrained optimal hook step, and applied it to the computation of periodic solutions in Couette flow. His method is matrix-free, and drastically reduces the computational time and memory required for the Newton iteration. Using it, he obtained five new three-dimensional periodic solutions that demonstrate the breakup and reorganization of near-wall coherent structures.

In plane Poiseuille flow, Toh and Itano (2003) identified a three-dimensional periodic-like solution that could originate from a heteroclinic connection between two equivalent Itano and Toh (2001) steady travelling waves that differ from each other by a spanwise shift of half a wavelength. Their solution is reminiscent of the heteroclinic cycle identified in a highly reduced dynamical-system approximation to near-wall turbulence by Aubry et al. (1988), who also observed a connection between two fixed points differing by a similar spanwise shift.

More recently, Duguet et al. (2008) found a three-dimensional periodic solution in Hagen–Poiseuille flow that bifurcates from the Pringle and Kerswell (2007) travelling wave without any discrete rotational symmetry.

Most of the unstable periodic orbits have been found for wall bounded shear flows, but there are a few exceptions. One is the periodic solution found by van Veen et al. (2006) in a Kida–Pelz highly-symmetric flow in a triply periodic domain (Kida, 1985), which reproduces the universal Kolmogorov (1941) energy spectrum with the right energy dissipation rate, even though its Reynolds number is too low to identify any real inertial range. Another periodic solution was obtained in the GOY (Gledzer–Ohkitani–Yamada) shell model by Kato and Yamada (2003), who observed that the solution not only exhibits the Kolmogorov inertial-range energy spectrum, but also intermittency.

It is curious that averaging along a single unstable periodic orbit of relatively short period, such as in these solutions or in those mentioned above for plane Couette flow (Kawahara and Kida, 2001; Viswanath, 2007), reproduces so well typical properties of turbulence, such as the energy spectra and the r.m.s. velocity profiles, which are long-time averages along true turbulent orbits. A possible interpretation has recently been proposed by Saiki and Yamada (2009, 2010), who showed that periodic orbits are really special. If one evaluates the statistical mean and variance of the time averages of dynamical variables along distinct periodic orbits with similar periods, the mean is nearly the same as that along chaotic segments with corresponding lengths, but the variance is significantly smaller than along the chaotic segments. This observation suggests that short periodic orbits might be better representations of the statistical behaviour of turbulent systems than comparable segments of their true temporal evolution in the sense that the statistical values along periodic orbits are confined to a narrower range around the exact one.

We now discuss the possible relevance of these simple invariant solutions to wall-bounded turbulence. All the solutions just described are unstable at the Reynolds numbers at which turbulence is observed, but the dimensions of their unstable manifolds in phase space are typically low (Waleffe, 2001; Kawahara, 2005; Wang et al., 2007; Kerswell and Tutty, 2007; Viswanath, 2009). Therefore, although we should not expect to observe them as such in real turbulence, a generic turbulent solution could approach them and spend a substantial fraction of its lifetime in their neighbourhood. In a nonlinear dynamical system the transverse intersections of the stable and unstable manifolds of the same or different periodic orbits lead to homoclinic or heteroclinic orbits, and to the appearance of chaotic behaviour through Smale's horseshoe (Guckenheimer and Holmes, 1986). Therefore, the tangle of stable and unstable manifolds of the above simple invariant solutions could represent turbulence dynamics, while the simple solutions themselves would represent coherent structures.

To ascertain whether that is really the case for near-wall turbulence, Jiménez et al. (2005) collected the simple solutions available at the time for Poiseuille and Couette flows, some of which we have seen to have velocity fluctuation profiles strongly reminiscent of real turbulence. The characteristics of those profiles are summarized in Figure 6.13, which has been adapted from that paper by adding some of the solutions that have been found since then. Each solution is represented by a single point whose coordinates are the maximum values, u'_{max} and v'_{max}, of its intensity profiles. Most solutions fall into one of two classes: a 'vortex-dominated' (upper) family, characterized by smaller u'_{max} and larger v'_{max}, and a 'streak-dominated' (lower) one, characterized by larger u'_{max} and smaller v'_{max}. As in §6.3 we take the streamwise and wall-normal velocity fluctuations as respectively representing the intensities of the streamwise velocity streaks and of the quasi-streamwise vortices.

In the upper-left corner of Figure 6.13 we find the vortex-dominated solutions, and in the lower-right corner the streak-dominated ones. The former are represented by Kawahara and Kida (2001) dynamic periodic solution, by Nagata (1990) upper-branch steady solutions, and by Jiménez and Simens (2001) autonomous travelling waves. The temporal averages of Viswanath (2007) periodic solutions in plane Couette flow are also part of this family. These are the solutions closer to real turbulence, and the full channel of Kim et al. (1987) is included in the figure as comparison.

The streak-dominated family includes Kawahara and Kida (2001) gentle periodic orbit, Nagata (1990) lower-branch steady solutions. the permanent wave obtained by Itano and Toh (2001) in Poiseuille flow, and the heteroclinic connection identified by Toh and Itano (2003). The figure includes the

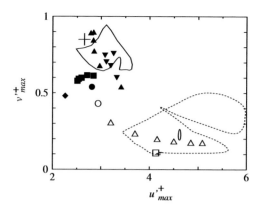

Figure 6.13 Classification into upper and lower families of the simple solutions, in terms of their maximum streamwise and wall-normal r.m.s. velocities, u'_{max} and v'_{max}. Solid symbols are classified as upper-branch, and open ones as lower-branch. The solid large and small loops represent the dynamic ($L_x^+ \times L_z^+ \times \delta^+ = 190 \times 130 \times 34$) and the gentle ($L_x^+ \times L_z^+ \times \delta^+ = 154 \times 105 \times 28$) periodic solutions at $Re = 400$ in Kawahara and Kida (2001). △, Nagata (1990) steady solution for several values of the spanwise wavelength at $Re = 400$ and $L_x/\delta = 2\pi$ (upper branch, $L_z^+ \times \delta^+ = 76 - 132 \times 35$; lower branch, $L_z^+ \times \delta^+ = 53 - 92 \times 24 - 25$). ■, autonomous solutions in Jiménez and Simens (2001) ($L_x^+ \times L_z^+ \times \delta_1^+ = 151 - 189 \times 180 \times 38 - 46$, where δ_1 is the filter height). ●, Waleffe (2003) upper-branch solution for $Q/\nu = 1303$, where Q is the volume flux per unit span. $L_x^+ \times L_z^+ \times \delta^+ = 387 \times 149 \times 123$; ○, Waleffe (2003) lower-branch solution for $Q/\nu = 1390$, $L_x^+ \times L_z^+ \times \delta^+ = 379 \times 146 \times 121$; □, Itano and Toh (2001) asymmetric wave for $Q/\nu = 4000$, $L_x/\delta \times L_z/\delta = \pi \times 0.4\pi$. The dotted loop is the periodic-like solution of Toh and Itano (2003). ▼, temporal averages of Viswanath (2007) periodic solutions (cases $P_2 - P_6$ in his table 1) for $Re = 400$ and $L_x/\delta \times L_z/\delta = 1.75\pi \times 1.2\pi$ in plane Couette flow. ♦, upper branch of Uhlmann et al. (2010) eight-vortex travelling-wave solution in square duct flow ($Re = 1371$, $L_x/\delta = 2\pi$). We have also shown the turbulent state computed by Kim et al. (1987) ($L_x^+ \times L_z^+ \times \delta^+ = 2300 \times 1150 \times 180$) by the symbol '+' for comparison.

two permanent waves by Waleffe (2003) for Poiseuille flow, which were classified as lower- or upper-branch in the original reference, but which appear here to be too close to the turning point to differ too much from each other.

The upper-branch of Uhlmann et al. (2010) eight-vortex travelling wave in a square duct is also included, and is relatively close to the upper-family solutions mentioned above, in spite of its significantly different boundary conditions, but the corresponding lower-branch solution has a weaker u'_{max} than the upper-branch one, and is thus quite different from the lower-family solutions discussed here.

Besides those statistical differences, the spatial structures of the upper and lower families are also quite different. The two flow fields in Figure 6.14 correspond to the upper and lower Nagata (1990) solutions at comparable Reynolds numbers and dimensions. They differ mainly in the location of the

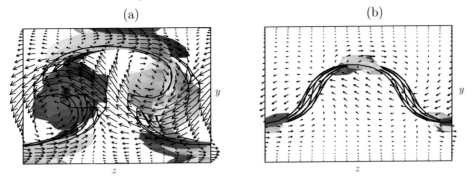

Figure 6.14 Projection of the streamwise vorticity, ω_x, and velocity fields in the cross-plane (z,y), for (a) the upper-branch ($L_z^+ = 99$), and (b) lower-branch ($L_z^+ = 67$) solutions of Nagata (1990). $Re = 400$, $L_x/\delta \times L_z/\delta = 2\pi \times 0.9\pi$. The thick solid lines are different sections of the surface $u = 0$, i.e, the critical layer. Arrows are the cross-plane velocities at $x = 0$, defined so that the first streamwise harmonic of u is proportional to $\cos(2\pi x/L_x)$. The arrows are uniformly scaled in wall units, and the longest arrows in (a) are roughly $1.9 u_\tau$. The dark- and light-gray objects are isosurfaces $\omega_x^+ = \pm 0.155$.

streamwise vorticity, which is concentrated in the form of sheets on a critical layer in the case of the lower-branch solution, but which has two concentrated tubular vortices in each flank of the streak in the case of the upper one. The spatial structure of the upper solution reproduces well the near-wall coherent structures, with a pair of counter-rotating staggered streamwise vortices flanking a wavy low-velocity streak (see §6.3). The streamwise vortical structures are responsible for the larger $v'_{\max}$ in the upper family, but note that there is no one-to-one correspondence between the strength of the vortices and that of the streak. The latter is a consequence of the action of the vortices over a period of time, and this time scale is as important in determining the streak intensity as the vortex strength. There is little doubt that, in the absence of viscosity or of other limiting factor, a long permanent vortex would eventually pump the mean velocity profile into uniform streamwise regions in which the streamwise velocity would be that of one or the other wall. Viscosity or instabilities limit that distortion, and the simplest explanation of the weaker streaks, and smaller $u'_{\max}$, of the upper-branch solutions is that the flank vortices shorten the effective time scale of the pumping by providing an effective eddy viscosity that homogenizes the streamwise-velocity profile.

The lower solution family is probably not directly related to turbulence, and it is now believed to play an important role in the subcritical transition to turbulence (Kerswell, 2005; Eckhardt et al., 2007). At Reynolds numbers low enough for laminar flow to be stable, there is a boundary in phase space

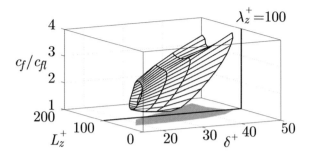

Figure 6.15 Friction coefficient, normalized by a corresponding laminar value, c_f/c_{fl}, of Nagata (1990) solution for plane Couette flow as a function of the spanwise wavelength L_z^+, and of the friction Reynolds number δ^+. Streamwise wavelength $L_x/\delta = 2\pi$

that separates the basins of attraction of the turbulent and laminar states. Some of the lower-family solutions for plane Poiseuille and Couette flow have only one unstable direction in phase space, so that those invariant sets and their stable manifolds form the laminar-to-turbulent separatrix (Itano and Toh, 2001; Waleffe, 2003; Kawahara, 2005; Wang et al., 2007; Schneider et al., 2008).

Figure 6.15 shows the friction coefficient of Nagata (1990) solutions normalized with that of the laminar state, for a fixed value of the streamwise wavelength $L_x/\delta = 2\pi$, as a function of the spanwise wavelength and of the friction Reynolds number. Both the upper and the lower branches of the solution have higher friction coefficients than the laminar state. The range of spanwise wavelengths for which the solutions exist is always around $L_z^+ = 100$, which is close to the observed mean separation of buffer-layer streaks, and the range itself, $L_z^+ \approx 50 - 150$, is in good agreement with the range of streak spacings found in experimental turbulent boundary layers by Smith and Metzler (1983), $\lambda_z^+ \approx 50 - 200$. The same range of spanwise wavelengths is also found in Kawahara and Kida (2001) stronger periodic solution in plane Couette flow (Kawahara, 2009), and is around the width of the narrowest minimal flow unit for sustained turbulence (Jiménez and Moin, 1991). Although its relation with the above nonlinear states is unclear, linear analysis of the system around the state point corresponding to the averaged turbulent flow demonstrates that a three-dimensional infinitesimal perturbation with $L_z^+ \approx 100$, also representing streaks and streamwise vortices, leads to optimal transient growth (del Álamo and Jiménez, 2006).

In pipe flow, on the other hand, Kerswell and Tutty (2007) performed a detailed analysis of the correlations between the velocity fields of a turbulent state and those of Wedin and Kerswell (2004) travelling waves with discrete rotational symmetry, and showed that turbulence only visits the close vicinity of the travelling waves in phase space for 10% of the time, suggesting that there should be other objects, such as periodic orbits, which turbulence approaches more frequently. More recently, Willis and Kerswell (2008) examined the relation of known travelling-wave solutions with flow states in a long pipe, and suggested that both the lower and upper branches of their solutions sit in an intermediate region of phase space, between laminar and turbulent states, rather than being embedded within the turbulent attractor itself.

We close this section by briefly discussing possible future areas of research in the nascent field of the possible relation between the coherent structures of turbulence and dynamical systems. Although it has turned out that the structures, the dynamics, and the statistics of buffer-layer coherent structures are tantalizingly close to those of some nonlinear equilibria, or to the finite-amplitude oscillations represented by known simple invariant sets in phase space, the dynamical interpretation of the large-amplitude intermittent events observed in the near-wall region remains unclear. One possibility is that those violent events correspond in phase space to the unstable or stable manifolds of the simple invariant solutions just described, or to their connections. Halcrow et al. (2009) have quantitatively identified heteroclinic connections between known equilibrium solutions of plane Couette flow, and van Veen et al. (2011) recently developed a numerical method for computing two-dimensional unstable manifolds in large systems, which they used to significantly extend the unstable manifold of Kawahara and Kida (2001) gentle periodic orbit in plane Couette flow and then to find homoclinic orbits from/to that periodic orbit (van Veen and Kawahara, 2011). These studies are the first steps in the exploration of global phase-space structures in near-wall turbulence.

Another possible research direction relates to the large-scale flow structures of transitional and turbulent wall flows, such as turbulent spots (Dauchot and Daviaud, 1995), oblique stripes (Prigent et al., 2002), puffs (Wygnanski and Champagne, 1973) and large-scale streaks (see §6.4). Most of the available invariant solutions have been obtained in more or less minimal flow units that cannot accommodate those large-scale structures. Spatially localized invariant solutions (including chaotic edge states) in a larger periodic box would be useful for the characterization of transitional large-scale structures (Cherhabili and Ehrenstein, 1997; Ehrenstein et al., 2008; Schneider et al., 2010; Duguet et al., 2009; Willis and Kerswell, 2009; Mellibovsky et al., 2009). We also saw in §6.4 that Flores and Jiménez (2010) have found quasi-periodic bursts in the logarithmic region of fully turbulent channel flow, strongly reminiscent of the regeneration cycle in the buffer layer, suggesting that the large-scale dynamics in the logarithmic layer might be related to invariant solutions in larger periodic domains.

6.6 Conclusions

In summary, our knowledge of wall-bounded turbulence has advanced a lot in the last thirty years, and we have argued that this has been due, in large part, to the availability of reasonably affordable direct numerical simulations. Although we have shown several instances of how that has happened in detail, it may be appropriate to summarize here the nature of the changes that simulations have brought to the study of turbulence.

Every science passes through several stages in its development. The most primitive one is characterized by the scarcity of suitable data, which leads to the appearance of theories attempting to explain the few data available, but whose further predictions cannot be easily tested. Planetary science was a typical example until the advent of space exploration, and turbulence spent much of the XXth century in this stage. Fluid flow is difficult to measure and to visualize, and the available data were relatively few and incomplete. It was possible to propose almost any theory about turbulence structure and dynamics, with relatively few constraints from observations.

The situation began to change in the 1950's with the advent of hot wires, which gave the first glimpses of the structure of the turbulent eddies, and it became a flood with direct numerical simulations. In the same way that it became impossible to speak about canals in Mars after the first space probes, the data generated by DNS, and to a lesser but important extent by experimental visualisations and PIV, have produced much tighter limits for our theories.

It is probably true that the data required to test almost any theory about the dynamics of canonical turbulent flows already exist in some numerical data base, or that, if it were known which data are missing, they could probably be easily computed in a relatively short time.

What this new development denies us is the freedom to speculate. Almost anything that can be proposed can be tested, and the nature of the turbulence problem is quickly changing from one of accumulating new data to one of interpreting existing ones. It is not necessarily a simpler problem, but it is different from much of what we have been doing up to now. The consequences for turbulence theory can only be guessed at this moment.

Acknowledgments The preparation of this article has been supported in part by the Spanish CICYT, under grant TRA2009-11498, by the European Research Council Advanced Grant ERC-2010.AdG-20100224, and a Grant-in-Aid for Scientific Research (B) from the Japanese Society for the Promotion of Science. The authors cordially thank D. Viswanath and M. Uhlmann for the original data of periodic solutions in Couette flow, and of travelling waves in a square-duct, and O. Flores for Figures 6.9(a) and 6.9(c).

References

Abe, H., Kawamura, H., and Matsuo, Y. 2004. Surface heat-flux fluctuations in a turbulent channel flow up to $Re_\tau = 1020$ with $Pr = 0.025$ and 0.71. *Int. J. Heat Fluid Flow*, **25**, 404–419.

Adrian, R. J. 2007. Hairpin vortex organization in wall turbulence. *Phys. Fluids.*, **19**, 041301.

Adrian, R. J., Meinhart, C. D., and Tomkins, C. D. 2000. Vortex organization in the outer region of the turbulent boundary layer. *J. Fluid Mech.*, **422**, 1–54.

Afzal, N., and Yajnik, K. 1973. Analysis of turbulent pipe and channel flows at moderately large Reynolds number. *J. Fluid Mech.*, **61**, 23–31.

Artuso, R., Aurell, E., and Cvitanović, P. 1990a. Recycling of strange sets: I. Cycle expansions. *Nonlinearity*, **3**, 325–359.

Artuso, R., Aurell, E., and Cvitanović, P. 1990b. Recycling of strange sets: II. Applications. *Nonlinearity*, **3**, 361–386.

Aubry, N., Holmes, P., Lumley, J. L., and Stone, E. 1988. The dynamics of coherent structures in the wall region of a turbulent boundary layer. *J. Fluid Mech.*, **192**, 115–173.

Bakewell, H. P., and Lumley, J. L. 1967. Viscous sublayer and adjacent wall region in turbulent pipe flow. *Phys. Fluids*, **10**, 1880–1889.

Barenblatt, G. I., Chorin, A. J., and Prostokishin, V. M. 2000. Self-similar intermediate structures in turbulent boundary layers at large Reynolds numbers. *J. Fluid Mech.*, **410**, 263–283.

Bradshaw, P. 1967. Inactive motions and pressure fluctuations in turbulent boundary layers. *J. Fluid Mech.*, **30**, 241–258.

Brown, G. L., and Roshko, A. 1974. On the density effects and large structure in turbulent mixing layers. *J. Fluid Mech.*, **64**, 775–816.

Butler, K. M., and Farrell, B. F. 1993. Optimal perturbations and streak spacing in wall-bounded shear flow. *Phys. Fluids A*, **5**, 774–777.

Cherhabili, A., and Ehrenstein, U. 1997. Finite-amplitude equilibrium states in plane Couette flow. *J. Fluid Mech.*, **342**, 159–177.

Christensen, K. T., and Adrian, R. J. 2001. Statistical evidence of hairpin vortex packets in wall turbulence. *J. Fluid Mech.*, **431**, 433–443.

Christiansen, F., Cvitanović, P., and Putkaradze, V. 1997. Spatiotemporal chaos in terms of unstable recurrent patterns. *Nonlinearity*, **10**, 55–70.

Clever, R. M., and Busse, F. H. 1992. Three-dimensional convection in a horizontal fluid layer subjected to a constant shear. *J. Fluid Mech.*, **234**, 511–527.

Clever, R. M., and Busse, F. H. 1997. Tertiary and quaternary solutions for plane Couette flow. *J. Fluid Mech.*, **344**, 137–153.

Cvitanović. 1987. Invariant measurement of strange sets in terms of cycles. *Phys. Rev. Lett.*, **61**, 2729–2732.

Darcy, H. 1854. Recherches expérimentales rélatives au mouvement de l'eau dans les tuyeaux. *Mém. Savants Etrang. Acad. Sci. Paris*, **17**, 1–268.

Dauchot, O., and Daviaud, F. 1995. Finite amplitude perturbation and spots growth mechanism in plane Couette flow. *Phys. Fluids*, **7**, 335–343.

Deardorff, J. W. 1970. A numerical study of three-dimensional turbulent channel flow at large Reynolds number. *J. Fluid Mech.*, **41**, 453–480.

Deardorff, J. W. 1973. The use of subgrid transport equations in a three-dimensional model of atmospheric turbulence. *J. Fluid Eng.*, **95**, 429–438.

deGraaff, D. B., and Eaton, J. K. 2000. Reynolds number scaling of the flat-plate turbulent boundary layer. *J. Fluid Mech.*, **422**, 319–346.

del Álamo, J. C., and Jiménez, J. 2003. Spectra of very large anisotropic scales in turbulent channels. *Phys. Fluids*, **15**, L41–L44.

del Álamo, J. C., and Jiménez, J. 2006. Linear energy amplification in turbulent channels. *J. Fluid Mech.*, **559**, 205–213.

del Álamo, J. C., Jiménez, J., Zandonade, P., and Moser, R. D. 2004. Scaling of the energy spectra of turbulent channels. *J. Fluid Mech.*, **500**, 135–144.

del Álamo, J. C., Jiménez, J., Zandonade, P., and Moser, R. D. 2006. Self-similar vortex clusters in the logarithmic region. *J. Fluid Mech.*, **561**, 329–358.

Duguet, Y., Pringle, C. C., and Kerswell, R. R. 2008. Relative periodic orbits in transitional pipe flow. *Phys. Fluids*, **20**, 114102.

Duguet, Y., Schlatter, P., and Henningson, D. S. 2009. Localized edge states in plane Couette flow. *Phys. Fluids*, **21**, 111701.

Eckhardt, B., Scheider, T. M., Hof., B., and Westerweel, J. 2007. Turbulence transition in pipe flow. *Ann. Rev. Fluid Mech.*, **39**, 447–468.

Ehrenstein, U., and Koch, W. 1991. Three-dimensional wavelike equilibrium states in plane Poiseuille flow. *J. Fluid Mech.*, **228**, 111–148.

Ehrenstein, U., Nagata, M., and Rincon, F. 2008. Two-dimensional nonlinear plane Poiseuille–Couette flow homotopy revisited. *Phys. Fluids*, **20**, 064103.

Faisst, H., and Eckhardt, B. 2003. Traveling waves in pipe flow. *Phys. Rev. Lett.*, **91**, 224502.

Feller, W. 1971. *An Introduction to Probability theory and its Applications.* third edn. Vol. 1. Wiley. pg. 446–448.

Flores, O., and Jiménez, J. 2005. Linear dynamics of turbulent structures in the log layer. Pages LR–1 of: *Proc. Div. Fluid Dyn. Am. Phys. Soc.*

Flores, O., and Jiménez, J. 2006. Effect of wall-boundary disturbances on turbulent channel flows. *J. Fluid Mech.*, **566**, 357–376.

Flores, O., and Jiménez, J. 2010. Hierarchy of minimal flow units in the logarithmic layer. *Phys. Fluids*, **22**, 071704.

Flores, O., Jiménez, J., and del Álamo, J. C. 2007. Vorticity organization in the outer layer of turbulent channels with disturbed walls. *J. Fluid Mech.*, **591**, 145–154.

Foias, C., Manley, O., Rosa, R., and Temam, R. 2001. *Navier–Stokes Equations and Turbulence*. Cambridge University Press.

Frederickson, P., Kaplan, J. L., Yorke, E. D., and Yorke, J. A. 1983. The Lyapunov dimension of strange attractors. *J. Diffl. Equat.*, **49**, 185–207.

Gaster, M., Kit, E., and Wygnanski, I. 1985. Large-scale structures in a forced turbulent mixing layer. *J. Fluid Mech.*, **150**, 23–39.

Gibson, J. F., Halcrow, J., and Cvitanović, P. 2008. Visualizing the geometry of state space in plane Couette flow. *J. Fluid Mech.*, **611**, 107–130.

Gibson, J. F., Halcrow, J., and Cvitanović, P. 2009. Equilibrium and travelling-wave solutions of plane Couette flow. *J. Fluid Mech.*, **638**, 243–266.

Guckenheimer, J., and Holmes, P. 1986. *Nonlinear Oscillations, Dynamical Systems, and Bifurcations of Vector Fields*, 2nd ed. Springer Verlag.

Hagen, G. H. L. 1839. Über den Bewegung des Wassers in engen cylindrischen Röhren. *Poggendorfs Ann. Physik Chemie*, **46**, 423–442.

Halcrow, J., Gibson, J. F., and Cvitanović, P. 2009. Heteroclinic connections in plane Couette flow. *J. Fluid Mech.*, **621**, 365–376.

Hamilton, J. M., Kim, J., and Waleffe, F. 1995. Regeneration mechanisms of near-wall turbulence structures. *J. Fluid Mech.*, **287**, 317–348.

Ho, C. H., and Huerre, P. 1984. Perturbed free shear layers. *Ann. Rev. Fluid Mech.*, **16**, 365–424.

Holmes, P., Lumley, J. L., and Berkooz, G. 1996. *Turbulence, Coherent Structures, Dynamical Systems and Symmetry*, 1st ed. Cambridge University Press.

Hoyas, S., and Jiménez, J. 2006. Scaling of the velocity fluctuations in turbulent channels up to $Re_\tau = 2003$. *Phys. Fluids*, **18**, 011702.

Hoyas, S., and Jiménez, J. 2008. Reynolds number effects on the Reynolds-stress budgets in turbulent channels. *Phys. Fluids*, **20**, 101511.

Hutchins, N., and Marusic, I. 2007. Evidence of very long meandering features in the logarithmic region of turbulent boundary layers. *J. Fluid Mech.*, **579**, 467–477.

Itano, T., and Generalis, S. C. 2009. Hairpin vortex solution in planar Couette flow: A tapestry of knotted vortices. *Phys. Rev. Lett.*, **102**, 114501.

Itano, T., and Toh, S. 2001. The dynamics of bursting process in wall turbulence. *J. Phys. Soc. Jpn.*, **70**, 703–716.

Jiménez, J. 1987. Coherent structures and dynamical systems. Pages 323–324 of: *Proc. CTR Summer Program 1987*. Stanford University.

Jiménez, J. 1998. The largest scales of turbulence. Pages 137–154 of: *CTR Ann. Res. Briefs*. Stanford University.

Jiménez, J. 2003. Computing high-Reynolds number flows: Will simulations ever substitute experiments? *J. of Turbulence*, **22**.

Jiménez, J. 2004. Turbulent flows over rough walls. *Ann. Rev. Fluid Mech.*, **36**, 173–196.

Jiménez, J. 2007. Recent developments in wall-bounded turbulence. *Rev. R. Acad. Cien. Serie A, Mat.*, **101**, 187–203.

Jiménez, J. 2012. Cascades in wall-bounded turbulence. *Ann. Rev. Fluid Mech.*, **44**, 27–45.

Jiménez, J., and Hoyas, S. 2008. Turbulent fluctuations above the buffer layer of wall-bounded flows. *J. Fluid Mech.*, **611**, 215–236.

Jiménez, J., and Moin, P. 1991. The minimal flow unit in near-wall turbulence. *J. Fluid Mech.*, **225**, 221–240.

Jiménez, J., and Moser, R. D. 2007. What are we learning from simulating wall turbulence? *Phil. Trans. Roy. Soc. A*, **365**, 715–732.

Jiménez, J., and Pinelli, A. 1999. The autonomous cycle of near wall turbulence. *J. Fluid Mech.*, **389**, 335–359.

Jiménez, J., and Simens, M. P. 2001. Low-dimensional dynamics in a turbulent wall flow. *J. Fluid Mech.*, **435**, 81–91.

Jiménez, J., Wray, A. A., Saffman, P. G., and Rogallo, R. S. 1993. The structure of intense vorticity in isotropic turbulence. *J. Fluid Mech.*, **255**, 65–90.

Jiménez, J., del Álamo, J. C., and Flores, O. 2004. The large-scale dynamics of near-wall turbulence. *J. Fluid Mech.*, **505**, 179–199.

Jiménez, J., Kawahara, G., Simens, M. P., Nagata, M., and Shiba, M. 2005. Characterization of near-wall turbulence in terms of equilibrium and 'bursting' solutions. *Phys. Fluids*, **17**, 015105.

Kato, S., and Yamada, M. 2003. Unstable periodic solutions embedded in a shell model turbulence. *Phys. Rev. E*, **68**, 025302(R).

Kawahara, G. 2005. Laminarization of minimal plane Couette flow: Going beyond the basin of attraction of turbulence. *Phys. Fluids*, **17**, 041702.

Kawahara, G. 2009. Theoretical interpretation of coherent structures in near-wall turbulence. *Fluid Dynamics Research*, **41**, 064001.

Kawahara, G., and Kida, S. 2001. Periodic motion embedded in plane Couette turbulence: regeneration cycle and burst. *J. Fluid Mech.*, **449**, 291–300.

Kawahara, G., Jiménez, J., Uhlmann, M., and Pinelli, A. 2003. Linear instability of a corrugated vortex sheet – a model for streak instability. *J. Fluid Mech.*, **483**, 315–342.

Keefe, L., Moin, P., and Kim, J. 1992. The dimension of attractors underlying periodic turbulent Poiseuille flow. *J. Fluid Mech.*, **242**, 1–29.

Kerswell, R. R. 2005. Recent progress in understanding the transition to turbulence in a pipe. *Nonlinearity*, **18**, R17–R44.

Kerswell, R. R., and Tutty, O. R. 2007. Recurrence of travelling waves in transitional pipe flow. *J. Fluid Mech.*, **584**, 69–102.

Kida, S. 1985. Three-dimensional periodic flows with high-symmetry. *J. Phys. Soc. Japan*, **54**, 3132–2136.

Kim, J. 1985. Turbulence structures associated with the bursting event. *Phys. Fluids*, **28**, 52–58.

Kim, J., and Hussain, F. 1993. Propagation velocity of perturbations in channel flow. *Phys. Fluids A*, **5**, 695–706.

Kim, J., Moin, P., and Moser, R. D. 1987. Turbulence statistics in fully developed channel flow at low Reynolds number. *J. Fluid Mech.*, **177**, 133–166.

Kim, K.C., and Adrian, R. J. 1999. Very large-scale motion in the outer layer. *Phys. Fluids*, **11**, 417–422.

Kline, S. J., Reynolds, W. C., Schraub, F. A., and Runstadler, P. W. 1967. Structure of turbulent boundary layers. *J. Fluid Mech.*, **30**, 741–773.

Kolmogorov, A. N. 1941. The local structure of turbulence in incompressible viscous fluids a very large Reynolds numbers. *Dokl. Akad. Nauk. SSSR*, **30**, 301–305. Reprinted in *Proc. Roy. Soc. London.* A **434**, 9–13 (1991).

Lee, S-H., and Sung, H. J. 2007. Direct numerical simulation of the turbulent boundary layer over a rod-roughened wall. *J. Fluid Mech.*, **584**, 125–146.

Lee, S-H., and Sung, H. J. 2011. Very-large-scale motions in a turbulent boundary layer. *J. Fluid Mech.*, **673**, 80–120.

Lozano-Durán, A., and Jiménez, J. 2010. Time-resolved evolution of the wall-bounded vorticity cascade. Pages EB-3 of: *Proc. Div. Fluid Dyn.* Am. Phys. Soc.

Marusic, I., and Adrian, R. J. 2012. The eddies and scales of wall-turbulence. *This volume*.

Mathis, R., Hutchins, N., and Marusic, I. 2009. Large-scale amplitude modulation of the small-scale structures in turbulent boundary layers. *J. Fluid Mech.*, **628**, 311–337.

Meinhart, C. D., and Adrian, R. J. 1995. On the existence of uniform momentum zones in a turbulent boundary layer. *Phys. Fluids*, **7**, 694–696.

Mellibovsky, F., Meseguer, A., Schneider, T. M., and Eckhardt, B. 2009. Transition in localized pipe flow turbulence. *Phys. Rev. Lett.*, **103**, 054502.

Moin, P., and Kim, J. 1982. Numerical investigation of turbulent channel flow. *J. Fluid Mech.*, **118**, 341–377.

Moin, P., and Mahesh, K. 1998. Direct numerical simulation: A tool in turbulence research. *Ann. Rev. Fluid Mech.*, **30**, 539–578.

Moin, P., and Moser, R. D. 1989. Characteristic-eddy decomposition of turbulence in a channel. *J. Fluid Mech.*, **200**, 471–509.

Morrison, W. R. B., Bullock, K. J., and Kronauer, R. E. 1971. Experimental evidence of waves in the sublayer. *J. Fluid Mech.*, **47**, 639–656.

Nagata, M. 1990. Three-dimensional finite-amplitude solutions in plane Couette flow: Bifurcation from infinity. *J. Fluid Mech.*, **217**, 519–527.

Nagata, M. 1997. Three-dimensional traveling-wave solutions in plane Couette flow. *Phys. Rev. E*, **55**, 2023–2025.

Oberlack, M. 2001. A unified approach for symmetries in plane parallel turbulent shear flows. *J. Fluid Mech.*, **427**, 299–328.

Obukhov, A. M. 1941. On the distribution of energy in the spectrum of turbulent flow. *Dokl. Akad. Nauk. SSSR*, **32**, 22–24.

Okino, S., Nagata, M., Wedin, H., and Bottaro, A. 2010. A new nonlinear vortex state in square-duct flow. *J. Fluid Mech.*, **657**, 413–429.

Österlund, J. M., Johansson, A. V., Nagib, H. M., and Hites, M. 2000. A note on the overlap region in turbulent boundary layers. *Phys. Fluids*, **12**, 1–4.

Pinelli, A., Uhlmann, M., Sekimoto, A., and Kawahara, G. 2010. Reynolds number dependence of mean flow structure in square duct turbulence. *J. Fluid Mech.*, **644**, 107–122.

Prigent, A., Grégorie, G., Chaté, H., Dauchot, O., and van Saarloos, W. 2002. Large-scale finite-wavelength modulation within turbulent shear flows. *Phys. Rev. Lett.*, **89**, 014501.

Pringle, C. C., and Kerswell, R. R. 2007. Asymmetric, helical, and mirror-symmetric traveling waves in pipe flow. *Phys. Rev. Lett.*, **99**, 074502.

Reddy, S. C., Schmid, P. J., Baggett, J. S., and Henningson, D. S. 1998. On stability of streamwise streaks and transition thresholds in plane channel flows. *J. Fluid Mech.*, **365**, 269–303.

Reynolds, O. 1894. On the dynamical theory of incompressible viscous fluids and the determination of the criterion. *Phil. Trans. Roy. Soc. London*, **186**, 123–164. Papers, ii, 535.

Richardson, L. F. 1920. The supply of energy from and to atmospheric eddies. *Proc. Roy. Soc. A*, **97**, 354–373.

Rogallo, R. S. 1981. *Numerical experiments in homogeneous turbulence*. Tech. Memo 81315. NASA.

Rogallo, R. S., and Moin, P. 1984. Numerical simulations of turbulent flows. *Ann. Rev. Fluid Mech.*, **16**, 99–137.

Saiki, Y., and Yamada, M. 2009. Time-averaged properties of unstable periodic orbits and chaotic orbits in ordinary differential equation systems. *Phys. Rev. E*, **79**, 015201(R).

Saiki, Y., and Yamada, M. 2010. Unstable periodic orbits embedded in a continuous time dynamical system – time averaged properties. *RIMS Kokyuroku*, **1713**, 111–123.

Schlatter, P., and Örlü, R. 2010. Assessment of direct numerical simulation data of turbulent boundary layers. *J. Fluid Mech.*, **659**, 116–126.

Schlatter, P., Örlü, R., Li, Q., Fransson, J.H.M., Johansson, A.V., Alfredsson, P. H., and Henningson, D. S. 2009. Turbulent boundary layers up to $Re_\theta = 2500$ through simulation and experiments. *Phys. Fluids*, **21**, 05702.

Schmiegel, A. 1999. *Transition to turbulence in linearly stable shear flows*. Ph.D. thesis, Faculty of Physics, Philipps-Universität Marburg.

Schneider, T. M., Gibson, J. F., Lagha, M., De Lillo, F., and Eckhardt, B. 2008. Laminar-turbulent boundary in plane Couette flow. *Phys. Rev. E*, **78**, 037301.

Schneider, T. M., Gibson, J. F., and Burke, J. 2010. Snakes and ladders: localized solutions of plane Couette flow. *Phys. Rev. Lett.*, **104**, 104501.

Schoppa, W., and Hussain, F. 2002. Coherent structure generation in near-wall turbulence. *J. Fluid Mech.*, **453**, 57–108.

She, Z.-S. 1993. Constrained Euler system for Navier–Stokes turbulence. *Phys. Rev. Lett.*, **70**, 1255–1258.

Siebesma, A. P., Bretherton, C. S., Brown, A., A.Chlond, Cuxart, J., Duynkerke, P. G., Jiang, H. L., Khairoutdinov, M., Lewellen, D., Moeng, C. H., Sanchez, E., Stevens, B., and Stevens, D. E. 2003. A large eddy simulation intercomparison study of shallow cumulus convection. *J. Atmos. Sci.*, **60**, 1201–1219.

Siggia, E. D. 1981. Numerical study of small scale intermittency in three-dimensional turbulence. *J. Fluid Mech.*, **107**, 375–406.

Sillero, J. A., Borrell, G., Gungor, A. G., Jiménez, J., Moser, R.D., and Oliver, T. A. 2010. Direct simulation of the zero-pressure-gradient boundary layer up to $Re_\theta = 6000$. Pages EB-4 of: *Proc. Div. Fluid Dyn. Am. Phys. Soc.*

Simens, M.P., Jiménez, J., Hoyas, S., and Mizuno, Y. 2009. A high-resolution code for turbulent boundary layers. *J. Comput. Phys.*, **228**, 4218–4231.

Smith, C. R., and Metzler, S. P. 1983. The characteristics of low speed streaks in the near wall region of a turbulent boundary layer. *J. Fluid Mech.*, **129**, 27–54.

Spalart, P. R., Coleman, G. N., and Johnstone, R. 2008. Direct numerical simulation of the Ekman layer: A step in Reynolds number, and cautious support for a log law with a shifted origin. *Phys. Fluids*, **20**, 101507.

Swearingen, J. D., and Blackwelder, R. F. 1987. The growth and breakdown of streamwise vortices in the presence of a wall. *J. Fluid Mech.*, **182**, 255–290.

Tennekes, H., and Lumley, J. L. 1972. *A First Course in Turbulence*. MIT Press.

Toh, S., and Itano, T. 2003. A periodic-like solution in channel flow. *J. Fluid Mech.*, **481**, 67–76.

Toh, S., and Itano, T. 2005. Interaction between a large-scale structure and near-wall structures in channel flow. *J. Fluid Mech.*, **524**, 249–262.

Townsend, A. A. 1961. Equilibrium layers and wall turbulence. *J. Fluid Mech.*, **11**, 97–120.

Townsend, A. A. 1976. *The Structure of Turbulent Shear Flow*, 2nd ed. Cambridge University Press.

Uhlmann, M., Kawahara, G., and Pinelli, A. 2010. Traveling-waves consistent with turbulence-driven secondary flow in a square duct. *Phys. Fluid*, **22**, 084102.

van Veen, L., and Kawahara, G. 2011. Homoclinic tangle on the edge of shear turbulence. *Phys. Rev. Lett.*, **107**, 114501.

van Veen, L., Kawahara, G., and Matsumura, A. 2011. On matrix-free computation of 2D unstable manifolds. *SIAM J. Sci. Comp.*, **33**, 25–44.

van Veen, L., Kida, S., and Kawahara, G. 2006. Periodic motion representing isotropic turbulence. *Fluid Dynamics Research*, **38**, 19–46.

Vincent, A., and Meneguzzi, M. 1991. The spatial structure and statistical properties of homogeneous turbulence. *J. Fluid Mech.*, **225**, 1–25.

Viswanath, D. 2007. Recurrent motions within plane Couette turbulence. *J. Fluid Mech.*, **580**, 339–358.

Viswanath, D. 2009. The critical layer in pipe flow at high Reynolds number. *Phil. Trans. Roy. Soc. A*, **367**, 561–576.

Waleffe, F. 1997. On a self-sustaining process in shear flows. *Phys. Fluids*, **9**, 883–900.

Waleffe, F. 1998. Three-dimensional coherent states in plane shear flows. *Phys. Rev. Lett.*, **81**, 4140–4143.

Waleffe, F. 2001. Exact coherent structures in channel flow. *J. Fluid Mech.*, **435**, 93–102.

Waleffe, F. 2003. Homotopy of exact coherent structures in plane shear flows. *Phys. Fluids*, **15**, 1517–1534.

Wang, J., Gibson, J., and Waleffe, F. 2007. Lower branch coherent states in shear flows: Transition and control. *Phys. Rev. Lett.*, **98**, 204501.

Wedin, H., and Kerswell, R. R. 2004. Exact coherent structures in pipe flow: travelling wave solutions. *J. Fluid Mech.*, **508**, 333–371.

Wedin, H., Bottaro, A., and Nagata, M. 2009. Three-dimensional traveling waves in a square duct. *Phys. Rev. E*, **79**, 065305(R).

Willis, A. P., and Kerswell, R. R. 2008. Coherent structures in localized and global pipe turbulence. *Phys. Rev. Lett.*, **100**, 124501.

Willis, A. P., and Kerswell, R. R. 2009. Turbulent dynamics of pipe flow captured in a reduced model: puff relaminarization and localized 'edge' states. *J. Fluid Mech.*, **619**, 213–233.

Wosnik, M., Castillo, L., and George, W. K. 2000. A theory for turbulent pipe and channel flows. *J. Fluid Mech.*, **421**, 115–145.

Wu, X., and Moin, P. 2010. Transitional and turbulent boundary layer with heat transfer. *Phys. Fluids.*, **22**, 085105.

Wygnanski, I. J., and Champagne, F. H. 1973. On transition in a pipe. Part 1. The origin of puffs and slugs and the flow in a turbulent slug. *J. Fluid Mech.*, **59**, 281–335.

Zhou, J., Adrian, R. J., S., Balachandar, and Kendall, T. M. 1999. Mechanisms for generating coherent packets of hairpin vortices in channel flow. *J. Fluid Mech.*, **387**, 353–396.

7

Recent Progress in Stratified Turbulence

James J. Riley and Erik Lindborg

7.1 Introduction

Stable density stratification can have a strong effect on fluid flows. For example, a stably-stratified fluid can support the propagation of internal waves. Also, at large enough horizontal scales, flow in a stably-stratified fluid will not have enough kinetic energy to overcome the potential energy needed to overturn; therefore flows at this horizontal scale and larger cannot overturn, greatly constraining the types of motions possible. Both of these effects were observed in laboratory experiments of wakes in stably-stratified fluids (see, e.g., (Lin and Pao, 1979)). In the wake experiments, generally the flow in the near wake of the source, e.g., a towed sphere or a towed grid, consisted of three-dimensional turbulence, little affected by the stable stratification. As the flow decayed, however, the effects of stratification became continually more important. After a few buoyancy periods, when the effects of stable stratification started to dominate, the flow had changed dramatically, and consisted of both internal waves and quasi-horizontal motions. Following Lilly (1983), we will call such motions, consisting of both internal waves and quasi-horizontal motions due to the domination of stable stratification, as "stratified turbulence". It has become clear that such flows, while being strongly constrained by the stable stratification, have many of the features of turbulence, including being stochastic, strongly nonlinear, strongly dispersive, and strongly dissipative.

A primary interest in stratified turbulence is how energy in such flows, strongly affected by stable stratification, is still effectively cascaded down to smaller scales and into three-dimensional turbulence, where it is ultimately dissipated.

More specifically we consider as stratified turbulence flows at a given horizontal scale ℓ_h with a corresponding characteristic velocity U having the following properties.

- The Reynolds number $R_h = U\ell_h/\nu$ is very large, where ν is the fluid kinematic viscosity.
- The Froude number $F_h = U/N\ell_h$ is very small, where N is the local buoyancy frequency, implying the the effects of the stably density stratification are in some sense dominant.
- The buoyancy Reynolds number $\mathcal{R} = F_h^2 R_h$ is very large, implying, as we will find in §7.2 below, that smaller-scale turbulence can exist.
- The Rossby number $Ro = U/f\ell_h$ is of the order of unity or larger, where f is the local inertial frequency, implying that the effects of the earth's rotation are not dominant.

Motions satisfying these assumptions generally occur at horizontal length scales above the Ozmidov scale, $\ell_o = (\epsilon/N^3)^{1/2}$ (Lumley (1964); Ozmidov (1965)), which is interpreted as characterizing the largest horizontal scale which can overturn (Riley and Lindborg, 2008). Here ϵ is the kinetic energy dissipation rate. Assuming a Taylor (1935) scaling for the energy dissipation rate, $\epsilon \sim U^3/\ell_h$, which is consistent with low Froude number scaling (Billant and Chomaz, 2001) and also with numerical simulations of stratified turbulence (Lindborg, 2006), then $\ell_o/\ell_h \sim F_h^{3/2}$, so that $\ell_o/\ell_h < 1$ implies that $F_h < 1$ and stable stratification effects become order 1.

Motions at horizontal length scales above the Ozmidov scale, and hence dominated by stable stratification, can often be found in the atmosphere and the oceans. For example, the Ozmidov scale in the oceans is typically of the order of 1 m (Gargett et al., 1981), so that stratified turbulence is expected to be found for horizontal scales greater than a few meters up to several hundred meters or more. In strongly stable atmospheric boundary layers, the Ozmidov scale has also been found to be of the order of 1 m (Frehlich et al., 2008), with stratified turbulence again expected to be in the range of a few meters up to several hundred meters or more. In the upper troposphere, fewer estimates of ϵ exist, but the Ozmidov scale is expected to be on the order of 10's of meters, so that the stratified turbulence range would be from scales of 10's of meters up to the kilometer range.

In the next section is given theoretical discussions of several aspects of stratified turbulence, including: a careful scaling of the governing equations; scaling arguments leading to predictions, e.g., of energy spectra; and a new derivation of the expression for the spectral energy flux in stratified turbulence, showing how this depends on the buoyancy Reynolds number. In the following section will be discussed some results of numerical simulations, including the simulation of decaying flows as well as of forced flows. Then in the next section results from some laboratory experiments will be reported, in particular some wake experiments. In the penultimate section results relevant to stratified turbulence from both the oceans and the atmosphere will be reported. Finally, in the last section, conclusions as well as open issues will be discussed.

It has not been our ambition to write a standard review paper covering the entire field of stratified turbulence, going back several decades in time, but rather to report on recent progress in the field, with some emphasis on problems which we have worked on ourselves. References going back more than ten years are made only when these are relevant in the context of more recent work which we discuss. For a more comprehensive review of research on this subject prior to 2000, the reader is referred to Riley and Lelong (2000). We have also had the ambition to identify some important problems in the field which are yet unsolved. In this context we have allowed ourselves to make some conjectures which we hope will stimulate further research in this field.

7.2 Scaling, cascade and spectra

7.2.1 Scaling

We consider the Navier-Stokes equations for a stratified incompressible flow under the Boussinesq approximation,

$$\frac{\partial \mathbf{u}'}{\partial t'} + \mathbf{u}' \cdot \nabla' \mathbf{u}' = -\frac{1}{\rho_0} \nabla' p' + \nu \nabla'^2 \mathbf{u}' - \frac{\rho' g}{\rho_0} \mathbf{e}_z \tag{7.1}$$

$$\nabla' \cdot \mathbf{u}' = 0 \tag{7.2}$$

$$\frac{\partial \rho'}{\partial t'} + \mathbf{u}' \cdot \nabla' \rho' = \kappa \nabla'^2 \rho' - \frac{\partial \overline{\rho}}{\partial z'} w', \tag{7.3}$$

where $\mathbf{e}_z$ is the unit vector in the vertical (z) direction, $\mathbf{u}'$ is the velocity, w' the vertical velocity component, p' the pressure, ν and κ are the kinematic viscosity and the molecular diffusivity, respectively, g the gravitational acceleration, ρ_0 a constant reference density, and ρ' is a perturbation from a linear density profile $\overline{\rho}$. The Brunt-Väisälä frequency, defined as $N = \sqrt{-(g/\rho_0)\partial \overline{\rho}/\partial z'}$, is assumed constant.

Following Riley et al. (1981), Lilly (1983), Billant and Chomaz (2001) and Brethouwer et al. (2007) we will now put these equations into nondimensional form. To do this, we introduce a horizontal length scale ℓ_h, a vertical length scale ℓ_v, a velocity scale U, the aspect ratio $\alpha = \ell_v/\ell_h$, and the horizontal Froude number $F_h = U/(\ell_h N)$. The length scale ℓ_h may be interpreted as the horizontal length scale of the largest structures of stratified turbulence, or, using a standard concept in turbulence theory, the "integral length scale". As we will see below, ℓ_v represents the vertical scale of instability of the horizontal motion, leading to three-dimensional turbulence. We shall also find that these length scales will be widely different in the two directions. Correspondingly, U may be interpreted as the typical horizontal velocity scale corresponding to the largest structures. The horizontal velocity is scaled by U, the vertical velocity by UF_h^2/α, the horizontal and vertical lengths by ℓ_h and ℓ_v, respectively, the time by ℓ_h/U, ρ' by $U^2\rho_0/(g\ell_v)$ and the pressure by $\rho_0 U^2$. This leads to the nondimensional equations

$$\frac{\partial \mathbf{u}_h}{\partial t} + \mathbf{u}_h \cdot \nabla_h \mathbf{u}_h + \frac{F_h^2}{\alpha^2} w \frac{\partial \mathbf{u}_h}{\partial z} = -\nabla_h p + \frac{1}{Re_h}\left[\frac{1}{\alpha^2}\frac{\partial^2 \mathbf{u}_h}{\partial z^2} + \nabla_h^2 \mathbf{u}_h\right] \quad (7.4)$$

$$F_h^2\left[\frac{\partial w}{\partial t} + \mathbf{u}_h \cdot \nabla_h w + \frac{F_h^2}{\alpha^2} w \frac{\partial w}{\partial z}\right] = -\frac{\partial p}{\partial z} - \rho + \frac{F_h^2}{Re_h}\left[\frac{1}{\alpha^2}\frac{\partial^2 w}{\partial z^2} + \nabla_h^2 w\right] \quad (7.5)$$

$$\nabla_h \cdot \mathbf{u}_h + \frac{F_h^2}{\alpha^2}\frac{\partial w}{\partial z} = 0 \quad (7.6)$$

$$\frac{\partial \rho}{\partial t} + \mathbf{u}_h \cdot \nabla_h \rho + \frac{F_h^2}{\alpha^2} w \frac{\partial \rho}{\partial z} = w + \frac{1}{Re_h Sc}\left[\frac{1}{\alpha^2}\frac{\partial^2 \rho}{\partial z^2} + \nabla_h^2 \rho\right], \quad (7.7)$$

where nondimensional variables are noted without a prime. Here $Re_h = U\ell_h/\nu$ is the Reynolds number, $Sc = \nu/\kappa$ is the Schmidt number, $\mathbf{u}_h$ and w are the horizontal and vertical velocities, respectively, and ∇_h is the horizontal gradient operator. In the limit $F_h = U/(l_h N) \to 0$ and $Re \gg 1$, with $F_h^2/\alpha^2 \leq \mathcal{O}(1)$, we obtain

$$\frac{\partial \mathbf{u}_h}{\partial t} + \mathbf{u}_h \cdot \nabla_h \mathbf{u}_h + \frac{F_h^2}{\alpha^2} w \frac{\partial \mathbf{u}_h}{\partial z} = -\nabla_h p + \frac{1}{Re\,\alpha^2}\frac{\partial^2 \mathbf{u}_h}{\partial z^2} \quad (7.8)$$

$$0 = -\frac{\partial p}{\partial z} - \rho \quad (7.9)$$

$$\nabla_h \cdot \mathbf{u}_h + \frac{F_h^2}{\alpha^2}\frac{\partial w}{\partial z} = 0 \quad (7.10)$$

$$\frac{\partial \rho}{\partial t} + \mathbf{u}_h \cdot \nabla_h \rho + \frac{F_h^2}{\alpha^2} w \frac{\partial \rho}{\partial z} = w + \frac{1}{ReSc\,\alpha^2}\frac{\partial^2 \rho}{\partial z^2} \quad (7.11)$$

Note that an interesting outcome of this scaling is that the flow must be hydrostatic to lowest order in F_h.

The perhaps most important assumption underlying recent progress in the theory of stratified turbulence (Billant and Chomaz, 2001) is that the aspect ratio, α, will adjust in such a way that the nonlinear terms including the vertical velocity will be of leading order in the equations (7.8) and (7.11), that is_a_

$$\alpha \sim F_h \Rightarrow l_v \sim \frac{U}{N}.\qquad(7.12)$$

It has been shown that this differential scale l_v develops even when a larger vertical scale is imposed upon the flow (Riley and deBruynKops, 2003). A necessary condition for equation (7.12) to be valid is that $\mathcal{R} \equiv F_h^2 Re > 1$. If $\mathcal{R} \ll 1$, on the other hand, the vertical length scale can be estimated as $\ell_v \sim \ell_h Re^{-1/2}$ (Godoy-Diana et al., 2004). In that case, we thus find that ℓ_v has the same $Re^{-1/2}$-dependence as the thickness of a laminar boundary layer and the dynamics will also be viscously dominated, just as in a laminar boundary layer. It is in the other limit, $\mathcal{R} \gg 1$, that the motion we call stratified turbulence will prevail. The most characteristic feature of stratified turbulence, as compared to classical Kolmogorov turbulence, is its strong degree of anisotropy. In geophyscial applications F_h may be as small as 10^{-3}; the aspect ratio, α, which is generally a little larger, although it scales with F_h, may be smaller than 10^{-2}. The small aspect ratio has given rise to the name "pancake turbulence".

In geophysical contexts the equations (7.8–7.11) are often known as the "primitive equations", here written in incompressible form and without including any system rotation. The primitive equations have been used in weather forecast models for decades. In such applications, however, the resolution has generally been too coarse to simulate stratified turbulence. To do this, it is required that a very wide range of scales is resolved.

7.2.2 Ranges of scales, cascade and spectra

Gage (1979) and Lilly (1983) conjectured that stratified turbulence may exhibit an inverse energy cascade, just as two-dimensional turbulence. The crucial assumption made by Lilly was that, as $F_h \to 0$, then also $F_h^2/\alpha^2 \to 0$ so that the terms involving the vertical velocity in equations (7.8) and (7.11) will become negligible in the limit of strong stratification. The resulting set of equations would then possess two quadratic inviscid invariants, total energy and the integral of the square of the vertical vorticity. Just as the two-dimensional Navier-Stokes equations, which also have two quadratic inviscid invariants, kinetic energy and enstrophy, the set of equations describing stratified turbulence would then exhibit an inverse energy cascade.

On the other hand, if the scaling assumption of Billant and Chomaz (2001) is correct – that the aspect ratio, α, will adjust in such a way that, as $F_h \to 0$,

---

[a] Lilly (1983) had suggested something similar when he argued that stratified turbulence would continually develop smaller vertical scales until the local Richardson number became of order 1.

the ratio F_h^2/α^2 remains of order 1, so that the terms including the vertical velocity will be of leading order – then the set of equations has only one inviscid quadratic invariant, which is total energy, that is, the kinetic plus potential energy. With a linear mean density profile, the potential energy is defined as $E_P = b'^2/2$, where $b' = \rho'g/\rho_0 N$ is the buoyancy, normalised in such a way that it has the same dimensions as velocity, rather than acceleration. With only one quadratic invariant, there is no reason to believe that stratified turbulence should exhibit an inverse energy cascade. On the contrary, it can be argued that the principal cascade is in the forward direction, just as in three-dimensional turbulence. It can also be argued that potential energy will be of the same order as the kinetic energy, or $b' \sim U$. Through the collective work of several investigators over recent years (e.g., Riley and deBruynKops (2003); Waite and Bartello (2004); Lindborg (2005, 2006); Brethouwer et al. (2007)) it has been demonstrated that stratified turbulence indeed exhibits a forward energy cascade in which both kinetic and potential energy are transferred from small wave numbers (large scales) towards large wave numbers (small scales). In this respect its dynamics are more similar to classical isotropic three-dimensional turbulence than to two-dimensional turbulence.

As pointed out in the introduction, the horizontal integral length scale of stratified turbulence can be estimated in the same way as the classical estimate of the integral length scale of isotropic turbulence (Taylor, 1935), that is

$$\ell_h \sim \frac{U^3}{\epsilon}, \qquad (7.13)$$

where ϵ is the energy dissipation rate. Note that this relation is consistent with the lowest order in F_h scaling of the equations (Billant and Chomaz (2001)) and also with the numerical simulations of stratified turbulence (Lindborg (2006); Brethouwer et al. (2007), Figure 1). It can be argued that the Ozmidov scale, $\ell_O = (\epsilon/N^3)^{1/2}$, is the largest horizontal scale possessing sufficient kinetic energy to overturn (Riley and Lindborg, 2008). Therefore the Ozmidov scale is the transition scale between stratified turbulence and classical three-dimensional turbulence. The ranges of length scales encompassing stratified turbulence can be estimated in terms of the Froude number as

$$\frac{\ell_h}{\ell_O} \sim F_h^{-3/2}, \qquad \frac{\ell_v}{\ell_O} \sim F_h^{-1/2}. \qquad (7.14)$$

The range of horizontal scales is thus wider by a factor of F_h^{-1} than the range of vertical scales. This is, again, a reflection of the strong degree of anisotropy of stratified turbulence. In addition, $\ell_O/\eta \sim \mathcal{R}^{3/4}$ with $\eta = (\nu^3/\epsilon)^{1/4}$, so that $\mathcal{R} \gg 1$ implies a large separation between the Ozmidov and the Kolmogorov scales.

In a similar way as the classical Kolmogorov (1941) $k^{-5/3}$-spectrum can be found from dimensional arguments together with matched asymptotic analysis (Lundgren, 2003) we can formulate power law hypotheses for the horizontal and vertical energy spectra. The classical Kolmogorov $k^{-5/3}$-spectrum can be viewed as an overlap range between a large scale range for which the two relevant parameters are a velocity scale U and the length scale $L \sim U^3/\epsilon$ and a small scale range where the two relevant parameters are ϵ and ν, alternatively ϵ and η. In the limit of large Reynolds number L and η become widely separated and there is an overlap range where the only common parameter is ϵ and in this range the spectrum must consequently be of the form $\sim \epsilon^{2/3} k^{-5/3}$. For the horizontal spectrum of stratified turbulence we can apply a very similar type of reasoning, with the difference that the relevant parameters in the small scale range are now ϵ and ℓ_O. In the limit of small F_h, ℓ_h and ℓ_O become widely separated and there is an overlap range where the only common parameter is ϵ and in this range the horizontal spectrum should therefore have a similar form as the Kolmogorov spectrum.

To be more specific, for stratified turbulence we should distinguish between the spectra of kinetic and potential energies. Likewise, we should distinguish between the mean dissipation of kinetic energy, ϵ_K, and the mean dissipation of potential energy, defined as

$$\epsilon_P = \kappa \langle \nabla b' \cdot \nabla b' \rangle, \tag{7.15}$$

where $\langle \rangle$ denotes a spacial mean value. The inertial range expressions for the horizontal spectra of kinetic and potential energy can be assumed to have the form (Lindborg, 2006)

$$E_K(k_h) = C_1 \epsilon_K^{2/3} k_h^{-5/3}, \quad E_P(k_h) = C_2 \frac{\epsilon_P}{\epsilon_K^{1/3}} k_h^{-5/3}, \tag{7.16}$$

where C_1 and C_2 are two constants. The expressions (7.16) may be interpreted as the two-dimensional horizontal spectra, in which case $k_h = \sqrt{k_x^2 + k_y^2}$, or may be interpreted as the one-dimensional horizontal energy spectra, in which case k_h is the magnitude of the wave number corresponding to the direction in which the spectra are measured. For example, if the measurement is carried out in the x-direction, then $k_h = |k_x|$. The constants C_1 and C_2 will differ depending on whether the expressions (7.16) are interpreted as the one-dimensional or two-dimensional spectra, with a converting factor approximately equal to 1.4 (Lindborg and Brethouwer, 2007).

For the vertical energy spectrum the large scale range parameters are U and N, while the small scale range parameters are ϵ and N. The only common parameter is N and in the overlap range, that is for vertical wave numbers satisfying $1/\ell_v \ll k_v \ll k_O$, the spectrum should therefore have the form

$$E(k_v) \sim N^2 k_v^{-3}. \tag{7.17}$$

This form should be valid for the kinetic as well as the potential energy spectrum. It should be emphasised, however, that in order to obtain a vertical spectrum of the form (7.17) the Froude number must be very small, since the the total range of vertical scales can be estimated as $F_h^{-1/2}$. The requirement on the Froude number for a horizontal $k_h^{-5/3}$-range to exist is much weaker, since the total range of horizontal scales can be estimated as $F_h^{-3/2}$. Thus, it must be expected that the horizontal $k_h^{-5/3}$-spectrum can be observed much more frequently in nature than the vertical k_v^{-3}-spectrum.

That the horizontal spectra do not depend on N in the overlap range is a consequence of the assumption that ℓ_h is independent of N. Likewise, that the vertical spectra do not depend on ϵ in the overlap range is a consequence of the assumption that ℓ_v is independent of ϵ. The dimensional arguments leading to (7.16) and (7.17) are by no means rigorous. Exactly the same type of criticism (Landau and Lifshitz, 1987) that can be raised against Kolmogorov's (1941) theory for not taking into account possible effects of intermittency can also be raised against the hypotheses formulated here. At best, they can be regarded as good approximations.

It should be pointed out that the contribution from the vertical velocity to the kinetic energy is very small in stratified turbulence, and that the power spectrum of the vertical velocity will scale differently than (7.16) and (7.17). The magnitude of the vertical velocity corresponding to the integral scales can be estimated from the continuity equation as

$$w' \sim \ell_v u'/\ell_h \sim \epsilon^{1/3} \ell_h^{-2/3} u'/N. \tag{7.18}$$

This estimate suggests that the prefactor of the power spectrum of vertical velocity should differ from the spectra (7.16) and (7.17) by a factor of $\epsilon^{2/3}/N^2$. The horizontal power spectrum of vertical velocity should thus scale as

$$E_w(k_h) \sim \frac{\epsilon^{4/3}}{N^2} k_h^{-1/3}. \tag{7.19}$$

An important consequence of (7.19) is that the vertical velocity is a small scale quantity, with dominant spectral contribution from the wave numbers corresponding to the Ozmidov length scale, rather than the integral length scale. The large scale spectral contribution to the mean square vertical velocity, $\langle w'w'\rangle$, can be estimated as $\sim U^2 F_h^2$. On the other hand, by integrating (7.19) between k_ℓ and k_O we find that the dominant contribution is the small scale contribution, which can be estimated as $\sim U^2 F_h$, which is larger by a factor of F_h^{-1} than the large scale contribution.

The mean rate of transfer of kinetic energy to potential energy can be written as

$$\mathcal{B} = N\langle w'b'\rangle, \qquad (7.20)$$

where we have used the symbol $\mathcal{B}$, since this quantity may also be called the buoyancy flux. Defined in this way, $\mathcal{B}$ has the dimension of energy transfer per unit mass, the same dimension as ϵ. An important, and yet unresolved, issue is whether or not the spectral contribution to the buoyancy flux is dominated by horizontal scales of the order of ℓ_h, or whether the scales of the order ℓ_O also makes a non-negligible contribution. A horizontal buoyancy flux spectrum steeper than k_h^{-1} would give a dominant contribution from scales which are much larger than the Ozmidov length scale, as in the model of Holloway (1988), while a spectrum more shallow than k_h^{-1} would give a dominant contribution from scales of the same order as the Ozmidov length scale, as in the model of Holloway (1986). Forced direct numerical simulations (Brethouwer et al., 2007) in which only the kinetic energy is forced at large scales, as well as decaying simulations (Riley and deBruynKops, 2003) in which the potential energy is initially set to zero, clearly indicate that potential energy will grow to almost the same order of magnitude as the kinetic energy at large scales. This suggest that the buoyancy flux is a large scale quantity. The small scale contribution may not be negligible, however. The estimate (7.18) together with the estimate $b' \sim U$ suggest that the prefactor of the buoyancy flux should be independent of N and only depend on ϵ. Such a spectrum should scale as $\sim \epsilon k_h^{-1}$. With such a spectrum, however, the total buoyancy flux would be larger than ϵ by a factor of $\ln(k_O/k_\ell)$, while it should be of the same order of magnitude. To correct for this factor, we hypothesise that the horizontal buoyancy flux spectrum may have the form

$$B(k_h) \sim \epsilon k_h^{-1} [\ln(k_O/k_\ell)]^{-1} \qquad (7.21)$$

in the inertial range. With such a spectrum the energy transfer to potential energy would be distributed throughout the inertial range. It should be emphasised that it remains to be investigated whether (7.21) is consistent with observational data and numerical simulations.

7.2.3 The spectral energy flux

In the Kolmogorov theory of three-dimensional, isotropic turbulence the concept of a forward energy cascade is intimately connected with the notion of a constant spectral energy flux through the inertial range. For forced isotropic turbulence, the leading order deviation from a constant energy flux can be derived. In this subsection, we will give a similar derivation for stratified turbulence, in which case the energy flux is the sum of the flux of kinetic energy and the flux of potential energy. We will also derive expressions for the separate fluxes of kinetic and potential energies.

As the reference example, we first consider forced isotropic turbulence, for which case the equation for the energy spectrum, $E(k)$, can be written as

$$\frac{\partial E}{\partial t} = T(k) - 2\nu k^2 E(k) + F(k), \tag{7.22}$$

where $T(k)$ is the energy transfer spectrum and $F(k)$ is the forcing spectrum. The spectral energy flux is defined as

$$\Pi(k) = -\int_0^k T(k)\,\mathrm{d}k. \tag{7.23}$$

We make the assumption of stationarity and that the forcing is zero for $k \gg k_\ell$. For such wave numbers the energy flux can be calculated as

$$\Pi(k) = \epsilon - \int_0^k 2\nu k^2 E(k)\,\mathrm{d}k. \tag{7.24}$$

where ϵ is the dissipation. Inserting into (7.24) a Kolmogorov spectrum, $E(k) = K\epsilon^{2/3}k^{-5/3}$, we find

$$\Pi(k) = \epsilon[1 - \frac{3}{2}K(k\eta)^{4/3}], \tag{7.25}$$

where $K \approx 1.5$ is the Kolmogorov constant and $\eta = \nu^{3/4}/\epsilon^{1/4}$ is the Kolmogorov length scale. For $k\eta \ll 1$, the second term in (7.25) is negligible and the energy flux is approximately constant and equal to ϵ.

For stratified turbulence, we instead consider the two-dimensional horizontal energy spectrum and the corresponding two-dimensional energy flux. Assuming axisymmetry with respect to the vertical axis the horizontal energy flux spectrum, the horizontal energy transfer spectrum, and the horizontal energy spectrum can be defined as

$$\Pi(k_h) = -\int_0^{k_h} T(k_h)\,dk_z\,,\quad T(k_h) = \int_{-\infty}^{\infty} 2\pi k_h T^{(3)}(k_h,k_z)\,dk_z\,, \quad (7.26)$$

$$E(k_h) = \int_{-\infty}^{\infty} 2\pi k_h E^{(3)}(k_h,k_z)\,dk_z\,, \quad (7.27)$$

where $E^{(3)}(k_h,k_z)$ is the three-dimensional spectral energy density and $T^{(3)}(k_h,k_z)$ is the corresponding spectral energy transfer density. In the appendix we prove that the two-dimensional spectral flux has the same form as the three-dimensional spectral energy flux (7.25) for isotropic turbulence, with the only difference that the numerical factor multiplying $(k\eta)^{4/3}$ is different. For stratified turbulence, we distinguish between the spectra of kinetic energy and potential energies, the transfer spectra of kinetic energy, T_K, and of potential energy, T_P, and the spectral flux of kinetic energy, Π_K, and of potential energy, Π_P.

Assuming that forcing is applied only in the kinetic energy equation, the equations for the kinetic and potential energy spectra can be written as

$$\frac{\partial E_K(k_h)}{\partial t} = T_K(k_h) - B(k_h) - D_K(k_h) - 2\nu k_h^2 E_K(k_h) + F(k_h)\,, (7.28)$$

$$\frac{\partial E_P(k_h)}{\partial t} = T_P(k_h) + B(k_h) - D_P(k_h) - 2\kappa k_h^2 E_P(k_h)\,. \quad (7.29)$$

Here, we have distinguished between the dissipation spectra corresponding to the viscous terms including only vertical gradients, denoted by $D_K(k_h)$ and $D_P(k_h)$, and the dissipation spectra corresponding to terms including only horizontal gradients, which are the terms including ν and κ explicitly. Assuming stationarity and integrating we find

$$0 = -\mathcal{B} - \epsilon_K + \int_0^\infty F(k_h)\,dk_h\,, \quad (7.30)$$

$$0 = \mathcal{B} - \epsilon_P\,, \quad (7.31)$$

where we have also included the range of classical Kolmogorov turbulence in the integration. However, in this analysis Kolmogorov turbulence is only important as an energy sink providing for the dissipation. If the buoyancy Reynolds number is large, the dissipation at scales larger than ℓ_O will be small, however not equal to zero. It can be argued that the dominant contribution to this small amount of dissipation comes from the viscous terms including vertical gradients, rather than from terms including horizontal gradients. For $k_h l_v \ll 1$ it can be assumed that the vertical shear is dominated by scales of the order of l_v and we can thus make the estimate

$$D_K(k_h) \sim \nu l_v^{-2} E_K(k_h) \sim \nu \frac{N^2}{U^2} C_1 \epsilon_K^{2/3} k_h^{-5/3}, \qquad (7.32)$$

and a corresponding estimate for D_P. An important consequence of (7.32) is that the small amount of energy which is dissipated at scales larger than ℓ_O is dissipated at the largest horizontal scales, of the order of ℓ_h. This can be seen by integrating (7.32) from k_ℓ to $k \gg k_\ell$,

$$\int_{k_\ell}^{k} \nu \frac{N^2}{U^2} C_1 \epsilon_K^{2/3} k_h^{-5/3} \, dk_h = \nu \frac{3N^2}{2U^2} C_1 \epsilon_K^{2/3} \ell_h^{2/3} [1 - (k\ell_h)^{-2/3}]$$

$$\sim \frac{\epsilon}{\mathcal{R}} [1 - (k\ell_h)^{-2/3}], \qquad (7.33)$$

where we have used Taylor's estimate (7.13) for ℓ_h. By this estimate the parameter $\mathcal{R} = Re_h F_h^2$ can also be written as

$$\mathcal{R} = \frac{\epsilon}{N^2 \nu}, \qquad (7.34)$$

which is the standard definition of the buoyancy Reynolds number. From (7.33) we can thus conclude that the dominant contribution to the dissipation at larger scales than ℓ_O comes from horizontal scales of the order of ℓ_h, and is smaller than the total dissipation by $\mathcal{R}^{-1}$. We can now deduce the form of total spectral energy flux in the inertial range of stratified turbulence. Assuming stationarity and that the forcing spectrum is zero for $k \gg k_\ell$, adding (7.27) and (7.26) and integrating, we find

$$\Pi(k_h) = \epsilon[1 - \mathcal{O}(\mathcal{R}^{-1})], \qquad (7.35)$$

where $\Pi = \Pi_K + \Pi_P$ and $\epsilon = \epsilon_K + \epsilon_P$. To arrive at (7.35) we have used (7.30), (7.31) and (7.33).

Thus, the leading order viscous term is independent of wave number k_h and scales as $\mathcal{R}^{-1}$. This is different from the flux relation (7.25) for isotropic turbulence, where the leading order viscous term decreases with decreasing wave number. At moderate buoyancy Reynolds number, $\mathcal{R} \sim \mathcal{O}(10)$, the energy flux from large to small scales is diminished by the combined influence of viscosity and stratification, while at $\mathcal{R} \sim \mathcal{O}(1)$ the energy flux is almost entirely shut off.

The spectral fluxes of kinetic energy and potential energy will have the form

$$\Pi_K(k_h) = \epsilon_K[1 - \mathcal{O}(\mathcal{R}^{-1})] + \int_{k_h}^{\infty} B(k_h) \, dk_h, \qquad (7.36)$$

$$\Pi_P(k_h) = \epsilon_P[1 - \mathcal{O}(\mathcal{R}^{-1})] - \int_{k_h}^{\infty} B(k_h) \, dk_h, \qquad (7.37)$$

in the inertial range. If the buoyancy flux has its dominant spectral contribution from large scales, then the last term in these expressions will be negligible in the inertial range and each of the fluxes will be approximately constant in the inertial range. If, on the other hand, our conjecture (7.21) is correct, the last term will give a non-negligible contribution. Using (7.21) we can make the estimate

$$\int_{k_h}^{\infty} B(k_h)\,dk_h \sim \int_{k_h}^{k_O} \epsilon k_h^{-1}[\ln(k_O/k_\ell)]^{-1}\,dk_h \sim \epsilon \frac{\ln(k_O/k_h)}{\ln(k_O/k_\ell)}, \tag{7.38}$$

where we have replaced the upper integration limit by k_O. If (7.38) is correct, then the kinetic energy flux decreases logarithmically through the inertial range, starting with $\Pi_K \approx \epsilon_K + \epsilon_P$ at $k_h \sim k_\ell$ and ending with $\Pi_K \approx \epsilon_K$ at $k_h \sim k_O$, while the potential energy flux increases correspondingly, starting with $\Pi_P \approx 0$ at $k_h \sim k_\ell$ and ending with $\Pi_P \sim \epsilon_P$ at $k_h \sim k_O$. The energy transfer, $\mathcal{B}$, from kinetic to potential energy, which in our idealised model is exactly equal to ϵ_P, would then be distributed throughout the inertial range.

7.2.4 Predictions of analytical theories of turbulence

Some recent research employing analytical theories of turbulence has been used to study stratified turbulence. For example, Cambon, Staquet and their colleagues (see Sagaut and Cambon (2008) for a summary of this work) used an anisotropic version of the EDQNM (eddy-damped, quasi-normal Markovian) closure theory (Orszag, 1970) in their study of turbulence in stable stratification. They expand their solutions in terms of the linear modes of the system, which they term toroidal modes (sometimes called vortical or vortex modes) and poloidal modes (internal wave modes in the linear case). (See also Bartello (1995) and Riley and Lelong (2000) for discussions of similar approaches to this decomposition.) They also performed direct numerical simulations, and compared simulation results with predictions from the EDQNM theory. Starting with initially isotropic turbulence (see Godeferd and Staquet (2003)) they find, in particular, that at low Froude number, strong anisotropization of the flow occurs with the toroidal mode energy moving towards vertical wave vectors. This in turn produces the collapse of the vertical motion and vertical layering, resulting in a quasi-horizontal pancake structure. Such structures are then consistent with the scaling arguments discussed earlier in s §7.2.1. They did not, however, pursue scaling arguments leading to the spectral behavior discussed in §7.2.2.

Galperin and his colleagues (see Galperin and Sukoriansky (2010) for a recent discussion of this work) use a quasi-normal scale elimination (QNSE) scheme to close the spectral form of the governing equations; the method is based upon a successive ensemble averaging of the velocity and temperature fields over the smallest scales and calculating a corresponding eddy viscosity and eddy diffusivity. They predict, in particular, vertical energy spectra proportional to $N^2 k_z^{-3}$, consistent with the predictions discussed earlier in §7.2.2. Furthermore, they find horizontal turbulence diffusivities proportional to $\epsilon^{1/3} k^{-4/3}$. This behavior of the horizontal diffusivities corresponds to a Richardson law for horizontal diffusion, which is consistent with a number of oceanic horizontal dispersion results of Okubo (1971). Furthermore they point out that this form for the horizontal diffusivity is consistent with a horizontal energy spectrum proportional to $\epsilon^{2/3} k^{-5/3}$, as also discussed earlier in §7.2.2. They have yet to directly predict, however, this form of the horizontal energy spectrum.

7.3 Numerical simulations

In order to perform direct numerical simulations of stratified turbulence, one needs a computational domain whose horizontal extent is at least of the order of $l_h \sim U^3/\epsilon$, and whose vertical extent is a least of the order of $l_v \sim U/N$. The smallest scale which is to be resolved in all three directions is the Kolmogorov scale (Pope, 2000). Given these requirements, the total numer of grid points will scale as $\mathcal{R}^{9/4} F_h^{-7/2}$. It is therefore extremely demanding to simulate a state for which $\mathcal{R} \gg 1$ and $F_h \ll 1$. One of the first studies coming close to such a state, and addressing the problem of stratified turbulence using numerical simulation, was that of Riley and deBruynKops (2003). They were interested in testing the suggestions of Lilly (1983), Babin et al. (1997), and Billant and Chomaz (2000, 2001) that strongly stratified turbulent at high enough Reynolds numbers would necessarily lead to smaller-scale turbulence. In addition, they were interested in the relevance to geophysical-scale motions of past laboratory experiments and numerical simulations which were conducted at low Froude number, but also at low buoyancy Reynolds numbers.

They chose to simulate the dynamics of a flow that initially consisted of Taylor-Green vortices, superimposed with low-level, broad-banded noise, i.e., the initial velocity field was given by

$$\mathbf{v}(\mathbf{x},0) = \mathcal{U} \cos(kz) \big[\cos(kx)\sin(ky), -\sin(kx)\cos(ky), 0\big] + \mathbf{v}_N(\mathbf{x}),$$

where $\mathcal{U}$ and k are constant velocity and wave number, respectively, and $\mathbf{v}_N$ is the added noise. The background density stratification was chosen with uniform buoyancy frequency N, and the initial density perturbation about the background was set to zero. Defining the Froude and Reynolds numbers here as $F_\ell = u'/N\ell$ and $R_\ell = u'\ell/\nu$, respectively, where u' in the volume-averaged rms velocity and $\ell = 1/k$, simulations were run for initial Froude numbers of 0.112 and 0.224 and initial Reynolds numbers in the range of 560 to 2,250. These initial Froude numbers were selected to be low enough, and the initial Reynolds number high enough, to put the flows in the stratified turbulence regime.

The Taylor-Green vortices, consisting of only horizontal motion but with well-defined, and equal, vertical and horizontal length scales, were chosen to mimic the low Froude number regimes in the laboratory wake studies of Lin and Pao (1979), Liu (1995), and Spedding (1997), but at higher Reynolds numbers. They also roughly mimic the intermediate scale motions of the atmosphere and oceans, ones with horizontal scales larger than the Ozmidov scale, but small enough that the effects of the earth's rotation are not important.

The evolution of the flows was assumed to satisfy the Navier-Stokes equations subject to the Boussinesq approximation, which were solved numerically on up to 512^3-point computational grids. Consistent with the Taylor-Green initialization, periodic boundary conditions were imposed in all three directions, and Fourier-spectral methods were used, along with Adams-Bashforth time stepping, and an integrating factor for the molecular viscosity and diffusion terms.

Riley and deBruynkops found that, indeed, strong shearing of the horizontal motion immediately started to develop, as suggested by Lilly (1983). For example, Figure 7.1 is a plot of the volume-averaged Richardson number Ri_V as a function of time, for simulations at a given Froude number ($F_\ell = 0.224$) but for a range of Reynolds numbers ($R_\ell = 560$ to 2250). (Here Ri_V is defined as $N^2 / \left\langle \left| \frac{\partial \mathbf{u}_h}{\partial z} \right|^2 \right\rangle_V$, where $\langle \cdot \rangle_V$ denotes a volume average.) It is seen that Ri_V initially has a value close to 9; minimum values of the Richardson number in the flow are above 3, so that the flow would be stable based upon a simplistic application of linear stability theory. The theory does not hold for this case, however, since the flow is not planar and not rectilinear. In fact it is the three-dimensional nature of the flow that leads to the stratified flow dynamics. As suggested by theoretical arguments (Lilly, 1983; Babin et al., 1997; Billant and Chomaz, 2000, 2001), strong shearing of the flow develops, and Ri_V rapidly drops to order 1, the value depending on the Reynolds number. For the highest Reynolds number case Ri_V is well

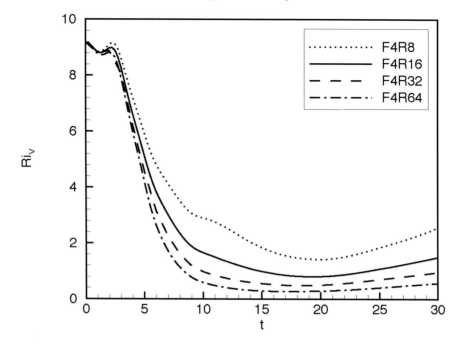

Figure 7.1 The volume averaged Richardson number as a function of time for cases at $F_\ell = 0.224$ and R_ℓ ranging from 560 to 2,250.

below 1, indicating that there must be some regions of potential instability and turbulence.

Such regions were discernible from various visualizations of the flow field. For example, the top panel in Figure 7.2 shows a horizontal slice through the vertical component of the velocity field at a time of about 20, a time when Ri_V had clearly become small (see Figure 7.1). Several oscillatory patterns in the vertical velocity are observed. The lower panel in the figure displays a vertical cut along the white dashed line in the upper panel, giving the total density field. What appears to be local Kelvin–Helmholtz-type rollers are clear visible. These subsequently breakdown into three-dimensional, turbulent-like motion. The flow appeared to consist of such intermittent patches of small-scale instabilities.

More insight into the flow dynamics can be found from examining the horizontal energy spectra of the horizontal velocity, as shown in Figure 7.3, for various times in the simulations of the case with $F_\ell = 0.22$, $R_\ell = 2250$. The solid line gives the initial conditions, consisting of the Taylor-Green modes plus noise. The striking feature of the figure is that, as the flow evolves in time, there is a strong transfer of energy to small scales, even though the

7: *Recent Progress in Stratified Turbulence* 285

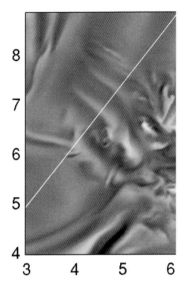

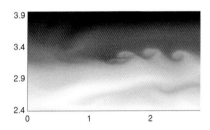

Figure 7.2 The top panel shows part of a horizontal slice through the vertical velocity field. The white dashed line gives the orientation of a vertical slice through the horizontal plane. The bottom panel shows the total density on that vertical slice. Both panels are for the case with $F_\ell = 0.112$, $R_\ell = 1,125$.

flow is decaying. The spectrum is the most broad at the time of 20, at which point the flow has lost over 50% of its energy, but a time near which the dissipation rates of potential and kinetic energies peak (not shown). An interesting point in this figure is that, at the time of 20, the spectrum exhibits a $k_h^{-5/3}$ behavior over almost one decade in wave number, hinting at an inertial range as discussed in §7.2. (Unpublished plots of the potential energy spectra exhibited similar behavior.) It was suggested that the larger-scale, quasi-horizontal motions behave almost inviscidly while transferring energy to smaller scales, implying that something analogous to Kolmogorov's inertial range ideas for high Reynolds number, three-dimensional turbulence might apply.

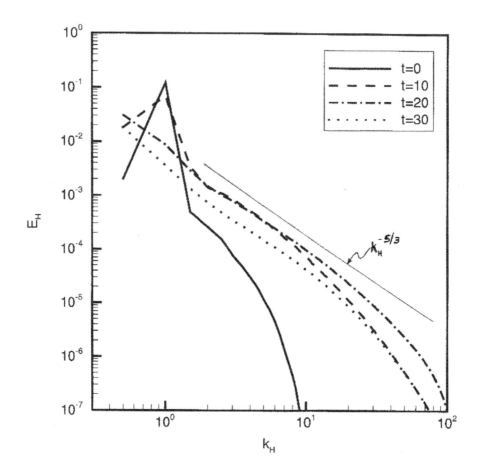

Figure 7.3 Horizontal energy spectra of the horizontal velocity at four different times for the case with $F_\ell = 0.224$, $R_\ell = 2250$.

Various statistics were computed. For example, a mixing efficiency defined as the ratio of the dissipation rate of potential energy to that of the total energy, i.e., $\epsilon_P/(\epsilon_K + \epsilon_P)$, was found to be about 0.29, in the range of many field and laboratory measurements. Furthermore it was found that the ratio of the partial dissipation rate $\nu \left\langle \left(\frac{\partial u}{\partial z}\right)^2 + \left(\frac{\partial v}{\partial z}\right)^2 \right\rangle$ to the kinetic energy dissipation rate ϵ was about 0.412, not far from the isotropic turbulence value of 0.323, indicating that smaller-scale turbulence was developing. This is in contrast to most laboratory wake experiments in the strongly stratified regime where, due to the low Reynolds number at that stage of the experiments, this partial dissipation rate carries almost the entire dissipation rate (see, e.g., Fincham et al. (1996); Praud et al. (2005)).

To estimate what conditions were required for instabilities and small-scale turbulence to develop, an argument was presented which used the local Richardson number and scaling the energy dissipation rate as $u_H'^3/\ell_h$, similar to that argued by Taylor for three-dimensional, non-stratified turbulence. The result was that the Richardson number would be locally less than 1, allowing for possible instabilities and turbulence, when the combination of the local Froude and Reynolds numbers, $F_h^2 Re_h$, was greater than 1. Using the Taylor scaling for the dissipation rate, this combination $F_h^2 Re_h$ is found to be approximately proportional to the buoyancy Reynolds number $\mathcal{R} = \epsilon/\nu N^2$. It was found from the simulations that, in order to maintain a strong cascade leading to inertial range behavior, $\mathcal{R}$ is required to be greater than about 10, which is consistent with the simulations of Brethouwer et al. (2007) discussed below. It has been found from laboratory and field experiments and from numerical simulations that, in order to maintain strong, active three-dimensional turbulence, the buoyancy Reynolds number should be greater than 20 (see, e.g., Smyth et al. (2005)). Therefore, the requirement on $\mathcal{R}$ to maintain a stratified turbulence inertial range is weaker than the requirement to maintain active, three-dimensional turbulence.

Based upon similar scaling arguments, Riley and deBruynKops also pointed out that, for the case of most of the laboratory experiments at low Froude number, the values of ℓ_v and ℓ_h are very similar, usually within a factor of 2 to 4. Therefore ℓ_v is determined by viscous effects and $\mathcal{R}$ is small, implying the spectral energy flux is almost entirely shut off, as discussed in §7.2. Thus the scaling of Lilly (1983) may be more appropriate than that of Billant and Chomaz (2000) for these flows, and most of the experiments are probably not in the range of stratified turbulence. See §7.4 for a more detailed discussion of this point.

Waite and Bartello have published as series of papers on stratified turbulence, addressing a number of key issues on this topic, including the effects of system rotation (Waite and Bartello, 2006b) and of forcing only internal wave modes (Waite and Bartello, 2006a).

In Waite and Bartello (2004) they presented results from forcing only the vortical mode. They solved the Navier-Stokes equations subject to the Boussinesq approximation, and utilized a hyperviscosity to treat the smaller-scale motions. Employing computational domains of up to 180^3 grid points, they performed simulations for a range of Froude numbers from very high, so that the solutions were similar to fully developed turbulent flows, to very low, where the stable stratification suppresses the smaller-scale turbulence and the flows were essentially laminar. In particular they examined energy spectra for these flows as a function of both the horizontal and the vertical wave numbers.

The vertical spectra tended to be rather flat out to wave numbers of the order of N/u' (see Figure 7.4), possibly indicating the decoupling of the motion in the vertical, as suggested by Lilly (1983). Unfortunately it is difficult to draw conclusions for the vertical spectra above the wave number of about $k_v \sim N/u'$, as the behavior is influenced by the hyperviscosity.

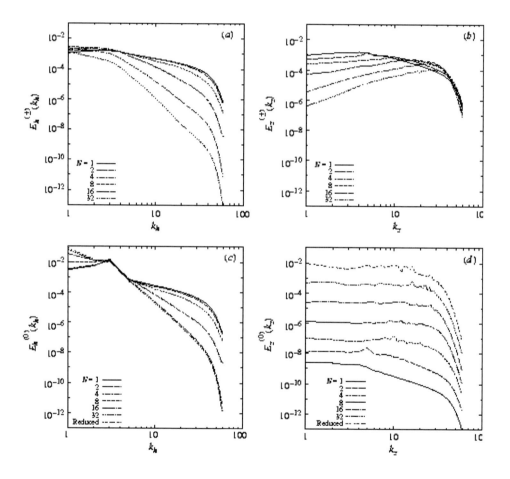

Figure 7.4 (a) Horizontal wave energy spectra, (b) vertical wave energy spectra, (c) horizontal vortical mode energy spectra, and (d) vertical vortical mode energy spectra from Waite and Bartello (2004). Note that in (d) the spectra have been offset from one another by factors of 10 for clarity. From Waite and Bartello (2004), reproduced with permission.

For small Froude numbers, the horizontal vortical mode energy spectra saturated as k_h^{-5}, with the visualizations of the flow indicating no small scale turbulence, consistent with the arguments presented in §7.2. One intermediate Froude number case, however, might be most relevant to stratified turbulence, i.e., the case with $F_h \equiv \frac{\sqrt{\langle \omega_v^2 \rangle}}{N} = 0.21$ (the case for $N = 4$ in the figure). From the visualizations this case appears to exhibit small-scale turbulence, much like the simulations of Riley and deBruynKops (2003) and Lindborg (2006). There is clearly a strong transfer of energy to smaller-scales, and a narrow range of the horizontal spectra not inconsistent with $k_h^{-5/3}$ behavior.

Of some interest, possibly related to the laboratory experiments (see §7.4) and the simulations of Riley and deBruynKops (2003), is the behavior of the energy spectra below the forcing wave number. There appears to be a weak upscale transfer of energy to larger scales, which is apparently enhanced as the strength of the stratification is increased (the Froude number is decreased).

Lindborg (2006) carried out a series of forced simulations; in order to resolve the finer and finer vertical scales that develop as $F_h \to 0$, he used stretched grids with successively higher resolution in the vertical with increasing strength of stratification. Inspired by the results of Waite and Bartello (2004), who confirmed the scaling relation (7.12) of Billant and Chomaz, Lindborg (2006) also used thin computational boxes whose thickness scaled as U/N. The aim of this method was to be able to use the same number of grid points for different runs with different degrees of stratification, and also for runs with very low F_h. To do this, it was also necessary to use different hyperdiffusion coefficients in the vertical and the horizontal directions. With this method it was possible to simulate a forward energy cascade at very low F_h. All the energy injected by the forcing at large scales was dissipated at small scales. Moreover, the kinetic and potential energy spectra agreed very well with the expression (7.16). Somewhat surprisingly, the same value 0.51 was obtained for the two constants C_1 and C_2 of the one-dimensional spectra, and these same values were also subsequently found from the simulation results of Riley and deBruynKops (2003).

Lindborg (2006) also showed that the method gave similar results in simulations including system rotation, provided that the Rossby number was not too low. With $Ro \sim 1$ the forward energy cascade was still present, a result which also was obtained in simulations by Waite and Bartello (2006b). In the inertial range the kinetic and potential energy spectra showed the same form in the case for $Ro \sim 1$ as in the case without system rotation. In Figure

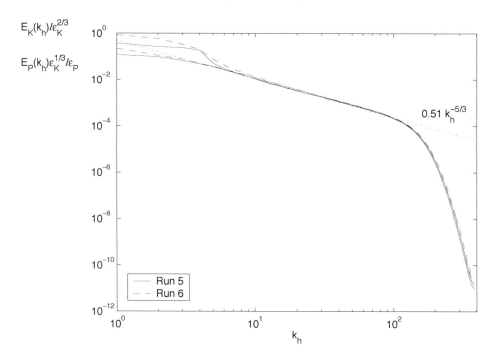

Figure 7.5 Scaled kinetic and potential energy spectra from simulations by Lindborg (2005). Run 5 is with system rotation and run 6 is without system rotation. The simulations were forced at wave number 4.

7.5 we have reproduced the kinetic and potential energy spectra from two of the runs presented in Lindborg (2005), run 5, which is with system rotation, and run 6, which is without system rotation. The only visible difference between the spectra from the two runs is that more energy is accumulated at small wave numbers in the run with system rotation as compared to the run without system rotation. In the inertial range the spectra from the two runs fall on top of each other.

The results of Lindborg build on several modeling assumptions, such as the use of hyperviscosity with different values of the vertical and horizontal coefficients, the use of stretched grids, and thin elongated boxes. They must therefore be interpreted with caution. The ultimate goal must, of course, be to test these results in highly resolved direct numerical simulations of the Navier-Stokes equations under the Boussinesq approximation. To obtain a stratified turbulence inertial range in such simulations, however, is extremely demanding in terms of the number of computational points which is required; it is possibly even more demanding than to obtain a classical inertial range of Kolmogorov turbulence. In fact, to obtain a stratified turbulence inertial range in a DNS, it is not only necessary to resolve the Ozmidov length scale, but also the Kolmogorov length scale, η, which by necessity is much smaller than the Ozmidov length scale; this can be seen from the relation

$$\mathcal{R} = \left(\frac{\ell_O}{\eta}\right)^{4/3}. \tag{7.39}$$

Since $\mathcal{R} \gg 1$ in stratified turbulence, it is thus necessary to resolve also a considerable range of classical Kolmogorov turbulence below the Ozmidov length scale. Brethouwer et al. (2007) carried out forced direct numerical simulations with $\mathcal{R} \sim 10$ and $F_h \sim 0.02$ and obtained a forward energy cascade of kinetic and potential energy. The energy spectra in the inertial range were in reasonable agreement with (7.16). In particular, the potential energy spectrum was quite close to displaying a $k^{-5/3}$-range. Larger simulations at higher buoyancy Reynolds numbers are needed, however, to further test these results. But note that these resolution requirements may be considerably relaxed, however, when hyperviscosity is used, as in the simulations of Lindborg (2006).

In order to study the energy transfer process from large to small scales in a more realistic, yet controlled, flow Molemaker and McWilliams (2010) simulated a forced horizontal jet with spatially uniform rotation, vertical stratification and vertical shear in a horizontally periodic domain. The large scale Rossby number was small enough, ~ 0.5, for the generation of geostrophically balanced eddies at large scales with an associated inverse energy cascade towards even larger scales. However, boundary-intensified buoyancy fronts through the straining by the larger eddies also initiated a downscale energy cascade at somewhat smaller scales, in which both kinetic and potential energy were transferred towards even smaller scales. Moreover, the kinetic and potential energy wave number spectra were in good agreement with (7.16). In figure (7.6) we have reproduced their kinetic and potential energy spectra. The straight line corresponds to $k^{-5/3}$.

The perhaps most interesting result of Molemaker and McWilliams was the reverse of sign of the energy transfer spectrum, $B(k)$, at approximately the same wave number where the forward cascade started. While other simulations, either through the initial conditions or through the forcing scheme, have been designed in such a way that the large scale energy transfer is in the direction from kinetic to potential energy, the large scale energy transfer in the simulation by Molemaker and McWilliams was in the other direction, which is typical for the process of baroclinic instability, governing the large scale dynamics of geophysical flows. Yet, at the smaller scale where the downscale energy started, kinetic energy was instead being transferred

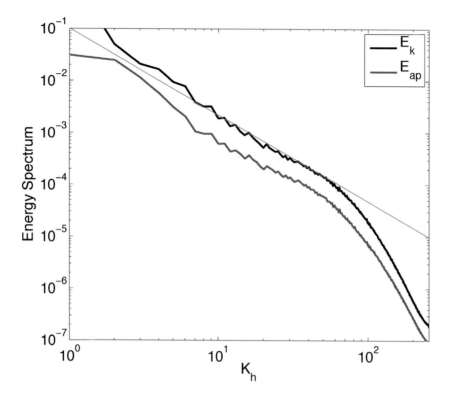

Figure 7.6 Kinetic and available potential energy spectra from a simulation of a forced horizontal jet by Molemaker and McWilliams (2010). The straight line represents $k^{-5/3}$. Reproduced with permission.

to potential energy. It seems that such a transfer is typical for stratified turbulence. The spontaneous reverse of sign can thus be seen as a strong indication that stratified turbulence is indeed generated in realistic geophysical flows.

7.4 Laboratory experiments

A number of laboratory experiments have been carried out to address turbulence in a stratified fluid, and which might have implications for stratified turbulence. In most of these experiments a towing tank was utilized, with water as the working fluid and salt used to create the stable stratification. Turbulence was often created by towing a sphere or an axisymmetric object (e.g., Lin and Pao (1979); Lin et al. (1992); Chomaz et al. (1993b,a); Spedding et al. (1996a,b); Bonnier et al. (2000)), although sometimes a towed grid was employed (e.g., Liu (1995); Praud et al. (2005)). Some experiments were conducted in the wake of a grid in a water tunnel (e.g., Stillinger et al. (1983); Itsweire et al. (1986)).

Typically in these experiments the local Froude number in the near wake of the source (e.g., the grid or the sphere) is large, so that initially the turbulence is not strongly affected by the stable stratification. For example, all three components of the turbulent kinetic energy are of the same order. The effect of stable stratification is first apparent in the rise in the (available) potential energy, E_P. In the approximately homogeneous turbulence grid experiments, this is defined as

$$E_P = -\frac{1}{2}\frac{g}{\rho_0}\left(\frac{d\bar{\rho}}{dz}\right)^{-1}\overline{\rho^2},$$

where ρ is the density fluctuation about the ambient, the overbar represents a time average, and $b'^2 = -\frac{g}{\rho_0}\left(\frac{d\bar{\rho}}{dz}\right)^{-1}\overline{\rho^2}$, so that this definition of potential energy is consistent with that given in §7.2.2. The potential energy grows from a value near zero to reach an appreciable fraction of the total kinetic energy (Liu, 1995), at which point the ratio of potential to total kinetic energy remains approximately constant. This latter behavior is very similar to results from numerical simulations (Riley and deBruynKops, 2003; Lindborg, 2006).

In most of the experiments a power law decay of the kinetic energy is found, i.e.,

$$\frac{u'^2}{U^2} = c\left(\frac{Ut}{M}\right)^{-n}, \quad \frac{\epsilon M}{U^3} = -\frac{3}{2}nc\left(\frac{Ut}{M}\right)^{-n-1},$$

where U is the speed of the object, u' is an rms turbulent velocity, M is a length scale of the object (e.g., the grid spacing or the sphere diameter), t is the time measured from the passage of the source, and c and n are empirical constants determined from the experiments. It is also useful to define an integral length scale L based upon u' and ϵ, following Taylor's estimate (equation (7.13)),

$$\frac{L}{M} = \frac{1}{M}\frac{u'^3}{\epsilon} = \frac{2c}{3n}\left(\frac{Ut}{M}\right)^{1-n/2}.$$

Note that in a stratified turbulence regime, this definition of the integral scale would be equivalent to the definition of ℓ_h given in §7.2.1.

In order to determine the parameter regime of the experiments, and their relevance to stratified turbulence, it is useful to establish the local Froude and buoyancy Reynolds numbers. As discussed in §7.1 and §7.2, for stratified turbulence the Froude number is small while the buoyancy Reynolds number is large. As mentioned in the previous section, estimates from ocean data and numerical simulations indicated that the buoyancy Reynolds number must be above about 10 to 20 for smaller-scale turbulence to be sustained (Smyth and Moum, 2000). Given the above power law behavior of u', ϵ, and L, then:

$$F_L = \frac{u'}{NL} = \frac{3n}{2}(Nt)^{-1}, \text{ and}$$

$$\mathcal{R} = \frac{\epsilon}{\nu N^2} = \frac{3}{2} nc R_M F_M^{1-n} (Nt)^{-n-1},$$

where $R_M = UM/\nu$ and $F_M = U/NM$ are the Reynolds and Froude numbers based upon the source characteristics. Note the strong decay in time of both quantities, but especially $\mathcal{R}$. As pointed out by Riley et al. (1981), using essentially the same power law arguments as above, at approximately one buoyancy period ($Nt \sim 1$), regardless of the initial Froude number, the Froude number becomes of order 1 and the effects of stable stratification become of first order. Therefore stratified turbulence might be expected in these experiments for Nt greater than about 1 if the buoyancy Reynolds number is sufficiently large.

It is therefore useful to examine typical values of F_L and $\mathcal{R}$ in the experiments. For example, Liu (1995) performed towed grid experiments in a salt-stratified water tank, at a grid Reynolds number of $R_M = 4.3 \cdot 10^3$, and at grid Froude numbers of $F_M = 12.7$ and 6.4. His results ranged over the time period of $Nt = 2.5$ to 37.7. He found, approximately, $c = 0.045$ and $n = 1.3$. For the case of $F_M = 12.7$, at one buoyancy period ($Nt/2\pi = 1$), when the effects of stable stratification have become important, then $F_L = 0.31$ while $\mathcal{R} = 25.7$. This latter value for $\mathcal{R}$ is slightly above the nominal value for smaller-scale turbulence to be sustained, so it is possible that his results might apply to stratified turbulence. For his experiments at $F_M = 6.4$, then at $Nt/2\pi = 1$, $F_L = 0.31$ while $\mathcal{R} = 31.6$, again above the nominal value for $\mathcal{R}$ for sustained smaller-scale turbulence. Therefore, in these experiments, as the flows decay, they become dominated by stable stratification, one aspect of stratified turbulence. By the time this happens, however, the local buoyancy Reynolds number has become near the critical value for sustaining small-scale turbulence; so it is not clear how well the results of these experiments are applicable to stratified turbulence.

This is also the case for the grid turbulence experiments of Praud et al. (2005). Their experiments were generally carried out at a much lower grid Froude number than those of Liu, with F_M ranging from 0.021 to 0.7. Their grid Reynolds numbers varied broadly over the range from 1,125 to 36,000. For a typical case presented, $R_M = 4,500$ and $F_M = 0.34$. Therefore, at a time of $Nt/2\pi = 1$, $F_L = 0.31$ while $\mathcal{R} = 63.8$. The time $Nt/2\pi = 1$ corresponds to a nondimensional of $Ut/M = 1.95$ Unfortunately measurements were not made at this early stage of decay. In fact, due to their experimental methodology, most of their measurements were made after the flows had decayed sufficiently for no overturning to occur, so that smaller-scale turbulence was not being sustained.

One of the first definitive studies of the wake of a localized object was that of Lin and Pao (1979), who examined the wake of a self-propelled slender body of aspect ratio of about 12 to 1. They did experiments with diameter Reynolds numbers of $R_D = 3 \cdot 10^4$, and for grid Froude numbers in the range of 16.4 to 89.9. Their empirical constants were found to be $c = 0.053$ and $n = 3/2$. Therefore, the corresponding local Froude and Reynolds numbers at $Nt/2\pi = 1$, for the case with $F_D = 16.4$, were $F_L = 0.36$ and $\mathcal{R} = 12.9$, probably on the margin to sustain smaller-scale turbulence. For the case with $F_D = 89.9$, $F_L = 0.36$ while $\mathcal{R} = 5.5$, the latter being so small that smaller-scale turbulence could not be sustained.

An intriguing result from the study of Lin and Pao was that, after a time of about $Nt/2\pi = 0.2$, the horizontal width of the wake began to grown somewhat faster (the growth rate changing from $t^{1/4}$ to $t^{0.4}$). From visualizations of the wake they found that quasi-horizontal vortices formed, continually growing in scale, consistent with their measurements of the increased growth rate of the wake width. At $Nt/2\pi = 0.2$, then $F_L = 1.8$ while $\mathcal{R} = 499$. It is possible that, for these values of the parameters, stratified turbulence has been achieved. This suggests that a feature of stratified turbulence could be a partial upscale transfer of kinetic energy in the horizontal, at least for situations where the flow is evolving, as in a turbulent wake, and with significant energy in vertical scales greater than U/N.

The most definitive set of wake measurements has been done by Spedding and his colleagues (e.g., Spedding et al. (1996a,b); Spedding (1997)). They towed spheres in a salt-stratified water tank for a range of diameter Reynolds numbers from $4.7 \cdot 10^3$ to $1.16 \cdot 10^4$, and diameter Froude numbers ranging from 10 to 240. In the early period of decay (see Figure 7.7), $0 \leq Nt \leq 2$, which they referred to as the 'three-dimensional turbulent initial wake', they found a power law decay with $c = 0.36$ and $n = 4/3$. At about $Nt = 2$ the kinetic energy decay rate decreased from $t^{-4/3}$ to about $t^{-1/2}$, indicating a possible change in the physics. For the case with $F_D = 10$ and $R_D = 5.1 \cdot 10^3$, at $Nt = 2$ the local Froude and buoyancy Reynolds numbers take on the values $F_L = 1.0$ and $\mathcal{R} = 346$, while for the case of $F_D = 240$, $R_D = 5.3 \cdot 10^3$,

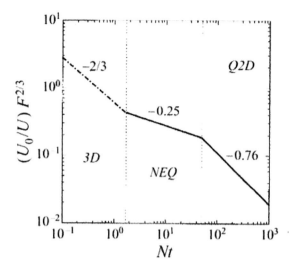

Figure 7.7 A universal curve for the evolution of the mean centerline axial velocity as a function of Nt. Three regimes are identified: a three-dimensional turbulent initial wake (3D); a transitional, non-equilibrium regime (NEQ) where wake collapse occurs and low kinetic energy decay rates are accompanied by conversion of potential to kinetic energy; and a quasi-horizontal state (Q2D) where vertical motions have become very small. Reproduced from Spedding (1997) with permission.

the corresponding values are $F_L = 1.0$ and $\mathcal{R} = 125$. In both cases F_L may be small enough and $\mathcal{R}$ large enough for stratified turbulence to be achieved, possibly explaining the change in decay rate. They refer to this second range as the 'non-equilibrium range' where stratification effects have become important. In this second range, $c = 0.25$ and $n = 1/2$. At the end of this range, at $Nt = 50$, the value of F_L has been reduced to 0.015, and the values of $\mathcal{R}$ have reduced to 8.6 and 43.5 for the cases with $F_L = 10$ and 240, respectively. It is possible that in this latter case smaller-scale three-dimensional turbulence could be sustained. Interestingly, at about $Nt = 50$ the decay rate changes again, to about $t^{-3/4}$, possibly indicating that the flow dynamics have again changed, as F_L has become sufficiently small.

Spedding also produced visual images of the vertical vorticity in the horizontal plane through the center of the wake out to very late times (see Figure 7.8). Clearly the horizontal scales are growing in the images, consistent with the results of Lin and Pao. This growth, however, is only marginally apparent at $Nt = 50$, and mainly occurs for much later times, when $\mathcal{R}$ is very low. It is possible that, for the case with $F_L = 240$, where $\mathcal{R} = 43.5$ at $Nt = 50$, this flow is in the stratified turbulence regime; however, it is

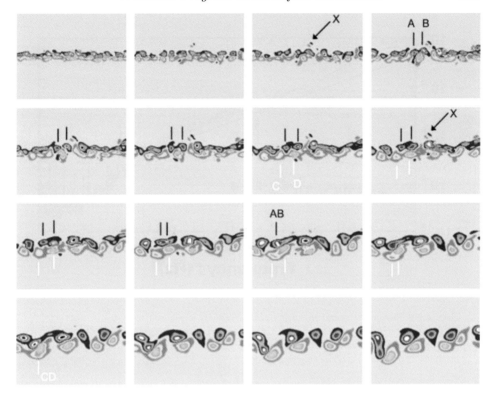

Figure 7.8 Color contour plots of vertical vorticity in the horizontal plane through the center of the wake of a towed sphere in a stratified fluid. The Froude and Reynolds numbers were 120 and 11.6×10^3, respectively, and the sequence of images were taken for times over the range $Nt \in [6, 314]$. Two separate pairing sequences are denoted by AB and CD. The motion of the sphere was from right to left. Reprinted with permission from "The streamwise spacing of adjacent coherent structures in stratified wakes", by G.R. Spedding, *Physics of Fluids*, **14**(11), © 2002, American Institute of Physics.

not definitive. Thus the applicability of these results for horizontal growth to stratified turbulence is unclear.

7.5 Field data

There is now a considerable amount of field data which is consistent with the conclusions of stratified turbulence scaling analysis as presented in §7.2.1 and with the numerical simulations presented in §7.3. We will first discuss meteorological data, both in stable atmospheric boundary layers and in the upper atmosphere. This will be followed with a discussion of ocean data.

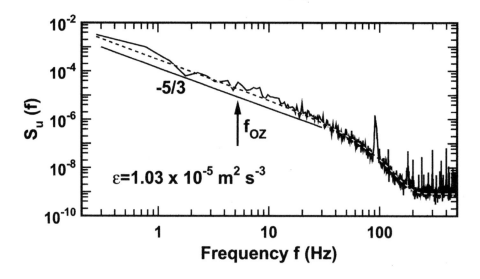

Figure 7.9 Frequency spectrum of horizontal velocity measured in a stable, nocturnal boundary layer. Both a line with a −5/3 slope, and the Ozmidov frequency, are indicated on the plot. The mean horizontal wind speed was about 9.2 m/s. Reproduced from Frehlich et al. (2008) with permission.

7.5.1 Stable atmospheric boundary layers

Some atmospheric boundary layer data are now becoming available to test the predictions of stratified turbulence theory. Frehlich et al. (2008) obtained velocity and temperature data in a stable, nocturnal boundary layer from *in situ*, fast response turbulence sensors mounted on a tethered lifting system. In an approximately 9.2 m/s wind, temperature frequency spectra could be interpreted as horizontal wave number spectra, and vice-versa. The Ozmidov scale was estimated to be 1.73 m, while the local Froude number was estimated to be 0.28, similar to that in the direct numerical simulations of Riley and deBruynKops (2003). Figure 7.9 is a plot of the frequency spectra of the horizontal velocity. Similar results were obtained for the temperature spectra. The results for the spectrum above the Ozmidov scale appear to be consistent with the scaling of Kolmogorov. At frequencies below the Ozmidov scale, however, Kolmogorov scaling would not be expected to hold, as the flow is probably not isotropic and stratification effects have become important. On the other hand, the results below the Ozmidov scale are consistent with the assumptions and predictions of stratified turbulence theory as discussed in §7.2, as a $k_h^{-5/3}$ spectra range is found.

7.5.2 Atmospheric mesoscale data

In a landmark paper Nastrom and Gage (1985) presented a comprehensive study of the atmospheric kinetic energy and potential temperature wave number spectra measured by commercial aircraft in the upper troposphere and lower stratosphere. The study very much confirmed results from previous measurements (e.g., Vinnichenko (1969)) of more limited scope. In Figure 7.10 we have reproduced a figure from Nastrom and Gage (1985) showing the compound spectra of meridional wind, zonal wind and potential temperature. The meridional wind spectrum and the potential temperature spectrum are shifted one and two decades to the right, respectively. The potential temperature spectrum can be transformed into a potential energy spectrum if the Brunt-Väisälä frequency is known. The figure shows a global average, including both tropospheric and stratospheric summer and winter data. Nastrom & Gage also investigated the variance of the spectra with respect to latitude and season as well as the difference between the tropospheric and stratospheric spectra and spectra taken over land and sea. All spectra showed the same general shape as in Figure 7.10, with minor differences in magnitude. At scales between 2.6 km and about 300–400 km the measured spectra have slopes very close to –5/3, and at larger scales the spectra steepen considerably to an approximate slope of –3.

The Nastrom & Gage spectrum has been the object of considerable debate. Until recently, the discussion has been confined to a rather limited group of theoretically oriented dynamicists. However, as the resolving capacity of numerical models has dramatically increased during the last ten years and the horizontal scale of resolution in forecast and climate models is entering into the mesoscale range, the problem has also become a matter of practical concern. In order to develop efficient subgrid models and accurately optimise the resolution, it has become more relevant to understand the dynamical origin of the spectrum. As a matter of fact, it has also been demonstrated (Koshyk and Hamilton (2001); Skamarock (2004); Hamilton et al. (2008)) that highly resolved numerical models are actually capable of reproducing the transition from a k^{-3} to a $k^{-5/3}$ spectrum range.

The rather limited k^{-3}-spectrum range at larger scales is consistent with the theory of quasi-geostrophic turbulence of Charney (1971), which explains this as arising from a downscale cascade of potential enstrophy, defined as half the square of potential vorticity. Although this theory is quite simplified, there is no reason to believe that it does not catch the essentials of the dynamics behind the k^{-3}-spectrum range. We will not question this interpretation, but focus on the dynamic origin of the $k^{-5/3}$-spectrum range.

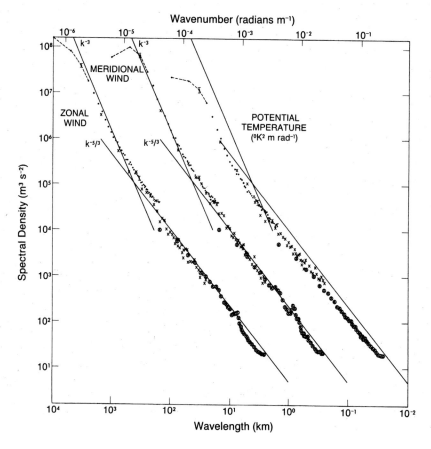

Figure 7.10 From left to right: variance power spectra of zonal wind, meridional wind, and potential temperature near the tropopause from Global Atmospheric Sampling Program aircraft data. The spectra for meridional wind and temperature are shifted one and two decades to the right, respectively. Reproduced from Nastrom and Gage (1985) with permission.

The different hypotheses can be categorised according to whether it is assumed that the $k^{-5/3}$-spectrum range is produced by gravity waves or by some type of turbulence, and whether it is assumed that there is a downscale or upscale energy cascade. In the early stage of the debate the assumption of the downscale energy cascade was combined with the gravity wave hypothesis (Dewan, 1979; VanZandt, 1982), while the assumption of an upscale energy cascade was combined with the turbulence hypothesis (Lilly, 1983). In recent work, however, the assumption of a downscale cascade has been combined with the hypothesis that the motion undergoing the cascade is best described as some type of turbulence, whether it may be quasigeostrophic turbulence (Tung and Orlando, 2003); surface quasigeostropic turbulence (Tulloch and Smith, 2006, 2009), or stratified turbulence (Lindborg, 2006). Here, we will argue that the stratified turbulence hypothesis is the only one proposed so far which is consistent with observations.

The inverse cascade hypothesis, which has stimulated a great deal of observational, theoretical and numerical work, is today mainly of historical interest. Early attempts to simulate an inverse energy cascade in a strongly stratified fluid (e.g., Herring and Métais (1989)) more or less failed, while subsequent numerical simulations (e.g., Riley and deBruynKops (2003); Lindborg (2006)) have shown that there is a forward energy cascade in such a fluid. There are also some direct observational evidence pointing against the inverse cascade hypothesis. Lindborg (1999) suggested that it should be possible to decide whether the energy cascade is upscale or downscale by calculating third order velocity structure functions from airplane data. The third order structure function law of Kolmogorov (1941) is maybe the most central theorem in turbulence theory (Frisch, 1995). The third-order longitudinal structure function D_{LLL} is defined as the third-order statistical moment of the difference, $\delta u_L = u'_L - u_L$, between the longitudinal velocity components at two points separated by a distance r. The longitudinal velocity component is the component in the same direction as the separation vector $\mathbf{r}$ between the two points. Kolmogorov's "four-fifth's' law" says that for three-dimensional isotropic turbulence,

$$D_{LLL} = -\frac{4}{5}\epsilon r \qquad (7.40)$$

in the inertial range of separation distances, where ϵ is the mean dissipation. For two-dimensional turbulence we instead find (Lindborg (1999))

$$D_{LLL} = \frac{3}{2}\epsilon r, \qquad (7.41)$$

where ϵ here is the upscale energy flux. The negative linear dependence on r in Kolmogorov (1941) is a reflection of the fact that there is downscale energy cascade in three-dimensional turbulence, while the positive linear dependence in (7.41) is a reflection of the fact that there is an upscale energy cascade in two-dimensional turbulence. If Lilly's hypothesis were correct, a measurement of the third-order structure function should give a positive linear dependence on r, in accordance with (7.41), at separations of the order of $10 - 100$ km. A negative linear dependence, on the other hand, would be a strong indication of a forward energy cascade. Cho and Lindborg (2001) calculated third-order structure functions from aircraft data reported in the MOZAIC data base (measurement of ozone and water vapour by airbus in-service aircraft; see Marenco et al. (1998)). Converged and unambiguous results were obtained only in the stratosphere, where the third-order structure function showed a negative linear range at mesoscale separations, in clear conflict with the prediction (7.41). The negative linear range indicated that there is, indeed, a downscale energy cascade.

The gravity wave hypothesis can be questioned on kinematical grounds. Linear internal gravity waves carry no vertical vorticity. All their kinetic energy is associated with the horizontal divergence of the horizontal velocity field. Inertia-gravity waves, on the other hand, do carry some vertical vorticity, whose magnitude depends on their frequency. For an inertia-gravity wave with frequency ω, the ratio between the energy associated with vertical vorticity and the energy associated with horizontal divergence is equal to $(f_0/\omega)^2$, where f_0 is the inertial frequency. Typical frequencies of inertia-gravity waves in the upper troposphere and lower stratosphere with horizontal wavelengths of the order of 100 km are considerably larger than f_0 (see Dewan (1997)) and references therein), with typical periods not greater than a few hours. If the dynamical origin of the mesoscale spectrum were a field of such waves, then the spectrum would therefore be totally dominated by the divergent part of the velocity field. Lindborg (2007) showed that it is possible to make a vortical-divergent decomposition of the kinetic energy of the mesoscale motions by analysing second order velocity structure functions calculated from aircraft data (Cho and Lindborg, 2001). It was found that the energy associated with the vortical part of the mesoscale velocity field is a little bit larger than the energy associated with the divergent part. Therefore, inertia-gravity waves can hardly be the dynamical origin of the mesoscale $k^{-5/3}$-spectrum. Simulation by Brethouwer et al. (2007)) of stratified turbulence show that kinetic energy is almost perfectly equipartioned between vortical and divergent modes. Unpublished analyses of the simulations by Riley and deBruynKops (2003) showed similar results. The structure function analysis by Lindborg (2007) indicated that divergent energy was typically about two thirds of the vortical energy in the atmospheric mesoscale motions. This is sufficiently close to the simulation results to be consistent with the results from the simulations.

The theory of quasigeostrophic turbulence was formulated by Charney (1971). The quasi-geostrophic equations are derived as a low Froude number and low Rossby number (strong rotation) limit of the equations describing atmospheric and oceanic motions. Just as the two-dimensional Navier-Stokes equations the quasi-geostrophic equations have two inviscid quadratic invariants, total energy and potential enstrophy, which is defined as half the square of potential vorticitiy. As shown by Charney, quasigeostrophic turbulence very much resembles two-dimensional turbulence. There is an inverse energy cascade and a forward cascade of potential enstrophy. As already pointed out, the limited approximate k^{-3}-range seen in the atmospheric spectrum in Figure 7.10 at synoptic scales, that is scales of the order of 1000 km, may be interpreted as a potential enstrophy cascade range.

Tung and Orlando (2003) suggested that the quasi-geostrophic equations may allow for a weak "shadow" forward energy cascade accompanying the forward enstrophy cascade. In the spectrum, such a forward energy cascade would become visible at the wave number, k_t, where the k^{-3}-spectrum of the enstrophy cascade is of equal magnitude as the $k^{-5/3}$-spectrum of the energy cascade, that is

$$k_t \sim \sqrt{\frac{\eta}{\epsilon}}, \qquad (7.42)$$

where η is the enstrophy flux and ϵ is the energy flux of the shadow cascade. According to the hypothesis of Tung & Orlando both the k^{-3}-range at synoptic scales and the $k^{-5/3}$-range at mesoscales can thus be explained within the theory of quasi-geostrophic turbulence. The hypothesis can be questioned on several grounds. First, it has not been shown that the quasigeostrophic equations allow for a sufficiently strong forward shadow cascade which would be required to explain the observed spectrum. It is true that there exists such weak shadow cascades in both two-dimensional turbulence (Davidson, 2008) and quasigeostrophic turbulence. However, the strength of these are rapidly decreasing with Reynolds number. In the atmosphere, the Reynolds number is very large and the quasi-geostrophic equations would therefore not be able to produce a sufficiently strong shadow cascade. Second, the Rossby number corresponding to the mesoscale motions can be estimated to be of the order of unity or larger, as pointed out by Lilly (1983) and more rigorously shown by Lindborg (2009). Since the quasigeostrophic equations are derived as a low Rossby number limit, it must therefore be questioned if any type of quasi-geostrophic theory can explain the mesoscale energy spectrum. To these two points can be added a third. Quasigeostrophic motions are exclusively built up by the vortical part of the velocity field. Observations (Lindborg, 2007) as well as simulations (Hamilton et al., 2008; Skamarock and Klemp, 2008) show that the energy in mesoscale motions which is associated with divergent modes is of the same order of magnitude as the energy associated with vortical modes.

The two last points of criticism can also be raised against the surface quasi-geostrophic (SQG) hypothesis formulated by Tulloch and Smith (2006). Apart from the low Froude number and Rossby number assumptions leading to the general quasigeostrophic model, the surface quasigeostrophic model is derived by introducing an assumption of zero or negligible potential vorticity everywhere except at a horizontal surface. The SQG model for the atmospheric energy spectrum proposed by Tulloch & Smith identifies this surface as the tropopause. The model explains the observed spectrum as arising under special dynamic conditions which are assumed to be prevalent in the region just below the tropopause. In particular, the $k^{-5/3}$-spectrum measured by Nastrom and Gage is supposed to be specific for "observations primarily collected near the tropopause" (Tulloch and Smith (2006)). As argued by Lindborg (2009) the SQG-model is not consistent with observations. A similar $k^{-5/3}$-spectrum as measured by Nastrom and Gage (1985) close to the tropopause has also been measured at considerable lower altitudes in the middle troposphere (Cho et al., 1999; Frehlich and Sharman, 2010). Moreover, not even the Nastrom & Gage spectrum seen in Figure 7.10 was measured sufficiently close to the tropopause to be explained by the SQG model. The hypothesis can therefore be rejected.

Thus, it seems that the stratified turbulence hypothesis is the only hypothesis presented so far which is consistent with observations. A number of questions must find their answers, however, before it can be regarded as a solid theory. Most importantly, the energetics of the forward energy cascade must be further investigated. In stratified turbulence, there is a kinetic to potential energy transfer at the largest scales of the cascade and possibly also throughout the $k^{-5/3}$-spectrum range, as discussed in §7.2. It remains to be tested whether there is such a transfer at scales of the order of 500 km in the atmosphere. To do this, one would need to make simultaneous measurements of the vertical velocity and the potential temperature at these scales.

The perhaps most important unanswered question is what the energy source is of the forward energy cascade. This question is connected to the problem of identifying the characteristic transition wave number, k_t, between the k^{-3}-spectrum range and the $k^{-5/3}$-spectrum range. Although it seems unreasonable that the $k^{-5/3}$-spectrum range can be fully explained within a quasigeostrophic model, the idea of a shadow forward energy cascade put forward by Tung & Orlando may turn out to be fruitful. As described in basic meteorological textbooks (e.g., Vallis (2006)) the energy source of large scale weather systems is baroclinic instability acting at a horizontal length scale $\mathcal{L}$, which is a few thousand kilometres and scales with the Rossby deformation radius, $\mathcal{L}_R \sim 500$ km. If η is the rate at which potential enstrophy is injected into the large scale weather systems, the corresponding energy injection rate can be estimated as

$$P \sim k_{\mathcal{L}}^{-2} \eta, \tag{7.43}$$

where $k_{\mathcal{L}}$ is the wave number corresponding to $\mathcal{L}$. If the atmospheric energetics could be completely described by the lowest order quasigeostrophic model, then the injected energy would almost exclusively go into an upscale cascade, while the potential enstrophy would go into a downscale cascade. However, quasi-geostrophic theory is constructed as a low Rossby number expansion theory and the upscale energy cascade is observed in the limit of zero Rossby number. At finite Rossby number, it is reasonable to assume that that there is a fraction of the injected energy which goes into a downscale shadow cascade and that this fraction is determined by the Rossby number rather than the Reynolds number, as in two-dimensional turbulence. The energy injection rate by baroclinic instability corresponds to ~ 3 W/m^2 (Oort, 1964), while the energy flux through the mesoscale range can be estimated as 4×10^{-5} m^2/s^3 (Cho and Lindborg, 2001), leading to a value of ~ 0.4 W/m^2 after integration over a 10 km column. With $Ro \sim 0.1$, a quantitatively realistic assumption would therefore be that the fraction of the injected energy which is going into a forward energy cascade can be estimated as

$$\frac{\epsilon}{P} \sim Ro. \tag{7.44}$$

According to such a model the energy source of the forward energy cascade would be large scale baroclinic instability. In contrast to the suggestion by Tung and Orlando (2003) there would be no downscale energy cascade in the limit of zero Rossby number. However, just as suggested by Tung and Orlando, it would be possible to estimate the transition wave number using their relation (7.42). Combining (7.42), (7.43) and (7.44) we obtain

$$k_t \sim Ro^{-1/2} k_{\mathcal{L}}. \tag{7.45}$$

This implies that the scale determining the transition between the k^{-3}-spectrum range and the $k^{-5/3}$-spectrum range would be possible to be estimated as $\mathcal{L}_t \sim Ro^{1/2} \mathcal{L}_R$. This suggestion should, of course, be seen as a conjecture which remains to be tested.

7.5.3 Ocean data

There are several sets of field data that are consistent with both of the principal assumptions of stratified turbulence, namely that the local Reynolds number is high and Froude number is low, and also are consistent with the predictions of an inertial range at the appropriate horizontal length scales.

Ewart (1976) was able to obtain horizontal spectra of ocean temperature along isobaric trajectories at various depths and geographical locations. He use the self-propelled underwater research vehicle SPURV, which recorded data at very high frequencies (corresponding to data samples at about every 0.18 m), and with high temperature accuracy. He presented data from 14 datasets from experiments off the coast of San Diego (at 30°N, 124°W), near the Cobb seamount (at 47°N, 131°W), off the coast of Mexico (at 21°N, 110°W), and off the Hawaiian chain (at 20°N, 156°W).

Figure 7.11 gives temperature spectra from data taken at depths of 300, 650, 1000, and 1600 m off the coast of San Diego. These data are consistent with 12 of the 14 data sets presented. A solid line at a slope of $k_h^{-5/3}$ has been added for reference. Below horizontal wave numbers of 10^{-2} cpm the data appear to follow the Garrett and Munk (1979) internal wave scaling of k_h^{-2}. For wave numbers above 10^{-2} cpm, however, a fairly consistent spectral range of $k_h^{-5/3}$ is found, down to scales of a few meters. Since the Ozmidov scale ℓ_O in the ocean is typically near 1 m (Gargett et al., 1981), the Froude number for these motions, given approximately as $F_h = u'/N\ell_h = (\ell_O/\ell_h)^{2/3}$, is small for ℓ_h in this power law range. So the conditions are consistent with the assumptions of stratified turbulence. They are clearly inconsistent with the assumptions of the inertial-convective range of Obukhov (1949) and Corrsin (1951), which implies isotropic turbulence with no effect of stably stratification. In addition, Ewart, in remarking about the higher wave number range of the spectra, noted that the data records show "intermittent patches of intense fluctuations", which he suggested are due to intermittent turbulence. This is also consistent with the behavior of stratified turbulence observed in the numerical simulations (e.g., Riley and deBruynKops (2003); Lindborg (2006)).

Klymak and Moum (2007a,b) measured temperature near the Hawaiian ridge from horizontal tows of a microstructure platform. Making measurements at depths from 700 to 3000 m, they obtained displacement spectra by interpreting temperature spectra in terms of vertical displacement. They also obtained direct measurements of the kinetic energy dissipation rate. From these experiments they were able to obtain horizontal displacement spectra over a range of several hundred meters down below one meter. Figure 7.12 gives typical spectra of the horizontal gradient of vertical displacement (slope) obtained from their experiments. (Note that k_x in this figure corresponds to k_h in this paper.) For wave numbers below a few tenths of a cycle per meter (horizontal length scales above a few meters), the spectra

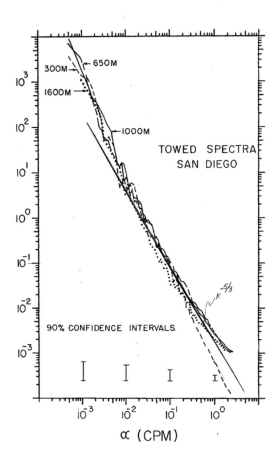

Figure 7.11 Power spectra of temperature versus horizontal wave number α computed from runs off the coast of San Diego ($30°$N, $124°$W). The spectra are normalized by $N/N_0 G$, where G is the temperature gradient averaged over several measurements, and the buoyancy frequency N has been fitted to the exponential profile $N = N_0\, e^{-z/b}$. The units of the spectra are $(\mathrm{m}^2/\mathrm{cpm})$. Reproduced from Ewart (1976) with permission.

approximately follow a $k_h^{1/3}$ curve, especially for the more energetic cases. (The $k_h^{1/3}$ spectra for slope correspond to $k_h^{-5/3}$ spectra for temperature.) Klymak and Moum remark that the "horizontal spectra exhibit a turbulent shape ($k_x^{1/3}$) to surprisingly low wave numbers," for the most intense turbulence "to scales exceeding 500 m".

Estimates can be made for the Ozmidov scale for this case. The mean buoyancy frequency was about $N^2 = 2 \cdot 10^{-5}\,\mathrm{s}^{-2}$, while the kinetic energy dissipation rate ϵ_K ranged from about $4.3 \cdot 10^{-10}$ to $1.2 \cdot 10^{-7}\,\mathrm{m}^2/\mathrm{s}^3$, corresponding to an Ozmidov scale ranging from about 7 cm to 1.2 m. Again the

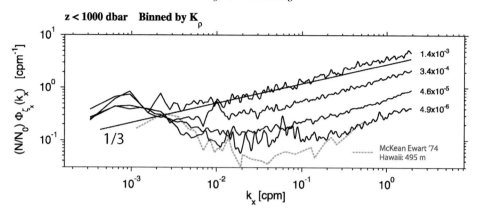

Figure 7.12 Average isopycnal slope spectra, binned in one-decade bins of the turbulent diffusivity $\mathbf{K}_\rho = 0.2\epsilon/N^2$ and normalized by (N_0/N) (Reproduced from Klymak and Moum (2007b) with permission). A line with slope $k_x^{1/3}$ has been added for comparison.

inertial range behavior is below the Ozmidov wave number of order 1 cpm, indicating consistency with stratified turbulence assumptions.

Several groups have reported horizontal spectra using seismic imaging (e.g., Holbrook and Fer (2005); Krahmann et al. (2008); Ménesguen et al. (2009); Sheen et al. (2009)). These studies all present results which are consistent with the assumptions and the predictions of stratified turbulence theory. It must be noted that the method is new and that there were still some unresolved issues regarding the interpretation of the measurements when these studies were presented. The raw signal gives a local deviation of the speed of sound from its background value and is related to local temperature. Only recently (Papenberg et al., 2010) has the exact relationship between the signal and the temperature field been established. Thus, there are now very good prospects that the method will be able to provide valuable new information on fine structure of ocean dynamics.

Sheen et al. (2009) present data taken close to the Subantarctic Front in the South Atlantic Ocean. They interpret their data in a manner similar to Klymak and Moum (2007a,b), and found horizontal displacement spectra consistent with an internal wave interpretation ($k_h^{-2.5}$, Garrett and Munk (1979)) in the wave number range of 0.0015 to 0.005 cpm, but consistent with a stratified turbulence interpretation in the wave number range of 0.015 to 0.04 cpm, referring to the latter as the 'turbulent bandwidth'. Holbrook and Fer (2005) used seismic imaging to measure vertical displacement in the Norwegian Sea. They obtained horizontal spectra from about 10 km down to about 30 m and noted that, for horizontal scales below about 300 m, 'the slope is close to that expected in the inertial-subrange of turbulence', i.e., close to $k_h^{-5/3}$ behavior. Dissipation rates were not measured in their study. The local buoyancy frequency, however, was measured to be about $N = 4 \cdot 10^{-3}\,\mathrm{s}^{-1}$. Interpreting the displacement spectra as potential energy spectra using the potential energy coefficient $C_p = 0.5$ found from numerical simulations (Lindborg, 2006), and using the approximate relationship of ϵ_p/ϵ_k obtained from the numerical simulations of Riley and deBruynKops (2003), allows the estimation of the kinetic energy dissipation rate, which is found to be $\epsilon_k = 3 \cdot 10^{-8}\,\mathrm{m^2/s^3}$. Thus the estimate for the Ozmidov scale is 0.7 m. Of course the interpretation of the results as classical inertial-subrange turbulence is entirely inconsistent with this theory, but it is consistent with the predictions of stratified turbulence theory.

Ménesguen et al. (2009) used seismic imaging to obtain the horizontal dependence of geoseismic sections in the Gulf of Cadiz. In particular, they studied the energetic, long-lived anticyclonic, lens-shaped vortices known as Meddies, which are particularly abundant in the outflow from the Mediterranean Sea in the depth range of about 700 to 1500 m. Instead of horizontal spectra they obtained second-order horizontal structure functions. The inertial range prediction from stratified turbulence theory is that these structure functions should vary as $r^{2/3}$. They consistently found $r^{2/3}$ behavior for the structure functions in the horizontal range of about 300 m to 2.8 km. They also performed numerical simulations of these Meddies and found results consistent with the field studies. In particular they found a strong downscale transfer of potential and kinetic energy, consistent with previous stratified turbulence simulations. Since the downscale transfer had a strong signature of potential vorticity, they concluded that there were nonlinear mechanisms other than internal waves, and stated that 'these show an anisotropy and a spectral slope consistent with the framework of stratified turbulence, which differs from that of Garrett and Munk of internal waves'. In addition, they suggested that results from Krahmann et al. (2008), taken during the same field campaign, can better be interpreted, for at least their more energetic regions, with a slope of $k_h^{-5/3}$ than the k_h^{-2} internal wave slope.

7.6 Conclusions

From scaling arguments, numerical simulations, and field data, it is becoming increasingly apparent that the dynamics of stratified turbulence play an important role in many geophysical flows. The requirements of stratified turbulence are that the Froude number is low (i.e., the effects of stratification are strong) and Reynolds number high (the effects of viscosity weak), and such that the buoyancy Reynolds number ($\epsilon/\nu N^2$) is above a critical value; these requirements are often met, e.g., in stably-stratified regions of the oceans and the atmosphere.

Stratified turbulence is typified as being extremely anisotropic, with horizontal, energy-containing scales (say ℓ_h) being much larger than vertical scales of importance, such that its motions are sometimes referred to as 'pancake eddies'. Horizontal scales range from the energy-containing scale down to the Ozmidov scale ($\epsilon^{1/2}/N^{3/2}$), roughly the largest horizontal scale at which overturning can occur. The ratio of these scales, ℓ_h/ℓ_O, was shown in §7.2 to be of the order of $F_h^{-3/2}$, where the Froude number $F_h = U/N\ell_h$ is expected to be very small; therefore the scale separation is expected to be very large. The scale of importance in the vertical is $\ell_v = U/N$; if the buoyancy Reynolds number is large enough, then instabilities occur at this scale, resulting locally in smaller-scale turbulence. The vertical range of scales for stratified turbulence is from U/N down to the Ozmidov scale, resulting in a scale ratio of $F_h^{-1/2}$.

A most important characteristic of stratified turbulence is that there is a tendency for strong vertical shearing of the horizontal flow to continually develop, leading to a strong cascade of energy from horizontal scales of ℓ_h down to the Ozmidov scale; the resulting instabilities and turbulence imply a further, more isotropic cascade of energy into the Kolmogorov range. Scaling arguments suggest, and numerical simulations and field data support, the existence of a stratified turbulence inertial range with both the kinetic and potential energy spectra scaling as $k_h^{-5/3}$. In §7.2 we further speculated on spectra for both the buoyancy flux and for spectral energy transfer.

There are a number of unresolved issues regarding stratified turbulence which suggest further research. For example, what are the specific physical mechanisms that cause the strong energy transfer in stratified turbulence? Are the suggestions for the buoyancy spectra and spectral energy transfer given in §7.2 accurate? What are the dynamics of vertical scales larger than ℓ_v? Do they tend to be uncorrelated, as suggested by the simulations and Waite and Bartello, and the arguments of Lilly? Do they lead to the apparently upscale transfer of energy in the horizontal, as observed in the laboratory experiments and in some numerical simulations? Scaling arguments suggest that for two Lagrangian particles separated by a horizontal distance $d(t)$, but in the same vertical (or isopycnal) plane, the separation distance scale as $d(t) \sim \epsilon t^3$, which is consistent with some ocean data. Is in fact this generally true for stratified turbulence? For mesoscale simulations for both the oceans and the atmosphere, much of stratified turbulence lies in the subgrid scales in both the horizontal and the vertical. Of interest is how best should the dynamics of stratified turbulence be parameterized in larger-scale simulations.

The research area of stratified turbulence lies somewhere between the more applied fields of meteorology and oceanography on the one hand and the more basic field of theoretical turbulence research on the other hand. It is the authors' hope that the area will be able to attract young talented researchers from both these sides – researchers who are ready to solve the many interesting problems that still remain.

Acknowledgements The authors would like to thank Peter Bartello, Peter Davidson, Mike Waite, and Jim McWilliams for their very helpful discussions during the course of writing this chapter. JJR acknowledges support from the National Science Foundation and from the Army Research Office.

Appendix

In this appendix we show that the two-dimensional energy flux, defined in equation (7.26), has the same form as the three-dimensional spectral energy flux (equation (7.25)) for forced isotropic turbulence, with the only difference that the numerical factor multiplying $(k\eta)^{4/3}$ is different. This may not be obvious, since the two-dimensional energy flux as defined in equation (7.26) includes an integration out to infinity in the k_z-direction, while the three-dimensional energy flux does not include any contribution from large wave numbers. To clearly see the difference between the three-dimensional and the two-dimensional energy flux we write them as

$$\Pi(k) = -\int_0^k 4\pi k^2 T^{(3)}(k')\,\mathrm{d}k', \tag{7.46}$$

$$\Pi^{(2)}(k) = -\int_0^k \int_0^\infty 4\pi k_h T^{(3)}(k')\,\mathrm{d}k_z \mathrm{d}k_h, \tag{7.47}$$

where $T^{(3)}$ is the three-dimensional spectral energy transfer density, which under the assumption of isotropy is a function of $k' = \sqrt{k_h^2 + k_z^2}$, where $k_h = \sqrt{k_x^2 + k_y^2}$. For forced isotropic turbulence, the difference between equations (7.46) and (7.47) can be written as

$$\Pi(k) - \Pi^{(2)}(k) = \int_0^k \int_k^\infty 4\pi k_h D^{(3)}(k')\,\mathrm{d}k_z \mathrm{d}k_h + \int\int\int_\Omega D^{(3)}(k')\,\mathrm{d}k_x \mathrm{d}k_y \mathrm{d}k_z, \tag{7.48}$$

where

$$D^{(3)}(k') = \frac{\nu}{2\pi} E(k') \qquad (7.49)$$

is the three-dimensional dissipation spectral density and Ω is defined as the domain in Fourier space confined between the cylinder of radius k and height $2k$ and the sphere with radius k, both centred at the origin. Assuming that the energy spectrum, $E(k')$, is a monotonic decreasing function for $k' > k$, where k is in the inertial range, and that $E(k') = K\epsilon^{2/3} k'^{-5/3}$ in the inertial range, upper bounds of the two integrals in equation (7.48) can be estimated as

$$\int_0^k \int_k^\infty 4\pi k_h D^{(3)}(k') \, dk_z dk_h < \int_0^k 2\nu k_h \, dk_h \int_k^\infty E(k_z) \, dk_z$$

$$< \nu k^2 K \epsilon^{2/3} \int_k^\infty k_z^{-5/3} \, dk_z = \frac{3}{2} K \epsilon (\eta k)^{4/3}, \qquad (7.50)$$

$$\int \int \int_\Omega D^{(3)}(k') \, dk_x dk_y dk_z < \left(2\pi k^3 - \frac{4\pi k^3}{3}\right) \frac{\nu}{2\pi} K \epsilon^{2/3} k^{-5/3} \qquad (7.51)$$

$$= \frac{1}{3} K \epsilon (\eta k)^{4/3}, \qquad (7.52)$$

where $\eta = \nu^{3/4}/\epsilon^{1/4}$ is the Kolmogorov scale. To make the estimate (7.51) we have multiplied the volume of the integration domain, Ω, by the maximum value of the integrand.

Comparing the expression (7.48) with the expression (7.25) for the three-dimensional energy flux and using the estimates (7.50) and (7.51), we can conclude that the inertial range expression for the two-dimensional energy flux for forced isotropic turbulence can be written as

$$\Pi^{(2)}(k) = \epsilon[1 - cK(k\eta)^{4/3}], \qquad (7.53)$$

where $3/2 < c < 10/3$. The two-dimensional energy flux has thus the same form as the three-dimensional energy flux (equation (7.25)), with the only difference is that the numerical factor in front of $(k\eta)^{4/3}$ is a little bit larger.

References

Babin, A., Mahalov, A., Nicolaenko, B., and Zhou, Y. 1997. On the asymptotic regimes and the strongly stratified limit of rotating Boussinesque equations. *Theor. Comput. Fluid Dyn.*, **9**, 223.

Bartello, P. 1995. Geostrophic adjustment and inverse cascades in rotating stratified turbulence. *J. Atmos. Sci.*, **52**, 4410–4428.

Billant, P., and Chomaz, J.-M. 2000. Three-dimensional stability of a columnar vortex pair in a stratified fluid. *J. Fluid Mech.*, **419**, 65.

Billant, P., and Chomaz, J.-M. 2001. Self-similarity of strongly stratified inviscid flows. *Phys. Fluids*, **13**, 1645.

Bonnier, M., Eiff, O., and Bonneton, P. 2000. On the density structure of far-wake vortices in a stratified fluid. *Dyn. Atmos. Oceans*, **31**, 117.

Brethouwer, G., Billant, P., Lindborg, E., and J.-M., Chomaz. 2007. Scaling analysis and simulations of strongly stratified turbulent flows. *J. Flud. Mech*, **585**, 343–368.

Charney, J.G. 1971. Geostrophic turbulence. *J. Atmos. Sci.*, **28**, 1087–1095.

Cho, J.Y.N., and Lindborg, E. 2001. Horizontal velocity structure functions in the upper troposphere and lower stratosphere 1. Observerations. *J. Geophys. Res.*, **106**, 10223–10232.

Cho, J.Y.N., Zhu, Y., Newell, R.E., Anderson, B.E., Barrick, J.D., Gregory, G.L., Sachse, G.W., Carroll, M.A., and Albercook, G.M. 1999. Horizontal wavenumber spectra of winds, temperature, and trace gases during the Pacific Exploratory Missions: 1. Climatology. *J. Geophys. Res.*, **104**, 5697–5716.

Chomaz, J.-M., Bonneton, P., and Hopfinger, E.J. 1993a. The structure of the near wake of a sphere moving horizontally in a stratified fluid. *J. Fluid Mech.*, **254**, 1–21.

Chomaz, J.-M., Bonneton, P., Butet, A., and Hopfinger, E.J. 1993b. Vertical diffusion of the far wake of a sphere moving in a stratified fluid. *Phys. Fluids*, **A5**, 2799–2806.

Corrsin, S. 1951. On the spectrum of isotropic temperature fluctuations in isotropic turbulence. *J. Appl. Phys.*, **22**, 469.

Davidson, P.A. 2008. Cascades and fluxes in two-dimensional turbulence. *Phys. Fluids*, **20**, 025106.

Dewan, E. 1979. Stratospheric spectra resembling turbulence. *Science*, **204**, 832–835.

Dewan, E. 1997. Saturated-cascade similitude theory of gravity wave spectra. *J. Geophys. Res.*, **102**, 29779–29817.

Ewart, T.E. 1976. Observations from straightline isobaric runs of SPURV. *Proc. IAPSO/IAMAP PSII*, Edinburgh, United Kingdom, Joint Oceanographic Assembly, 1–18.

Fincham, A.M., Maxworthy, T., and Spedding, G.R. 1996. Energy dissipation and vortex structure in freely decaying stratified grid turbulence. *Dyn. Atmos. Oceans*, **23**, 155–169.

Frehlich, F., and Sharman, R. 2010. Climatology of velocity and temperature turbulence statistics determined from rawinsonde and ACARS/AMDAR data. *J. Appl. Meteorol. Climatology*, **49**, 1149–1169.

Frehlich, R., Meillier, Y., and Jensen, M.L. 2008. Measurements of boundary layer profiles with *in situ* sensors and Doppler lidar. *J. Atmos. Oceanic Technol.*, **25**, 1328–1340.

Frisch, U. 1995. *Turbulence*. Cambridge University Press.

Gage, K.S. 1979. Evidence for a $k^{-5/3}$ law inertial range in mesoscale two-dimensional turbulence. *J. Atmos. Sci.*, **36**, 1950–1954.

Galperin, B., and Sukoriansky, S. 2010. Geophysical flows with anisotropic turbulence and dispersive waves: flows with stable stratification. *Ocean Dyn.*, **60**, 1319–1337.

Gargett, A.E., Hendricks, P.J., Sanford, T.B., Osborn, T.R., and Williams, A.J. 1981. A composite spectrum of vertical shear in the open ocean. *J. Phys. Ocean.*, **11**, 339–369.

Garrett, C., and Munk, W. 1979. Internal waves in the ocean. *Annu. Rev. Fluid Mech.*, **11**, 339–369.

Godeferd, F.S., and Staquet, C. 2003. Statistical modeling and direct numerical simulation of decaying stably stratified turbulence. Part 2. Large-scale and small-scale anisotropy. *J. Fluid Mech.*, **486**, 115–159.

Godoy-Diana, R., Chomaz, J.M., and Billant, P. 2004. Vertical length scale selection for pancake vortices in strongly stratified viscous fluids. *J. Fluid Mech.*, **504**, 229–238.

Hamilton, K., Takahashi, Y.O., and Ohfuchi, W. 2008. Mesoscale spectrum of atmospheric motionos investigated in a very fine resolution global general circulation model. *J. Geophys. Res.*, **113**, D18110.

Herring, J.R., and Métais, O. 1989. Numerical experiments in forced stably stratified turbulence. *J. Fluid. Mech.*, **202**, 97–115.

Holbrook, W.S., and Fer, I. 2005. Ocean internal wave spectra inferred from seismic reflection transects. *Geophys. Res. Lett.*, **32**, L15606, doi:10.1029/2005GL023733.

Holloway, G. 1986. Considerations on the Theory of Temperature Spectra in Stably Stratified Turbulence. *J. Phys. Ocean*, **16**, 2179–2183.

Holloway, G. 1988. The buoyancy flux from internal gravity wave breaking. *Dyn. Atmos. Oceans*, **12**, 107–125.

Itsweire, E.C., Helland, K.N., and van Atta, C.W. 1986. The evolution of grid-generated turbulence in stably stratified fluid. *J. Fluid Mech.*, **162**, 299–338.

Klymak, J.M., and Moum, J.N. 2007a. Oceanic isopycnal slope spectra. Part I: Internal waves. *J. Phys. Oceanogr.*, **37**, 1215–1231.

Klymak, J.M., and Moum, J.N. 2007b. Oceanic isopycnal slope spectra. Part II: Turbulence. *J. Phys. Oceanogr.*, **37**, 1232–1245.

Kolmogorov, A.N. 1941. Dissipation of energy in the locally isotropic turbulence. *Dokl. Adad. Nauk SSSR*, **32(1)**.

Koshyk, J.N., and Hamilton, K. 2001. The horizontal kinetic energy spectrum and spectral budget simulated by a high-resolution troposphere-stratosphere-mesosphere GCM. *J. Atmos. Sci.*, **58**, 329–348.

Krahmann, G., Klaeschen, D., and Reston, T. 2008. Mid-depth internal wave energy off the Iberian Peninsula estimated from seismic reflection data. *J. Geophys. Res.*, **113**, C12016,doi:10.1029/2007/JC004678.

Landau, L.D., and Lifshitz, E.M. 1987. *Fluid Mechanics, Second Edition*. Pergamon.

Lilly, D.K. 1983. Stratified turbulence and the mesoscale variability of the atmosphere. *J. Atmos. Sci.*, **40**, 749.

Lin, J.-T., and Pao, Y.-H. 1979. Wakes in stratified fluids: a review. *Ann. Rev. Fluid Mech.*, **11**, 317–338.

Lin, Q., Lindberg, W.R., Boyer, D.L., and Fernando, H.J.S. 1992. Stratified flow past a sphere. *J. Fluid Mech.*, **240**, 315–354.

Lindborg, E. 1999. Can the atmospheric kinetic energy spectrum be explained by two-dimensional turbulence? *J. Fluid Mech.*, **388**, 259–288.

Lindborg, E. 2005. The effect of rotation on the mesoscale energy cascade in the free atmosphere. *Geophys. Res. Lett.*, **32**, L01809.

Lindborg, E. 2006. The energy cascade in a strongly stratified fluid. *J. Fluid Mech.*, **550**, 207–242.

Lindborg, E. 2007. Horizontal wavenumber spectra of vertical vorticity and horizontal divergence in the upper troposphere and lower stratosphere. *J. Atmos. Sci.*, **64**, 1017–1025.

Lindborg, E. 2009. Two comments on the surface quasigeostrophic model for the atmospheric kinetic energy spectrum. *J. Atmos. Sci.*, **66**, 1069–1072.

Lindborg, E., and Brethouwer, G. 2007. Stratified turbulence forced in rotational and divergent modes. *J. Fluid Mech.*, **586**, 83–108.

Liu, H.-T. 1995. Energetics of grid turbulence in a stably stratified fluid. *J. Fluid Mech.*, **296**, 127–157.

Lumley, J.L. 1964. The spectrum of nearly inertial turbulence in a stably stratified fluid. *J. Atmos. Sci.*, **21**, 99–102.

Lundgren, T.S. 2003. Kolmogorov turbulence by matched asymptotic expansions. *Phys. Fluids*, **15**, 1074–1081.

Marenco, A., Thouret, V., Nédélec, P., Smith, H., Helten, M., Kley, D., Karcher, F., Simon, P., Law, K., Pyle, J., Pashmann, G., Von Wrede, R., Hume, C., and Cook, T. 1998. Measurements of ozone and water vapor by airbus in-service aircraft: the MOZAIC airborne program, an overview. *J. Geophys. Res.*, **103**, 25631–25642.

Ménesguen, C., Hua, B.L., Papenberg, C., Klaeschen, D., Géli, L., and Hobbs, R.W. 2009. Effect of bandwidth on seismic imaging of rotating stratified turbulence surrounding an anticyclonic eddy from field data and numerical simulations. *Geophys. Res. Lett.*, **36**, L00D05,doi:10.1029/2009GL039951.

Molemaker, J.M., and McWilliams, J.C. 2010. Local balance and cross-scale flux of available potential energy. *J. Fluid Mech.*, **645**, 295–314.

Nastrom, G.D., and Gage, K.S. 1985. A climatology of Atmospheric Wavenumber Spectra of Wind and Temperature Observed by Commmercial Aircaraf. *J. Atmos. Sci.*, **42**, 950–960.

Obukhov, A.M. 1949. Structure of the temperature field in turbulent flows. *Izv. Akad. Nauk SSSR, Ser. Geofiz.*, **13**, 58–69.

Okubo, A. 1971. Oceanic diffusion diagrams. *Deep-Sea Res.*, **18**, 789–802.

Oort, A.H. 1964. On estimates of the atmospheric energy cycle. *Month. Weath. Rev.*, **92**, 483–493.

Orszag, S.A. 1970. Analytical theories of turbulence. *J. Fluid Mech.*, **41**, 363–386.

Ozmidov, R.V. 1965. On the turbulent exchange in a stably stratified ocean. *Bull. Acad. Sci. U.S.S.R.*, **1**, 493–497.

Papenberg, C., Klaeschen, D., Krahmann, G., and Hobbs, R.W. 2010. Ocean temperature and salinity inverted from combined hydrographic and seismic data. *Geophys. Res. Lett.*, **37**, L04601.

Pope, S.B. 2000. *Turbulent Flows*. Cambridge University Press.

Praud, O., Fincham, A.M., and Sommeria, J. 2005. Decaying grid turbulence in a strongly stratified fluid. *J. Fluid Mech.*, **522**, 1–33.

Riley, J.J., and deBruynKops, S.M. 2003. Dynamics of turbulence strongly influenced by buoyancy. *Phys. Fluids*, **15**(7), 2047–2059.

Riley, J.J., and Lelong, M.-P. 2000. Fluid Motions in the Presence of Strong Stable Stratification. *Ann. Rev. Fluid Mech.*, **32**, 613–657.

Riley, J.J., and Lindborg, E. 2008. Stratified turbulence: a possible interpretation of some geophysical turbulence measurements. *J. Atmos. Sci.*, **65**, 2416–2424.

Riley, J.J., Metcalfe, R.W., and Weissman, M.A. 1981. Direct numerical simulations of homogeneous turbulence in density stratified fluids. *Nonlinear Properties of Internal Waves, ed. B.J. West*, 79–112.

Sagaut, P., and Cambon, C. 2008. *Homogeneous Turbulence Dynamics*. Cambridge University Press.

Sheen, K.L., White, N.J., and Hobbs, R.W. 2009. Estimating mixing rates from seismic images of oceanic structure. *Geophys. Res. Lett.*, **36**, L00D04,doi:10.1029/2009GL040106.

Skamarock, W.C. 2004. Evaluating Mesoscale NWP models using kinetic energy spectra. *Month. Weath. Rev.*, **132**, 3019–3032.

Skamarock, W.C., and Klemp, J.B. 2008. A time-split nonhydrostatic atmospheric model for weather research and forecasting applications. *J. Comp. Phys.*, **227**, 3465–3485.

Smyth, W.D., and Moum, J.N. 2000. Length scales of turbulence in stably stratified mixing layers. *Phys. Fluids*, **12**, 1327–1342.

Smyth, W.D., Nash, J.D., and Moum, J.N. 2005. Differential diffusion in breaking Kelvin-Helmholtz Billows. *J. Phys. Ocean.*, **35**, 1004–1022.

Spedding, G.R. 1997. The evolution of initially turbulent bluff-body wakes at high internal Froude number. *J. Fluid Mech.*, **337**, 283–301.

Spedding, G.R. 2002. The streamwise spacing of adjacent coherent structures in stratified wakes. *Phys. Fluids*, **14**, 3820–3828.

Spedding, G.R., Browand, F.K., and Fincham, A.M. 1996a. The long-time evolution of the initially turbulent wake of a sphere in a stable stratification. *Dyn. Atmos. Oceans*, **23**, 171–182.

Spedding, G.R., Browand, F.K., and Fincham, A.M. 1996b. Turbulence, similarity scaling and vortex geometry in the wake of a towed sphere in a stably stratified fluid. *J. Fluid Mech.*, **314**, 53–103.

Stillinger, D.C., Helland, K.N., and van Atta, C.W. 1983. Transition of homogeneous turbulence to internal waves in stratified fluid. *J. Fluid Mech.*, **131**, 91–122.

Taylor, G.I. 1935. Statistical theory of turbulence. *Proc. Roy. Soc. A*, **151**, 421–478.

Tulloch, R., and Smith, K.S. 2006. A theory for the atmospheric energy spectrum: Depth-limited temperature anomalies at the tropopause. *Proc. Nat. Ac. Sci.*, **103**, 14690–14694.

Tulloch, R., and Smith, K.S. 2009. Quasigeostrophic turbulence with explicit surface dynamics: Application to the atmospheric energy spectrum. *J. Atmos. Sci.*, **66**, 450–467.

Tung, K.K., and Orlando, W.W. 2003. The k^{-3} and the $k^{-5/3}$ energy spectrum of atmospheric turbulence: Quasigeostrophic two-level simulation. *J. Atmos. Sci.*, **60**, 824–835.

Vallis, G.K. 2006. *Atmospheric and Oceanic Fluid Dynamics*. Cambridge University Press.

VanZandt, T.E. 1982. A universal spectrum of buoyancy waves in the atmosphere. *Geophys. Res. Lett.*, **9**, 575–578.

Vinnichenko, V.K. 1969. Empirical studies of atmospheric structure and spectra in the free atmosphere. *Radio Sci.*, **4**, 1115–1125.

Waite, M.L., and Bartello, P. 2004. Stratified turbulence dominated by vortical motions. *J. Fluid Mech.*, **517**, 281–308.

Waite, M.L., and Bartello, P. 2006a. Stratified turbulence generated by internal gravity waves. *J. Fluid Mech.*, **546**, 313–339.

Waite, M.L., and Bartello, P. 2006b. Transition from geostrophic to stratified turbulence. *J. Fluid Mech.*, **568**, 89–108.

8
Rapidly-Rotating Turbulence: An Experimental Perspective

P.A. Davidson

8.1 The evidence of the early experiments

We consider rapidly-rotating turbulence; that is, turbulence in which the fluctuating velocity in the rotating frame, $\mathbf{u}$, is smaller than, or of the order of, $|\mathbf{\Omega}|\ell$, where $\mathbf{\Omega} = \Omega\hat{\mathbf{e}}_z$ is the bulk rotation vector and ℓ a suitably defined integral scale. The Navier–Stokes equation in the rotating frame is

$$\frac{\partial \mathbf{u}}{\partial t} + (\mathbf{u} \cdot \nabla)\mathbf{u} = -\nabla(p/\rho) + 2\mathbf{u} \times \mathbf{\Omega} + \nu\nabla^2\mathbf{u}, \qquad (8.1)$$

where p is the so-called reduced pressure, which incorporates the irrotational centrifugal force. It is conventional to introduce the Rossby number, $\mathrm{Ro} = u/\Omega\ell$, to measure the relative importance of the inertial and Coriolis forces, and so our primary interest is turbulence in which $\mathrm{Ro} = O(1)$, or smaller. It is well known that such turbulence is characterised by the presence of long, columnar eddies aligned with the rotation axis, and there has been much discussion as to the mechanism by which these columnar structures form. The various theories differ in detail, but all agree that inertial waves play an important role. (We shall review the properties of inertial waves in §8.2.)

Several, now classic, experiments set the scene in the 1970s and 80s. Ibbetson & Tritton (1975) looked at freely-decaying turbulence in a rotating annulus in which $\mathrm{Ro} = O(1)$. Like all subsequent researchers, they observed that rotation causes the eddies to grow rapidly along the rotation axis, forming columnar structures. They also noted that inertial waves are important for transporting energy across the flow. Unlike most later studies, however, they observed an increase in the rate of energy dissipation as a result of rotation, though this is almost certainly an artefact of the shallowness of the annulus used, which meant that the columnar eddies could rapidly span the domain and so induce dissipative Ekman layers on the upper and lower surfaces.

8: Rapidly-Rotating Turbulence: An Experimental Perspective

A rather different configuration was investigated by Hopfinger *et al* (1982) who continually forced the flow with a vertically oscillating grid. The resulting turbulence was strongly inhomogeneous, with Ro large close to the grid, but order unity (or less) away from the grid. In regions where Ro dropped below $O(1)$, the flow became quasi-two-dimensional, characterised by the presence of long-lived, columnar eddies aligned with the rotation axis. These vortices were predominantly cyclonic, a finding that was to be confirmed by many later experiments. Also, finite-amplitude waves were seen to travel along the core of the vortices, waves which looked very like the vortex–soliton solution of Hasimoto (1972).

A similar vertically-oscillating grid experiment is reported in Dickenson & Long (1983) where, once again, the flow is continually forced and highly inhomogeneous. The emphasis, however, was rather different. They focussed on times for which the turbulence created by the grid was still spreading through the tank. They found that, when Ro was large and the rotation weak, the cloud of turbulence created by the grid spread along the rotation axis at a rate $\sqrt{\nu_t t}$, where ν_t is a turbulent diffusivity. They attributed this to conventional turbulent diffusion. However, for Ro $\leq O(1)$, the cloud grew in length at a rate proportional to t, $\ell_{\text{cloud}} \sim t$, suggesting a wave-like dispersion of energy. This is consistent with the suggestion of Ibbetson & Tritton (1975) that energy is transported by inertial waves, and indeed, in variations of the Dickenson & Long experiment, Davidson *et al* (2006) and Kolvin *et al* (2009) confirmed that the scaling $\ell_{\text{cloud}} \sim t$ is almost certainly a direct result of the dispersal of energy by low-frequency inertial waves propagating along the rotation axis (see §8.4).

These various experiments, while certainly stimulating in their own right, are a long way from the idealised flow preferred by many theoreticians; i.e. statistically homogeneous turbulence. Perhaps the first experimental investigation of freely-decaying, homogeneous, rotating turbulence, free from the influence of boundaries, is that of Wigeland & Nagib (1978). They looked at turbulence in a wind tunnel, generated by a rotating honeycomb. This configuration was later refined by Jacquin *et al* (1990), who reported many detailed measurements, including the observations that the rate of decay of energy is inhibited by rotation and that, for Ro $\leq O(1)$, the integral scale of the turbulence parallel to the rotation axis grows linearly with time, $\ell_{//} \sim t$. These important results were reconfirmed much later in the homogeneous experiments of Staplehurst *et al* (2008), but we shall defer our discussion of the more recent experiments until §8.5.

Summarizing the results of the early experiments, we might observe:

(i) long-lived, columnar vortices form on a fast time scale, of order Ω^{-1};
(ii) these vortices are predominantly cyclonic;
(iii) there is evidence from the inhomogeneous experiments that energy is transported predominantly by inertial waves when $\mathrm{Ro} \leq O(1)$;
(iv) in homogeneous turbulence at $\mathrm{Ro} \leq O(1)$, the integral scale parallel to the rotation axis grows linearly with time, $\ell_{//} \sim t$;
(v) rotation influences the energy decay rate, with the dissipation rate reduced in homogeneous turbulence, and increased in overly confined domains due to the formation of Ekman layers on the horizontal surfaces.

Shortly after these early experiments were reported, numerical simulations of homogeneous turbulence began to become common place, many of which are reviewed in Cambon & Scott (1999) and Cambon (2001). However, the relationship between the simulations and experiments is often far from clear-cut. For example, some simulations impose $\mathrm{Ro} = u/\Omega\ell \ll 1$, while nearly all the experiments on rotating turbulence are necessarily around $\mathrm{Ro} \sim 1$, or a bit lower. That is, a physical experiment intended to investigate the generation of inertial waves by turbulence, and the subsequent interplay of the waves and turbulence, cannot be forced at $\mathrm{Ro} \ll 1$, since in such cases the inertial waves which dominate the flow come straight off the grid, and so are an artefact of the forcing, rather than of the turbulence itself. So it is common practice in turbulence experiments to force briefly at $\mathrm{Ro} > 1$, and then let Ro drift down to $\mathrm{Ro} \sim 1$ as the energy of the turbulence decays. This is rather different to the numerical simulations which are often continually forced at small Ro, or else are unforced but have $\mathrm{Ro} = u/\Omega\ell \ll 1$ imposed as an initial condition. (We shall return to this distinction between the experiments and simulations in §8.5.) Moreover, some simulations employ an artificial body force which has no experimental counterpart, and most are confined to periodic computational domains, so that periodicity almost certainly plays a significant role in the long-term behaviour of the flow. One exception to this are the decay simulations of Liechtenstein *et al*, 2005, which used a reasonably large computational domain, though it has to be said that such large-scale simulations are still relatively rare.

In any event, columnar structures are also observed in the numerical simulations, just like in the experiments, though often these appear over timescales much longer than that required for columnar eddies to emerge in the laboratory experiments. The origin of this slow emergence of columns in the low-Ro simulations has been the subject of much discussion, but nearly all theories suggest that the anisotropy results from a weak *non-linear* coupling of the inertial waves (so-called resonant triad interactions), operating over long periods of time, $\Omega t \gg 1$. (See, for example, Waleffe, 1993, Smith & Waleffe, 1999, Smith & Lee, 2005, and §8.4 for a discussion of anisotropy arising from resonant triad interactions.)

In the light of these observations we might pose the following questions:

(i) what is the mechanism by which columnar vortices form in laboratory experiments?
(ii) given that the columnar vortices form rapidly in these experiments, on a time scale of Ω^{-1}, do the non-linear resonant triad interactions, which are usually invoked to explain the results of numerical simulations, play a key role in the laboratory experiments?
(iii) why is there a dominance of cyclones over anticyclones?
(iv) can we characterise the influence of rotation on energy dissipation?

Perhaps it is worth saying something about the structure of this chapter. As we shall see, the recurring themes of rotating turbulence are the emergence of columnar vortices, the dispersal of energy by inertial waves, and the interplay between the waves and the columnar eddies. Consequently, before we discuss the experimental results, it seems appropriate to summarise what is known about inertial waves and their relationship to columnar vortices. We do this in §8.2 and §8.3 in the context of linearised (low Ro) theory, before touching on non-linear theories in §8.4. We focus initially on low-Ro dynamics because the few rigorous statements that can be made about inertial waves (and their relationship to columnar eddies) occurs in the context of linear theory. Moreover, when confronted with a bewildering range of experimental observations, it is natural to ask which can be explained in terms of simple linear theory, and which require the heavy guns of non-linear phenomenology. So we start, in §8.2 and §8.3, with some simple linear dynamics.

8.2 Background: inertial waves and the formation of Taylor columns

Let us review briefly the main properties of non-turbulent rotating flows, i.e. introduce the idea of Taylor columns and inertial waves. As we shall see, both of these phenomena play an important role in our interpretation of the experimental evidence. When $\text{Ro} \ll 1$, the Euler equation in a rotating frame of reference simplifies to the linear equation

$$\frac{\partial \mathbf{u}}{\partial t} = 2\mathbf{u} \times \mathbf{\Omega} - \nabla(p/\rho), \tag{8.2}$$

whose curl yields

$$\frac{\partial \boldsymbol{\omega}}{\partial t} = 2\left(\boldsymbol{\Omega} \cdot \nabla\right) \mathbf{u}, \tag{8.3}$$

where $\boldsymbol{\omega}$ is the vorticity. If the motion is steady, or quasi-steady, then $\partial \boldsymbol{\omega}/\partial t$ may be neglected and we obtain,

$$\left(\boldsymbol{\Omega} \cdot \nabla\right) \mathbf{u} = 0. \tag{8.4}$$

Evidently, rapidly rotating, quasi-steady motion is subject to the powerful constraint that $\mathbf{u}$ is two-dimensional, in the sense that it is independent of the coordinate parallel to $\boldsymbol{\Omega}$. This is, of course, the Taylor–Proudman theorem and it has many curious consequences. Consider, for example, the experiment shown schematically in Figure 8.1. A small object is towed across the base of a rotating tank which is filled with water. It is observed that, as the object moves, the column of fluid located between it and the surface of the fluid also moves, as if rigidly attached to the object. This column is known as a Taylor column. Thus, for example, a fluid particle initially at point A will move across the tank, always centred above the object.

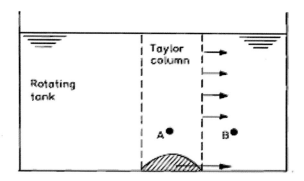

Figure 8.1 A small object is slowly towed across the base of a rotating tank. As the object moves it carries with it the column of fluid located between it and the upper surface of the liquid.

The existence of this Taylor column may be rationalised with the aid of (8.4) which demands that $\partial u_z/\partial z = 0$ and so forbids any axial straining of fluid elements. Since a vertical column of fluid cannot be stretched or compressed, there can be no flow over the object as it drifts across the tank.

Rather, the fluid, such as that at point B, must flow around the vertical cylinder which circumscribes the object, as if the Taylor column were rigid. Of course, it is natural to enquire as to how the fluid lying within the Taylor column knows to move with the object, which brings us to inertial waves.

The Coriolis force endows a rotating, incompressible fluid with a remarkable property: it can support internal wave motion. Consider the application of the operator $\nabla \times (\partial/\partial t)$ to (8.3). This yields the wave-like equation

$$\frac{\partial^2}{\partial t^2} \left(\nabla^2 \mathbf{u} \right) + 4 \left(\mathbf{\Omega} \cdot \nabla \right)^2 \mathbf{u} = 0, \tag{8.5}$$

which supports plane waves of the form

$$\mathbf{u} = \hat{\mathbf{u}} \exp\left[j \left(\mathbf{k} \cdot \mathbf{x} - \varpi t \right) \right], \quad \varpi = \pm 2 (\mathbf{k} \cdot \mathbf{\Omega}) / |\mathbf{k}|. \tag{8.6}$$

These are known as inertial waves. The group velocity of inertial waves, which is the velocity at which energy propagates away from a disturbance in the form of wave packets, is

$$\mathbf{c}_g = \frac{\partial \varpi}{\partial k_i} = \pm 2 \mathbf{k} \times (\mathbf{\Omega} \times \mathbf{k}) / |\mathbf{k}|^3. \tag{8.7}$$

Evidently, inertial waves have the unusual property that their group velocity is perpendicular to their phase velocity. Moreover, (8.7) tells us that

$$\mathbf{c}_g \cdot \mathbf{\Omega} = \pm 2 k^{-3} \left[k^2 \Omega^2 - (\mathbf{k} \cdot \mathbf{\Omega})^2 \right], \tag{8.8}$$

so that the positive signs in (8.6) and (8.7) correspond to wave energy travelling upward, while the negative signs correspond to energy propagating downward. Note also that inertial waves are helical, since (8.3) and (8.6) demand $\hat{\boldsymbol{\omega}} = \mp |\mathbf{k}| \hat{\mathbf{u}}$, where $\hat{\boldsymbol{\omega}}$ is the amplitude of the vorticity. It follows that the vorticity and velocity fields are parallel and in phase, so that inertial waves have maximum helicity density, $h = \mathbf{u} \cdot \boldsymbol{\omega}$, with the $+$ sign in (8.8) corresponding to negative helicity and vice versa. Thus wave packets with negative helicity propagate upward, while wave packets with positive helicity travel downward.

The frequency of inertial waves is independent of $|\mathbf{k}|$ but does depend on the relative orientation of $\mathbf{\Omega}$ and $\mathbf{k}$, varying from $\varpi = 0$ to $\varpi = 2|\mathbf{\Omega}|$. Low-frequency waves have $\mathbf{k} \cdot \mathbf{\Omega} \approx 0$ and a group velocity of $\mathbf{c}_g = \pm 2\mathbf{\Omega}/|\mathbf{k}|$. Such waves can be generated in a rotating tank by slowly oscillating an object, say a disc. Wave energy then propagates away from the object and along the rotation axis with a speed $c_g \sim 2\Omega R$, where R is the characteristic size of object (the disc radius, for example).

Now suppose that, instead of oscillating the disc, we move it slowly along

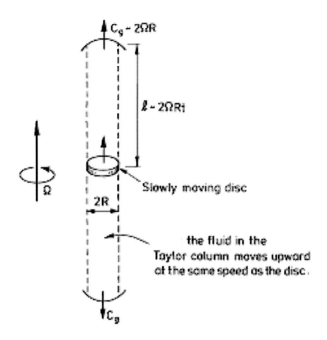

Figure 8.2 The formation of a Taylor column by inertial waves generated by a slowly moving disc (from Davidson, 2004)

the axis of rotation with a speed V, starting at time $t = 0$. Low-frequency waves propagate in the $\pm\Omega$ directions, carrying energy from the disc at a speed $c_g \approx 2\Omega/|\mathbf{k}|$. Since the largest wavelengths travel fastest, and these have a magnitude of $|\mathbf{k}| \approx \pi/R$, we would expect to find wave-fronts located a distance $\sim (2/\pi)\Omega Rt$ above and below the disc, as shown in Figure 8.2. An exact solution to this problem is given in Greenspan (1968, §4.3) and it turns out that the naive picture shown in Figure 8.2 is surprisingly accurate. At time t the inertial waves generated by the disc fill a column of radius R and half-length $\ell \approx (2/\pi)\Omega Rt$. The fluid which lies within this column has the same axial velocity as the disc, while that lying outside the column is quiescent in the rotating frame. Evidently the inertial waves have created a Taylor column, whose length grows at the rate $\ell \sim c_g t$.

We can now understand how the Taylor column shown in Figure 8.1 forms. As the object is towed slowly across the base of the tank it emits inertial waves. These travel upward (low-frequency waves propagate parallel to $\boldsymbol{\Omega}$) with a velocity of $\sim 2\Omega R$, and so reach the free surface on a time scale which is virtually instantaneous by comparison with the slow timescale of the movement of the object. As the inertial waves propagate upward they carry the information that tells the fluid within the column to move horizontally, keeping pace with the towed object. Thus the Taylor column is continually formed and re-formed by a train of inertial waves emitted by the object. Indeed, in some respects, the column *is* just a superposition of inertial waves. When we suppress the time derivative in (8.2) to give the Taylor–Proudman theorem, we filter out these waves. However, their long-term effect, which is the formation of the Taylor column, is still captured by the quasi-steady solution.

8.3 The spontaneous growth of Taylor columns from compact eddies at low Ro

So far we have considered inertial wave generation in the context of boundary value problems. The hallmark of a such problem is that there is an imposed time scale set by the movement of the boundary. For example, slowly moving boundaries excite low frequencies and hence (8.6) demands that **k** is more-or-less horizontal. Once the direction of **k** is fixed we can work out the direction of propagation of wave energy from (8.7), which is along the rotation axis in the case of low-frequency inertial waves. However, when thinking about the evolution of a rotating cloud of turbulence, it turns out that the initial value problem is much more relevant. This is more subtle than a boundary value problem, because there is no imposed time-scale to fix the direction of **k**, and hence $\mathbf{c}_g$. Rather, the direction of propagation of wave energy is determined, via (8.7), by the distribution of energy amongst the various **k**-vectors in the initial velocity field. That is, when we Fourier transform the initial condition we find a certain amount of energy associated with each wavevector, **k**, and that energy then propagates in accordance with (8.7). The act of taking the inverse transform, which is tantamount to superimposing all of these waves, determines where the wave energy gets to. For an arbitrary initial condition, then, we might expect the energy to disperse every which way. It turns out, however, that there is still a tendency for wave energy to propagate preferentially along the rotation axes to form columnar structures. In particular, we shall see that vortices (blobs of vorticity) spontaneously evolve into columnar eddies, i.e. the vortices create transient Taylor columns (Davidson *et al*, 2006).

Consider the initial value problem consisting of a localised blob of vorticity sitting in an otherwise quiescent, rapidly-rotating fluid (Figure 8.3). Let the characteristic scale of the blob be δ and a typical velocity be u. If Ro $= u/\Omega\delta \ll 1$ then the subsequent motion consists of a spectrum of linear inertial waves whose group velocity is dictated by the initial distribution of $\mathbf{k}$ in accordance with (8.7). Thus the energy will disperse in all directions with a typical speed of $|\mathbf{c}_g| \sim \Omega\delta$. However, this radiation of energy is subject to a powerful constraint, which systematically favours dispersion of energy along the rotation axis. Let V_R be a cylindrical volume of radius R and infinite length that encloses the vorticity field $\boldsymbol{\omega}$ at $t = 0$ (the tangent cylinder). Then it is readily shown that the axial component of the angular impulse, $\frac{1}{3}\int [\mathbf{x} \times (\mathbf{x} \times \boldsymbol{\omega})]_z dV$, which is equal to the angular momentum held in the tangent cylinder, $\int_{V_R} [\mathbf{x} \times \mathbf{u}]_z dV$, is conserved for all time. In short, angular momentum can disperse along the rotation axis only. To show that this is so, we first take the cross product of (8.2) with $\mathbf{x}$, which yields

$$\frac{\partial}{\partial t}(\mathbf{x} \times \mathbf{u}) = 2\mathbf{x} \times (\mathbf{u} \times \boldsymbol{\Omega}) + \nabla \times (p\mathbf{x}/\rho), \qquad (8.9)$$

the axial component of which can be rewritten as

$$\frac{\partial}{\partial t}(\mathbf{x} \times \mathbf{u})_z = -\nabla \cdot \left[\left(\mathbf{x}^2 - z^2\right)\Omega\mathbf{u}\right] + [\nabla \times (p\mathbf{x}/\rho)]_z. \qquad (8.10)$$

Integrating this over V_R we find that the pressure term drops out, since there is no net pressure torque acting on the cylinder, while the first term on the right-hand side converts to a surface integral which integrates to zero because of continuity. It follows that the axial component of angular momentum, $H_z = \int (\mathbf{x} \times \mathbf{u})_z dV$, is conserved in V_R, as suggested above. (A similar proof is readily constructed in terms of the angular impulse of the eddy, $\frac{1}{3}\int [\mathbf{x} \times (\mathbf{x} \times \boldsymbol{\omega})]_z dV$.)

The constraint imposed by the conservation of H_z systematically biases the dispersion of energy. For example, as the energy radiates to fill a three-dimensional volume whose size grows as $V_{3D} \sim (c_g t)^3 \sim (\delta\Omega t)^3$, we would expect that conservation of energy, $u^2 V_{3D} \sim$ constant, requires the velocity outside the tangent cylinder to fall as $|\mathbf{u}| \sim |\mathbf{u}_0|(\Omega t)^{-3/2}$. However, inside the tangent cylinder the axial component of angular momentum is confined to a cylindrical region whose volume grows as $V_{1D} \sim c_g t \delta^2 \sim \Omega t \delta^3$. Conservation of H_z, i.e. $V_{1D} u \delta \sim$ constant, then suggests that the characteristic velocity inside the cylinder falls more slowly, as $|\mathbf{u}| \sim |\mathbf{u}_0|(\Omega t)^{-1}$. This is illustrated in Figure 8.3. It follows that the energy density inside the tangent cylinder is significantly greater than that outside, and so the dominant influence of inertial wave radiation is to spread the energy of the vortex along the rotation axis. (If the eddy happens to have zero angular impulse,

8: Rapidly-Rotating Turbulence: An Experimental Perspective   327

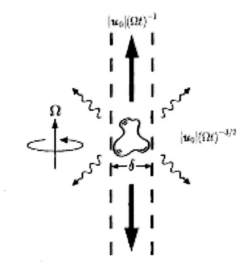

Figure 8.3 While the energy of a disturbance can propagate in any direction (wiggly arrows), the angular momentum can disperse along the rotation axis only (solid arrows). This biases the dispersion of energy, with the energy density within the tangent cylinder higher than that outside.

we can arrive at exactly the same conclusions by considering the conservation of linear impulse.) These two scalings, $|\mathbf{u}| \sim |\mathbf{u}_0|(\Omega t)^{-1}$ inside V_R and $|\mathbf{u}| \sim |\mathbf{u}_0|(\Omega t)^{-3/2}$ outside, can be confirmed by exact calculation using more detailed mathematical methods, such as the method of stationary phase.

In summary, then, a vortex confined to the region $|\mathbf{x}| < \delta$ is obliged to disperse energy in all directions, but the energy density is always greatest on the rotation axis, creating a pair of columnar vortices as shown schematically in Figure 8.4. These columnar structures are, in effect, transient Taylor columns.

A simple example taken from Davidson et al. (2006) illustrates the point. Suppose our initial condition consists of a spherical region of swirling fluid, $\mathbf{u} = \Lambda r \exp\left[-\left(r^2 + z^2\right)/\delta^2\right] \hat{\mathbf{e}}_\theta$ in cylindrical polar coordinates (r, θ, z), where Λ is a measure of the initial vortex strength. Then equation (8.5) yields the axisymmetric wave equation

$$\frac{\partial^2}{\partial t^2}\left[r\frac{\partial}{\partial r}\frac{1}{r}\frac{\partial \Gamma}{\partial r} + \frac{\partial^2 \Gamma}{\partial z^2}\right] + (2\Omega)^2 \frac{\partial^2 \Gamma}{\partial z^2} = 0, \tag{8.11}$$

where $\Gamma = r u_\theta$. This may be solved using a Hankel-cosine transform. The solution is complicated, but it is readily confirmed that a good approximation to the exact solution is

$$u_\theta \approx \Lambda\delta \int_0^\infty \kappa^2 e^{-\kappa^2} J_1\left(\frac{2\kappa r}{\delta}\right) \left\{\exp\left[-\left(\frac{z}{\delta} - \frac{\Omega t}{\kappa}\right)^2\right] + \exp\left[-\left(\frac{z}{\delta} + \frac{\Omega t}{\kappa}\right)^2\right]\right\} d\kappa, \tag{8.12}$$

$$u_z \approx \Lambda\delta \int_0^\infty \kappa^2 e^{-\kappa^2} J_0\left(\frac{2\kappa r}{\delta}\right) \left\{-\exp\left[-\left(\frac{z}{\delta} - \frac{\Omega t}{\kappa}\right)^2\right] + \exp\left[-\left(\frac{z}{\delta} + \frac{\Omega t}{\kappa}\right)^2\right]\right\} d\kappa, \tag{8.13}$$

$$\omega_\theta \approx \Lambda\delta \int_0^\infty \kappa^2 e^{-\kappa^2} \frac{2\kappa}{\delta} J_1\left(\frac{2\kappa r}{\delta}\right) \left\{-\exp\left[-\left(\frac{z}{\delta} - \frac{\Omega t}{\kappa}\right)^2\right] + \exp\left[-\left(\frac{z}{\delta} + \frac{\Omega t}{\kappa}\right)^2\right]\right\} d\kappa, \tag{8.14}$$

$$\omega_z \approx \Lambda\delta \int_0^\infty \kappa^2 e^{-\kappa^2} \frac{2\kappa}{\delta} J_0\left(\frac{2\kappa r}{\delta}\right) \left\{\exp\left[-\left(\frac{z}{\delta} - \frac{\Omega t}{\kappa}\right)^2\right] + \exp\left[-\left(\frac{z}{\delta} + \frac{\Omega t}{\kappa}\right)^2\right]\right\} d\kappa, \tag{8.15}$$

where J_1 and J_0 are the usual Bessel functions, $\kappa = k_r \delta/2$, and k_r is the radial wavenumber. Note that the motion in not purely azimuthal, as in the initial condition, but is helical with $u_\theta \sim u_z$ and $\omega_\theta \sim \omega_z$. It is clear from the form of the exponentials in (8.12)–(8.15) that the kinetic energy and enstrophy disperse along the z-axis forming two columnar structures of radius δ and length $\ell_z \sim \delta \Omega t$, and whose centres are located at $z = \pm \delta \Omega t$. This is illustrated schematically in Figure 8.4. Of course, these two lobes of vorticity are just a form of transient Taylor column.

The precise form of (8.12) for $\Omega t \to \infty$ may be found by insisting that the arguments in the exponentials remain of order unity as $\Omega t \to \infty$. At

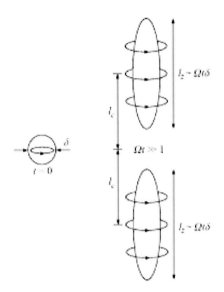

Figure 8.4 An initial blob of vorticity converts itself into a pair of columnar eddies (transient Taylor columns) via inertial wave propagation, Ro ≪ 1.

location $z = \delta\Omega t$, for example, we find

$$u_\theta(\Omega t \to \infty) \approx \Lambda\delta \left[\sqrt{\pi}/e\right] J_1(2r/\delta)(\Omega t)^{-1}, \qquad (8.16)$$

yielding $u_\theta \sim \Lambda\delta(\Omega t)^{-1}$ inside the tangent cylinder and

$$u_\theta \sim \Lambda\delta(\Omega t)^{-3/2}(r/z)^{-1/2}$$

outside the cylinder, in line with the discussion above. Thus the off-axis energy density is significantly weaker than that near the axis, as predicted.

Notice that, at late times, the helicity is given by $h \approx u_\theta \omega_\theta + u_z \omega_z$ due to the elongated form of the eddies, and that (8.12)–(8.15) confirm that $h < 0$ for the upward propagating vortex while $h > 0$ for its downward propagating partner.

In this simple example, then, we find that two transient Taylor columns form spontaneously as a result of linear wave propagation. Note in particular that nonlinear resonant triads interactions are not required. Whether nor not this kind of spontaneous generation of columnar structures lies behind the elongated vortices seen in laboratory experiments remains a controversial question. This is a key issue to which we shall return.

It turns out that the columnar vortex formation shown in Figure 8.4 is not limited to Ro $\ll 1$, but rather persists up to Ro ~ 1.4 for cyclonic initial conditions, and Ro ~ 0.4 for an anticyclonic vortex, as shown in the numerical simulations of Sreenivasan & Davidson (2008). Here Ro is defined by Ro $= |u_{\max}|/2\Omega\delta$, where $u_{\max}$ is the maximum velocity in the rotating frame at $t = 0$. This surprising observation is possibly a manifestation of the fact that there exists certain classes of finite-amplitude inertial waves. The idea is the following. One looks for solutions of the nonlinear equation

$$\frac{\partial \boldsymbol{\omega}}{\partial t} = \nabla \times (\mathbf{u} \times \boldsymbol{\omega}) + 2\left(\boldsymbol{\Omega} \cdot \nabla\right)\mathbf{u} \qquad (8.17)$$

in the form of Beltrami flows in the rotating frame, $\boldsymbol{\omega} = \mp k\mathbf{u}$, whose spatial structure is governed by $\nabla^2 \mathbf{u} + k^2 \mathbf{u} = 0$. Equation (8.17) then simplifies to the linear expression

$$2\left(\boldsymbol{\Omega} \cdot \nabla\right)\mathbf{u} \pm k\frac{\partial \mathbf{u}}{\partial t} = 0. \qquad (8.18)$$

Just such finite-amplitude solutions, in the form of axisymmetric waves, are reported in, for example, Greenspan (1968) and they have the same phase velocity as linear inertial waves. However, these solutions cannot be superimposed unless they share a common value of k, and so it is far from clear that they lie behind the finite-amplitude waves seen in the numerical simulations of Sreenivasan *et al* (2008).

Let us now take a step a little closer to turbulence and consider an initial condition consisting of many eddies (discrete blobs of vorticity) randomly but uniformly distributed in space. We take Ro $\ll 1$, so the problem is again linear and superposition tells us that every eddy in the initial condition will behave like our vortex in Figure 8.3. However, because of the off-axis radiation emitted by the initial vortices, each eddy is now immersed in a sea of waves emitted by its neighbours, and this could mask the Taylor column formation shown in Figure 8.4. So the question is: in this more complex situation, would we still expect to see $\ell_z \sim \delta\Omega t$, where ℓ_z is now defined in some statistical manner?

Because of the random nature of the initial condition we adopt the conventional approach in turbulence and use two-point correlations as our primary diagnostic tool. It should be kept in mind, however, that these are imperfect diagnostics for the task in hand, as autocorrelations strip out nearly all the phase information present in a signal, and it is precisely this phase information that allows energy to disperse by linear wave propagation. Thus autocorrelations are virtually blind to the very thing we seek to characterise. Luckily, however, some residual phase information is retained. Indeed, it is readily confirmed (see Staplehurst et al, 2008) that, for Ro ≪ 1, a statistically homogeneous initial condition composed of a random sea of vortices of the form $\mathbf{u} = \pm \Lambda r \exp\left[-2(r^2 + z^2)/\delta^2\right] \hat{\mathbf{e}}_\theta$ leads to

$$Q_\perp \approx Q_\perp^{(s)}(\mathbf{r}) + \tfrac{1}{2}\langle \mathbf{u}^2 \rangle \int_0^\infty \kappa^3 e^{-\kappa^2} J_0\left(\frac{2\kappa r_\perp}{\delta}\right) \left\{ \exp\left[-\left(\frac{r_{//}}{\delta} - \frac{2\Omega t}{\kappa}\right)^2\right] \right.$$
$$\left. + \exp\left[-\left(\frac{r_{//}}{\delta} + \frac{2\Omega t}{\kappa}\right)^2\right] \right\} d\kappa, \quad (8.19)$$

where $Q_\perp(\mathbf{r}) = \langle \mathbf{u}_\perp(\mathbf{x}) \cdot \mathbf{u}_\perp(\mathbf{x}+\mathbf{r}) \rangle$ is the two-point velocity correlation in the transverse plane. In (8.19) we use the notation $r_{//} = r_z$, $r_\perp^2 = r_x^2 + r_y^2$, $k_\perp^2 = k_x^2 + k_y^2$, $\kappa = k_\perp \delta/2$, and $Q_\perp^{(s)}(\mathbf{r})$ is a steady contribution to $Q_\perp$ which need not concern us for the moment. Note that the time-dependent part of correlation (8.19) is centred on $r_{//} \sim \pm 2\delta\Omega t$ and its characteristic axial length-scale grows as $\ell_z \sim \delta\Omega t$. Evidently $Q_\perp$ mimics, albeit in a statistical sense, the behaviour of the individual eddies. Moreover, for $\Omega t \to \infty$ and at location $r_{//} = 2\delta\Omega t$ (where $Q_\perp - Q_\perp^{(s)}(\mathbf{r})$ is centred), we have the analogue of (8.16),

$$\frac{Q_\perp(r_{//} = 2\delta\Omega t)}{\tfrac{1}{2}\langle \mathbf{u}^2 \rangle} \approx \left[\sqrt{\pi}/e\right] J_0(2r_\perp/\delta)(2\Omega t)^{-1}, \quad \Omega t \to \infty. \quad (8.20)$$

Evidently $Q_\perp(r_{//} = 2\delta\Omega t) \sim (\Omega t)^{-1}$ for $r_\perp \sim \delta$ and, for $r_\perp \gg \delta$, $Q_\perp(r_{//} = 2\delta\Omega t) \sim (\Omega t)^{-3/2}(r_\perp/r_{//})^{-1/2}$, so that the off-axis contribution to $Q_\perp$ falls off more quickly than the contribution near $r_\perp = 0$. In short, as $\Omega t \to \infty$, so $Q_\perp$ becomes increasingly centred near the $r_{//}$ axis.

Note that the autocorrelation $Q_\perp - Q_\perp^{(s)}(\mathbf{r})$ is self-similar for $\Omega t \gg 1$ and $r_{//} \sim 2\delta\Omega t$ provided $r_{//}$ is normalised by $2\delta\Omega t$. In particular

$$\frac{Q_\perp(\mathbf{r},t) - Q_\perp^{(s)}(\mathbf{r})}{\tfrac{1}{2}\langle \mathbf{u}^2 \rangle} \approx \sqrt{\pi}\frac{\exp\left(-\hat{r}_{//}^{-2}\right)}{\hat{r}_{//}^5} J_0\left(\frac{2r_\perp/\delta}{\hat{r}_{//}}\right)\frac{1}{2\Omega t}, \quad \Omega t \gg 1, \quad (8.21)$$

where $\hat{r}_{//} = r_{//}/2\delta\Omega t$. Moreover, $Q_\perp^{(s)}(\mathbf{r})$ is centred near $\mathbf{r} = 0$, thus making a negligible contribution at $r_{//} \sim 2\delta\Omega t$ for $\Omega t \gg 1$, and so (8.21) reduces to the self-similar form

$$Q_\perp(\Omega t \to \infty) \approx \frac{1}{\Omega t} F\left(r_\perp/\delta, r_{//}/\delta\Omega t\right), \quad \Omega t \gg 1, \quad (8.22)$$

except near $r = 0$. This admits the simple physical interpretation that all of the eddies grow in the axial direction at the rate $\ell_z \sim \delta\Omega t$. It is readily confirmed that the parallel velocity correlation $\langle \mathbf{u}_{//}(\mathbf{x}) \cdot \mathbf{u}_{//}(\mathbf{x}+\mathbf{r}) \rangle = \langle u_z u_z' \rangle$ behaves in exactly the same way, as do the equivalent two-point vorticity correlations, as we would expect from (8.13)–(8.15). In Staplehurst et al. (2008) it is argued that this behaviour is not particular to the model eddy used in the calculation, but rather generalises to any kind of eddy; i.e. in all cases we would expect the autocorrelations to be self-similar at large times when $r_{//}$ is normalised by $2\delta\Omega t$. We may think of this as the statistical signature of energy dispersion by linear inertial waves propagating from a sea of compact vortex blobs.

It is worth noting that the time dependent term in (8.19) relies crucially on the fact that the initial condition is anisotropic. In fact, as is well known, when Ro $\ll$ 1 and nonlinearity is neglected, any isotropic initial condition leaves $Q_{ij} = \langle u_i u_j' \rangle$ independent of t. However, in the linear regime it is hard to believe that the imposition of isotropy changes the way that individual eddies behave. So the vanishing of the time-dependent contribution to Q_{ij} in such cases is possibly a reflection of the fact that Q_{ij} is a poor diagnostic, rather than an indication that the dynamical processes have changed.

It is natural to use (8.22) as a criterion that may be deployed in the laboratory to test if columnar structure formation in homogeneous turbulence is the result of quasi-linear wave propagation, as described above, or else the result of nonlinear processes, such as resonant triads. However, one must exercise some caution here. For example, we have already seen that Q_{ij} is a poor diagnostic in the case of isotropic initial conditions. Moreover, it is entirely possible that certain nonlinear theories also lead to the self-similar solution $Q_\perp(\Omega t \to \infty) \approx Q_\perp\left(r_{//}/\delta\Omega t\right)$, and indeed some numerical calculations using an EDQNM closure model exhibit what looks close to a linear growth in $\ell_{//}$ (Cambon, 2001). So all-in-all, the issue is far from clear cut.

8.4 Anisotropic structuring via nonlinear wave interactions: resonant triads

We now move from linear to nonlinear considerations and discuss so-called resonant triad theory which seeks to describe weakly nonlinear interactions between waves, in our case inertial waves. We follow the development of Waleffe (1993) and Smith & Waleffe (1999), adopting their notation for the most part. The first step is to find a way of decomposing the velocity field into a superposition of helical inertial waves. Once this is achieved it is possible to examine the exchange of energy between waves promoted by the nonlinear inertial force $\mathbf{u} \cdot \nabla \mathbf{u}$. Let us start with a single inertial wave.

Plane inertial waves have a helical structure of the form

$$\mathbf{u} = \hat{\mathbf{u}} \exp\left[j\left(\mathbf{k} \cdot \mathbf{x} - \varpi_s t\right)\right] = b_s(\mathbf{k}) \mathbf{h}_s(\mathbf{k}) \exp\left[j\left(\mathbf{k} \cdot \mathbf{x} - \varpi_s t\right)\right],$$

where

$$\mathbf{h}_s = \hat{\mathbf{e}}_1 \times \hat{\mathbf{e}}_2 + sj\hat{\mathbf{e}}_2, \quad s = \pm 1, \quad \varpi_s = 2s\mathbf{\Omega} \cdot \hat{\mathbf{e}}_1.$$

Here $\hat{\mathbf{e}}_1 = \mathbf{k}/|\mathbf{k}|$, $\hat{\mathbf{e}}_2 = \mathbf{k} \times \mathbf{\Omega}/|\mathbf{k} \times \mathbf{\Omega}|$ and $\hat{\mathbf{e}}_3 = \hat{\mathbf{e}}_1 \times \hat{\mathbf{e}}_2$ are orthogonal unit vectors and we might think of $b_s(\mathbf{k}) = \hat{u}_3$ as the amplitude of the wave, while $\mathbf{h}_s$ defines its spatial structure. This is a circularly polarised wave with a velocity which is normal to $\mathbf{k}$, constant in magnitude, and rotates about $\mathbf{k}$ as the wave propagates. The wave satisfies $\hat{\boldsymbol{\omega}} = j\mathbf{k} \times \hat{\mathbf{u}} = \mp |\mathbf{k}| \hat{\mathbf{u}}$ and so the vorticity and velocity fields are in phase and everywhere parallel. Thus, as noted earlier, inertial waves have maximum helicity, with $s = 1$ corresponding to negative helicity and $s = -1$ to positive helicity. Moreover, expression (8.8) tells us that wave packets with negative helicity ($s = 1$) travel upward ($\mathbf{c}_g \cdot \mathbf{\Omega} > 0$), and those with positive helicity ($s = -1$) propagate downward ($\mathbf{c}_g \cdot \mathbf{\Omega} < 0$).

It follows that a sea of linear inertial waves can be expressed in the form

$$\mathbf{u} = \sum_{\mathbf{k}} \sum_{s=\pm 1} b_s(\mathbf{k}) \mathbf{h}_s(\mathbf{k}) \exp\left[j\left(\mathbf{k} \cdot \mathbf{x} - \varpi_s t\right)\right], \tag{8.23}$$

where the amplitudes, $b_s(\mathbf{k})$, are constants when nonlinearity and viscosity are ignored. The purpose of resonant triad theory is to try and predict the slow variation of the amplitudes $b_s(\mathbf{k})$ arising from the small but finite effects of inertia, assuming $\mathrm{Ro} \ll 1$.

A strictly kinematic statement, closely related to (8.23), but having nothing to do with rotating fluids or inertial waves, is the following. An instantaneous incompressible velocity field $\mathbf{u}(\mathbf{x})$, measured in an inertial frame, can be decomposed into helical modes according to

$$\mathbf{u} = \sum_{\mathbf{k}} \sum_{s=\pm 1} a_s(\mathbf{k}) \mathbf{h}_s(\mathbf{k}) \exp\left[j\mathbf{k} \cdot \mathbf{x}\right],$$

where the $a_s(\mathbf{k})$ are amplitudes. Moreover, substituting this expansion into the (non-rotating) Navier–Stokes equation yields (see e.g. Waleffe, 1993),

$$\left(\frac{\partial}{\partial t} + \nu k^2\right) a_{s_k} = \frac{1}{2} \sum_{\mathbf{k}+\mathbf{p}+\mathbf{q}=0} \sum_{s_p, s_q} C_{\mathbf{k}\mathbf{p}\mathbf{q}}^{s_k s_p s_q} a_{s_p}^* a_{s_q}^*, \tag{8.24}$$

where

$$C^{s_k s_p s_q}_{\mathbf{kpq}} = \tfrac{1}{2}(s_q q - s_p p)\left[\left(\mathbf{h}^*_{s_k} \times \mathbf{h}^*_{s_p}\right) \cdot \mathbf{h}^*_{s_q}\right], \tag{8.25}$$

and $*$ indicates a complex conjugate. Note that the nonlinear terms on the right of (8.24) are restricted to wavenumbers that satisfy the triad relationship $\mathbf{k} + \mathbf{p} + \mathbf{q} = 0$. This is simply a kinematic artefact of performing a Fourier-like decomposition. We shall revisit (8.24) shortly.

Let us now return to rotating fluids and inertial waves in the limit of weak inertia, Ro $\ll 1$. If the fluid is rotating as a whole with angular velocity $\mathbf{\Omega}$, and $\mathbf{u}$ is measured in the rotating frame of reference, then (8.24) becomes

$$\left(\frac{\partial}{\partial t} + \nu k^2 + j\varpi_{s_k}\right) a_{s_k} = \frac{1}{2} \sum_{\mathbf{k}+\mathbf{p}+\mathbf{q}=0} \sum_{s_p, s_q} C^{s_k s_p s_q}_{\mathbf{kpq}} a^*_{s_p} a^*_{s_q},$$

where $\varpi_s = 2s\mathbf{\Omega} \cdot \mathbf{k}/k$ is the inertial wave frequency. Writing

$$a_{s_k}(\mathbf{k}, t) = b_{s_k}(\mathbf{k}, t) \exp\left[-j\varpi_{s_k} t\right]$$

in accordance with (8.23), we find that the equivalent evolution equation for the amplitudes $b_{s_k}(\mathbf{k}, t)$ is

$$\left(\frac{\partial}{\partial t} + \nu k^2\right) b_{s_k} =$$
$$\frac{1}{2} \sum_{\mathbf{k}+\mathbf{p}+\mathbf{q}=0} \sum_{s_p, s_q} C^{s_k s_p s_q}_{\mathbf{kpq}} b^*_{s_p} b^*_{s_q} \exp\left[j\left(\varpi_{s_k} + \varpi_{s_p} + \varpi_{s_q}\right) t\right]. \tag{8.26}$$

Now the nonlinear terms in (8.26) are of order Ro. Since Ro is assumed small, the inertial waves oscillate rapidly on the fast time scale of Ω, while the amplitudes $b_{s_k}(\mathbf{k}, t)$ evolve slowly on the long time scale RoΩ. The oscillating term $\exp\left[j\left(\varpi_{s_k} + \varpi_{s_p} + \varpi_{s_q}\right) t\right]$ in (8.26) means that contributions in which $\varpi_{s_k} + \varpi_{s_p} + \varpi_{s_q} \neq 0$ oscillate rapidly on the fast time scale Ω, and these tend to average to zero upon integration over times of order (RoΩ)$^{-1}$. Thus, for small Ro, the dominant contribution to the right of (8.26) comes from waves which satisfy the resonance condition, $\varpi_{s_k} + \varpi_{s_p} + \varpi_{s_q} \approx 0$. Formally, for Ro $\to 0$, we may replace (8.26) by

$$\left(\frac{\partial}{\partial t} + \nu k^2\right) b_{s_k} = \frac{1}{2} \sum_{\substack{\mathbf{k}+\mathbf{p}+\mathbf{q}=0}}^{\varpi_{s_k}+\varpi_{s_p}+\varpi_{s_q}=0} \sum_{s_p, s_q} C^{s_k s_p s_q}_{\mathbf{kpq}} b^*_{s_p} b^*_{s_q}, \tag{8.27}$$

which is the same as (8.24), except that the nonlinear interactions are now restricted to resonant triads, which satisfy

$$\mathbf{k} + \mathbf{p} + \mathbf{q} = 0, \quad \varpi_{s_k} + \varpi_{s_p} + \varpi_{s_q} = 0. \tag{8.28}$$

We note in passing that, for a small but finite Ro, the exact resonance condition can be relaxed to that of near resonances,

$$\mathbf{k} + \mathbf{p} + \mathbf{q} = 0, \quad \varpi_{s_k} + \varpi_{s_p} + \varpi_{s_q} = O(\text{Ro}).$$

In any event, introducing $\cos \vartheta = k_{//}/k$, we may rewrite (8.28) in the form

$$k \cos \vartheta_k + p \cos \vartheta_p + q \cos \vartheta_q = 0, \quad s_k \cos \vartheta_k + s_p \cos \vartheta_p + s_q \cos \vartheta_q = 0, \quad (8.29)$$

where we have used only the axial component of the triad relationship. These may be rewritten as

$$\frac{\cos \vartheta_k}{s_p q - s_q p} = \frac{\cos \vartheta_p}{s_q k - s_k q} = \frac{\cos \vartheta_q}{s_k p - s_p k}. \quad (8.30)$$

The question now arises; does the evolution equation (8.27), subject to the constraints (8.30), predict the build up of anisotropy seen in laboratory experiments? The immediate answer has to be no, at least not without some additional constraint, since the nonlinear terms in (8.27) possess no arrow of time. That is, for every initial condition that pushes the turbulence to an axially elongated state, there will be another which drives the turbulence in the opposite direction. To break the deadlock, we need to make some assumption as to which of the nonlinear interactions contributing to (8.27) are statistically likely to occur, and which are statistically unlikely. This is where Waleffe's *instability assumption* enters the picture.

Let us temporarily abandon rapidly-rotating turbulence and return instead to conventional homogeneous turbulence governed by (8.24) and (8.25). We use the shorthand $C_k = C_{\mathbf{kpq}}^{s_k s_p s_q}$ and consider the idealised situation of a single triad evolving independently of the other helical modes. The three differential equations that describe this isolated triad are (omitting viscosity)

$$\frac{da_{s_k}}{dt} = C_k a_{s_p}^* a_{s_q}^*, \quad \frac{da_{s_p}}{dt} = C_p a_{s_q}^* a_{s_k}^*, \quad \frac{da_{s_q}}{dt} = C_q a_{s_k}^* a_{s_p}^*, \quad (8.31)$$

where (8.25) demands

$$C_k + C_p + C_q = 0, \quad s_k k C_k + s_p p C_p + s_q q C_q = 0, \quad (8.32)$$

which can be shown to be the consequences of energy and helicity conservation, respectively. Thus one of the three energy transfer coefficients has a sign opposite to the other two. This system possesses three steady solutions, each of which correspond to energy being held initially in only one mode. Two of these steady solutions turn out to be linearly stable and the third unstable, with the unstable mode corresponding to the energy transfer coefficient whose sign is opposite to the other two. Further, it is possible to

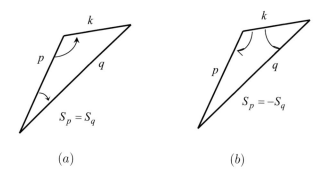

Figure 8.5 The directions of energy transfer within a triad that follow from Waleffe's *instability assumption*. In (a) the helicity of the two largest wavenumbers are of the same sign, while in (b) the helicity of the two largest wavenumbers is of opposite sign.

show that the smallest wavenumber in the triad is the unstable one whenever the two larger wavenumbers have helicities of opposite sign, whereas if the two largest wavenumbers have helicities of the same sign, then the unstable mode is the intermediate wavenumber.

Let us now consider homogeneous turbulence where we have many interacting triads. For each triad one of the three energy transfer coefficients has a sign opposite to the other two (the unstable mode of the isolated triad). There are two possibilities, either energy passes from this mode to the other two modes, or else this mode absorbs energy from the other two. In Waleffe (1993) it is proposed that, *on average*, we would expect energy to pass from the mode of opposite sign to the two modes of the same sign. This assumption, which is non-rigorous, is based on an appeal to the dynamical characteristics of an isolated triad. This is referred to as the *instability assumption* and is illustrated in Figure 8.5 where, without any loss of generality, we order the wavenumbers as $k < p < q$.

Numerical simulations of isotropic turbulence seem to be consistent with the instability assumption (Waleffe, 1993). Moreover, the assumption is consistent with statistical theories of two-dimensional turbulence. In Waleffe (1993) and Smith & Waleffe (1999) it is proposed that the instability assumption should also apply to rapidly-rotating turbulence. When this is combined with the resonant triad condition (8.30) we obtain some interesting results.

Let us return, then, to rotating turbulence and order the wavenumbers in a given triad as before, with $k < p < q$. When modes p and q have helicities of the same sign then the instability assumption asserts that mode p transfers energy to the other two, as in Figure 8.5(a). However, from (8.30)

we have
$$\cos\vartheta_p = \frac{s_q k - s_k q}{s_p q - s_q p}\cos\vartheta_k = \frac{s_q k - s_k q}{s_k p - s_p k}\cos\vartheta_q, \tag{8.33}$$

and when $s_p = s_q$ this requires $|\cos\vartheta_p| > |\cos\vartheta_k|, |\cos\vartheta_q|$. So in this situation energy transfers from the mode most closely aligned with the rotation axis to modes whose wave-vectors are closer to the transverse k_x-k_y plane. On the other hand, when modes p and q have helicities of opposite sign then the instability assumption asserts that the k mode transfers energy to the other two, as in Figure 8.5(b). However,

$$\cos\vartheta_k = \frac{s_p q - s_q p}{s_q k - s_k q}\cos\vartheta_p = \frac{s_p q - s_q p}{s_k p - s_p k}\cos\vartheta_q, \tag{8.34}$$

and when $s_p = -s_q$ this requires $|\cos\vartheta_k| > |\cos\vartheta_p|, |\cos\vartheta_q|$. Once again, energy transfers to the wave-vectors which are closer to the transverse k_x-k_y plane. In either case, then, *if* the instability assumption is applicable, resonant triad interactions tend to increase the anisotropy, which is not inconsistent with the experiments. So it seems that there are both linear and nonlinear reasons to expect rotating turbulence to become anisotropic, characterised by axially elongated structures.

8.5 Recent experimental evidence on inertial waves and columnar vortex formation

Let us now turn to the more recent experiments of Davidson *et al.* (2006), Morize *et al.* (2005, 2006), Bewley *et al.* (2007), Staplehurst *et al.* (2008), and Kolvin *et al.* (2009), starting with the first. Here the turbulence is created near the top a large tank by dragging a grid part way through the fluid and then back out. The turbulence is then left to itself. Initially Ro is large, but it drifts down towards unity as the energy of the flow decays. The high initial value of Ro ensures that no waves are generated by the grid used to create the turbulence, and so those waves which do appear arise from the turbulence itself.

When Ro reaches a value of order unity, columnar eddies emerge from the turbulent cloud, propagating in the axial direction, as shown schematically in Figure 8.6. Measurements show that these columnar vortices elongate at a constant rate (see Figure 8.7a), and that the growth rate is proportional to Ω and to the transverse scale of the vortices, δ, (Figure 8.7b). Thus we have

$$\ell_z \sim \Omega t \delta.$$

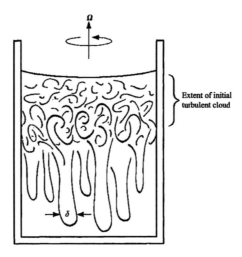

Figure 8.6 Schematic of the experiment of Davidson *et al.* (2006)

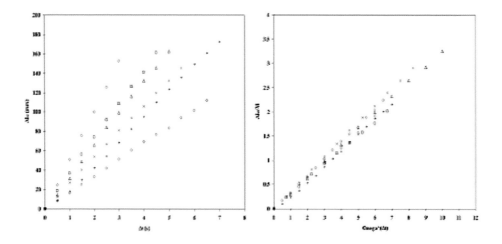

Figure 8.7 The variation of length, ΔL, versus time of the dominant columnar structure in six distinct experiments which had different rotation rates and different mesh sizes, M. The width of the columnar eddies scale with M. The graph on the left (ΔL versus t) shows that each columnar vortex grows at a constant rate, while that on the right ($\Delta L/M$ versus Ωt) confirms that the columnar eddies grow as $\ell_z \sim \Omega t \delta$.

This suggests that the vortices elongate by quasi-linear inertial wave propagation, at least in this particular experiment. (By quasi-linear waves we mean waves which propagate with the group velocity of linear waves, but have a finite amplitude.) A remarkably similar result was subsequently obtained by Kolvin *et al.* (2009), who performed a variation of the Dickenson & Long (1983) experiment and showed that a Fourier decomposition of the spreading turbulent cloud revealed $c_g \approx 2\Omega/k$, in line with (8.7). These authors particularly emphasise that it is surprising to see linear behaviour when Ro ~ 1, as this represents the propagation of finite-amplitude waves. The same authors also emphasise that, while the early development of the turbulence seems to be dominated by energy transport by quasi-linear inertial waves, the late-time development is dominated by nonlinear interactions.

One weakness of these experiments, however, is that the turbulence is inhomogeneous, with the columnar vortices growing into a quiescent region. It is not immediately apparent that the same thing will happen within the interior of a homogeneous field of turbulence, particularly when Ro ~ 1, so that linear and nonlinear processes compete. Nevertheless, Davidson *et al.* (2006) speculate that, in such a case, inertial waves might continue to form columnar structures on the time scale Ω^{-1}, and that these elongated eddies will provide a catalyst for nonlinear interactions as they spiral up the surrounding vorticity field. If such a situation did indeed arise, we would expect to see the longitudinal integral scale, $\ell_{//}$, to grow as $\ell_{//} \sim \ell_\perp \Omega t$, as the nonlinear dynamics shadow the growth of the columnar eddies.

This hypothesis was put to the test in the homogeneous, freely-decaying experiments of Staplehurst *et al.* (2008). In this case the grid is dragged once through the entire tank and the fluid then left to itself. As before, the initial value of Ro is large, but as the turbulence decays, Ro falls. The experiments were performed in a large vessel, 35 integral scales across, and any mean flow carefully suppressed so that the resulting motion is a good approximation to homogeneous turbulence (Figure 8.8). Four robust phenomena were observed:

(1) when Ro $= u/\Omega\ell$ reaches a value close to unity, columnar eddies start to form and these eventually dominate the large, energy-containing scales;
(2) during the formation of these columnar eddies, the integral scale parallel to $\boldsymbol{\Omega}$ grows linearly in time (as in Jacquin *et al.* 1990), i.e. $\ell_z \sim \ell_0 \Omega t$ where ℓ_0 is the initial integral scale;
(3) more cyclones than anticyclones are observed, as in many other experiments;
(4) the rate of energy decay is reduced by rotation, as in Jacquin *et al.* (1990).

We shall discuss (3) and (4) in §8.6 and §8.7, respectively. Here we focus on the first two observations. The fact that the integral scale grows as $\ell_z \sim \ell_0 \Omega t$ tentatively suggests that the columnar eddies might form by

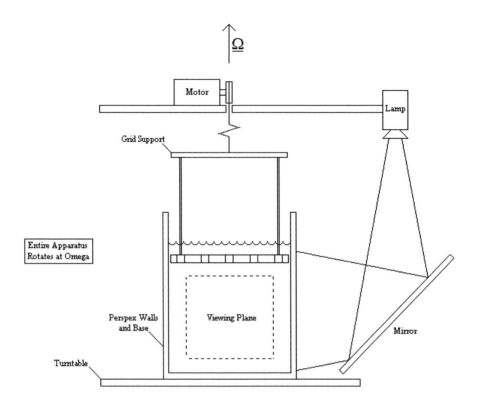

Figure 8.8 Experimental apparatus used in Staplehurst *et al.* (2008)

quasi-linear wave propagation, rather than by resonant triad interactions. This is perhaps surprising, since Ro ~ 1. In order to test this hypothesis, two-point autocorrelations where measured as a function of $r_{//}$ and t. This was done for four different experiments which had varying values of Ω and ℓ_0, but were otherwise similar. According to the linear theory of §8.3, the autocorrelations from all four experiments and at all times in each experiment should collapse onto a single universal curve, provided $r_{//}$ is normalised by $\Omega \ell_0 t$, i.e. linear theory demands $\left[Q_\perp / u_\perp^2\right] (\Omega t \to \infty) \approx \left[Q_\perp / u_\perp^2\right] \left(r_{//} / \ell_0 \Omega t\right)$. This is exactly what was found, and so it is possible, but by no means certain, that the columnar vortices seen in Staplehurst *et al.* (2008) were formed by quasi-linear wave propagation. However, as noted earlier, one cannot rule out the possibility that certain nonlinear processes might also lead to $\left[Q_\perp / u_\perp^2\right] (\Omega t \to \infty) \approx \left[Q_\perp / u_\perp^2\right] \left(r_{//} / \ell_0 \Omega t\right)$, so one must exercise considerable caution when interpreting these experimental results. Still, the observation that $Q_\perp$ is self-similar in this way needs to be explained, and Occam's razor suggests that wave propagation is a natural starting point for discussion.

The possibility that quasi-linear behaviour occurs at Ro ~ 1 is consistent with the computations of Sreenivasan & Davidson (2008). Here a sequence of numerical simulations were undertaken to investigate the influence of a finite Ro on columnar vortex formation. In each simulation a single eddy was allowed to evolve in the presence of background rotation. In the rotating frame the eddy had an initial condition very similar to that discussed in §8.3, and the initial Rossby number, defined as Ro $= |u_{\max}|/2\Omega\delta$, was varied over a wide range. The computations showed that the transition from almost purely linear to fully nonlinear behaviour is abrupt. In the case of cyclones the transition occurs in the narrow range $1.4 <$ Ro < 3.0, with columnar vortex formation (via inertial waves) below Ro $= 1.4$ and nonlinear centrifugal bursting of the vortex blob for Ro > 3.0. The equivalent range for anticyclones is also surprisingly narrow, $0.4 <$ Ro < 1.6. Of course these findings relate to the particular initial conditions used in the simulations, but they do at least illustrate how quasi-linear inertial wave propagation can persist up to Rossby numbers of Ro ~ 1.

Perhaps it is worth emphasising that it is only the large, energy-containing vortices in these experiments which form columnar vortices. Most of the enstrophy, on the other hand, is held in much smaller vortices. Moreover, since Ro ~ 1 based on the large scales, the effective Rossby number for these smaller enstrophy-containing eddies will be large. This suggests that we have two types of dynamics occurring simultaneously in the turbulence: the large eddies form columnar structures at Ro ~ 1 (possibly through quasi-linear wave propagation, or possibly by some other process), but these columnar eddies are immersed in a sea of smaller vortices which continue to evolve in a highly nonlinear way. Of course, the anisotropy created at the large scales can then feed nonlinearly down to the smaller eddies, because these sit in the shadow of the large columnar vortices and so are subject to a strain field whose axial length scale grows as $\ell_z \sim \ell_0 \Omega t$.

We now turn to the experiments of Morize *et al.* (2005, 2006) and Bewley *et al.* (2007), starting with Morize *et al.* (2005, 2006). The set-up here is similar to that of Staplehurst *et al* (2008), though the total depth of the water (44cm) is limited to 11 mesh lengths. Consequently the axial confinement in these experiments is significant and the increase in axial length scale quickly limited by the domain size. Thus the initial linear growth, $\ell_z \sim \ell_0 \Omega t$, seen in Jacquin *et al.* (1990), Staplehurst *et al.* (2008) and Kolvin (2009), cannot be properly checked since the large eddies rapidly span the domain. Nevertheless, in common with the other experiments, it is observed that the large scales are dominated by long-lived columnar eddies aligned with Ω. Three phenomena were investigated in detail:

(1) the asymmetry between cyclones and anticyclones;
(2) the influence of rotation on the inertial-range energy spectrum; and
(3) the influence of rotation on the rate of decay of energy.

Let us start with (1). Visualisations of the axial vorticity in horizontal planes showed that long-lived cyclonic vortices ($\omega_z > 0$) rapidly form, while the anticyclonic vorticity ($\omega_z < 0$) remains more disorganised and filamentary. There is also evidence that the columnar cyclones tend to organise the surrounding vorticity field, spiralling up the adjacent vortex filaments, as suggested by Davidson et al. (2006). The skewness of ω_z, $S_\omega = \langle \omega_z^3 \rangle / \langle \omega_z^2 \rangle^{3/2}$, was used as a measure of the dominance of cyclones and was found to grow initially as $S_\omega \sim (\Omega t)^{0.7}$, before saturating at around $S_\omega \sim 1$ and then decaying. Similar behaviour is reported in the experiments of Staplehurst et al. (2008) and the numerical simulations of Bokhoven et al. (2008), though the saturation of S_ω occurs at the lower value of $S_\omega \sim 0.35$ in both of these later studies. Morize et al. (2005) speculate that the growth of this asymmetry is driven by inertial waves, and we shall return to this idea in §8.7. Turning now to the energy spectrum, $E(k)$, the exponent m in the inertial range scaling $E \sim k_\perp^{-m}$ is found to increase from Kolmogorov's $m = 5/3$ at short times to $m \approx 2.3$ at larger values of Ωt. At no point does it reach $m = 3$, nor is there any sustained period in which $m = 2$. (Various phenomenological models have been proposed that predict either $m = 2$ or $m = 3$, the latter being typical of two-dimensional turbulence.) Finally, Morize et al. (2006) examined the rate of energy decay and found $u^2 \sim t^{-n}$, where the exponent n various from 2 at low rotation rates to $n = 1$ at higher values of Ω. We shall return to this topic in §8.7 where we shall see that, for Ro ~ 1, the exponent $n = 1$ is found in other experiments. Morize et al. (2006) speculate that their higher values on n are due to confinement effects.

Let us now turn to the experiment of Bewley et al. (2007). This is particularly interesting as it is one of the few rotating turbulence experiments where the initial forcing is at low Ro. This has a dramatic effect on the subsequent dynamics, producing a flow quite unlike any of those discussed so far. Inertial waves are produced directly from the grid (rather than from the turbulence) and rapidly fill the domain. Subsequent reflections from the boundaries produce large-scale standing waves which are as energetic as the turbulence itself. These waves, in turn, produce an oscillating mean flow and large-scale inhomogeneities in the turbulence. As these authors note, the dynamics of such turbulent flows are intimately linked to the overall geometry of the flow domain and will exhibit few universal characteristics. The dramatic differences between this experiment and the others discussed above emphasise the fact that there is not just one class rotating turbulence, but rather many classes, and due attention must be paid to the value of Ro, the nature of the forcing, and the boundary conditions.

As discussed earlier, a similar difficulty can arise when trying to compare laboratory experiments with certain numerical simulations. In a wave bearing system, which is what rapidly-rotating turbulence is, the nature of the forcing (if there is any), and of the boundary conditions, will play a strong role in what is seen. (If you beat a drum, what you hear depends on how you beat it and on how it is constructed.) So it is, perhaps, difficult to make a quantitative comparison between most laboratory experiments and, say, an artificially forced numerical simulation in a periodic cube at low Ro. That is not to say that there have been no attempts to make quantitative comparisons between rotating turbulence experiments and numerical simulations (see, for example, Godeferd & Lollini, 1999), but these are still relatively rare.

8.6 The cyclone–anticyclone asymmetry: speculative cartoons

Most experiments and numerical simulations of rotating turbulence show that there is an asymmetry between cyclones and anticyclones, with a dominance of cyclones (see, for example, Bartello et al., 1994, Morize et al., 2005, Bokhoven et al., 2008, and Staplehurst et al., 2008). There have been a number of attempts to explain this asymmetry, though the entire issue remains largely unresolved.

One particularly elegant explanation was offered by Gence & Frick (2001). They considered fully-developed, isotropic turbulence which is suddenly subjected to a background rotation. They showed that, at the moment at which the rotation is applied,

$$\frac{\partial}{\partial t} \langle \omega_z^3 \rangle = 0.4\Omega \langle \omega_i \omega_j S_{ij} \rangle,$$

where S_{ij} is the rate-of-strain tensor. Since $\langle \omega_i \omega_j S_{ij} \rangle > 0$ in fully-developed, isotropic turbulence, the vorticity skewness, $S_\omega = \langle \omega_z^3 \rangle \big/ \langle \omega_z^2 \rangle^{3/2}$, must become positive, which is taken to be indicative of the generation of cyclones. However, this only tells us what happens immediately after rotation is somehow 'switched on'. A quite different proposal is that of Bartello *et al.* (1994) who considered axisymmetric, two-dimensional columnar vortices and showed that, when viewed in an inertial frame of reference, the cyclones are typically more stable by Rayleigh's criterion than the anticyclones. However, the evidence of the experiments is not so much that columnar anticyclones form and then go unstable, but rather that they tend not to form in the first place. So this too seems somehow incomplete.

The numerical simulations of Sreenivasan & Davidson (2008) are interesting in this respect. Recall that, as Ro is increased, the transition from columnar vortex formation (via quasi-linear inertial waves) to fully non-linear centrifugal bursting is different for cyclonic and anticyclonic vortex blobs. In particular, the anticyclonic blobs require a lower value of Ro to form columnar structures. For example, in the simulations Sreenivasan *et al.* (2008), cyclonic eddies develop into columnar vortices provided Ro < 1.4, whereas anticyclonic eddies require the lower value of Ro < 0.4. It is argued in Sreenivasan & Davidson that this asymmetry is not just an artefact of the particular initial condition used, but is in fact a generic feature of axisymmetric cyclonic and anticyclonic vortices. The point is this: the centrifugal bursting which characterises nonlinear dynamics is driven by regions in which, in an inertial frame of reference, Rayleigh's discriminant is negative, $\partial \widehat{\Gamma}^2 / \partial r < 0$. (Here $\widehat{\Gamma}$ is the angular momentum in the inertial frame: $\widehat{\Gamma} = r u_\theta + \Omega r^2$.) Loosely speaking, the numerical simulations in Sreenivasan & Davidson show that columnar vortices form via inertial wave propagation provided that, everywhere in the initial condition, we satisfy $\partial \widehat{\Gamma}^2 / \partial r > 0$. Conversely, the vortex blob will burst radially outward (rather than grow axially) if there is a significant region in which $\partial \widehat{\Gamma}^2 / \partial r < 0$. Moreover, it is shown that, if we consider a sequence of initial conditions in which Ro increases progressively from Ro $\ll$ 1 up to Ro $\sim$ 1, anticyclonic eddies almost always exhibit regions of negative $\partial \widehat{\Gamma}^2 / \partial r$ before the cyclonic ones. In short, the fact that the transition range of Ro noted above is lower for anticyclones than cyclones is not particular to the initial condition used in the simulations, but rather is a generic feature of axisymmetric vortices.

Let us now return to the experiments of Hopfinger *et al.* (1982) and Staplehurst *et al.* (2008). Initially, we have a high value of Ro, but as Ro drifts down towards Ro $\sim$ 1 (either in space or time), columnar eddies start to appear. From the asymmetry described above, we might expect the first columnar structures to be cyclones, with significantly lower values of Ro needed to generate anticyclonic columns, which is consistent with the observations. Note, however, that this is little more than a superficial cartoon,

8: *Rapidly-Rotating Turbulence: An Experimental Perspective* 345

and that we are still a long way from really understanding why the cyclones are dominant.

8.7 The rate of energy decay

We now turn to the rate of energy decay in rotating turbulence which, in the case of homogeneous turbulence, is known to be suppressed by the rotation (e.g. Jacquin *et al.*, 1990.) The likely physical origin of this phenomenon is the observation that a significant percentage of the kinetic energy is held in columnar eddies, and these tend to be long-lived structures.

Perhaps the earliest attempt to quantify this was the analysis of Squires *et al.* (1994), to which we shall return shortly. In a related paper Zeman (1994) develops a phenomenological argument which suggests $u_\perp^2 \sim t^{-10/21}$, though the exponent 10/21 seems somewhat removed from the experimental observations which, as we shall see, are closer to $u^2 \sim t^{-1}$.

More recently, it is shown in Davidson (2010) that certain homogeneous turbulent flows which are statistically axisymmetric possess the Saffman-like invariant

$$L_\perp = \int Q_\perp(\mathbf{r}) d\mathbf{r} = \text{constant}. \tag{8.35}$$

This includes MHD, stratified and rotating turbulence. If the large scales in such a flow are self-similar, and they often are, then this requires $u_\perp^2 \ell_\perp^2 \ell_{//} = $ constant, where $\ell_\perp$ and $\ell_{//}$ are suitably defined transverse and longitudinal integral scales, and $u_\perp^2 = \langle u_x^2 + u_y^2 \rangle$. In such a situation conservation law (8.35) can be used to estimate the rate of decay of energy in rapidly-rotating turbulence. The argument, which is somewhat speculative, is developed in Davidson (2010) and goes as follows. We have already seen that, once the Rossby number falls below Ro ~ 1, $\ell_{//}$ grows as $\ell_{//} \sim \ell_0 \Omega t$ where ℓ_0 is the initial value of $\ell_{//}$. It follows that, in freely-decaying, rapidly-rotating turbulence, $u_\perp^2 \ell_\perp^2 \ell_0 \sim (\Omega t)^{-1}$. More precisely, we have

$$u_\perp^2 \ell_\perp^2 = u_0^2 \ell_0^2 \left(1 + \kappa \Omega t\right)^{-1}, \tag{8.36}$$

where $\kappa \sim 1$, $u_0 = u_\perp(t=0)$, and $t=0$ corresponds to the time at which columnar eddies first appear and $\ell_{//}$ starts to grow linearly. Note that u_0 and ℓ_0 are constrained to satisfy $u_0/\Omega \ell_0 \sim 1$, since Ro ~ 1 at $t=0$. Let us now suppose that the rate of dissipation of energy, $du_\perp^2/dt$, is uniquely determined by the integral scales, $u_\perp$ and $\ell_\perp$, and by Ω:

$$\frac{du_\perp^2}{dt} = F(u_\perp, \ell_\perp, \Omega).$$

Then dimensional analysis gives

$$\frac{du_\perp^2}{dt} = -G\left(u_\perp/\Omega\ell_\perp\right)\frac{u_\perp^3}{\ell_\perp}, \tag{8.37}$$

where G is some unknown function. Now we know that rotation increasingly inhibits energy dissipation for Ro ≤ 1, so G is likely to be an increasing function $u_\perp/\Omega\ell_\perp$ for Ro ≤ 1. The simplest option for G, which is consistent with the experimental evidence, is $G(\chi) \sim \chi$ for $\chi \leq 1$ and $G(\chi) \sim 1$ when $\chi > 1$, which is tantamount to requiring $du_\perp^2/dt \sim \Omega^{-1}$ when Ro ≤ 1, and indeed precisely this scaling has been advocated by Squires et al. (1994). Adopting this simple functional form for G, (8.37) becomes

$$\frac{du_\perp^2}{dt} = -\alpha\frac{u_\perp^4}{\Omega\ell_\perp^2}, \quad \text{Ro} \leq 1, \tag{8.38}$$

for some dimensionless constant α of order unity. Of course, it is a big step from (8.37) to (8.38), and this really needs to be independently assessed. Still, if we make this leap of faith, integrating (8.38) subject to the constraint of (8.36) yields, for large $u_0 t/\ell_0$,

$$u_\perp^2 \sim \frac{\Omega^2 \ell_o^2}{1+\kappa\Omega t} \sim (\Omega t)^{-1}, \quad \text{Ro} \leq 1. \tag{8.39}$$

This is interesting because the measurements of Jacquin et al. (1990) show $u^2 \sim t^{-n}$, with $n \approx 1.4$ for $\Omega = 0$, and $n \approx 0.81 \rightarrow 1.08$ for Ro ~ 1, which is close to (8.39). Moreover, the decay data of Staplehurst et al. (2008), which is shown in Figure 8.9, seems to follow a power law reasonably close to $u_\perp^2 \sim (\Omega t)^{-1}$. It should be emphasised, however, that it is notoriously difficult to get accurate measurements of energy decay exponents from experiments, so that any correspondence between (8.39) and the experimental data may be coincidental. Indeed Morize & Moisy (2006) found energy decay exponents considerably higher than $n = 1$, in the range $1 < n < 2$. However, their experiment was of limited vertical extent, so that columnar eddies rapidly span the domain. In such a situation homogeneity is quickly lost and Ekman layers form on the horizontal boundaries, which augment the dissipation of energy.

In this respect it is also interesting to look at the data of the numerical simulations. Teitelbaum & Mininni (2009) report $u^2 \sim t^{-1}$ in numerical simulations of rotating turbulence, though Bokhoven et al. (2008) and Thiele & Muller (2009) report significantly lower decay exponents in their numerical simulations. It should be noted, however, that all of these simulations were performed in periodic domains of modest size, and so the influence of periodicity cannot be ruled out.

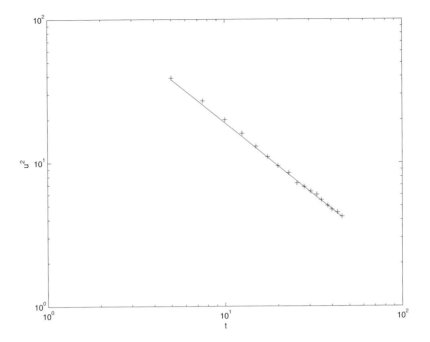

Figure 8.9 Plot of energy versus Ωt from the experiments of Staplehurst et al., 2008. The solid line is $u^2 \sim (\Omega t)^{-1}$

Finally we note that prediction (8.39) conflicts with the suggestions of Squires et al.(1994) who, on the basis of large-eddy simulations in periodic domains, and on the assumption that $L_\perp$ = constant, propose $u_\perp^2 \sim t^{-3/5}$, $\ell_\perp \sim t^{1/4}$ and $\ell_{//} \sim t$. While the linear growth in $\ell_{//}$ is consistent with experimental observations, the other estimates seem paradoxical as they require $L_\perp = u_\perp^2 \ell_\perp^2 \ell_{//} \sim t^{9/10}$, which is certainly not constant.

8.8 Concluding remarks

We have focussed on three issues:

(1) What is the mechanism by which columnar vortices form?
(2) Why is there a dominance of cyclones?
(3) How does rotation influence the rate of decay of energy?

While the experimental evidence from various sources is beginning to converge, we still do not have satisfactory answers to any of these questions.

Let us start with the mechanism of columnar vortex formation. It seems likely from the experiments of Davidson et al. (2006) and Kolvin et al. (2009)

that a localised cloud of turbulence spreads into adjacent, quiescent fluid by quasi-linear inertial wave propagation at Ro ~ 1. This creates columnar eddies in the form of transient Taylor columns. It is tempting to conclude, therefore, that the same thing happens within the interior of a field of homogeneous turbulence, and indeed the experiments of Staplehurst et al. (2008) lend some tentative support to this view. However, the fact that the autocorrelations from these experiments all collapse onto a single universal curve when $r_{//}$ is normalised by $\Omega \ell_0 t$, while suggestive of linear inertial waves, is far from conclusive. That is to say, it is perfectly conceivable that nonlinear phenomenology could also yield the prediction $Q_\perp(\Omega t \to \infty) \approx Q_\perp\left(r_{//}/\ell_0 \Omega t\right)$. So the jury is still out on the homogeneous case.

Similarly, while almost everyone sees a dominance of cyclonic vortices over anticyclones, there is still no satisfactory explanation for this. The various possible mechanisms discussed in §8.6 are mere cartoons and not terribly predictive. For example, we might ask: why does the skewness $S_\omega = \langle \omega_z^3 \rangle \Big/ \langle \omega_z^2 \rangle^{3/2}$ initially grow and then saturate? Evidently, there is a lot more work to be done here.

Finally, there is growing experimental evidence that the energy decay law for homogeneous rotating turbulence may be something close to $u_\perp^2 \sim (\Omega t)^{-1}$ for Ro ≤ 1. However, this too is subject to considerable uncertainty and is poorly understood. First, it is notoriously difficult to get accurate energy decay exponents from experimental data, so $n \approx 1$ is simply a best guess. Moreover, decay exponents extracted from numerical simulations in periodic cubes are suspect if the eddies straddle the cube, and there are relatively few simulations to date in which the computational domain remains an order of magnitude greater than $\ell_{//}$. So the evidence for $u_\perp^2 \sim (\Omega t)^{-1}$ is somewhat tenuous. And even if we were to agree that something like $u_\perp^2 \sim (\Omega t)^{-1}$ is correct, we have no good explanation for this particular power law. For example, the argument of §8.7 rest crucially on the decay law

$$\frac{du_\perp^2}{dt} = -\alpha \frac{u_\perp^4}{\Omega \ell_\perp^2}, \quad \text{Ro} \leq 1,$$

originally proposed by Squires et al. (1994). To date, this has received little independent support.

In summary, then, the experiments have come a long way, but there is still little we understand and a great deal that we do not.

Acknowledgement The author would like to thank Philip Staplehurst for many interesting discussions on rotating turbulence.

References

Bartello, P., Metais, O., Lesieur, M., 1994. Coherent structures in rotating three-dimensional turbulence. *J Fluid Mech.*, **237**, 1–29.

Bewley, G.P., Lathrop, D.P., Mass, L.R.M. & Sreenivasan, K.R., 2007. Inertial waves in rotating grid turbulence. *Phys. Fluids*, **19**, 071701.

Bokhoven, L.J.A., Cambon, C., Liechtenstein, L., Godeferd, F.S. & Clercx, H.J.H., 2008. Refined vorticity statistics in decaying rotating three-dimensional turbulence. *J. Turbulence*, **9**(6).

Cambon, C., 2001. Turbulence and vortex structures in rotating and stratified flows. *Eur. J. Mech. B*, **20**, 489–510.

Cambon, C. & Scott J.F., 1999. Linear and nonlinear models of anisotropic turbulence. *Ann. Rev. Fluid Mech.*, **31**, 1–53.

Davidson, P.A., Staplehurst, P.J. & Dalziel S.B., 2006. On the evolution of eddies in a rapidly rotating system. *J Fluid Mech.*, **557**, 135–144.

Davidson, P.A., 2010. On the decay of Saffman turbulence subject to rotation, stratification or an imposed magnetic field. *J. Fluid Mech.*, Published online doi:10.1017/S0022112010003496

Davidson, P.A., 2004, *Turbulence: an Introduction for Scientists and Engineers*, Oxford University Press.

Dickenson, S.C. & Long, R.R., 1983. Oscillating-grid turbulence including effects of rotation. *J. Fluid Mech.*, **126**, 315–333.

Gence, J.-N. & Frick C., 2001. Naissance des correlations triple de vorticite dans une turbulence homogene soumise á une rotation. *C. R. Acad. Sc. Paris*, Ser. IIB, **329**, 351–362.

Godeferd, F.S. & Lollini, L., 1999. Direct numerical simulation of turbulence with rotation and confinement. *J. Fluid Mech.* **393**, 257–308.

Greenspan, H.P., 1968, *The Theory of Rotating Fluids*. Cambridge University Press.

Hasimoto, H., 1972. A soliton on a vortex filament. *J. Fluid Mech.* **51**, 477–485.

Hopfinger, E.J., Browand, F.K. & Gagne, Y., 1982. Turbulence and waves in a rotating tank. *J. Fluid Mech.*, **125**, 505–534.

Ibbetson, A. & Tritton, D.J., 1975. Experiments on turbulence in a rotating fluid. *J. Fluid Mech.*, **68**(4), 6639–672.

Jacquin, L., Leuchter, O., Cambon, C. & Mathieu, J., 1990. Homogenous turbulence in the presence of rotation. *J. Fluid Mech.*, **220**, 1–52.

Kolvin, I., Cohen, K., Vardi, Y., & Sharon, E., 2009. Energy transfer by inertial waves during the buildup of turbulence in a rotating system. *Phy. Rev. Lett.*, **102**, 014503.

Liechtenstein, L., Godeferd, F. & Cambon, C., 2005. Nonlinear formation of structures in rotating, stratified turbulence. *J. Turbulence*, **6**, 1-18.

Morize, C., Moisy, F. & Rabaud, M., 2005. Decaying grid-generated turbulence in a rotating tank. *Phys. Fluids* **17**, 095105.

Morize, C. & Moisy, F., 2006. Energy decay of rotating turbulence with confinement effects. *Phys. Fluids* **18**, 065107.

Squires K.D., Chasnov, J.R., Mansour, N.N. & Cambon, C., 1994. The asymptotic state of rotating turbulence at high Reynolds number. AGARD-CP-551, 4.1–4.9.

Smith, L.M. & Waleffe, F., 1999. Transfer of energy to two-dimansional large scales in forced, rotating three-dimensional turbulence. *Phys. Fluids* **11**(6), 1608–1622.

Smith, L.M. & Lee, Y., 2005. On near resonances and symmetry breaking in forced rotating flows at moderate at moderate Rossby number. *J. Fluid Mech.*, **535**, 111–142.

Sreenivasan, B. & Davidson, P.A., 2008. On the formation of cyclones and anticyclones in a rotating fluid. *Phys. Fluids*, **20**(8), 085104.

Staplehurst P.J., Davidson P.A., Dalziel S.B., 2008. Structure formation in homogeneous freely decaying rotating turbulence. *J. Fluid Mech.*, **598**, 81–105.

Teitelbaum, I. & Mininni, P.D., 2009. Effect of helicity and rotation on the free decay of turbulent flows, *Phys. Rev. Lett.* **103**, 014501

Thiele, M. & Muller, W.-C., 2009. Structure and decay of rotating homogeneous turbulence. *J. Fluid Mech.* **637**, 425–442

Waleffe, F., 1993, Inertial transfers in the helical decomposition. *Phys. Fluids A*, **5**(3), 667–685.

Wigeland, R.A. & Nagib, H. M., 1978. Grid generated turbulence with and without rotation about the streamwise direction. *IIT Fluid & Heat Transfer Rep.* R. 78-1.

Zeman, O. 1994. A note on the spectra and decay of rotating homogeneous turbulence. *Phys. Fluids*, **6**, 3221

9
MHD Dynamos and Turbulence

S.M. Tobias, F. Cattaneo and S. Boldyrev

9.1 Introduction

Magnetic fields are ubiquitous in the universe (Parker (1979); Zeldovich *et al.* (1983)). Their interaction with an electrically conducting fluid gives rise to a complex system–a magnetofluid–whose dynamics is quite distinct from that of either a non conducting fluid, or that of a magnetic field in a vacuum (Cowling (1976)). The scales of these interactions vary in nature from metres to megaparsecs and in most situations, the dissipative processes occur on small enough scales that the resulting flows are turbulent. The purpose of this review is to discuss a small fraction of what is currently known about the properties of these turbulent flows. We refer the reader to several recent reviews for a broader view of the field (Biskamp (2003); Galtier (2008, 2009); Lazarian (2006); Lazarian & Cho (2005); Müller & Busse (2007); Kulsrud & Zweibel (2008), Bigot *et al.* 2008, Sridhar 2010, Brandenburg & Nordlund 2010). The electrically conducting fluid most commonly found in nature is ionized gas, i.e. a plasma, and its description in terms of all its fundamental constituents is extremely complex (see e.g. Kulsrud (2005)). In many circumstances, however, these complexities can be neglected in favour of a simplified description in term of a single fluid interacting with a magnetic field. Formally, this approach is justifiable when the processes of interest occur on timescales long compared with the light-crossing time, and on spatial scales much larger that any characteristic plasma length. Despite its simplified nature, the resulting Magneto-Hydrodynamic (MHD) approximation is of great general applicability and can adequately describe many astrophysical systems ranging from galaxy clusters, to the interstellar medium, to stellar and planetary interiors, as well as laboratory experiments in liquid metals. The dynamics of turbulent flows under this approximation will form the basis for our discussion.

There are many similarities between turbulence in a magnetofluid and in a non-electrically conducting fluid – hereinafter a regular fluid. The most fundamental is the idea of a turbulent cascade whereby energy is transferred from a large injection scale by nonlinear interactions through an inertial range to small scales where it is converted into heat by dissipative processes (see e.g. Davidson (2004)). In analogy with regular turbulence, one of the objectives of theories of MHD turbulence is to predict the general form of the inertial range and dissipative subrange, and to identify the important processes therein. There are, however, many significant differences. First of all, MHD describes the evolution of two vector fields, the velocity and the magnetic field, and hence the specification of the state of the system is more involved. A further crucial difference concerns the influence of global constraints. In regular turbulence global constraints become progressively weaker at small scales, and in general it is always possible to find a scale below which the motions become unconstrained. For example, in a rotating fluid scales smaller than the Rossby radius hardly feel the rotation at all; in a stratified fluid, motions on a scale much smaller than a representative scale height hardly feel the stratification. In contrast, there is no scale in MHD turbulence below which the fluid becomes unmagnetized. This has an important consequence: the cascade in regular turbulence approaches an isotropic state at small scales, whereas the cascade for MHD turbulence becomes progressively more anisotropic at small scales, this being true even if there is no large-scale field. Moreover, an important effect in MHD turbulence that has no direct correspondence in regular turbulence is the so-called dynamo process whereby a turbulent electrically conducting fluid becomes self magnetized. Finally, there are several differences between regular and MHD turbulence that are social in origin. We are constantly immersed in regular turbulence. We have a direct experience of it in our everyday life. Thus our development of models and theories of regular turbulence is both strongly guided and strongly constrained by experimental data and intuitions. Not so for MHD turbulence. Even though MHD turbulence is very widespread in the universe, we have practically no direct experience of it in our daily pursuits. Laboratory experiments can be conducted but they are, in general difficult to perform, and difficult to diagnose (Verhille *et al.* (2010)). As a result, MHD theorists have enjoyed a greater creative freedom and have expressed it by generating an impressive array of competing theories; computer simulations being the only obstacle standing in the way of the expression of unbounded imagination. As a testament to this, we note that whereas the form of the energy spectrum in isotropic homogeneous (regular) turbulence was very much a settled issue by the late forties (for a historical perspective see Frisch (1995)), the corresponding problem in MHD turbulence is still very much open to debate today.

9.1.1 Formulation and equations

In this review we shall restrict our discussion to incompressible MHD. In terms of the fluid velocity this is appropriate for subsonic flows. In MHD an additional constraint on the magnitude of the magnetic fluctuations is also required for the applicability of incompressibility. It is customary to define the plasma β as the ratio of the gas pressure to the magnetic pressure – then incompressibility is appropriate for situations in which $\beta \gg 1$. This condition is satisfied, for instance, in planetary and stellar interiors, but not in dilute plasmas such as those found in the solar corona and in many fusion devices (Kulsrud (2005)). We note here also that in some specific circumstances the incompressible equations provide a good approximation even if $\beta \ll 1$ (Biskamp (2003)). The evolution of the magnetofluid is described by the momentum equation together with the induction equation and the requirements that both the velocity and magnetic fields be solenoidal. In dimensionless form, these can be written as

$$\partial_t \mathbf{v} + \mathbf{v} \cdot \nabla \mathbf{v} = -\nabla p + \mathbf{J} \times \mathbf{B} + Re^{-1}\nabla^2 \mathbf{v} + \mathbf{f}, \tag{9.1}$$

$$\partial_t \mathbf{B} = \nabla \times (\mathbf{v} \times \mathbf{B}) + Rm^{-1}\nabla^2 \mathbf{B}, \tag{9.2}$$

$$\nabla \cdot \mathbf{B} = \nabla \cdot \mathbf{v} = 0, \tag{9.3}$$

where $\mathbf{v}$ is the fluid velocity, $\mathbf{B}$ is the magnetic field intensity, p the pressure, $\mathbf{J}$ is the current density, and $\mathbf{f}$ is the total body force, that could include rotation, buoyancy, etc. Here, $Re = v_o \ell / \nu$ in the momentum equation is the familiar Reynolds number and Rm in the induction equation is the corresponding magnetic Reynolds number defined analogously by $Rm = v_o \ell / \eta$, where η is the magnetic diffusivity; it measures the stength of inductive processes relative to diffusion. As is customary, we have assumed that the density ρ_o is constant and uniform, and we measure $\mathbf{B}$ in units of the equivalent Alfvén speed $(B/\sqrt{4\pi\rho_o})$. The ratio of the magnetic to the kinetic Reynolds number is a property of the fluid and is defined as the magnetic Prandtl number Pm. In most naturally occurring systems, Pm is either extremely large, like in the interstellar medium, the intergalactic medium and the solar wind, or extremely small like in the dense plasmas found in stellar interiors or liquid metals. Interestingly, it is never close to unity except in numerical simulations. This has some important consequences in terms of what we can compute as opposed to what we would like to understand; we shall return to this issue later. In this review, we shall concentrate on the cases in which Re is large, so that the flows are turbulent, and Rm can be small, as in some liquid metal experiments, moderate, as in planetary interiors, or large as in most astrophysical situations.

For an ideal fluid, i.e. one in which Rm is infinite, magnetic field lines move with the fluid as if they were frozen in. This is analogous to the behaviour of vortex lines in an inviscid fluid. Indeed, Alfvén's Theorem states that the magnetic flux through a material surface is conserved in the same way as Kelvin's theorem asserts that the circulation of a material contour is conserved (Cowling (1976); Moffatt (1978)). In fact, there is a formal analogy between the induction and vorticity equations. It should be noted however that, whereas the vorticity is the curl of the velocity, no such relationship exists between **v** and **B**. As a result, arguments based on the formal analogy between the two equations can sometimes be useful, and sometimes be misleading. In general, as a rule of thumb, if the magnetic field is weak compared with the velocity it behaves analogously to the vorticity; if it is comparable it behaves like the velocity. This point will be discussed more fully later.

As in hydrodynamic turbulence, much insight can be gained from examining conservation laws. In the ideal limit there are three quadratic conserved quantities: the total energy

$$E = \frac{1}{2} \int_V \left(v^2 + B^2 \right) d^3x, \qquad (9.4)$$

the cross-helicity

$$H^c = \int_V \mathbf{v} \cdot \mathbf{B}\, d^3x, \qquad (9.5)$$

and the magnetic helicity

$$H = \int_V \mathbf{A} \cdot \mathbf{B}\, d^3x, \qquad (9.6)$$

where **A** is the vector potential satisfying $\mathbf{B} = \nabla \times \mathbf{A}$. Here the volume V is either bounded by a material flux surface, or is all space provided that the fields decay sufficiently fast at infinity (Moffatt (1978))[1]. It should be noted that the limit $\mathbf{B} \to 0$ is a delicate one; only the energy survives as a conserved quantity and a new conserved quantity the kinetic helicity ($H^k = \int_V \mathbf{v} \cdot [\nabla \times \mathbf{v}]\, d^3x$) appears. This emphasises the fundamental difference between hydrodynamics and MHD.

It can be argued analytically and verified numerically that, in the presence of small dissipation, energy decays faster than magnetic helicity and cross-helicity (Biskamp (2003)). Therefore, in a turbulent state, energy cascades toward small scales (analogous to the energy cascade in 3D hydrodynamics), with the magnetic fluctuations approximately in equipartition with

---

[1] Many numerical simulations utilise periodic boundary conditions, for which care must be taken in defining the magnetic helicity.

the velocity fluctuations. The magnetic helicity cascades toward large scales (analogous to the energy cascade in 2D hydrodynamics); the inverse cascade of magnetic helicity may lead to the formation of large-scale magnetic fields, which are not in equipartition with the velocity fluctuations (Pouquet et al. (1976)). The role of cross-helicity is more subtle. Having the dimension of energy, it also cascades towards small scales (Biskamp (2003)). However, cross-helicity dissipation is not sign-definite: cross-helicity may be either amplified or damped locally. As we shall see, this leads to local self-organization in the turbulent inertial interval.

We can now discuss the most significant difference between hydrodynamic and MHD turbulence. In hydrodynamic turbulence, mean flows or large-scale eddies advect small-scale fluctuations without affecting their dynamics (Batchelor (1953)). This is a consequence of the Galilean invariance of the Navier–Stokes equation. The situation is quite the opposite in MHD turbulence. While the large-scale velocity field can be removed by Galilean transformation, the large-scale magnetic field cannot. Such a large-scale magnetic field could arise owing to a number of different processes. It could be either generated by large-scale eddies, as in the interstellar medium, or imposed by external sources, as in the solar wind or plasma fusion devises. The large-scale magnetic field mediates the energy cascade at all scales in the inertial interval. As a consequence, as we pointed out earlier, weak small-scale turbulent fluctuations become anisotropic, since it is much easier to shuffle strong magnetic field lines than to bend them. The smaller the scale, the stronger the anisotropy caused by the large-scale field. This is in a stark contrast with hydrodynamic turbulence where large-scale anisotropic conditions get "forgotten" by smaller eddies, so that the turbulent fluctuations become isotropic as their scale decreases.

In this review, we shall organise our discussion into two main sections describing the dynamics of turbulence with and without a significant mean field. If the mean field is unimportant then we neglect it altogether and consider the "dynamo" case, whereas if it is important we shall assume that it is strong and shall not concern ourselves about its origin. One should note that this distinction may depend on scale and the same system may be well described by the "dynamo" case at some scales and the "guide-field" case at others. Finally, we shall mostly be concerned with driven turbulence, that eventually evolves to a stationary state. There is a substantial body of literature considering turbulent decay (see e.g. Biskamp (2003)) that will not be discussed here.

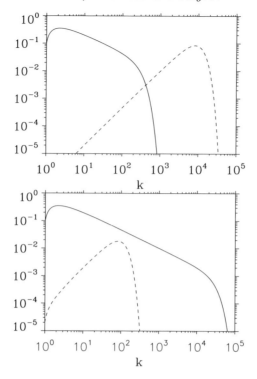

Figure 9.1 Possible schematic spectra for kinetic energy (solid line) and magnetic energy (dashed line) for dynamo action at (a) high and (b) low Pm. When Pm is large the resistive scale for the magnetic field is much smaller than a typical eddy and so the flow appears large-scale and smooth. When Pm is small, the resistive scale lies in the inertial range of the turbulence where the velocity is rough.

9.2 Dynamo

In this section we consider the case where there is no externally imposed magnetic field. It is well known that the turbulent motion of an electrically conducting fluid can lead to the amplification of a seed magnetic field (Parker (1979); Moffatt (1978)). This generation process is termed dynamo action and can lead to a substantial level of magnetisation. Two questions naturally arise. Under what conditions does dynamo action take place and what is the final state of the turbulence and magnetic field in the magneto-fluid? We discuss these questions in turn.

9.2.1 The onset of turbulent dynamo action

We consider a situation in which an electrically conducting fluid, confined within some bounded region of space, is in a state of turbulent motion, driven by stationary forces. At some time, a very weak magnetic perturbation is introduced into the fluid. The turbulent motions will stretch the magnetic field and will therefore typically lead either to its amplification or its decay. In the case of growth the field will eventually become strong, and the nonlinear Lorentz force term in the momentum equation will become important. This force will modify the turbulence in such a way that the field no longer grows and the system will settle down to a stationary state. However, initially, as the initial field is small, the Lorentz force is negligible and the question of growth versus decay of the field can be addressed by assuming that the velocity is given and by solving the induction equation alone. This defines the kinematic dynamo problem.

For a prescribed velocity field, this is a linear problem for the magnetic field $\mathbf{B}$. There are three cases that are typically considered within the literature: for a steady velocity one can seek solutions of the form $\mathbf{B} = \mathbf{B}(\mathbf{x})\exp\sigma t$ and the induction equation becomes a classical eigenvalue problem for the dynamo growth-rate σ. The value of Rm for which the real part of the growth-rate ($R_\lambda(\sigma)$) becomes positive is termed the critical magnetic Reynolds number Rm^{crit} for the onset of dynamo action. Another commonly considered case is one where the velocity is periodic in time, then the induction equation defines a Floquet problem and solutions consist of a periodic and an exponentially growing part and again one can define Rm^{crit} in analogy with the steady case. Finally there is the case in which the velocity is a stationary random process. Here one utilises a statistical description of the field, and seeks conditions under which the moments of the probability distribution function for $B = |\mathbf{B}|$ grow exponentially (Zeldovich et al. (1990)). In all of these cases Rm^{crit} corresponds to that value at which the induction processes overcome diffusion. Naïvely one would expect that once $Rm > Rm^{crit}$ dynamo action would be guaranteed, but this, actually, is not necessarily so. In fact establishing dynamo action in the limit $Rm \to \infty$, the so-called fast-dynamo problem, is technically very difficult (see for instance Childress & Gilbert (1995)). For instance it has been shown that flows that do not have chaotic streamlines can not be fast dynamos (Klapper & Young (1995)). This exemplifies the somewhat paradoxical role of diffusion in dynamo action. On the one hand too much diffusion suppresses dynamo action, on the other hand not enough diffusion also makes dynamo action impossible. The reason is that reconnection is required in order to change the magnetic topology to allow for the growth of the field (see e.g. Dormy & Soward (2007)). In fact, as we shall see, the diffusive scale at which reconnection occurs plays an important role in determining the properties of dynamo action.

Once a growing solution has been identified, it is of interest to determine properties of the eigenfunction. The detailed properties depend on the precise form of the velocity, but typically most of the energy of the growing eigenfunction is concentrated at the reconnection scales. However there is considerable interest, mostly astrophysically motivated, in cases where a significant fraction of the energy is found on scales larger than a typical scale for the velocity. This is called the large-scale dynamo problem and is most commonly discussed within the framework of mean field electrodynamics (Moffatt (1978); Krause & Raedler (1980)). One of the early successes of this kinematic theory was to establish that a lack of reflectional symmetry of the underlying flow is a necessary condition for the generation of large-scale magnetic fields. There is a substantial body of literature that discusses the large-scale dynamo problem. It is fair to say that currently there is considerable controversy as to whether mean-field electrodynamics can be applied in cases where Rm is large (see e.g. Brandenburg & Subramanian (2005); Boldyrev & Cattaneo (2004); Cattaneo & Hughes (2009)). In this present review we shall not discuss at length the large-scale problem, but instead focus on the amplification and saturation of magnetic fields of any scale.

Although anti-dynamo theorems exclude the possibility of dynamo action for certain flows and magnetic fields that possess too much symmetry, for example two-dimensional flows or axisymmetric fields (Cowling (1933); Zel'dovich (1956)), it is well established that most sufficiently complicated laminar flows do lead to growing fields at sufficiently high Rm. However here we are interested in the corresponding question for turbulent flows. In order to address this issue we need to discuss two distinct possibilities corresponding to the cases of small and large magnetic Prandtl number Pm.

9.2.2 High Pm versus low Pm, smooth versus rough

For the purposes of this discussion, we shall assume that the fluid Reynolds number is high so that the flow is turbulent with a well-defined inertial range that extends from the integral scale l_0 to the dissipative scale l_ν, and a dissipative sub-range for the scales $l < l_\nu$. It is useful to define the second order structure function $\Delta_2(r) = \langle |(\mathbf{v}(\mathbf{x},t) - \mathbf{v}(\mathbf{x}+\mathbf{r},t)).\mathbf{e}_r|^2 \rangle$, where $r = |\mathbf{r}|$ and $\mathbf{e}_r = \mathbf{r}/r$, where we have assumed homogeneity and isotropy. We can then characterise the inertial and dissipative ranges by the scaling exponents of $\Delta_2(r)$ with $\Delta_2(r) \sim r^{2\alpha}$, where α is the roughness exponent. In the dissipative sub-range we expect the velocity to be a smooth function of position and therefore $\alpha = 1$, whilst for the inertial range the velocity fluctuates rapidly in space, i.e. it is *rough* and $\alpha < 1$; for example, for Kolmogorov turbulence $\alpha = 1/3$. The inertial range can also be characterised by the slope of the energy spectrum $E_k \sim k^{-p}$ where p is related to the roughness exponent by $p = 2\alpha + 1$. The dissipative scale is also related to α and given by $l_\nu = l_0 Re^{-1/1+\alpha}$.

For this given velocity field the scale at which magnetic dissipation becomes important and reconnection occurs can be analogously defined by $l_\eta = l_0 Rm^{-1/1+\alpha}$. The ratio of the two dissipative scales is given by $l_\nu/l_\eta = Pm^{1/1+\alpha}$. Clearly, irrespective of the value of α, if $Pm \gg 1$ then the resistive scale l_η is much smaller than the viscous scale l_ν, and therefore reconnection and dissipation of magnetic energy occur in the dissipative sub-range where the velocity is spatially smooth, as in Figure 9.1(a). By contrast, if $Pm \ll 1$ the dissipative scale is much greater than the viscous scale and therefore reconnection occurs in the inertial range where the velocity is rough and therefore fluctuates rapidly, as shown in Figure 9.1(b). Therefore $Pm = 1$ defines the boundary between dynamo action driven by rough and smooth velocities.

It will become apparent that in general it is harder to drive a dynamo with a rough velocity than with a smooth velocity. We now examine some specific examples.

9.2.3 Random dynamos - the Kazantsev formulation

We now look at the simplest model of (kinematic) dynamo action driven by a random flow, the so-called Kazantsev model (Kazantsev 1968). This is the only known solvable model for dynamo action and, despite its simplicity, it captures many of the relevant features of turbulent field generation.

The model is based on a prescribed Gaussian, delta correlated (in time) velocity field, with zero mean and covariance

$$\langle v_i(\mathbf{x}+\mathbf{r},t)v_j(\mathbf{x},\tau)\rangle = \kappa_{ij}(\mathbf{x},\mathbf{r})\delta(t-\tau). \tag{9.7}$$

Isotropy and homogeneity imply that the velocity correlation function has the form

$$\kappa_{ij}(\mathbf{r}) = \kappa_N(r)\left(\delta_{ij} - \frac{r_i r_j}{r^2}\right) + \kappa_L(r)\frac{r_i r_j}{r^2}. \tag{9.8}$$

Further, incompressibility gives $\kappa_N = \kappa_L + (r\kappa'_L)/2$, where the primes denote differentiation with respect to r, and now the velocity statistics can be characterized by the single scalar function $\kappa_L(r)$. A corresponding expression for the magnetic correlator can be defined by

$$\langle B_i(\mathbf{x}+\mathbf{r},t)B_j(\mathbf{x},t)\rangle = H_{ij}(\mathbf{x},\mathbf{r},t), \tag{9.9}$$

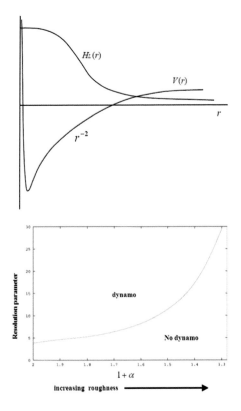

Figure 9.2 (a) Schematic representation of the potential $V(r)$ and the spatial part of the longitudinal magnetic correlator $H_L(r)$ in the Kazantsev model. The potential has a $1/r^2$ behaviour in the inertial range. Its overall height is determined by the velocity roughness, and its large r behaviour by the boundary conditions. The magnetic correlator is peaked at small scales and decays exponentially for large values of r. (b) Critical value of the size parameter L for dynamo action in the Kazantsev model as a function of the velocity roughness $1 + \alpha$. As the velocity becomes rougher the effort required to drive a dynamo increases

where, analogously to (9.8), the H_{ij} function can be represented as

$$H_{ij} = H_N(r,t)\left(\delta_{ij} - \frac{r_i r_j}{r^2}\right) + H_L(r,t)\frac{r_i r_j}{r^2}. \tag{9.10}$$

Similarly, the condition $\nabla \cdot \mathbf{B} = 0$ gives $H_N = H_L + (rH_L')/2$.

It now remains to determine the equation governing the evolution of $H_L(r,t)$ in terms of the input function $\kappa_L(r)$. This equation is derived by differentiating equation (9.9) with respect to time, using the dimensional version of the induction equation and expressions (9.7), (9.8) and (9.10).

Straightforward (and lengthy) manipulation yields

$$\partial_t H_L = \kappa H_L'' + \left(\frac{4}{r}\kappa + \kappa'\right) H_L' + \left(\kappa'' + \frac{4}{r}\kappa'\right) H_L, \qquad (9.11)$$

where the the 'renormalized' velocity correlation function $\kappa(r) = 2\eta + \kappa_L(0) - \kappa_L(r)$ has been introduced, and η is the magnetic diffusivity.

Equation (9.11) was originally derived by Kazantsev (1968), and can be rewritten in a related form that formally coincides with the Schrödinger equation in imaginary time. This is done by effecting the change of variable, $H_L = \psi(r,t) r^{-2} \kappa(r)^{-1/2}$, to obtain

$$\partial_t \psi = \kappa(r)\psi'' - V(r)\psi, \qquad (9.12)$$

which describes the wave function of a quantum particle with variable mass, $m(r) = 1/[2\kappa(r)]$, in a one-dimensional potential $(r > 0)$:

$$V(r) = \frac{2}{r^2}\kappa(r) - \frac{1}{2}\kappa''(r) - \frac{2}{r}\kappa'(r) - \frac{(\kappa'(r))^2}{4\kappa(r)}. \qquad (9.13)$$

Different authors have investigated the solutions of equation (9.12) for various choices of $\kappa(r)$ (see Zeldovich *et al.* (1990); Arponen & Horvai (2007); Chertkov *et al.* (1999)). Here we restrict attention to the two extreme cases in which l_η is in the deep dissipative subrange ($Pm \gg 1$) and the velocity is smooth, and the case where l_η is in the inertial range ($Pm \ll 1$) and the velocity is rough, so that $\alpha < 1$. For the smooth case, exponentially growing solutions of equation (9.11) can be found and the magnetic energy spectrum E_M is peaked at the dissipative scale. The spectrum in the range $1/l_\eta < k < 1/l_\nu$ has a power law behaviour with $E_M \sim k^{3/2}$, irrespective of the spectral index for the velocity in the inertial range (Kulsrud & Anderson (1992); Schekochihin *et al.* (2002)). This regime for a smooth velocity is also referred to as the Batchelor regime, since this is the regime studied by Batchelor (1959a) for passive-scalar advection.

For the rough case $\kappa(r) \sim r^{1+\alpha}$ in the inertial range. This scaling follows from the fact that Eq. (9.12) depends only on the time-integral of the velocity correlation function (9.8), that is, on the turbulent diffusivity. When matching the Kazantsev model with a realistic velocity field, one therefore needs to match the turbulent diffusivities. In the Kazantsev model the diffusivity is given by $\kappa(r)$, while for a realistic velocity field it is estimated as $\Delta_2(r)\tau(r)$, where $\tau(r) \sim r/[\Delta_2(r)]^{1/2} \sim r^{1-\alpha}$ is the typical velocity decorrelation time. This leads to the inertial-interval scaling $\kappa(r) \sim \Delta_2(r)\tau(r) \sim r^{1+\alpha}$ presented above. The Schrödinger equation (9.12) therefore has the effective potential $U_{\text{eff}}(r) = V(r)/\kappa(r) = A(\alpha)/r^2$ in the inertial range, where $A(\alpha) = 2 - 3(1+\alpha)/2 - 3(1+\alpha)^2/4$. At small scales $(l < l_\eta)$ this effective potential is regularised by magnetic diffusion as shown in Figure 9.2(a). Growing dynamo solutions correspond to bound states for the wave-function ψ, which are guaranteed to exist if $A(\alpha) < -1/4$. Since this is always the case for $0 < \alpha < 1$, this demonstrates that dynamo action is always possible even in the case of a rough velocity (Boldyrev & Cattaneo (2004)). The corresponding wave-function will be concentrated around the minimum of the potential at $r \sim l_\eta$ and it will decay exponentially for $r > l_\eta$ (see Figure 9.2). We stress at this point that the effective potential always remains $1/r^2$ in the inertial range with its depth decreasing as the roughness increases ($\alpha \to 0$). This justifies our previous statement that it is harder to drive dynamo action the rougher the velocity. An asymptotic analysis of the solution to equation (9.12) demonstrates that the growth-time, τ, is of the order of the turnover time of the eddies at the diffusive scale (l_η); that is, $\tau \sim l_\eta^{1-\alpha} \sim Rm^{(\alpha-1)/(\alpha+1)}$. This makes good physical sense since these are the eddies that have the largest shear rate. Furthermore the spatial part of the wave-function for large $r \gg l_\eta$ decays exponentially as $\exp\left[-2\tau^{-1/2}r^{(1-\alpha)/2}/(1-\alpha)\right]$. The presence of the factor $1-\alpha$ in the denominator implies that if α is close to unity (smooth velocities) the wave-function is localised close to the resistive scale. On the other hand, if α is close to zero (rough velocity) the wavefunction is more spread out. This can be used to determine the requirements for the onset of dynamo action, which can be expressed either in terms of a critical magnetic Reynolds number or in terms of a "size parameter" L. The latter is a more useful measure, since it relates directly to the effort, computational or experimental, that is needed to achieve dynamo action. Consider equation (9.12) as a two-point boundary value problem in a finite domain of size l_0 (see e.g. Boldyrev & Cattaneo (2004)), with boundary conditions $\psi(0) = \psi(l_0) = 0$. As the velocity becomes rougher the wave-function spreads out and a larger domain (i.e. larger l_0) is required to contain the wave-function. Figure 9.2(b) shows the minimum value of l_0 measured in units of l_η (which is the size parameter L) for which a growing solution can be found as a function of $1+\alpha$. Clearly there is a dramatic increase in L as the velocity becomes rougher ($1+\alpha$ gets smaller). Hence the effort required to drive a dynamo also increases.[2]

We now return to the question posed at the start of this section, what is the effect of changing Pm for the onset of dynamo action in a random

---

[2] An earlier attempt to solve the Kazantsev model for small Pm was made in (Rogachevskii & Kleeorin 1997). However, this paper employed an incorrect asymptotic matching procedure for deriving the dynamo threshold and the dynamo growth rate. The correct analysis was presented in (Boldyrev & Cattaneo (2004); Malyshkin & Boldyrev (2010)).

flow? At high Pm the dynamo operates in the dissipative sub-range where the velocity is smooth and so the effort necessary to drive a dynamo is modest. This state of affairs continues until Pm decreases through unity. At that point, the viscous scale becomes smaller than the resistive scale and the dynamo begins to operate in the inertial range, where the velocity is rough, with a corresponding increase in the effort required to sustain dynamo action. Once the dynamo is operating in the inertial range, further decreases in Pm do not make any difference to the effort required. In terms of critical Rm as a function of increasing Pm the curve takes the form of a low plateau and a high plateau joined by a sharp increase around $Pm = 1$ – see for example the curves in Figure 9.3. On the other hand, if a sequence of calculations is carried out at fixed Rm and increasing Re a sharp drop in the growth-rate σ will be observed when $Re \approx Rm$ (Boldyrev & Cattaneo (2004)). If the initial Rm is moderate, this could lead to the loss of dynamo action at some Re, which could be misinterpreted as a critical value of Pm at which dynamo action becomes impossible (Christensen *et al.* (1999); Yousef *et al.* (2003); Schekochihin *et al.* (2004a, 2005)). We shall return to this theme later.

The Kazantsev model is, of course, based on a number of somewhat restrictive assumptions, one of which is that the flow considered is completely random and has no coherent part. We discuss in the next section the role of coherent structure in dynamo problems. Within the statistical framework there has been substantial effort to extend the basic model to more general cases. One common criticism is that delta-correlated velocities are artificial, with real turbulence having correlation times of order the turnover time. However, one should remember that in most turbulent flows the turnover time decreases as one goes to small scales. Near onset the dynamo growth time can be very large compared with the turnover time – and hence the correlation time – of the small eddies that typically participate in the dynamo process. Noting that the growing magnetic fluctuations are predominantly concentrated at the resistive scales, where the relative motion of magnetic-field lines is affected by magnetic diffusion and, therefore, these lines do not separate with the eddy turnover rate, one expects that the assumption of zero correlation time is not as restrictive as may first appear. Once this is realised, it is to be expected that the dynamo behaviour near onset should be similar for cases with short but finite correlation time to that with zero correlation time, and indeed this is confirmed by models in which the correlation time of the turbulence is finite (Vainshtein & Kichatinov (1986); Kleeorin *et al.* (2002)). A second possible extension is to flows that lack reflectional symmetry. In this case the velocity and magnetic correlators are

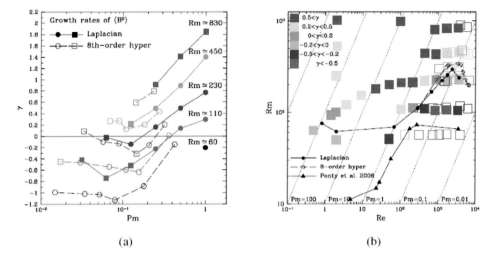

Figure 9.3 Onset of dynamo action at moderate Pm, from Schekochihin et al. (2007). (a) Growth/decay rate of magnetic energy as a function of Pm for five values of Rm. (b) Growth/decay rates in the parameter space (Re, Rm). Also shown are the interpolated stability curves Rm^{crit} as a function of Re based on the Laplacian and hyperviscous runs are shown separately. For an in-depth discussion of the reason for the apparent discrepancy between the curves with and without a mean flow see Schekochihin et al. (2007)

each defined by two functions, one as before related to the energy density (either kinetic or magnetic), the other related to the helicity (either kinetic or magnetic) (Vainshtein & Kichatinov (1986)). In this case similar analysis leads to the derivation of a pair of coupled Schroedinger-like equations for the two parts of the magnetic correlator (Vainshtein & Kichatinov (1986); Berger & Rosner (1995); Kim & Hughes (1997)). Although the analysis is now more involved, it is possible to show that the system remains self-adjoint (Boldyrev et al. (2005)). It can also be shown that for sufficient kinetic helicity there is the possibility of extended states – these are states that do not decay exponentially at infinity and can be interpreted as large-scale dynamo solutions, in direct analogy with the solutions of the well-known alpha-dynamo model by Steenbeck et al. (1966). However, at large Rm the largest growth-rates remain associated with the localised bound states, so that the overall dynamo growth-rate remains controlled by small-scale dynamo action (Boldyrev et al. (2005); Malyshkin & Boldyrev (2007, 2008, 2009)). Anisotropic versions of the Kazantsev model have also been constructed by Schekochihin et al. (2004b).

Finally there are extensions of the model that take into account departures from Gaussianity. Again on physical grounds one does not expect this to introduce fundamental changes from the Gaussian case. The reason is similar to the one given above; near marginality the growth time is long compared with the eddy turnover time and so the dynamo process feels the sum of many uncorrelated events, which approaches a Gaussian even if the individual events themselves are not. This situation arises in numerical simulations of dynamo action based on the full solution of the induction equation for velocities derived from solving the Navier–Stokes equations. Many such efforts are summarised in Figure 9.3, which shows the critical magnetic Reynolds number as a function of Reynolds number for a collection of such calculations (Schekochihin et al. (2007)). It is clear that at large and moderate Pm these results are consistent with the predictions of the Kazantsev model above – the plateau at high Pm is succeeded by a jump in the critical Rm when Pm approaches unity – such a curve is visible in Figure 9.3. Even the size of the jump is consistent with the analysis above. However, since the numerical cost of resolving the very thin viscous boundary layers rapidly becomes prohibitive at small Pm, this regime is not really accessible to direct numerical simulation (DNS). One scheme to alleviate this problem is to use large eddy simulations (LES) to generate the velocity (Ponty et al. (2007)). We would advise great caution when using this approach. As the discussion above shows, small changes in the roughness exponent can lead to huge changes in the critical Rm or alternatively the dynamo growth-rate. In the case of LES, one would need to be able to control very delicately how the velocity roughness is related to the smoothing scale. We also note that the jump in critical Rm is captured by shell models of dynamo action (see e.g. Frick & Sokoloff (1998)) for which it is possible to achieve a large separation of the viscous and resistive dissipation scales. We are not certain what, if anything, to make of this.

One of the interesting features of the Kazantsev model is that *for random flows* the only thing that matters for the onset of dynamo action is the roughness exponent of the spectrum in the neighbourhood of the dissipative scale. Therefore, within this framework, features associated with high-order moments, or large-scale boundary conditions do not matter. Thus from the point of view of dynamo onset a velocity derived from solving the Navier–Stokes equations, with random forcing should yield qualitatively similar results to a synthesised random velocity with the same spectrum. So far this seems to be broadly consistent with the results of numerical simulations.[3]. However, for flows with a substantial non-random component, outside of the range of validity of the Kazantsev model it may be that characteristics other

---

[3] A critical discussion of this question can be found in (Mason et al. (2011))

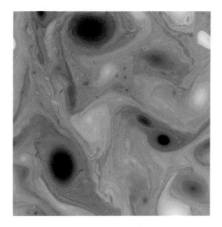 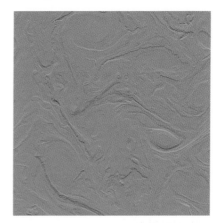

 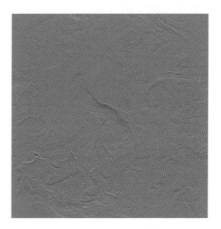

Figure 9.4 Dynamos with and without coherent structures. Out-of-plane velocity (left) and B_x (right) for flows with coherent structures (top) and purely random structure (bottom) – after Tobias & Cattaneo (2008b). The velocity with coherent structures is a better dynamo and generates more filamentary magnetic fields.

than the spectral slope of the velocity do play a key role in determining dynamo onset, and it is this possibility that we address in the next section. The results of the next section should therefore act as a caveat when considering the applicability of the results of this section.

9.2.4 Coherent Structure dynamos

We now turn to the case where the flow consists of two components, a random component as described above and a more organised component whose coherence time is long compared with the turnover time (see e.g. Mininni et al. (2005)). There are many examples of naturally occuring flows that can be characterised in this way; for example, flows in an accretion disks that consists of small-scale turbulence and coherent long-lived vortices (see e.g. Bracco et al. (1999); Godon & Livio (2000); Balmforth & Korycansky (2001), Taylor columns in rapidly rotating turbulent planetary interiors (see e.g. Busse (2002)) or the large-scale flows driven by propellors in laboratory experiments (see e.g. Monchaux et al. (2007)). In all these cases the turbulence arises because the Reynolds number is huge, whilst the coherent part is associated with some constraint, such as rotation, stratification or large-scale forcing. The natural question then is what determines the dynamo growth-rate?

In the last section we gained some understanding for the case of entirely random flows – we know that if the dynamo were entirely random then the dynamo growth-rate would largely be determined by the spectral index of the flow at the dissipative scale. Conversely, if the flow were laminar again we could appeal to laminar dynamo theory – but would not try to characterise the flow in terms of spectra. Here we need to consider the case where the flow is the sum of the two components, bearing in mind that the dynamo of the sum is not the sum of the dynamos (see e.g. Cattaneo & Tobias (2005)). We address these issues by examining a specific case. We consider a flow that has both long-lived vortices and a random component with a well-defined spectrum. For those who are interested, such flows arise naturally, for instance, as solutions to the active scalar equations (Tobias & Cattaneo (2008b)). In tandem we also generate a second flow with the same spectrum but no coherent structures obtained by randomising the phases. Figure 9.4 compares the out-of plane velocity for the two flows. We consider the dynamo properties for large Rm so we anticipate that dynamo action would occur for either flow. Therefore, if the considerations of the previous section were to apply, these two flows would exhibit very similar dynamo properties. However, this is actually not so – the flow with the coherent structures is a more efficient dynamo in the sense that at the same Rm the dynamo growth-rate is higher. Moreover the structure of the resulting magnetic field (shown in Figure 9.4) indicates that the coherent structures play the major role in field generation.

Given the above considerations, it is reasonable to ask what controls dynamo action in a flow with a superposition of coherent structures with different spatial scales and different turnover times? This is equivalent to a turbulent cascade, but here each of the eddies are long-lived with a coherence time greater than their local turnover time. For the general case this is a very difficult question to answer. However, there is a case in which some statements can be made – namely when each of the dynamo eddies are 'quick dynamos'. A 'quick dynamo' (as defined by Tobias & Cattaneo (2008a)) is one that reaches a neighbourhood of its maximal growth-rate quickly as a function of Rm. In that case it can be shown that the dynamo is driven by the coherent eddy which has the shortest turnover time τ and has $Rm \geq \mathcal{O}(1)$. Since both the turnover time of the eddy and the local Rm are functions of the spatial scale of the eddy and the slope of the spectrum, the location in the spectrum of the eddy responsible for dynamo action also depends on the spectral slope as well as the magnetic Reynolds number on the integral scale (Tobias & Cattaneo (2008a)).

Some of these ideas become particularly germane in explaining the behaviour of dynamos in liquid metal experiments. The typical laboratory dynamo device consists of a confining vessel and propellors, designed to drive a large-scale flow with desirable laminar dynamo properties; i.e. with low Rm^{crit}. Because the magnetic Prandtl number of liquid metals is tiny $\mathcal{O}(10^{-6})$ the corresponding Reynolds numbers are huge. Thus the actual flow consists of the large-scale flow planned by the designers plus turbulent fluctuations that can be comparable in magnitude to the mean flow. Invariably, it has been found that the actual critical magnetic Reynolds number for the mean flow and turbulence is significantly higher than that envisaged for the laminar flow alone. It is important to realise that these devices operate in a regime where at best the magnetic Reynolds number is close to the marginal one for the laminar flow; it is definitely below the critical Rm for the turbulent part. In all the cases discussed the role of turbulence is to hinder the dynamo, and this can be understood in terms a renormalised diffusivity that increases with the degree of randomness. This increase in diffusivity in astrophysical situations is irrelevant, since Rm is so far above critical that increasing the diffusivity makes little difference. However in the case of the experiments, this increase has a catastrophic effect on the chances of success for the experiment[4].

9.2.5 Saturation of turbulent dynamos

The exponential growth of the magnetic field described in the last section is accompanied by an exponential growth of the magnetic forces, which will eventually become comparable with those driving the turbulence. In this second nonlinear phase the exponential growth of the magnetic field

---

[4] Therefore we conclude that all funding for experiments should immediately be channeled to theorists studying dynamos at large Rm (or writing reviews).

will become saturated and the magneto-turbulence will settle down to some stationary level of magnetization. It is of interest to speculate about the nature of the saturation process, and about the general properties of the final state, although in general the saturation mechanism may be subtle (see e.g. Cattaneo & Tobias (2009))

As before there is a large difference between high and low magnetic Prandtl number regimes. In the high Pm regime the dynamo is operating at scales in the sub-inertial range of the velocity for which the Reynolds number is small. Therefore the inertial term in the momentum equation can be neglected and the velocity can be split into two components; one driven by the external forces, which is the original velocity of the kinematic dynamo problem and has a characteric scale $\gg l_\eta$, the other, driven by the Lorentz force, encodes the back-reaction of the magnetic field and is mostly concentrated around the diffusive scale. In this high Pm regime saturation can be successfully addressed by semi-analytical models, phenomenological models and numerical experiments. The semi-analytical models are ultimately based on some closure of the MHD equations. Within the framework of the Kazantsev equation the magnetically driven velocity produces a change in the velocity correlator, which leads to the nonlinear saturation (Subramanian (1999, 2003)). A similar approach can be taken by constructing a Fokker–Planck equation for the probability distribution function for magnetic fluctuations rather than an equation for the magnetic correlator. It can be shown that the coefficients of this equation again are determined by the velocity correlation function which can be modified nonlinearly in a similar manner to above (Boldyrev (2001)).

An interesting phenomenological model has been proposed by Schekochihin and collaborators (Schekochihin et al. (2002)). The model begins with the kinematic growth of fields at the resistive scale, which is much smaller than the viscous scale. In this kinematic phase the eddies driving the dynamo are the ones at the viscous scale. The authors anticipate that nonlinear effects begin to be important when $\mathbf{v} \cdot \nabla \mathbf{v} \approx \mathbf{B} \cdot \nabla \mathbf{B}$ at that scale. The left hand side is easily estimated to be v^2/l_ν. To estimate the right hand side the authors use the foliated structure of the magnetic field. To have a geometrical understanding of what this means, it is useful to distinguish between the orientation and the direction of a vector field. For example a change from horizontal to vertical is a change in orientation, whereas a change from up to down is a change in direction. For a typical foliated field the orientation changes slowly on the scale of the velocity, whereas the direction changes rapidly on the scale l_η (Finn & Ott (1988)). Hence the regions of high curvature occupy practically none of the volume and the right hand side is estimated to be b^2/l_ν, where b is a typical field strength at the scale l_η. Thus the nonlinear saturation begins when the magnetic energy comes into equipartition with the energy at the viscous scale. The effect is to suppress the dynamo growth associated with the eddies at the viscous scale. This suppression is not necessarily achieved by a dramatic reduction of the amplitude of the eddies but rather by a subtle modification of their geometry. In particular if the velocity becomes more aligned with the local magnetic field then it cannot distort that field and therefore contribute to its amplification. However, slightly larger eddies can still sustain growth, albeit at a slightly slower rate. Growth will continue until the magnetic energy comes into equipartition with the energy of these eddies. The process will continue to larger and larger eddies until the magnetic energy reaches equipartition with the energy of the flow at the integral scale. This nonlinear adjustment is characterised by a growth of the magnetic energy on an algebraic (rather than exponential) time and there is a corresponding shift of the characteristic scale of the magnetic field to larger scales. This idea of scale-by-scale modification of the velocity to reach some form of global equipartition can be formalised in terms of either Fokker–Planck equations (Schekochihin *et al.* (2002)) or a Kazantsev model which is appropriately modified to take account of the growing degree of anisotropy and finite correlation time (Schekochihin *et al.* (2004b)). There are even models constructed where the final state does not reach global equipartition but only a fraction of equipartition. According to Schekochihin *et al.* (2002), this occurs when Pm is large but $Pm \leq Re^{1/2}$ with $B^2/U^2 \approx Pm/Re^{1/2}$. In this scenario, however, by the time the final state is reached the characteristic scale of the magnetic field is still smaller than the viscous scale. Schekochihin *et al.* (2002) estimate this to be the case when $Pm \gg Re^{1/2}$.

There have been a number of simulations at moderate to high Pm, and within the normal restrictions of numerical simulations the results seem to conform to this general picture (Maron *et al.* (2004)). There is good evidence that in the kinematic phase the magnetic spectrum is compatible with the $k^{3/2}$ prediction of the Kazantsev model. The appearance of the magnetic field is indeed that of a foliated structure. In the saturated state, the magnetic and kinetic energies are comparable, although for moderate values of Pm the magnetic energy increases with Pm. As the saturation progresses the magnetic spectrum grows and flattens, which is compatible with the creation of magnetic structures larger than the resistive scale. It is always the case that the magnetic energy exceeds the kinetic energy at small scales (see e.g. Carati *et al.* (2006)). Moreover the magnetic field is more intermittent than the velocity – indeed the pdf for the velocity field remains close to

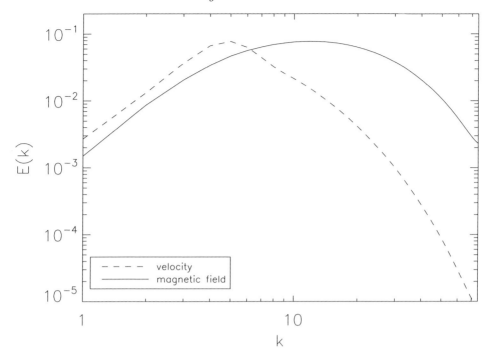

Figure 9.5 Velocity and magnetic spectra in a numerical simulation of magnetised turbulence; courtesy of Mason and Malyshkin (private communication)

that of a Gaussian whilst that for the magnetic field is better described by an exponential – although the degree of intermittency is reduced in the saturated state as compared with the kinematic state (Cattaneo (1999)).

Remarkably many of these features of the saturated state persist in the $Pm \approx \mathcal{O}(1)$ regime, which is much more accessible to numerical simulations. These address various issues ranging from the degree of intermittency to the nature of the kinetic and magnetic energy spectrum. Although there is no universal agreement there are a number of features that are robust. The magnetic fluctuations are always more intermittent than the velocity fluctuations. Furthermore there is a statistical anisotropy along and transverse to the local magnetic field, with a stronger degree of intermittency transverse (Müller et al. (2003)). These authors have argued that the intermittency results are consistent with a She-Leveque scaling in which the more singular structures are current sheets. Concerning energy spectra, results show that the kinetic and magnetic energy spectra track each other, with the magnetic energy spectra exceeding the kinetic one at all but the largest scales, as in Figure 9.5. The existence of an inertial range, characterised by a well-defined

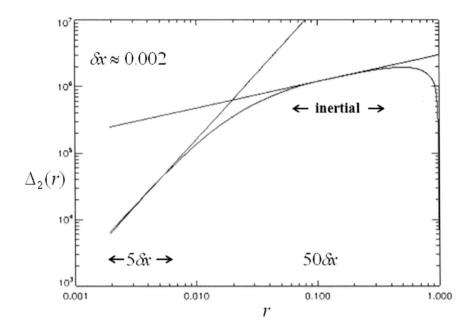

Figure 9.6 Second order longitudinal velocity structure function $\Delta_2(r)$ measured in a numerical simulation of magnetised turbulence as a function of separation r. At small scales $\Delta_2(r) \sim r^2$; at larger separations its power law behaviour defines the velocity roughness. Clearly, in order to represent numerically a rough velocity, at least of the order of a few tens of gridpoints is required. Calculations courtesy of Gianluigi Bodo & Cattaneo.

power law, is still a matter for debate. There are various reasons for this. One, of course, is that the resolution of the numerical simulations is limited. Another one is the presence of bottleneck effects, whose intensity is linked to the use of hyperdiffusivities. This, coupled with the limited resolution, can have effects that propagate into the putative inertial range. Finally there is a tendency to present results sometimes in terms of the shell-averaged spectrum, sometimes in terms of the one-dimensional spectrum relevant to a direction chosen for computational convenience and sometimes in terms of a one-dimensional spectrum perpendicular to the local mean field. All these issues notwithstanding, there are two things that can be said with certainty. Most researchers agree that the spectrum, if it exists, has a slope somewhere between -1.5 and -1.7 – we shall return to this issue in the next section. The second is that people have a tendency to find in their results confirmation of their expectations, and to attribute deviations to other effects, such as intermittency or bottlenecks. For example it is not unheard of for people to interpret a spectrum compatible with a spectral slope of $-3/2$ as a spectrum of $-5/3$ with corrections induced by a bottleneck (Haugen et al. (2004)) or intermittency (Beresnyak & Lazarian (2006)). No doubt as computers get bigger and numerics get better some kind of consensus will emerge.

At present the low Pm regime remains largely unchartered. In this regime the dynamo is operating in the inertial range of the turbulence. Inertial terms are certainly important and the velocity can no longer be thought of as consisting of two components. In this case analytical progress can only be made by resorting to standard closure models such as EDQNM (Pouquet et al. (1976)). One of the basic ingredients of these theories is the decorrelation time of triple moments. Whereas in hydrodynamic turbulence there is overwhelming evidence that this quantity is of the order of the turnover time, there is no consensus for of what this quantity should be in MHD theory. Indeed it may depend sensitively on the field strength and magnetic Reynolds number of the magnetised turbulence. For this reason it is not clear that utilising EDQNM is a reliable way to make analytical progress. One of the problems is that, whereas in hydrodynamic turbulence there is a vast body of experimental evidence, there is no such thing in MHD, and therefore one is forced to rely almost entirely on numerical experiments, with all of their limitations. These limitations become extremely severe in the low Pm regime. There is a generic expectation that for sufficiently low Pm an asymptotic regime is reached where the level of saturation becomes independent of Pm (Pétrélis & Fauve (2008)). Some steps have been taken to move in the direction of simulating dynamos at low Pm, but these are still in the $Pm \sim \mathcal{O}(1)$ regime, with all the associated dynamics (see e.g. Mininni (2006); Iskakov et al. (2007)). It is important to appreciate the difficulties inherent in reaching the low Pm regime numerically, and why caution should be exercised when interpreting the results of numerical simulations that are described as being in this regime. The essence of the problem is that the velocity field driving the dynamo is rough. It is possible to ask what sort of resolution is required to represent such a velocity. Figure 9.6 shows the second order structure function for a turbulent velocity; the distance here is expressed in units of the grid spacing. At small distances the slope approaches two, implying that on a few grid-points the function becomes *smooth*. It is important to realise that this feature persists even if the dissipation is entirely numerical. For larger distances (i.e. more gridpoints) the slope becomes less than two and eventually approaches a power law behaviour, characteristic of a rough velocity. So for example, in this case the range of scales over which the function becomes rough is of the order of 20-30 gridpoints. In the low Pm regime we should think of this size as the thickness of a magnetic boundary layer that is able to recognise the velocity as rough. Inspection of

Figure 9.2b gives an estimate of the size of the domain required to contain a growing eigenfunction in units of the magnetic boundary layer thickness. For a Kolmogorov-like velocity $1+\alpha = 1.333$ and so this is of the order of 30. This would give of the order 600–900 gridpoints simply to capture the magnetically growing eigenfunction. If one then wants a reasonable description of the velocity inertial range one probably requires at least a decade below this, giving a requirement in terms of gridpoints of several thousands (or tens of thousands). Possibly one could save a factor of ten by not matching the velocity inertial range to a resolved viscous sub-range by using LES. This is indeed the approach used by Ponty et al. (2008), but again we caution the reader that playing this game requires exceptional control of the LES, and even with the LES, simulations of the order of 1000^3 are unavoidable. Again, as machines get larger, the task of exploring this regime may not appear so daunting.

9.3 Mean field

We now turn our attention to the case in which there is an externally imposed magnetic field. As mentioned earlier, here the motivations are that in many astrophysically significant situations the regions of interest are threaded by a large scale magnetic field, or that we are interested in small scales whose dynamics is influenced by a magnetic field at larger scales. It is reasonable, and customary, to concentrate on the idealized case in which the large scale field is uniform. We denote the uniform background magnetic field by $\mathbf{B}_0$ and choose a coordinate frame with the z-axis along $\mathbf{B}_0$. The magnetic field is then given by $\mathbf{B}(\mathbf{x},t) = \mathbf{B}_0 + \mathbf{b}(\mathbf{x},t)$, where $\mathbf{b}(\mathbf{x},t)$ is the fluctuating part. Instead of the guide field $\mathbf{B}_0$ we will often use the Alfvén velocity $\mathbf{v}_A = \mathbf{B}_0/\sqrt{4\pi\rho_0}$, where ρ_0 is a uniform fluid density. We begin by discussing small perturbations to a uniform field; the magnetohydrodynamic equations support linear waves that propagate on the stationary and uniformly magnetized background. We seek wave-like solutions of the MHD equations for the velocity and magnetic fluctuations in the form $\mathbf{v}(\mathbf{x},t) = \mathbf{v}_k \exp(-i\omega t + i\mathbf{k}\cdot\mathbf{x})$ and $\mathbf{b}(\mathbf{x},t) = \mathbf{b}_k \exp(-i\omega t + i\mathbf{k}\cdot\mathbf{x})$. Recall that we are considering the incompressible case, $\mathbf{k}\cdot\mathbf{v}_k=0$, which rules out acoustic modes.

The remaining waves can be classified according to their polarizations. There are two so-called "shear-Alfvén" waves. Their dispersion relation is $\omega = |k_\parallel| v_A$, where $k_\parallel$ is the guide-field parallel component of the wavevector, and their polarization vector $\mathbf{b}_k$ is normal to both the wavevector $\mathbf{k}$ and the guide field $\mathbf{B}_0$ (i.e. $\mathbf{b}_k \cdot \mathbf{k} = \mathbf{b}_k \cdot \mathbf{B}_0 = 0$). If the group velocity of a shear-Alfvén wave is in the direction of the guide field, its velocity polarisation is anti-parallel to the magnetic polarisation, i.e. $\mathbf{v}_k = -\mathbf{b}_k$. If the group velocity of such a wave is in the direction opposite to the guide field, its velocity polarisation is parallel to the magnetic polarisation, i.e. $\mathbf{v}_k = \mathbf{b}_k$.

The other two modes are the so-called "pseudo-Alfvén" waves. Their dispersion relation is also $\omega = |k_\parallel| v_A$, however, the polarisations are different. The magnetic polarisation is normal to the wavevector (i.e. $\mathbf{b}_k \cdot \mathbf{k} = 0$) and lies in the plane of $\mathbf{k}$ and $\mathbf{B}_0$. If the group velocity of a pseudo-Alfvén wave is in the direction of the guide field, its velocity polarisation is anti-parallel to the magnetic polarisation, i.e., $\mathbf{v}_k = -\mathbf{b}_k$. If the group velocity of such a wave is in the direction opposite to the guide field, its velocity polarisation is parallel to the magnetic polarisation, i.e., $\mathbf{v}_k = \mathbf{b}_k$.

For those familiar with magnetoacoustic waves, it is helpful to note that the shear-Alfvén and pseudo-Alfvén linear waves are the limiting cases of the so-called "Alfvén" and "slow" waves, respectively, in the general picture of compressible MHD in the limit of low compressibility; that is $v_A \ll v_s$, where v_s is the speed of sound. An arbitrarily small perturbation of a uniformly magnetised fluid is a superposition of non-interacting MHD waves. In the purely incompressible case, however, both the shear-Alvfén and pseudo-Alfvén modes become rather significant. As we show in the next section, shear-Alfvén and pseudo-Alfvén wave packets even at finite amplitudes become *exact* solutions of the incompressible nonlinear MHD equations. Any perturbation of incompressible magnetized fluid can be expanded in these modes.

9.3.1 Formulation

The MHD equations take an especially simple form when written in the so-called Elsasser variables, $\mathbf{z}^\pm = \mathbf{v} \pm \mathbf{b}$,

$$\left(\frac{\partial}{\partial t} \mp \mathbf{v}_A \cdot \nabla\right) \mathbf{z}^\pm + \left(\mathbf{z}^\mp \cdot \nabla\right) \mathbf{z}^\pm = -\nabla P + \frac{1}{2}(\nu+\eta)\nabla^2 \mathbf{z}^\pm + \frac{1}{2}(\nu-\eta)\nabla^2 \mathbf{z}^\mp + \mathbf{f}^\pm, \tag{9.14}$$

where $\mathbf{v}$ is the fluctuating plasma velocity, $\mathbf{b}$ is the fluctuating magnetic field normalized by $\sqrt{4\pi\rho_0}$, $\mathbf{v}_A = \mathbf{B}_0/\sqrt{4\pi\rho_0}$ is the contribution from the uniform magnetic field $\mathbf{B}_0$, $P = (p/\rho_0 + b^2/2)$ includes the plasma pressure p and the magnetic pressure $b^2/2$, and the forces $\mathbf{f}^\pm$ mimic possible driving mechanisms. In the case of incompressible fluid, the pressure term is not an independent function, but it ensures incompressibility of the $\mathbf{z}^+$ and $\mathbf{z}^-$ fields. Turbulence can be excited in a number of ways, for example by driving velocity fluctuation (the "dynamo-type" driving), or by launching colliding Alfvén waves, etc. The steady-state inertial interval should be independent of the details of the driving.

In the incompressible case, it follows from the MHD equations that for

$\mathbf{z}^{\mp}(\mathbf{x},t) \equiv 0$, neglecting dissipation and driving, there is an exact nonlinear solution that represents a non-dispersive wave packet propagating along the direction $\mp\mathbf{v}_A$, i.e. $\mathbf{z}^{\pm}(\mathbf{x},t) = F^{\pm}(\mathbf{x}\pm\mathbf{v}_A t)$ where $F^{\pm}$ is an arbitrary function. A wave packet $\mathbf{z}^{\pm}$ therefore propagates without distortion until it reaches a region where $\mathbf{z}^{\mp}$ does not vanish. Nonlinear interactions are thus solely the result of collisions between counter-propagating Alfvén wave packets. These exact solutions may therefore be thought of as typical nonlinear structures in incompressible MHD turbulence, somewhat analogous to eddies in incompressible hydrodynamic turbulence.[5] Any perturbations of the velocity and magnetic fields can be rewritten as perturbations of the Elsasser fields. The conservation of energy and cross-helicity by the ideal incompressible MHD equations discussed earlier is equivalent to the conservation of the Elsasser energies $E^+ = \int (\mathbf{z}^+)^2 \, d^3x$ and $E^- = \int (\mathbf{z}^-)^2 \, d^3x$. The energies E^+ and E^- are independently conserved and they cascade in a turbulent state towards small scales owing to nonlinear interactions of oppositely moving z^+ and z^- Alfvén packets.

Incompressible MHD turbulence therefore consists of Alfvén wave packets, which are distorted and split into smaller ones owing to nonlinear interaction, until their energy is converted into heat by dissipation. However, when the amplitudes of the wave packets are small they can survive many independent interactions before getting destroyed, in which case MHD turbulence can be considered as an ensemble of weakly interacting linear waves.

To estimate the strength of the nonlinear interaction, we need to compare the linear terms, $(\mathbf{v}_A \cdot \nabla)\mathbf{z}^{\pm}$, which describe advection of Alfvén wave packets along the guide field, with the nonlinear terms, $(\mathbf{z}^{\mp} \cdot \nabla) \mathbf{z}^{\pm}$, which are responsible for the distortion of wave packets and for the energy redistribution over scales. Denote b_λ as the rms magnetic fluctuation at the field-perpendicular scale $\lambda \propto 1/k_\perp$, and assume that the typical field-parallel wavenumber of such fluctuations is $k_\parallel$. For Alfvén waves, magnetic and velocity fluctuations are of the same order, $v_\lambda \sim b_\lambda$, so one estimates $(\mathbf{v}_A \cdot \nabla)\mathbf{z}^{\pm} \sim v_A k_\parallel b_\lambda$ and $(\mathbf{z}^{\mp} \cdot \nabla) \mathbf{z}^{\pm} \sim k_\perp b_\lambda^2$. Then the turbulence is called "weak" when the linear terms dominate, i.e.

$$k_\parallel v_A \gg k_\perp b_\lambda, \qquad (9.15)$$

and it is called "strong" otherwise. Depending on the driving force, turbulence exhibits either a weak or strong regime in a certain range of scales.

---

[5] For other types of turbulence, e.g. rotating or stratified, there are no corresponding exact nonlinear solutions, and the interactions do not occur via collisions in this manner. Care must therefore be taken in applying the ideas and techniques of MHD turbulence more generally.

In the next section we describe early approaches to MHD turbulence that essentially treated the turbulence as weak.

9.3.2 Early Approaches

The first model of MHD turbulence was proposed by Iroshnikov (1963) and Kraichnan (1965) who invoked the picture of interacting Alfvén packets propagating along a strong large-scale magnetic field. Iroshnikov and Kraichnan considered two wave packets of field-perpendicular size λ propagating with the Alfvén velocity in opposite directions along a magnetic-field line. They further assumed that the wave packets are isotropic; that is, their field-parallel scale is also λ. Assuming that significant interaction occurs between eddies of comparable sizes[6], one then estimates from equations (9.14) that during one collision, that is one crossing time λ/v_A, the distortion of each wave packet is $\delta v_\lambda \sim (v_\lambda^2/\lambda)(\lambda/v_A)$. The distortions add up randomly, therefore, each wave packet will be distorted relatively strongly after $N \sim (v_\lambda/\delta v_\lambda)^2 \sim (v_A/v_\lambda)^2$ collisions with uncorrelated wave packets moving in the opposite direction. The energy transfer time is, therefore,

$$\tau_{IK}(\lambda) \sim N\lambda/v_A \sim \lambda v_A/v_\lambda^2 = \lambda/v_\lambda (v_A/v_\lambda). \tag{9.16}$$

It is important that in the Iroshnikov-Kraichnan interpretation, a wave packet has to experience *many* uncorrelated interactions with oppositely moving wave packets (since $\tau_{IK} \gg \lambda/v_A$) before its energy is transferred to a smaller scale. The requirement of constant energy flux over scales $\epsilon = v_\lambda^2/\tau_{IK}(\lambda) = \mathsf{const}$ immediately leads to the scaling of the fluctuating fields $v_\lambda \propto b_\lambda \propto \lambda^{1/4}$, which results in the energy spectrum (the Iroshnikov-Kraichnan spectrum),

$$E_{IK}(k) \sim |v_k|^2 4\pi k^2 \propto k^{-3/2}. \tag{9.17}$$

This expression follows from the fact that if the second-order structure function of the velocity field scales as $\lambda^{1/2}$, then its Fourier transform, the spectrum of the velocity field, scales as $|v_k|^2 \propto k^{-7/2}$. Substituting the derived scaling of the fluctuating fields into formula (9.15) one concludes that, since $k_\parallel \sim k_\perp \sim 1/\lambda$, turbulence becomes progressively *weaker* as the scale of the fluctuations decreases. Therefore, the Iroshnikov-Kraichnan picture is the picture of weak MHD turbulence.

Iroshnikov and Kraichnan did not consider anisotropy of the spectrum, so $E_{IK}(k)$ was assumed to be three-dimensional and isotropic. As discussed

---

[6] This is a standard assumption of "locality" of turbulence, see, e.g., (Aluie & Eyink (2010)).

earlier, over the years the assumption of isotropy was shown to be incorrect: even if large-scale isotropy is ensured by the driving force, it will break at small scales due to nonlinear interaction. The phenomenological picture of anisotropic weak MHD turbulence was proposed by Ng & Bhattacharjee (1996) and Goldreich & Sridhar (1997), and it was put on more formal mathematical ground by Galtier et al. (2000). This theory is discussed in the next section.

9.3.3 Weak turbulence

When the condition (9.15) is satisfied, one may assume that turbulence consists of weakly interacting shear-Alfvén and pseudo-Alfvén waves. The main results of the weak turbulence theory can be explained by using the following dimensional arguments, which were proposed by Ng & Bhattacharjee (1996) and Goldreich & Sridhar (1997) before the more rigorous treatment by Galtier et al. (2000) was developed. First, let us note that two Alfvén waves with wavevectors $\mathbf{k}_1$ and $\mathbf{k}_2$ interact with a third one if the following resonance conditions are satisfied: $\omega(k) = \omega(k_1) + \omega(k_2)$ and $\mathbf{k} = \mathbf{k}_1 + \mathbf{k}_2$. Only the waves propagating in opposite directions along the guide field interact with each other, therefore, $k_{1\|}$ and $k_{2\|}$ should have opposite signs. Since the Alfvén waves have a linear dispersion relation, $\omega = |k_\||v_A$, the solution of the resonance equations is nontrivial only if either $k_{1\|} = 0$ or $k_{2\|} = 0$ (Shebalin et al. (1983)). The turbulence dynamics therefore depend on the fluctuation spectrum at $k_\| = 0$. As the Alfvén waves interact with the fluctuations at $k_\| = 0$, the $k_\|$ components of their wavevectors do not change, and the energy of waves with given $k_\|$ cascade in the direction of large $k_\perp$.

In the region $k_\perp \gg k_\|$, the polarization of the pseudo-Alfvén waves is almost parallel to the guide field $\mathbf{B}_0$. Therefore, the pseudo-Alfvén modes ($\mathbf{z}_p$) influence the shear-Alfvén ones ($\mathbf{z}_s$) through the term $(\mathbf{z}_p^\pm \cdot \nabla)\mathbf{z}_s^\mp \sim z_p^\pm k_\| z_s^\mp$, whilst the shear-Alfvén modes, whose polarization is normal to the guide field, interact with each other as $(\mathbf{z}_s^\pm \cdot \nabla)\mathbf{z}_s^\mp \sim z_s^\pm k_\perp z_s^\mp$. Since $k_\|$ does not change in the energy cascade, in the inertial interval we have $k_\perp \gg k_\|$, and we obtain the important property of MHD turbulence that the dynamics of the shear-Alfvén modes decouple from the dynamics of the pseudo-Alfvén ones (Goldreich & Sridhar (1995)). Pseudo-Alfvén modes are passively advected by the shear-Alfvén ones. The spectrum of a passive scalar is the same as the spectrum of the advecting velocity field (Batchelor (1959)), therefore the spectra of pseudo- and shear-Alfvén waves should be the same.

These spectra can be derived from dimensional arguments similar to those of Iroshnikov and Kraichnan, if one takes into account the anisotropy of wave packets with respect to the guide field. As before, denote the field-perpendicular scale of the interacting wave packets as λ, and their field-parallel scale as l; but now the field-parallel scale l does not change during interactions. The collision time (crossing time) is now given by l/v_A. During one collision, the counter-propagating packets get deformed by $\delta v_\lambda \sim (v_\lambda^2/\lambda)(l/v_A)$. The number of uncorrelated interactions before the wave packets get destroyed is $N \sim (v_\lambda/\delta v_\lambda)^2 \sim \lambda^2 v_A^2/(l^2 v_\lambda^2)$, and the energy cascade time is $\tau_w \sim Nl/v_A$. From the condition of constant energy flux $\epsilon = v_\lambda^2/\tau_w = \text{const}$, one derives $v_\lambda \propto b_\lambda \propto \lambda^{1/2}$, and the field-perpendicular energy spectrum

$$E(k_\perp) \propto k_\perp^{-2}, \tag{9.18}$$

where the two-dimensional Fourier transform is used in the (x, y) plane in order to calculate the spectrum.

These results can be put on a more rigorous mathematical ground by using the approach of weak turbulence theory. The basic assumption of this theory is that in the absence of nonlinear interaction, the waves have random phases, and that the Gaussian rule can be applied to express their higher order correlation functions in terms of the second-order ones.[7] A perturbative theory of weak MHD turbulence has also been developed (Galtier et al. (2000), see also Galtier et al. (2002)). By expanding the Elsasser form of the MHD eq. (9.14) up to the second order in the nonlinear interaction and using the Gaussian rule to split the fourth-order correlators, the authors derived a closed system of kinetic equations governing the wave energy spectra[8].

These equations confirm the important fact that wave energy cascades in the Fourier space in the direction of large $k_\perp$, and the universal spectrum of wave turbulence is established in the region $k_\perp \gg k_\parallel$. The equations also demonstrate that in this region the dynamics of the shear-Alfvén waves decouple from the dynamics of the pseudo-Alfvén waves, and the pseudo-Alfvén waves are passively advected by the shear-Alfvén ones, in agreement with the previous qualitative consideration. The main objects of interest in this theory are the correlation functions of the shear-Alfvén

---

[7] Many papers contributed over the years to the development of fundamental ideas on MHD turbulence (see e.g., the reviews in Biskamp (2003); Ng et al. (2003)). The general methods of weak turbulence theory have been extensively reviewed (Zakharov et al. (1992); Newell et al. (2001)).

[8] It should be noted that the Gaussian assumption is, in fact, not necessary to derive the kinetic equatons. The same closure for the long time behavior of the spectral cumulants can be derived without a priori assumptions on the statistic of the process, e.g., Newel et al. (2001), Elskens & Escande (2003).

waves, $\langle \mathbf{z}^\pm(\mathbf{k},t)\mathbf{z}^\pm(\mathbf{k}',t)\rangle = e^\pm(\mathbf{k},t)\delta(\mathbf{k}+\mathbf{k}')$. In addition, it was assumed that the waves propagating in the opposite directions are not correlated, that is, $\langle \mathbf{z}^\pm(\mathbf{k},t)\mathbf{z}^\mp(\mathbf{k}',t)\rangle = 0$. The kinetic equations for the Elsasser energy spectra $e^\pm(\mathbf{k},t)$ then have the form:

$$\partial_t e^\pm(\mathbf{k}) = \int M_{\mathbf{k},\mathbf{pq}} e^\mp(\mathbf{q})[e^\pm(\mathbf{p}) - e^\pm(\mathbf{k})]\delta(q_\parallel)d_{\mathbf{k},\mathbf{pq}}, \qquad (9.19)$$

where the interaction kernel is given by

$$M_{\mathbf{k},\mathbf{pq}} = (\pi/v_A)(\mathbf{k}_\perp \times \mathbf{q}_\perp)^2(\mathbf{k}_\perp \cdot \mathbf{p}_\perp)^2/(k_\perp^2 q_\perp^2 p_\perp^2), \qquad (9.20)$$

and the shorthand notation $d_{\mathbf{k},\mathbf{pq}} \equiv \delta(\mathbf{k}-\mathbf{p}-\mathbf{q})\,d^3p\,d^3q$ is adopted. It is also customary to use the phase-volume compensated energy spectrum calculated as $E^\pm(\mathbf{k},t)dk_\parallel\,dk_\perp = e^\pm(\mathbf{k},t)2\pi k_\perp dk_\parallel\,dk_\perp$. In this section we consider only statistically balanced turbulence, that is, we assume $E^+ = E^-$. The balanced stationary solution of equation (9.19) was found (Galtier et al. (2000)) to have the general form $E^\pm(\mathbf{k}) = g(k_\parallel)k_\perp^{-2}$, where $g(k_\parallel)$ is an arbitrary function that is smooth at $k_\parallel = 0$.

It should be noted, however, that the derivation of (9.19) based on the weak interaction approximation is not rigorous. It follows from the wave resonance condition, and as is evident from equation (9.19), that only the $q_\parallel = 0$ components of the energy spectrum $e(\mathbf{q})$ are responsible for the energy transfer. However, if we apply equation (9.19) to these dynamically important components themselves, that is, if we set $k_\parallel = 0$ in (9.19), we observe an inconsistency. Indeed, the perturbative approach implies that the linear frequencies of the waves are much larger than the frequency of their nonlinear interaction. The nonlinear interaction in (9.19) remains nonzero as $k_\parallel \to 0$ while the linear frequency of the corresponding Alfvén modes, $\omega_k = k_\parallel v_A$, vanishes. Therefore, as shown by Galtier et al. (2000), an additional assumption that goes beyond the theory of weak turbulence, should be made. Namely, one has to assume the smoothness of the function $g(k_\parallel)$ at $k_\parallel = 0$; this assumption is crucial for deriving the spectrum $E(\mathbf{k}) \propto k_\perp^{-2}$ since, according to (9.19) the wave dynamics essentially depend on the energy spectrum at $k_\parallel \to 0$.

A definitive numerical verification of such a spectrum therefore seems desirable. It is however quite difficult to perform direct numerical simulations of weak MHD turbulence based on equation (9.14). The major problem faced by such simulations is to satisfy simultaneously the two conditions, $k_\perp \gg k_\parallel$, which is necessary to reach the universal regime where the dynamically unimportant pseudo-Alfvén mode decouples, and $k_\parallel B_0 \gg k_\perp b_\lambda$, which is

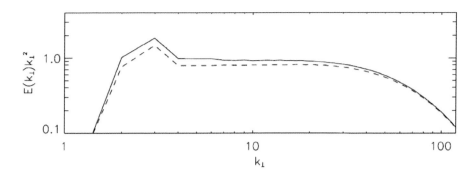

Figure 9.7 Compensated spectra of E^+ (solid line) and E^- (dash line) in numerical simulations of weak MHD turbulence based on reduced MHD equations, with numerical resolution $1024^2 \times 256$, and Reynolds number $Re = Rm = 6000$ (from Boldyrev & Perez (2009)).

the condition of weak turbulence. These two conditions are hard to achieve with present-day computing power.

The first numerical tests (Bhattacharjee & Ng (2001) and Ng et al. (2003)), used a scattering model based on MHD equations expanded up to the second order in nonlinear interaction rather than the full MHD equations. Integration of such a model did reproduce the '−2 exponent'. Recently, direct numerical simulations of the so-called reduced MHD (RMHD) equations (see later), which explicitly neglect pseudo-Alfvén fluctuations and present a good approximation of the full MHD equations in the case of strong guide field, were conducted (Perez & Boldyrev (2008); Boldyrev & Perez (2009)); more details on numerical simulations are given in Section (9.3.5). They also confirmed the −2 exponent, see Figure 9.7

Weak turbulence theory predicts that as $k_\perp$ increases, the turbulence should eventually become strong. Indeed, owing to the obtained scaling of turbulent fluctuations, $b_\lambda \propto \lambda^{1/2}$, the linear terms in equation (9.14) scale as $(\mathbf{v}_A \cdot \nabla)\mathbf{z}^\pm \sim v_A b_\lambda/l \propto \lambda^{1/2}$, whilst the nonlinear ones are independent of λ as $(\mathbf{z}^\mp \cdot \nabla)\mathbf{z}^\pm \sim b_\lambda^2/\lambda \propto \text{const}$. One therefore observes that the condition of weak turbulence (9.15) should break at small enough λ, which means that at large $k_\perp$ turbulence becomes strong. In the next section we describe strong MHD turbulence.

9.3.4 Strong turbulence

As noted at the end of the previous section, weak MHD turbulence eventually becomes strong as the $k_\perp$ increases so that condition (9.15) is no longer satisfied. If this is the case then one can argue that the linear and nonlinear terms should be approximately balanced at all scales; in equation (9.15) this would mean that "$k_\| v_A \sim k_\perp b_\lambda$". This is the so-called critical balance condition (Goldreich & Sridhar (1995)). Before discussing the consequences of critical balance, let us give its more precise definition, and discuss its physical meaning.

In contrast with weak turbulence, in strong turbulence the magnetic field lines are relatively strongly bent by velocity fluctuations. The wave packets following these lines in opposite directions are strongly distorted in only one interaction, that is, one crossing time. During such an interaction small wave packets are guided not by the mean field obtained through averaging over the whole turbulent domain, but rather by a local guide field whose direction is deviated from the direction of the mean field by larger wave packets (Cho & Vishniac (2000)). Therefore, it would be incorrect to verify the critical balance condition by using the wavenumber $k_\|$ defined through the Fourier transform in the global z-direction, as sometimes proposed in the literature. Rather, the critical balance condition means that the nonlinear interaction time λ/b_λ should be on the order of the linear crossing time, which can be represented as l/v_A with some length l along the *local* guide field, at all scales.

The physical meaning of the critical balance can then be understood from the following causality principle (Boldyrev (2005)). Suppose that owing to nonlinear interaction the wave packets are deformed on the time scale $\tau_N \sim \lambda/v_\lambda$. During this time, the information about the deformation cannot propagate along the guide field further than a distance $l \sim v_A \tau_N$, and the fluctuations cannot be correlated at a larger distance along the guide field. This coincides with the statement of the critical balance. We also note that the condition of critical balance in the GS picture has a useful geometric property. Noting that the individual magnetic field lines in an eddy of size λ are locally deviated by the small angle b_λ/v_A, one derives that the Alfvén wave packet of length l displaces the magnetic field lines in the field perpendicular direction by a distance $\xi \sim b_\lambda l/v_A$. In the GS picture this displacement happens to be equal to the perpendicular wave-packet size λ.

As a consequence of critical balance, oppositely moving Alfvén wave-packets are significantly deformed during only one interaction (one crossing time). This is in contrast with weak turbulence discussed earlier, where a large number of crossing times was required to deform a wave packet. The nonlinear interaction time therefore assumes the Kolmogorov form $\tau_{GS} \sim \lambda/v_\lambda$, and the energy spectrum attains the Kolmogorov scaling in the field-perpendicular direction (the Goldreich-Sridhar spectrum):

$$E_{GS}(k_\perp) \propto k_\perp^{-5/3}, \qquad (9.21)$$

where the spectrum is calculated by using the two-dimensional Fourier transform in the (x,y) plane, in analogy with the anisotropic weak turbulence spectrum (9.18). As an important consequence of critical balance, strong MHD turbulence becomes progressively more anisotropic at smaller scales. Indeed, with the GS scaling $v_\lambda \propto \lambda^{1/3}$ and the critical balance condition it follows that $l \propto \lambda^{2/3}$, and the "eddies" or wave packets get progressively elongated along the local guide field as their scale decreases.

Such scale-dependent anisotropy and the scaling (9.21) seemed to find some support in early numerical simulations, which did not have strong enough guide field, i.e. they had $v_A \sim v_{rms}$, and did not have large enough inertial interval, (see e.g., Cho & Vishniac (2000); Cho et al. (2002)). Recent high resolution direct numerical simulations with a strong guide field, $v_A \geq 5v_{rms}$, verify the strong anisotropy of the turbulent fluctuations, supporting the argument that the original IK picture is incorrect. However, they consistently reproduce the field-perpendicular energy spectrum as close to $E(k_\perp) \propto k_\perp^{-3/2}$, (see, e.g., Maron & Goldreich (2001); Müller et al. (2003); Müller & Grappin (2005), Mason et al. 2006, Mason et al. (2008); Perez & Boldyrev (2008)), thus contradicting the GS model, see Figure 9.8.

The flattening of the spectrum compared with the GS theory means that the energy transfer time becomes progressively longer than the Goldreich–Sridhar time $\tau_{GS}(\lambda)$ at small λ. This can happen if the nonlinear interaction in the MHD equations is depleted by some mechanism. A theory explaining such depletion of nonlinear interaction has recently been proposed (Boldyrev (2005, 2006)). In addition to the elongation of the eddies in the direction of the guiding field, it is proposed that at each field-perpendicular scale λ ($\sim 1/k_\perp$) in the inertial range, typical shear-Alfvén velocity fluctuations, $\mathbf{v}_\lambda$, and magnetic fluctuations, $\pm\mathbf{b}_\lambda$, tend to align the directions of their polarizations in the field-perpendicular plane, and the turbulent eddies become anisotropic in that plane. In these eddies the magnetic and velocity fluctuations change significantly in the direction almost perpendicular to the directions of the fluctuations themselves, $\mathbf{v}_\lambda$ and $\pm\mathbf{b}_\lambda$. This reduces the strength of the nonlinear interaction in the MHD equations by $\theta_\lambda \ll 1$, which is the angle between the direction of the fluctuations and the direction of the gradient: $(\mathbf{z}^\mp \cdot \nabla)\mathbf{z}^\pm \sim v_\lambda^2 \theta_\lambda / \lambda$. One can argue that the alignment and anisotropy are stronger for smaller scales, with the alignment angle decreasing with scale as $\theta_\lambda \propto \lambda^{1/4}$. This 'dynamic alignment' process progressively reduces the strength of the nonlinear interactions as the scale of the fluc-

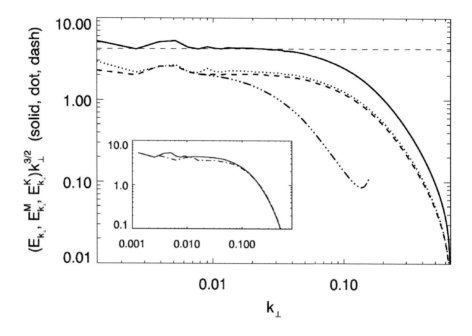

Figure 9.8 Compensated field-perpendicular total (solid), kinetic (dashed), and magnetic (dotted) energy spectra. The guide field is $B_0 = 5v_{rms}$, resolution $1024^2 \times 256$, Reynolds number $Re = Rm = 2300$ (based on field-perpendicular fluctuations). Dash-dotted curve: high-k part of field-parallel total energy spectrum. Insert: comparison of the field-perpendicular energy spectra for field-perpendicular resolutions of 512^2 (dash-dotted) and 1024^2 (solid) (from Müller & Grappin (2005)).

tuations decreases, which leads to the velocity and magnetic fluctuations $v_\lambda \sim b_\lambda \propto \lambda^{1/4}$, and to the field-perpendicular energy spectrum

$$E(k_\perp) \propto k_\perp^{-3/2}. \tag{9.22}$$

Dynamic alignment is a well-known phenomenon of MHD turbulence, (e.g., Dobrowolny *et al.* 1980, Biskamp (2003)), but the term has taken on a number of meanings. In early studies it essentially meant that *decaying* MHD turbulence asymptotically reaches the so-called Alfvénic state where either $\mathbf{v}(\mathbf{x}) \equiv \mathbf{b}(\mathbf{x})$ or $\mathbf{v}(\mathbf{x}) \equiv -\mathbf{b}(\mathbf{x})$, depending on initial conditions. Such behaviour is a consequence of cross-helicity conservation, (see e.g., Biskamp (2003)): energy decays faster than cross-helicity, and such selective decay leads asymptotically to alignment of magnetic and velocity fluctuations. Regions of polarized fluctuations have also been previously detected in numerical simulations of *driven* turbulence (Maron & Goldreich (2001)). The essense of the phenomenon that we discuss here is that in randomly driven

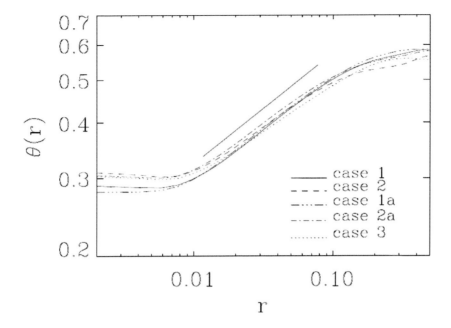

Figure 9.9 The alignment angle between shear-Alfvén velocity and magnetic fluctuations as a function of fluctuation scale r. The results are plotted for five independent simulations that differ by the large-scale driving mechanisms. The solid line has the slope of 1/4, predicted by the theory (from Mason et al. (2008)).

MHD turbulence the fluctuations $\mathbf{v}_\lambda$ and $\pm\mathbf{b}_\lambda$ tend to align their directions in such a way that the alignment angle becomes *scale-dependent* (Boldyrev (2005, 2006); Mason et al. (2006); Boldyrev et al. (2009)).

The first numerical observations of the scale-dependent dynamic alignment were presented in Mason et al. (2006) and Mason et al. (2008), (see also Beresnyak & Lazarian 2006). The results are shown in Figure 9.9. The effect was also observed in magnetized turbulence in the solar wind (Podesta et al. (2009)). It has been also argued (Boldyrev et al. 2009) that the phenomenon of scale-dependent dynamic alignment, and the corresponding energy spectrum $k_\perp^{-3/2}$ are consistent with the exact relations known for MHD turbulence (Politano & Pouquet (1998)).

As previously discussed, there are two possibilities for dynamic alignment: the velocity fluctuation $\mathbf{v}_\lambda$ can be aligned either with $\mathbf{b}_\lambda$ (positive alignment) or with $-\mathbf{b}_\lambda$ (negative alignment). This implies that the turbulent domain can be fragmented into regions (eddies) of positive and negative alignment. If no overall alignment is present, that is, the total cross-helicity of turbulence is zero, the numbers of positively and negatively aligned eddies are equal on average. In the case of nonzero total cross-helicity those numbers may be not equal. However, there is no essential difference between overall balanced (zero total cross-helicity) and imbalanced (non-zero total cross-helicity) strong turbulence (Perez & Boldyrev (2009)). Therein it is argued that strong MHD turbulence, whether balanced overall or not, has the characteristic property that at each scale it is locally imbalanced. Overall, it can be viewed as a superposition of positively and negatively aligned eddies. The scaling of the turbulent energy spectrum depends on the change in degree of alignment with scale, not on the amount of overall alignment.

9.3.5 Numerical frameworks

In the absence of a rigorous theory, the understanding of MHD turbulence largely relies on phenomenological arguments and numerical simulations. Numerical experiments are sometimes able to resolve controversies among different phenomenological assumptions. In this section we discuss the specific requirements for the numerical simulations imposed by the special features of MHD turbulence. In particular, the anisotropic nature of MHD turbulence requires the optimisation of the computation domain and the driving force in order to capture the dynamics of strongly anisotropic eddies. First, one needs to choose the forcing as to ensure that large-scale eddies, driven by the forcing, are critically balanced. Second, the simulation box should be expanded in the direction of the guide field (z-direction) so as to fit the anisotropic eddies, otherwise, residual box size effects can spoil the inertial interval if the Reynolds number if not large enough.

Numerical simulations, which have an inertial interval of a limited extent, demonstrate that when the guide field is not strong enough, $v_A \leq v_{rms}$, the observed turbulence spectrum is not very different from the Kolmogorov spectrum (e.g., Cho & Vishniac (2000); Cho et al. (2002); Mason et al. (2006)). As v_A exceeds $\sim 3v_{rms}$, the guide field becomes important and the spectral exponent changes to $-3/2$, (e.g., Müller et al. (2003), Mason et al. 2006). In the latter case, the simulation box should have the aspect ratio $L_\parallel/L_\perp \sim v_A/v_{rms}$. Current state-of-the-art high-resolution direct numerical simulations of MHD turbulence (Müller & Grappin (2005); Perez & Boldyrev (2009, 2010a)) discuss in detail the aforementioned effects of anisotropy related to the strong guide field.

We now demonstrate that anisotropic MHD turbulence can, in fact, be effectively simulated using simplified MHD equations. An obvious simplification stems from the fact that in the case of strong guide field the gradients of fluctuating fields are much smaller along the guide field than across the field. Such a system of equations, the so-called Reduced MHD (RMHD) equations, was derived in context of fusion devices (Strauss (1976); Kadomtsev & Pogutse (1974), see also Biskamp (2003)). In this system the following self-consistent ordering is used $L_\perp/L_\parallel \sim b_\perp/B_0 \sim b_\parallel/b_\perp \ll 1$, where $L_\parallel$ and $L_\perp$ are typical scales of magnetic perturbations, i.e., large scales of turbulent fluctuations in our case. As noticed by Goldreich & Sridhar (1995) this is precisely the ordering following from the critical balance condition for strong MHD turbulence. Therefore, the Reduced MHD system is suitable for studying strong turbulence. The Reduced MHD equations have the form:

$$\left(\tfrac{\partial}{\partial t} \mp \mathbf{v}_A \cdot \nabla\right) \tilde{\mathbf{z}}^\pm + (\tilde{\mathbf{z}}^\mp \cdot \nabla)\tilde{\mathbf{z}}^\pm = -\nabla_\perp P + \\ + \tfrac{1}{2}(\nu+\eta)\nabla^2 \tilde{\mathbf{z}}^\pm + \tfrac{1}{2}(\nu-\eta)\nabla^2 \tilde{\mathbf{z}}^\mp + \tilde{\mathbf{f}}_\perp, \qquad (9.23)$$

where the fluctuating fields and the force have only two vector components, e.g., $\tilde{\mathbf{z}}^\pm = \{\tilde{z}_1^\pm, \tilde{z}_2^\pm, 0\}$, but depend on all three spatial coordinates. Since both the velocity and the magnetic fields are divergence-free, they can be represented through the scalar potentials, $\mathbf{v} = \hat{\mathbf{z}} \times \nabla\phi$ and $\mathbf{b} = \hat{\mathbf{z}} \times \nabla\psi$, where $\hat{\mathbf{z}}$ is the unit vector in the z-direction. Then, the system (9.23) yields two scalar equations,

$$\partial_t \omega + (\mathbf{v}\cdot\nabla)\omega - (\mathbf{b}\cdot\nabla)j = B_0\partial_z j + \nu\nabla^2\omega + f_\omega, \\ \partial_t \psi + (\mathbf{v}\cdot\nabla)\psi = B_0\partial_z \phi + \eta\nabla^2\psi + f_\psi, \qquad (9.24)$$

where $j = \nabla_\perp^2 \psi$ is the current density and $\omega = \nabla_\perp^2 \phi$ is the vorticity. Note that it would be incorrect to suggest that system (9.24) describes quasi-two-dimensional turbulence, since the linear terms describing the field-parallel dynamics, that is, the terms containing $B_0\partial_z$, are of the same order as the nonlinear terms.

The system (9.24) is the best known form of the RMHD equations. The less common symmetric Elsasser form (9.23), on the other hand, has advantages for analytic study, for example, it allows one to derive relations analogous to those by Politano & Pouquet (1998) for the case of anisotropic MHD turbulence (see Perez & Boldyrev (2008)). Both full and RMHD systems can be used for numerical simulations of strong MHD turbulence, however, the reduced MHD system has half as many independent variables as the full MHD equations, allowing one to speed up the numerical computations by a factor of two.

Recently, it has been demonstrated that the system (9.23) can also be effectively used for numerical simulations of the universal regime of *weak* MHD turbulence (Perez & Boldyrev 2008). This may sound somewhat surprising and requires an explanation. As we discussed in Section 9.3.3, the weak turbulence spectrum becomes universal at $k_\perp \gg k_\|$, when the pseudo-Alfvén modes decouple from the cascade dynamics. Simultaneously, one needs to satisfy the condition of weak turbulence (9.15). In order to satisfy both conditions, one needs quite a low ratio of b/B_0, which implies quite a short Alfvén time and an increased computational cost. It turns out that the first condition can be relaxed if one uses the system where the pseudo-Alfvén modes are explicitly removed. This system is obtained from the full MHD system (9.14) if one sets $z_\|^\pm \equiv 0$. The resulting restricted system then formally coincides with (9.23), with the exception that it now can be used out of the region of validity of the RMHD equations, that is, $k_\perp \gg k_\|$. The restricted system explicitly neglects the pseudo-Alfvén modes, and, therefore, it describes the universal regime of weak MHD turbulence of shear-Alfvén waves as long as the driving force ensures the weak turbulence condition (9.15). The condition $k_\perp \gg k_\|$ is therefore not required.

To support this argument, one can demonstrate that the weak turbulence derivation based on the reduced system (9.23) leads to exactly the same kinetic equations (9.19), as the full MHD system (Galtier et al. (2002); Galtier & Chandran (2006), Perez & Boldyrev 2008, Galtier 2009). Moreover, direct numerical simulations of system (9.23) with the broad $k_\|$-spectrum of the driving force, necessary to ensure the weak turbulence condition (9.15), reproduce the energy spectrum of weak turbulence $k_\perp^{-2}$ (Perez & Boldyrev 2008), see Figure 9.7. To date, this has been the only available direct numerical study of the universal regime of weak MHD turbulence. We conclude that the reduced MHD system provides an effective framework for simulating the universal regimes of MHD turbulence, both weak and strong, providing the forcing is chosen accordingly. Depending on the $k_\|$ spectral width of the forcing, the driven turbulence is weak if the fluctuations satisfy the inequality (9.15), and it is strong if approximate equality holds in (9.15) instead. Further examples of numerical simulations of various regimes of MHD turbulence in the RMHD framework are available (e.g., Oughton et al. (2004); Rappazzo et al. (2007, 2008), Dmitruk et al. 2003).

It should be reiterated that even if the turbulence is driven in a "weak" fashion, it remains weak only for a certain range of scales, and it becomes strong below the field-perpendicular scales at which the condition (9.15) breaks down. With rapidly increasing capabilities of numerical simulations it should become possible to reproduce a large enough inertial interval, and to observe the predicted transition from weak to strong MHD turbulence.

9.3.6 Unbalanced turbulence

In the preceding sections we assumed that the MHD turbulence was statistically balanced, that is, the energies of counter-propagating Elsasser modes E^+ and E^- were the same. This, however, is by no means guaranteed, since the Elsasser energies are independently conserved by the ideal MHD equations. In nature and in the laboratory (e.g., the solar wind, interstellar medium, or fusion devices) MHD turbulence is typically unbalanced as it is generated by spatially localized sources or instabilities, so that the energy has a preferred direction of propagation. Moreover, there is a reason to believe that imbalance is an inherent fundamental property of MHD turbulence. As argued earlier, numerical simulations and phenomenological arguments indicate that even when the turbulence is balanced overall, it is unbalanced locally, creating patches of positive and negative cross-helicity (e.g., Matthaeus et al. (2008); Perez & Boldyrev (2009); Boldyrev et al. (2009)). We have already encountered this phenomenon when discussing dynamic alignment in Section 9.3.4. Unbalanced MHD turbulence is also called "cross-helical," as the normalized cross helicity, $H^C/E = \frac{1}{2}(E^+ - E^-)/(E^+ + E^-)$, is a natural measure of the imbalance.

In unbalanced MHD turbulence the energies of counter-propagating Elsasser modes are not equal, and, *a priori* might not have same scalings. This raises an interesting question as to whether the MHD turbulence is universal and scale-invariant. Indeed, if the energy spectra in the unbalanced domains are different, the overall energy spectrum does not have to be scale-invariant and universal, but rather may depend on the driving mechanism. Below we consider the cases of strong and weak unbalanced turbulence separately.

Strong unbalanced turbulence. Phenomenological treatments of strong unbalanced MHD turbulence are complicated by the fact one can formally construct two time scales for nonlinear energy transfer. The MHD system (9.14) suggests that the times of nonlinear deformation of the $z^\pm$ packets are $\tau^\pm \sim \lambda/z_\lambda^\mp$. These time scales are essentially different in the unbalanced case, which is hard to reconcile with the assumption that most effectively interacting counter-propagating eddies have comparable field-parallel and field-perpendicular dimensions, respectively. Several phenomenological models attempting to accommodate this time difference have been proposed recently. These, however, have led to conflicting predictions for the turbulent spectra of E^+ and E^- (Lithwick et al. (2007); Beresnyak & Lazarian (2008); Chandran (2008a); Perez & Boldyrev (2009)).

A possible resolution has been proposed (Perez & Boldyrev (2009)). Here the phenomenon of scale-dependent dynamic alignment was invoked to estimate the interaction times. In an unbalanced eddy, the field-perpendicular fluctuations of $\mathbf{v}_\lambda$ and $\mathbf{b}_\lambda$ are aligned in the field-perpendicular plane within a small angle θ_λ. In the phenomenology of scale-dependent dynamic alignment, these fluctuations are almost normal to the direction of their gradient, which is also in the field-perpendicular plane. In the case of strong imbalance, $z^+ \gg z^-$, $\mathbf{z}^+$ and $\mathbf{z}^-$ then form different angles with the direction of the gradient, and a geometric constraint requires that $z_\lambda^+ \theta_\lambda^+ \sim z_\lambda^- \theta_\lambda^-$. In the aligned case the nonlinear interaction time is increased by the alignment angle, $\tau^\pm \sim \lambda/z_\lambda^\mp \theta_\lambda^\mp$, leading to the conclusion that the nonlinear interaction times are the *same* for both z^+ and z^- packets. Therefore, z_λ^+ has the same scaling as z_λ^-, and θ_λ^+ has the same scaling as θ_λ^-.

As a result, in strong unbalanced MHD turbulence the spectra of the two sets of modes should have the same scaling, $E^+(k_\perp) \sim E^-(k_\perp) \propto k_\perp^{-3/2}$, but different amplitudes. This result is consistent with the picture that overall balanced MHD turbulence consists of regions of local imbalance of various strengths. In each of the unbalanced regions the fluctuations are dynamically aligned and the discussed phenomenology applies. The scaling of the spectrum of strong MHD turbulence is therefore universal, and it does not depend on the degree of overall imbalance of the turbulence. These results seem to be supported by numerical simulations (Perez & Boldyrev (2009, 2010a)).

To conclude this section we note that the above consideration allows one to predict the scalings of the Elsasser fields, however, it does not allow one to specify their amplitudes. Since the E^+ energy is mostly concentrated in positively aligned eddies, while E^- energy in negatively aligned ones, the amplitudes of the $E^\pm$ spectra averaged over the whole turbulent domain depend on the numbers of eddies of each kind. In particular, for each eddy one can estimate $(z_\lambda^+/z_\lambda^-)^2 \sim \epsilon^+/\epsilon^-$, however, this relation should not generally hold for the quantities $\langle (z_\lambda^\pm)^2 \rangle$ and $\langle \epsilon^\pm \rangle$ averaged over the whole turbulent domain (e.g., Perez & Boldyrev (2010b)). In principle, such an average may be not universal, but rather may depend on the structure of the large-scale driving. A possible refinement of the theory, which takes into account different populations of positively and negatively aligned eddies has been proposed (Podesta & Bhattacharjee (2010)).

Weak unbalanced turbulence. Weak unbalanced MHD turbulence allows for a more detailed consideration. A good starting point is provided by the equations (9.19) describing the evolution of the Elsasser energies (Galtier et al. (2000)). We recall that in deriving these equations one assumes that in the

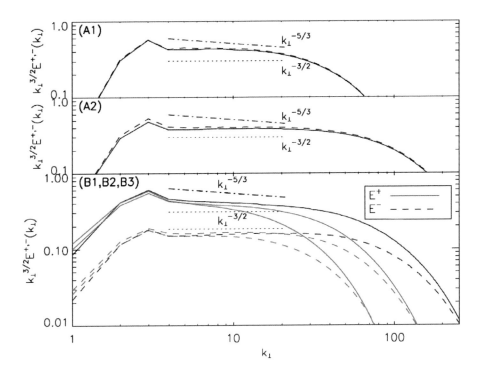

Figure 9.10 Compensated field-perpendicular spectra of strong balanced MHD turbulence (A1–A2), and strong unbalanced MHD turbulence (B1–B3) in numerical simulations of reduced MHD system. Runs A1, A2 have Reynolds numbers $Re = Rm = 2400$ and 6000, correspondingly. Runs B1, B2, and B3 have Reynolds numbers 900, 2200, and 5600. In runs B1–B3, the slopes of the Elsasser spectra become progressively more parallel and close to -3/2 as the Reynolds number increases (from Perez & Boldyrev (2010a)).

zeroth-order approximation, only the correlation functions $e^{\pm}(\mathbf{k}) \propto \langle \mathbf{z}^{\pm}(\mathbf{k}) \cdot \mathbf{z}^{\pm}(-\mathbf{k}) \rangle$ are non-zero. So far we have discussed the "balanced" solution of this system, $e^{+}(\mathbf{k}) = e^{-}(\mathbf{k}) \propto k_{\perp}^{-3}$. Galtier et al. (2000) realized that the system can also describe unbalanced MHD turbulence. They noticed that the system has a broader range of steady solutions: the right hand side integrals of equation (9.19) vanish for a one-parameter family of power-like solutions

$$e^{\pm}(\mathbf{k}) = g^{\pm}(k_{\parallel}) k_{\perp}^{-3\pm\alpha}, \qquad (9.25)$$

with arbitrary functions $g^{\pm}(k_{\parallel})$, which are smooth at $k_{\parallel} = 0$, and $-1 < \alpha < 1$. For $\alpha \neq 0$ these solutions correspond to unbalanced MHD turbulence. The different energy spectra correspond to different energy fluxes over scales, which in a steady state are equal to the rates of energy provided

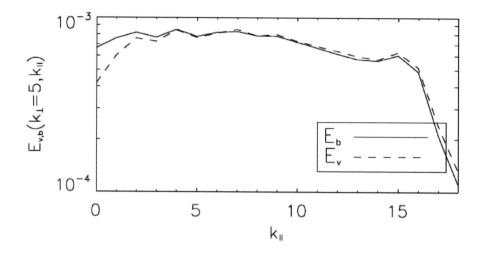

Figure 9.11 Field-parallel magnetic and kinetic energy spectra at given $k_\perp = 5$ in the inertial interval. The condensate of the residual energy appears at small $k_\parallel$. The simulations correspond to moderately imbalanced weak MHD turbulence $z^+/z^- \sim 2$, resolution $512^2 \times 256$, Reynolds number $Re = Rm = 2500$ (from Boldyrev & Perez (2009)).

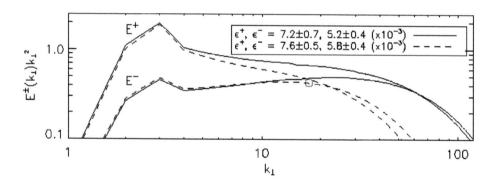

Figure 9.12 Compensated field-perpendicular spectra of the Elsasser fields in unbalanced weak MHD turbulence simulations based on reduced MHD equations. Resolution $1024^2 \times 256$, Reynolds numbers $Re = Rm = 2000$ and 4500. The spectra are anchored at large scales and pinned at the dissipation scales. As the Reynolds number increases, the spectral slopes should become progressively more parallel thus converging to -2 (from Boldyrev & Perez (2009)).

by the large-scale forcing to the Elsasser fields; we denote these rates as $\epsilon^\pm$. One can demonstrate directly from eq. (9.19) that α is uniquely determined if the ratio of the fluxes ϵ^+/ϵ^- is specified. One can further show that the solution with the steeper spectrum corresponds to the larger energy flux, and vice versa (Galtier et al. (2000); Lithwick & Goldreich (2003)).

Once the energy fluxes provided by the large-scale forcing are specified, the slopes of the energy spectra are specified but their amplitudes are not. Fully to remove the degeneracy, one further argues that at the dissipation scale the balance should be restored, that is, $e^+(k)$ should converge to $e^-(k)$. This so-called "pinning" effect was first pointed out in Grappin et al (1983), and has been elaborated upon in greater detail (see e.g. Galtier et al. (2000); Lithwick & Goldreich (2003); Chandran (2008a)). The pinning can be physically understood, if one notes that the $e^+(k)$ and $e^-(k)$ energy spectra are not expected to intersect in the inertial interval, as this would contradict the universality of the turbulence. The alignment of velocity and magnetic fluctuations, preserved by the nonlinear terms, can be broken only by the dissipation. This pinning effect is indeed observed in the simulations presented below.

According to the above picture, if the rates of energy supply are fixed, then the *slopes* of the energy spectra $e^\pm(k)$ are fixed as well. If the dissipation scale is now changed, the *amplitudes* of the spectra should change so as to maintain the specified slopes and to make them converge at the dissipation scale. This conclusion, although consistent with equations (9.19), seems however to be at odds with common intuition about turbulent systems, which suggests that small-scale dissipation should not significantly affect large-scale fields subject to the same large-scale driving.

This seeming contradiction was recently addressed (Boldyrev & Perez (2009)). It was proposed that driven weak MHD turbulence generates a residual energy condensate

$$\langle \mathbf{z}^+(\mathbf{k}) \cdot \mathbf{z}^-(-\mathbf{k}) \rangle = v^2 - b^2 \neq 0 \quad \text{at } k_\parallel = 0. \tag{9.26}$$

This condensate has been assumed to be zero in the standard derivations (Ng & Bhattacharjee 1996, Goldreich & Sridhar (1997); Galtier et al. (2000)), however, this assumption is not necessary and apparently incorrect, which has especially important consequences for the unbalanced case. The presence of the condensate can be explained as follows. Alfvén wave fluctuations at $k_\parallel \neq 0$ obey $\mathbf{v} = \pm \mathbf{b}$, in which case the residual energy vanishes. However, at $k_\parallel = 0$ fluctuations are not waves, and the Alfvénic condition should not be necessarily satisfied. The presence of the condensate (9.26) implies that magnetic and kinetic energies are not in equipartition at $k_\parallel = 0$.

The generation of the condensate may be a consequence of the breakdown of the mirror symmetry in unbalanced turbulence. Indeed, non-balanced MHD turbulence is not mirror-invariant, as it possesses non-zero cross-helicity, i.e. $H^c = \int (\mathbf{v} \cdot \mathbf{b}) d^3x = (E^+ - E^-)/4 \neq 0$. Non-mirror invariant turbulence can also possess nonzero magnetic helicity, $H^m = \int (\mathbf{A} \cdot \mathbf{B}) d^3x \neq 0$.

Helical magnetic fields should not necessarily be in equipartition with the velocity field, as is obvious from the fact that the corresponding Lorentz force may be equal to zero (consider, for example, a Beltrami field $\mathbf{B} = \alpha \nabla \times \mathbf{B}$, which is maximally helical, but exerts no force on the velocity field). In the case of weak MHD turbulence, the only region of phase space where the equipartition between the magnetic and velocity fluctuations may be broken is the region $k_\parallel = 0$, where such fluctuations are not Alfvén waves. In the case of a uniform guide field $\mathbf{B}_0$, which is relevant for the numerical set-up, these arguments can have further support if one notes that the magnetic helicity of *fluctuations* is not conserved, rather, it is generated by the term $\int [\mathbf{z}^+ \times \mathbf{z}^-]_\parallel d^3x$, (see e.g., Galtier et al. (2000)). Interestingly, the magnetic field of the condensate is generated by the $k_\parallel = 0$ component of the same term, $[\mathbf{z}^+ \times \mathbf{z}^-]_\parallel$. Such a condensate is indeed observed in the numerics, see Figure 9.11.

The conclusion of this analysis are the following. When the turbulence is balanced, the energy spectra are $E^\pm(k_\perp) \propto k_\perp^{-2}$, in agreement with the analytic prediction (Ng & Bhattacharjee 1996, Goldreich & Sridhar (1997); Galtier et al. (2000)). In the balanced case the evolution of $E^\pm$ fields is not affected by the condensate. In the unbalanced case the interaction with the condensate becomes essential, and the universal power-law spectra may not exist in an inertial interval of limited extent. Both spectra $E^\pm(k_\perp)$ have the large-scale amplitudes fully specified by the external forcing, and they converge at the dissipation scale. As the dissipation scale decreases, however, the spectral scalings (but not amplitudes) approach each other at large $k_\perp$ such that the universal scaling $k_\perp^{-2}$ is recovered for both the $E^\pm(k_\perp)$ spectra as $k_\perp \to \infty$, see Figure 9.12.

Finally, we would like to point out that unbalanced MHD turbulence is numerically more challenging than balanced turbulence. Indeed, in the case $z^+ \gg z^-$ the formal Reynolds number associated with the z^+ mode is much smaller than the Reynolds number associated with z^-. The stronger the imbalance the larger the Reynolds number and the resolution required to reproduce the inertial intervals for both fields. For example, numerical simulations with the resolution of 1024^2 points in the field-perpendicular direction do not allow one to address turbulence with imbalance stronger than $z^+/z^- \sim 2\text{--}3$, (see e.g., Perez & Boldyrev (2010a)).

9.4 Conclusions

The last two decades have seen many developments in the field of MHD turbulence. There is little doubt that the driving force behind many of these has been the remarkable increase in computing power. In a field in which it so difficult to carry out experiments or detailed observations, researchers have relied almost entirely on numerical simulations for the underpinning of their theoretical speculations and modelling. It is now a matter of routine to carry out simulations with in excess of one billion grid-points; and the expectation is that in the near future simulations with 10^{12} grid-points will become feasible. With these resolutions it is possible to explore fully three-dimensional configurations with moderately high Reynolds numbers. In other words, the turbulent regime has become available to computations. Given these advances and these possibilities, it is natural to conclude this review with some assessment of what has been learned – and is now considered fairly certain – and what still remains unclear or controversial, and will probably occupy the minds of researchers, and the CPU's of their computers, in the years to come.

Within the framework of dynamo theory it is now widely accepted that given a random velocity, dynamo action is always possible provided the magnetic Reynolds number is large enough. In the case of reflectionally symmetric turbulence (i.e. non-helical) the effort required to drive a dynamo appears to be determined by the slope of the energy spectrum of the velocity at spatial scales comparable to those on which magnetic reconnection occurs. The flatter the spectrum the harder it is to drive a dynamo. If this property is expressed in terms of the velocity roughness exponent, a marked difference emerges between the mechanisms leading to dynamo action for a smooth velocity as opposed to those in which the velocity is rough. This distinction between smooth and rough velocities then maps very naturally in two distinct regimes corresponding to large and small values of the magnetic Prandtl number. What is also clear now, is that this stochastic theory must be modified if coherent structures are present in the turbulence. These may, depending on their properties, take over the control of the dynamo growth rate. We are now only beginning to understand which properties of coherent structures contribute to the dynamo process, and certainly this will be an active area of future research. In the nonlinear regime, there are now compelling models for the saturation process leading to the establishment of a stationary turbulent state for dynamos at large Magnetic Prandtl number. These models, in general, rely on some specific geometrical property of the magnetic field, like foliation, and seem to be borne out by the existing simulations. It remains to be seen if the predictions of these, mostly phenomenological models, continue to hold at higher values of Rm. By contrast, the low magnetic Prandtl number regime remains largely unexplored. The difficulties are both conceptual, since the velocity correlator has a non-analytic behaviour at the reconnection (or resistive dissipation) scales, and numerical, since reproducing a wide separation of scales between the magnetic and velocity boundary layers is extremely computationally expensive; in this review we have determined the size of the calculation required to begin to settle the issue. The general expectation is, however, that the turbulence should become independent of of the magnetic Prandtl number provided the latter is small enough. The existence of this asymptotic regime remains conjectural pending the availability of either much bigger computers, or much better analytical approaches.

When the flows are affected by the large scale component of the magnetic field, irrespective of whether the latter is self generated or imposed externally, the techniques utilised in analysing the turbulence are different. Two important ideas have emerged that have strongly influenced our understanding of the turbulent process in this case. The first is that depending on the dominant mechanism responsible for energy transfer MHD turbulence can be either weak or strong. In the weak case, mean-fluctuation interactions dominate, whereas in the strong case, mean-fluctuation and fluctuation-fluctuation interactions are comparable. The common wisdom is that the strength of the fluctuation-fluctuation interactions increases at small scales, so that if the inertial range is sufficiently extended there will always be a transition from weak to strong turbulence. At present there are no cases in which this transition has been convincingly demonstrated numerically. The problem is associated with difficulties in reproducing a deep inertial range. The situation will no doubt improve with the next generation of supercomputers. The second important idea is that MHD turbulence, unlike its unmagnetised counterpart, has a richly geometrical structure. There is now universal agreement that the energy cascade is anisotropic with most of the energy being transferred transverse to the (local) mean field. The exact geometry of this process is controlled by the requirement of a "critical balance" between the crossing time of counter propagating wavepackets and the time for nonlinear transfer in the field-perpendicular direction. Furthermore, there are strong indications, both analytical and numerical, that the polarization vectors of magnetic and velocity fluctuations tend to align in the transverse plane. Although this phenomenon is intimately connected with the forward cascade of cross-helicity it has measurable consequences for the slope of the energy spectrum. This appears shallower than predicted in the absence of alignment. Both the elongation of the wavepackets and the degree of alignment increase at small scales, and hence the picture that emerges of MHD turbulence in the deep inertial range is that of a collection of interacting ribbon-like structures. Another problem that has recently emerged, and is at the moment the focus of considerable interest, is that of unbalanced cascades. This occurs when there is an excess of wavepackets propagating in one direction relative to those propagating in the opposite direction. The basic question here is whether unbalanced MHD turbulence is fundamentally different from balanced turbulence, or whether both are the same, with even balanced turbulence being made up of interwoven unbalanced patches that appear balanced only on average. Again, a definite resolution of these issues may have to wait for a further increase in computing resources.

Finally, we should remark about the areas of active research in MHD turbulence that are not discussed in this review. In dynamo theory, the most important problem remains that of the generation of large scale fields by turbulence lacking reflectional symmetry. At the moment, there appears to be a fracture between the conventional wisdom, mostly grounded in mean-field theory, and an increasing body of numerical experiments. It is possible that large-scale dynamos are "essentially nonlinear" rather than "essentially kinematic" and a major revision of the theory might be in order. It is also possible that the whole categorisation of dynamos into large and small-scale is misleading and that it is more useful to seek "system-scale" dynamo solutions to the MHD equations (Tobias *et al.* (2011)). In general, the lack of reflectional symmetry in astrophysical turbulence owes its origins to the presence of rotation. Strong rotation and shear flows are known to modify the nature of MHD turbulence and indeed may lead to the generation of new instabilties (Chandrasekhar (1961); Balbus & Hawley (1991)). Such magnetorotational turbulence is of great interest owing to its importance for the accretion process, and the nature of the turbulence may be very different to that discussed in this review (see e.g. Balbus & Hawley (1998) and the references therein).

For reflectionally symmetric turbulence, the most natural extension of what has been presented is to incorporate compressibility. This bring in one more set of linear waves – the fast magnetosonic waves – and modifies the nature of the pseudo-Alfvén waves. The resulting turbulence is extremely rich and is only now beginning to be explored in the nonlinear regime (see e.g. Beresnyak *et al.* (2005); Li *et al.* (2008); Chandran (2008b)). Finally, we should note that all of our considerations have been discussed within the framework of classical MHD. There is an almost bewildering variety of new effects that become important when the particular nature of a plasma is taken into account (see e.g. Kulsrud (2005); Howes *et al.* (2006)). Again, the full richness of this system will be a major preoccupation in the near future.

References

Aluie, H. and Eyink, G. L. 2010 Scale locality of magnetohydrodynamic turbulence. *Physical Review Letters* **104**, 081101.

Arponen, H. & Horvai, P. 2007 Dynamo effect in the Kraichnan magnetohydrodynamic turbulence. *Journal of Statistical Physics* **129**, 205–239.

Balbus, S. A. & Hawley, J. F. 1991 A powerful local shear instability in weakly magnetized disks. I – Linear analysis. II – Nonlinear evolution. *Astrophysical Journal* **376**, 214–233.

Balbus, S. A. & Hawley, J. F. 1998 Instability, turbulence, and enhanced transport in accretion disks. *Reviews of Modern Physics* **70**, 1–53.

Balmforth, N. J. & Korycansky, D. G. 2001 Non-linear dynamics of the corotation torque. *Mon. Not. Roy. Ast. Soc.* **326**, 833–851.

Batchelor, G. K. 1953 *The Theory of Homogeneous Turbulence*. Cambridge: Cambridge University Press.

Batchelor, G. K. 1959 Small-scale variation of convected quantities like temperature in turbulent fluid. Part 1. General discussion and the case of small conductivity. *Journal of Fluid Mechanics* **5**, 113–133.

Beresnyak, A. & Lazarian, A. 2006 Polarization intermittency and its influence on MHD turbulence. *Astrophysical Journal Letters* **640**, L175–L178.

Beresnyak, A. & Lazarian, A. 2008 Strong imbalanced turbulence. *Astrophys. J.* **682**, 1070–1075.

Beresnyak, A., Lazarian, A. & Cho, J. 2005 Density scaling and anisotropy in supersonic magnetohydrodynamic turbulence. *Astrophysical Journal Letters* **624**, L93–L96.

Berger, M. & Rosner, R. 1995 The evolution of helicity in the presence of turbulence. *Geophysical and Astrophysical Fluid Dynamics* **81**, 73–99.

Bhattacharjee, A. & Ng, C. S. 2001 Random scattering and anisotropic turbulence of shear Alfvén wave packets. *Astrophys. J.* **548**, 318–322.

Bigot, B., Galtier, S. & Politano, H. 2008 Development of anisotropy in incompressible magnetohydrodynamic turbulence. *Physical Review E* **78** 066301.

Biskamp, D. 2003 *Magnetohydrodynamic Turbulence*. Cambridge: Cambridge University Press.

Boldyrev, S. 2001 A solvable model for nonlinear mean field dynamo. *Astrophysical Journal* **562**, 1081–1085.

Boldyrev, S. 2005 On the spectrum of magnetohydrodynamic turbulence. *Astrophys. J. Lett.* **626**, L37–L40.

Boldyrev, S. 2006 Spectrum of magnetohydrodynamic turbulence. *Physical Review Letters* **96**, 115002.

Boldyrev, S. & Cattaneo, F. 2004 Magnetic field generation in Kolmogorov turbulence. *Physical Review Letters* **92**, 144501.

Boldyrev, S. & Perez, J. C. 2009 Spectrum of weak magnetohydrodynamic turbulence. *Physical Review Letters* **103**, 225001.

Boldyrev, S., Cattaneo, F. & Rosner, R. 2005 Magnetic field generation in helical turbulence. *Physical Review Letters* **95**, 255001.

Boldyrev, S., Mason, J. & Cattaneo, F. 2009 Dynamic alignment and exact scaling laws in magnetohydrodynamic turbulence. *Astrophys. J. Lett.* **699**, L39–L42.

Bracco, A., Chavanis, P. H., Provenzale, A. & Spiegel, E. A. 1999 Particle aggregation in a turbulent Keplerian flow. *Physics of Fluids* **11**, 2280–2287.

Brandenburg, A. & Nordlund, A. 2010 Astrophysical turbulence. *Reports on Progress in Physics* **74**, 046901.

Brandenburg, A. & Subramanian, K. 2005 Astrophysical magnetic fields and nonlinear dynamo theory. *Physics Reports* **417**, 1–209.

Busse, F. H. 2002 Convective flows in rapidly rotating spheres and their dynamo action. *Physics of Fluids* **14**, 1301–1314.

Carati, D., Debliquy, O., Knaepen, B., Teaca, B. & Verma, M. 2006 Energy transfers in forced MHD turbulence. *Journal of Turbulence* **7**, 51.

Cattaneo, F. 1999 On the origin of magnetic fields in the quiet Photosphere. *Astrophys. J. lett.* **515**, L39–L42.

Cattaneo, F. & Hughes, D. W. 2009 Problems with kinematic mean field electrodynamics at high magnetic Reynolds numbers. *Mon. Not. Roy. Ast. Soc.* **395**, L48–L51.

Cattaneo, F. & Tobias, S. M. 2005 Interaction between dynamos at different scales. *Phys. Fluids* **17**, 127105.

Cattaneo, F. & Tobias, S. M. 2009 Dynamo properties of the turbulent velocity field of a saturated dynamo, *JFM* **621**, 205–214.

Chandran, B. D. G. 2008a Strong Anisotropic MHD Turbulence with Cross Helicity. *Astrophys. J.* **685**, 646–658.

Chandran, B. D. G. 2008b Weakly Turbulent Magnetohydrodynamic Waves in Compressible Low-β Plasmas. *Physical Review Letters* **101** 235004.

Chandrasekhar, S. 1961 *Hydrodynamic and Hydromagnetic Stability*. International Series of Monographs on Physics, Oxford: Clarendon Press.

Chertkov, M., Falkovich, G., Kolokolov, I. & Vergassola, M. 1999 Small-scale turbulent dynamo. *Physical Review Letters* **83**, 4065–4068.

Childress, S. & Gilbert, A. D. 1995 *Stretch, Twist, Fold: The Fast Dynamo*. Berlin: Springer-Verlag.

Cho, J., Lazarian, A. & Vishniac, E. T. 2002 Simulations of magnetohydrodynamic turbulence in a strongly magnetized medium. *Astrophys. J.* **564**, 291–301.

Cho, J. & Vishniac, E. T. 2000 The anisotropy of Magnetohydrodynamic alfvénic turbulence. *Astrophys. J.* **539**, 273–282.

Christensen, U., Olson, P. & Glatzmaier, G. A. 1999 Numerical modelling of the geodynamo: a systematic parameter study. *Geophysical Journal International* **138**, 393–409.

Cowling, T. G. 1933 The magnetic field of sunspots. *Mon. Not. Roy. Ast. Soc.* **94**, 39–48.

Cowling, T. G. 1976 *Magnetohydrodynamics.*. Bristol: Adam Hilger.

Davidson, P. A. 2004 *Turbulence : an Introduction for Scientists and Engineers*. Oxford: Oxford University Press.

Dmitruk, P., Gómez, D. O. & Matthaeus, W. H. 2003 Energy spectrum of turbulent fluctuations in boundary driven reduced magnetohydrodynamics. *Physics of Plasmas* **10**, 3584–3591.

Dobrowolny, M., Mangeney, A. & Veltri, P. 1980 Fully developed anisotropic hydromagnetic turbulence in interplanetary space. *Physical Review Letters* **45**, 144–147.

Dormy, E. & Soward, A. M., ed. 2007 *Mathematical Aspects of Natural Dynamos*. CRC Press/Taylor & Francis.

Elskens, Y. & Escande, D. F. 2003 *Microscopic Dynamics of Plasmas and Chaos*. Bristol: Institute of Physics Publishing.

Finn, J. M. & Ott, E. 1988 Chaotic flows and fast magnetic dynamos. *Physics of Fluids* **31**, 2992–3011.

Frick, P. & Sokoloff, D. 1998 Cascade and dynamo action in a shell model of magnetohydrodynamic turbulence. *Physical Review E* **57**, 4155–4164.

Frisch, U. 1995 *Turbulence: the Legacy of A. N. Kolmogorov*. Cambridge: Cambridge University Press.

Galtier, S. 2008 Magnetohydrodynamic turbulence. In *SF2A-2008* ed. C. Charbonnel, F. Combes, & R. Samadi, pp. 313–326.

Galtier, S. 2009 Wave turbulence in magnetized plasmas. *Nonlinear Processes in Geophysics* **16**, 83–98.

Galtier, S. & Chandran, B. D. G. 2006 Extended spectral scaling laws for shear Alfvén wave turbulence. *Physics of Plasmas* **13** 11, 114505.

Galtier, S., Nazarenko, S. V., Newell, A. C. & Pouquet, A. 2000 A weak turbulence theory for incompressible magnetohydrodynamics. *Journal of Plasma Physics* **63**, 447–488.

Galtier, S., Nazarenko, S. V., Newell, A. C. & Pouquet, A. 2002 Anisotropic turbulence of shear Alfvén waves. *Astrophys. J. Lett.* **564**, L49–L52.

Godon, P. & Livio, M. 2000 The formation and role of vortices in protoplanetary disks. *Astrophys. J.* **537**, 396–404.

Goldreich, P. & Sridhar, S. 1995 Toward a theory of interstellar turbulence. 2: Strong alfvenic turbulence. *Astrophys. J.* **438**, 763–775.

Goldreich, P. & Sridhar, S. 1997 Magnetohydrodynamic turbulence revisited. *Astrophys. J.* **485**, 680–688.

Grappin, R., Leorat, J. & Pouquet, A. 1983 Dependence of MHD turbulence spectra on the velocity field-magnetic field correlation. *Astron. Astrophys.* **126**, 51–58.

Haugen, N. E., Brandenburg, A. & Dobler, W. 2004 Simulations of nonhelical hydromagnetic turbulence. *Phys. Rev. E* **70**, 016308.

Howes, G. G., Cowley, S. C., Dorland, W., Hammett, G. W., Quataert, E. & Schekochihin, A. A.2006 Astrophysical gyrokinetics: basic equations and linear theory. *Astrophysical Journal* **651**, 590–614

Iroshnikov, P. S. 1963 Turbulence of a conducting fluid in a strong magnetic field. *Astronomicheskii Zhurnal* **40**, 742–750.

Iskakov, A. B., Schekochihin, A. A., Cowley, S. C., McWilliams, J. C. & Proctor, M. R. E. 2007 Numerical demonstration of fluctuation dynamo at low magnetic Prandtl numbers. *Physical Review Letters* **98** 20, 208501.

Kadomtsev, B. B. & Pogutse, O. P. 1974 Nonlinear helical perturbations of a plasma in the tokamak. *Soviet Journal of Experimental and Theoretical Physics* **38**, 283.

Kazantsev, A. P. 1968 Enhancement of a magnetic field by a conducting fluid. *Soviet Journal of Experimental and Theoretical Physics* **26**, 1031–1034.

Kim, E. & Hughes, D. W. 1997 Flow helicity in a simplified statistical model of a fast dynamo. *Physics Letters A* **236**, 211–218.

Klapper, I. & Young, L. S. 1995 Rigorous bounds on the fast dynamo growth-rate involving topological entropy. *Comm. Math. Phys.* **175**, 623–646.

Kleeorin, N. & Rogachevskii, I. 1997 Intermittency and anomalous scaling for magnetic fluctuations. *Phys. Rev. E* **56** 417.

Kleeorin, N., Rogachevskii, I. & Sokoloff, D. 2002 Magnetic fluctuations with a zero mean field in a random fluid flow with a finite correlation time and a small magnetic diffusion. *Phys. Rev. E* **65**, 036303.

Kraichnan, R. H. 1965 Inertial-Range Spectrum of Hydromagnetic Turbulence. *Physics of Fluids* **8**, 1385–1387.

Krause, F. & Raedler, K. 1980 *Mean-Field Magnetohydrodynamics and Dynamo Theory*. Oxford, Pergamon Press, Ltd., 1980.

Kulsrud, R. M. 2005 *Plasma Physics for Astrophysics*. Princeton: Princeton University Press.

Kulsrud, R. M. & Anderson, S. W. 1992 The spectrum of random magnetic fields in the mean field dynamo theory of the galactic magnetic field. *Astrophysical Journal* **396**, 606–630.

Kulsrud, R. M. & Zweibel, E. G. 2008 On the origin of cosmic magnetic fields. *Reports on Progress in Physics* **71**, 046901.

Lazarian, A. 2006 Intermittency of magnetohydrodynamic turbulence: an astrophysical perspective. *International Journal of Modern Physics D* **15**, 1099–1111.

Lazarian, A. & Cho, J. 2005 Scaling, Intermittency and decay of MHD turbulence. *Physica Scripta* **T116**, 32–37.

Li, P. S., McKee, C. F., Klein, R. I. & Fisher, R. T. 2008 Sub-Alfvénic nonideal MHD turbulence simulations with ambipolar diffusion. I. Turbulence statistics. *Astrophysical Journal* **684**, 380–394.

Lithwick, Y. & Goldreich, P. 2003 Imbalanced Weak Magnetohydrodynamic Turbulence. *Astrophys. J.* **582**, 1220–1240.

Lithwick, Y., Goldreich, P. & Sridhar, S. 2007 Imbalanced Strong MHD Turbulence. *Astrophys. J.* **655**, 269–274.

Malyshkin, L. & Boldyrev, S. 2007 Magnetic dynamo action in helical turbulence. *Astrophysical Journal Letters* **671**, L185–L188.

Malyshkin, L. & Boldyrev, S. 2008 Amplification of magnetic fields by dynamo action in Gaussian-correlated helical turbulence. *Physica Scripta* **T132**, 014028.

Malyshkin, L. & Boldyrev, S. 2009 On magnetic dynamo action in astrophysical turbulence. *Astrophys. J.* **697**, 1433–1438.

Malyshkin, L. & Boldyrev, S. 2010 Magnetic Dynamo action at Low Magnetic Prandtl Numbers. *Physical Review Lett.* **105**, 215002

Maron, J. & Goldreich, P. 2001 Simulations of Incompressible Magnetohydrodynamic Turbulence. *Astrophys. J.* **554**, 1175–1196.

Maron, J., Cowley, S. & McWilliams, J. 2004 The Nonlinear Magnetic Cascade. *Astrophysical Journal* **603**, 569–583.

Mason, J., Cattaneo, F. & Boldyrev, S. 2006 Dynamic Alignment in Driven Magnetohydrodynamic Turbulence. *Physical Review Letters* **97**, 255002.

Mason, J., Cattaneo, F. & Boldyrev, S. 2008 Numerical measurements of the spectrum in magnetohydrodynamic turbulence. *Physical Review E* **77**, 036403.

Mason, J., Malyshkin, L., Boldyrev, S. & Cattaneo, F. 2011 Magnetic Dynamo Action in Random Flows with Zero and Finite Correlation Times. *Astrophysical Journal* **730**, 86.

Matthaeus, W. H., Pouquet, A., Mininni, P. D., Dmitruk, P. & Breech, B. 2008 Rapid Alignment of Velocity and Magnetic Field in Magnetohydrodynamic Turbulence. *Physical Review Letters* **100**, 085003.

Mininni, P. D. 2006 Turbulent magnetic dynamo excitation at low magnetic Prandtl number. *Physics of Plasmas* **13**, 056502.

Mininni, P. D., Ponty, Y., Montgomery, D. C., Pinton, J.-F., Politano, H. & Pouquet, A. 2005 Dynamo Regimes with a Nonhelical Forcing. *Astrophys. J.* **626**, 853–863.

Moffatt, H. K. 1978 *Magnetic Field Generation in Electrically Conducting Fluids*. Cambridge: Cambridge University Press.

Monchaux, R., Berhanu, M., Bourgoin, M., Moulin, M., Odier, P., Pinton, J.-F., Volk, R., Fauve, S., Mordant, N., Pétrélis, F., Chiffaudel, A., Daviaud, F., Dubrulle, B., Gasquet, C., Marié, L. & Ravelet, F. 2007 Generation of a Magnetic Field by Dynamo Action in a Turbulent Flow of Liquid Sodium. *Physical Review Letters* **98** 4, 044502.

Müller, W. & Busse, A. 2007 Recent developments in the theory of magnetohydrodynamic turbulence. In *Turbulence and Nonlinear Processes in Astrophysical Plasmas*, D. Shaikh & G. P. Zank, (eds.) *American Institute of Physics Conference Series*, vol. 932, pp. 52–57.

Müller, W. & Grappin, R. 2005 Spectral Energy Dynamics in Magnetohydrodynamic Turbulence. *Physical Review Letters* **95**, 114502.

Müller, W., Biskamp, D. & Grappin, R. 2003 Statistical anisotropy of magnetohydrodynamic turbulence. *Phys. Rev. E* **67**, 066302.

Newell, A. C., Nazarenko, S. & Biven, L. 2001 Wave turbulence and intermittency. *Physica D Nonlinear Phenomena* **152**, 520–550.

Ng, C. S. & Bhattacharjee, A. 1996 Interaction of Shear-Alfven Wave Packets: Implication for Weak Magnetohydrodynamic Turbulence in Astrophysical Plasmas. *Astrophys. J.* **465**, 845–854.

Ng, C. S., Bhattacharjee, A., Germaschewski, K. & Galtier, S. 2003 Anisotropic fluid turbulence in the interstellar medium and solar wind. *Physics of Plasmas* **10**, 1954–1962.

Oughton, S., Dmitruk, P. & Matthaeus, W. H. 2004 Reduced magnetohydrodynamics and parallel spectral transfer. *Physics of Plasmas* **11**, 2214–2225.

Parker, E. N. 1979 *Cosmical Magnetic Fields: their Origin and their Activity*. Oxford, Clarendon Press; New York, Oxford University Press..

Perez, J. C. & Boldyrev, S. 2008 On Weak and Strong Magnetohydrodynamic Turbulence. *Astrophys. J. Lett.* **672**, L61–L64.

Perez, J. C. & Boldyrev, S. 2009 Role of Cross-Helicity in Magnetohydrodynamic Turbulence. *Physical Review Letters* **102**, 025003.

Perez, J. C. & Boldyrev, S. 2010a Numerical simulations of imbalanced strong magnetohydrodynamic turbulence. *Astrophys. J. Lett.* **710**, L63-L66.

Perez, J. C. & Boldyrev, S. 2010b Strong magnetohydrodynamic turbulence with cross helicity. *Physics of Plasmas* **17**, 055903.

Pétrélis, F. & Fauve, S. 2008 Chaotic dynamics of the magnetic field generated by dynamo action in a turbulent flow. *Journal of Physics Condensed Matter* **20**, 494203.

Podesta, J. J. & Bhattacharjee, A. 2010 Theory of incompressible MHD turbulence with cross-helicity. *Astrophys. J.* **718**, 1151–1157.

Podesta, J. J., Chandran, B. D. G., Bhattacharjee, A., Roberts, D. A. & Goldstein, M. L. 2009 Scale-dependent angle of alignment between velocity and magnetic field fluctuations in solar wind turbulence. *Journal of Geophysical Research Space Physics* **114**, A01107.

Politano, H. & Pouquet, A. 1998 Dynamical length scales for turbulent magnetized flows. *Geophys. Res. Lett.* **25**, 273–276.

Ponty, Y., Mininni, P. D., Pinton, J.-F., Politano, H. & Pouquet, A. 2007 Dynamo action at low magnetic Prandtl numbers: mean flow versus fully turbulent motions. *New Journal of Physics* **9**, 296–305.

Pouquet, A., Frisch, U. & Leorat, J. 1976 Strong MHD helical turbulence and the nonlinear dynamo effect. *Journal of Fluid Mechanics* **77**, 321–354.

Rappazzo, A. F., Velli, M., Einaudi, G. & Dahlburg, R. B. 2007 Coronal Heating, Weak MHD Turbulence, and Scaling Laws. *Astrophysical Journal Letters* **657**, L47–L51.

Rappazzo, A. F., Velli, M., Einaudi, G. & Dahlburg, R. B. 2008 Nonlinear Dynamics of the Parker Scenario for Coronal Heating. *Astrophys. J.* **677**, 1348–1366.

Schekochihin, A. A., Cowley, S. C., Hammett, G. W., Maron, J. L. & McWilliams, J. C. 2002 A model of nonlinear evolution and saturation of the turbulent MHD dynamo. *New Journal of Physics* **4**, 84.

Schekochihin, A. A., Boldyrev, S. A. & Kulsrud, R. M. 2002 Spectra and Growth Rates of Fluctuating Magnetic Fields in the Kinematic Dynamo Theory with Large Magnetic Prandtl Numbers. *Astrophys. J.* **567**, 828–852.

Schekochihin, A. A., Cowley, S. C., Maron, J. L. & McWilliams, J. C. 2004 Critical Magnetic Prandtl Number for Small-Scale Dynamo. *Physical Review Letters* **92** 5, 054502.

Schekochihin, A. A., Cowley, S. C., Taylor, S. F., Hammett, G. W., Maron, J. L. & McWilliams, J. C. 2004 Saturated State of the Nonlinear Small-Scale Dynamo. *Physical Review Letters* **92** 8, 084504.

Schekochihin, A. A., Haugen, N. E. L., Brandenburg, A., Cowley, S. C., Maron, J. L. & McWilliams, J. C. 2005 The Onset of a Small-Scale Turbulent Dynamo at Low Magnetic Prandtl Numbers. *Astrophysical Journal Letters* **625**, L115–L118.

Schekochihin, A. A., Iskakov, A. B., Cowley, S. C., McWilliams, J. C., Proctor, M. R. E. & Yousef, T. A. 2007 Fluctuation dynamo and turbulent induction at low magnetic Prandtl numbers. *New Journal of Physics* **9**, 300.

Shebalin, J. V., Matthaeus, W. H. & Montgomery, D. 1983 Anisotropy in MHD turbulence due to a mean magnetic field. *Journal of Plasma Physics* **29**, 525–547.

Sridhar, S. 2010 Magnetohydrodynamic turbulence in a strongly magnetised plasma. *Astronomische Nachrichten* **331**, 93–100.

Strauss, H. R. 1976 Nonlinear, three-dimensional magnetohydrodynamics of non-circular tokamaks. *Physics of Fluids* **19**, 134–140.

Steenbeck, M. and Krause, F. & Rädler, K.-H. 1966 *Zeitschrift Naturforschung Teil A* **21**, 369.

Subramanian, K. 1999 Unified Treatment of Small- and Large-Scale Dynamos in Helical Turbulence. *Physical Review Letters* **83**, 2957–2960.

Subramanian, K. 2003 Hyperdiffusion in Nonlinear Large- and Small-Scale Turbulent Dynamos. *Physical Review Letters* **90** 24, 245003.

Tobias, S. M. & Cattaneo, F. 2008a Dynamo action in complex flows: the quick and the fast. *Journal of Fluid Mechanics* **601**, 101–122.

Tobias, S. M. & Cattaneo, F. 2008b Limited Role of Spectra in Dynamo Theory: Coherent versus Random Dynamos. *Physical Review Letters* **101** 12, 125003.

Tobias, S. M., Cattaneo, F. & Brummell, N.H. 2010 On the Generation of Organised Magnetic Fields. *Astrophysical Journal* **728**, 153.

Vainshtein, S. I. & Kichatinov, L. L. 1986 The dynamics of magnetic fields in a highly conducting turbulent medium and the generalized Kolmogorov–Fokker–Planck equations. *Journal of Fluid Mechanics* **168**, 73–87.

Verhille, G., Plihon, N., Bourgoin, M., Odier, P. & Pinton, J. 2010 Laboratory Dynamo Experiments. *Space Science Reviews* **152**, 543–564.

Yousef, T. A., Brandenburg, A. & Rüdiger, G. 2003 Turbulent magnetic Prandtl number and magnetic diffusivity quenching from simulations. *Astronomy & Astrophysics* **411**, 321–327.

Zakharov, V. E., L'Vov, V. S. & Falkovich, G. 1992 *Kolmogorov spectra of turbulence I: Wave turbulence.*

Zel'dovich, Y. B. 1956 The magnetic field in the two dimensional motion of a conducting turbulent fluid. *Zh. Exp. Teor. Fiz. SSSR* **31**, 154–155.

Zeldovich, I. B., Ruzmaikin, A. A. & Sokolov, D. D., eds. 1983 *Magnetic Fields in Astrophysics*, , vol. 3. Heidelrberg: Springer-Verlag.

Zeldovich, Y. B., Ruzmaikin, A. A. & Sokoloff, D. D. 1990 *The Almighty Chance.* World Scientific Lecture Notes in Physics, Singapore: World Scientific Publishers, 1990.

10
How Similar is Quantum Turbulence to Classical Turbulence?

Ladislav Skrbek and Katepalli R. Sreenivasan

10.1 Introduction

Superfluids can flow without friction and display two-fluid phenomena. These two properties, which have quantum mechanical origins, lie outside common experience with classical fluids. The subject of superfluids has thus generally been relegated to the backwaters of mainstream fluid dynamics. The focus of low-temperature physicists has been the microscopic structure of superfluids, which does not naturally invite the attention of experts on classical fluids. However, perhaps amazingly, there exists a state of superfluid flow that is similar to classical turbulence, qualitatively and quantitatively, in which superfluids are endowed with quasiclassical properties such as effective friction and finite heat conductivity. This state is called superfluid or quantum turbulence (QT) [Feynman (1955); Vinen & Niemela (2002); Skrbek (2004); Skrbek & Sreenivasan (2012)]. Although QT differs from classical turbulence in several important respects, many of its properties can often be understood in terms of the existing phenomenology of its classical counterpart. We can also learn new physics about classical turbulence by studying QT. Our goal in this article is to explore this interrelation. Instead of expanding the scope of the article broadly and compromising on details, we will focus on one important aspect: the physics that is common between decaying vortex line density in QT and the decay of three-dimensional (3D) turbulence that is nearly homogeneous and isotropic turbulence (HIT), which has been a cornerstone of many theoretical and modeling advances in hydrodynamic turbulence. A more comprehensive discussion can be found in a recent review by Skrbek & Sreenivasan (2012).

We shall discuss QT in two cryogenic fluids (see Figure 10.1): superfluid phase of liquid ^{4}He, called He II for historical reasons, and superfluid B-phase of the ^{3}He, which is a rarer helium isotope. QT has been extensively studied

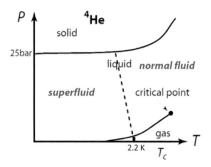

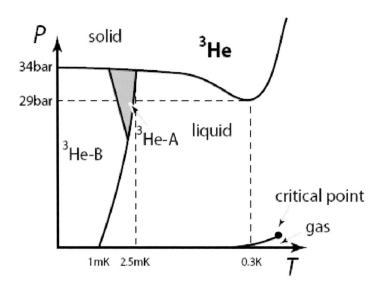

Figure 10.1 Schematic equilibrium phase diagrams for ^{4}He and ^{3}He, showing parameter spaces of the existence of working fluids, He II and ^{3}He-B, discussed in the text.

in both these quantum liquids. We set aside the underlying microscopic description [see, e.g. Tilley & Tilley (1986)] that explains superfluidity in these two systems with very different character – bosonic versus fermionic – and limit ourselves to widely accepted phenomenological description within the so-called two-fluid model, introduced with different emphases by Tisza (1938) and Landau (1941, 1947). Within the two-fluid model, both He II

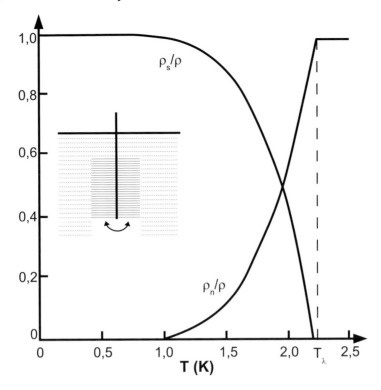

Figure 10.2 Temperature dependence of the normal and superfluid density ratio in He II. The ρ_n/ρ_s ratio was first experimentally established in the Andronikashivili (1946) experiment with a pile of thin closely spaced discs arranged as a torsional pendulum hanging in He II on a thin torsional fiber. If the discs are suspended in a classical viscous fluid of finite kinematic viscosity ν and the vertical distance between them is much less than the penetration depth $\delta = 2\nu/\Omega$, where Ω is the angular frequency of small torsional oscillations, the fluid is firmly trapped between the discs and contributes to the moment of inertia $\mathbf{J}$ of the pendulum. This is the case of He I. As the temperature is lowered below T_λ, the period of oscillations increases, as the superfluid, being inviscid, stays at rest and does not contribute to $\mathbf{J}$.

and ^{3}He can be regarded as consisting of two interpenetrating components: the normal component (obeying the Navier–Stokes equations and carrying all the entropy and the thermal capacity of the liquid) and the inviscid superfluid component. The two interpenetrating components flow without interaction (but see the paragraph immediately below), and support two independent velocity fields. We shall indicate the velocity of the normal component of density ρ_n by $\mathbf{v}_n$ and that of the superfluid component of density ρ_s by $\mathbf{v}_s$. The total density of the liquid is given by $\rho = \rho_n + \rho_s$, with the fraction ρ_n/ρ_s depending on the temperature. At the superfluid transition temperature (denoted by $T_\lambda \approx 2.17$ K for He II and of order

1 mK for ^{3}He-B) the entire liquid is normal and $\rho_n/\rho_s = 1$. This ratio drops with decreasing temperature until at 0 K the liquid is entirely superfluid, so that $\rho_n/\rho_s = 0$; see Figure 10.2. In practice, He II is nearly all superfluid below temperatures of the order of 1 K; ^{3}He-B below about $0.2T_c$, where the superfluid transition temperature T_c for ^{3}He varies between 1–2 mK depending on externally applied pressure.

When the superfluid component of He II flows faster than a limit (with respect to solid surfaces or the normal component), usually of order of a few cm/s, energy considerations suggest that it is advantageous for the flow to create line vortices. The core of these vortices is of atomic size (of the order of an Angstrom) in He II [Feynman (1955); Donnelly (1991)]. In ^{3}He-B this critical velocity is even lower, of order mm/s, and vortex cores (depending on pressure) are about 10–70 μm in size [Vollhardt & Wölfle (1992); Dobbs (2000)]. Around these cores, there is an irrotational circulation equal to $\kappa = 2\pi\hbar/m$, where m denotes the mass of one superfluid particle, i.e., ^{4}He atom or a Cooper pair of two ^{3}He atoms, and $\hbar$ is Planck's constant. The circulation is quantized instead of being arbitrary. This quantization of line vortices is an important characteristic of quantum liquids. In analogy with classical viscous flow, the flow of the superfluid component can be characterized by the 'superfluid Reynolds number' $Re_s = \ell v/\kappa$, where ℓ and v have their usual fluid dynamical meaning. The limit of very large Re_s is equivalent (in the non-dimensional sense) to a vanishing Planck's constant, $\hbar \propto \kappa \to 0$, so that the vorticity becomes a continuous variable as in the classical case. The Feynman criterion $Re_s \sim 1$ gives the velocity at which it becomes energetically favorable to form a quantized vortex. It does not mean, however, that upon exceeding this critical velocity quantized vortices necessarily appear in the flow, because of the nucleation barrier. Nucleation of quantized vortices is a complex problem whose thorough discussion falls outside our present scope. In short, quantized vortices can be nucleated intrinsically (i.e., in a completely vortex-free sample of fluid), or extrinsically (i.e., from already existing vortex loops which act as seeds). In any macroscopic sized He II sample, the seeds are always present in practice – it is hardly possible to create any macroscopic He II sample that would be entirely free of remnant vortices, and extrinsic vortex nucleation usually dominates. In ^{3}He-B, both intrinsic and extrinsic nucleation mechanisms are possible.

QT can be defined loosely as the most general motion of quantum fluids involving the dynamics of a complex tangle of the quantized vortices just defined. It is perhaps not surprising that the zero-temperature properties of QT are for the most part independent of whether it is generated in He II or

^{3}He. Here the differences appear at the smallest scales only. A key variable concerning the tangle is the vortex line density, L, which is the total length of the vortex core in a unit volume of quantum fluids. On scales much greater than the average spacing between vortex lines in this tangle ("inter-vortex spacing"), the superfluid flow closely mimics a classical fluid, as we shall elaborate; on length scales of the order of a few vortex line diameters, the two flows are indeed quite different because of the quantization condition on circulation. For temperatures at which normal and superfluid components are both present, any vortex line moving relative to the normal fluid is subject to a drag force, called the mutual friction force [Donnelly (1991); Vinen & Niemela (2002)]. This is the source of coupling between the two fluids, which otherwise, within the phenomenological two-fluid model, flow independently. It is reasonable to think that, as a result of this coupling, the normal fluid forces the superfluid component to behave quasi-classically at scales larger than the inter-vortex spacing [Vinen (2000)]. It turns out that even at very low temperature where there is little normal fluid, the decay of the vortex tangle often preserves a striking similarity with its classical counterpart. We shall discuss this feature as well.

The particular scope of this article is the decay of turbulence in the absence of sustained production. For classical turbulence in viscous fluids this situation arises in homogeneous and isotropic turbulence, as has been discussed in many reviews [e.g., Lesieur (1997)] since the pioneering work by Taylor (1935), von Kármán & Howarth (1938) and Kolmogorov (1941). A theoretical result of Kolmogorov is that the kinetic energy of turbulence, given by $E = \overline{\mathbf{v}^2}/2$, where $\mathbf{v} = \mathbf{v}(\mathbf{r},t)$ is the velocity vector, decays as $E(t) \propto t^{-\gamma}$, with $\gamma = 10/7$. The experimental data display large variability in the decay exponent γ; as discussed, among others, by Skrbek & Stalp (2000), at least some of this variation is due to the strong influence of the so-called virtual origin chosen for time. Numerical studies also yield substantially varying results for γ; as pointed out recently by Yakhot (2004), these differences could be effects of finite computational domains or of initial conditions. While discussing these aspects, we will particularly distinguish between the two cases: (i) the energy-containing scale of turbulence, ℓ (or the integral scale ℓ_i), is equal to the size of the apparatus, D, and stays approximately so during the decay; and (ii) ℓ is small enough compared to the size of the apparatus that it grows unconstrained during decay. We shall argue that carefully planned experiments on decaying QT could shed new light on an old puzzle concerning the so-called Loitsianskii invariant [Loitsianskii (1939)].

The remainder of this chapter is organized as follows. We review the experimental techniques and the relevant experimental data sets on decaying QT in Section 10.2, emphasizing the recent data in the limit of zero temperature. In Section 10.3 we consider aspects of possible comparisons and consider in Section 10.4 the general problem of temporal decay of energy in a universal setting; in Section 10.5, we discuss various scenarios of decay and relate them to energy spectrum characteristics. Section 10.6 considers effective viscosity and presents a simple phenomenological model to estimate it. We summarize the main points in Section 10.7.

10.2 Preliminary remarks on decaying QT

10.2.1 Experimental techniques

Three different experimental techniques have been used so far for obtaining *quantitative* data on decaying QT in He II and ^{3}He-B, all of them measuring the vortex line density, L, under the assumptions to be discussed below. We will explain in Section 10.4 the central role played by L.

(1) Second sound attenuation A powerful and historically the oldest technique is the second sound attenuation suitable for experiments in superfluid ^{4}He [Vinen (1957)]. The technique usually assumes that the vortex tangle is homogeneous and isotropic. In all the experiments discussed in this article, second sound has been generated and detected using two gold-plated porous sensors placed opposite each other across a flow channel, the width of which serves as an acoustic resonator. The DC biased second sound transducer is driven by the AC signal at one of the standing second sound resonances, and the signal from the receiver tracks the information on second sound attenuation providing a direct measure of the vortex line density in the channel. This is not a local measurement and roughly provides L averaged over the volume between transducers, typically of the order of 0.1–1 cm^3.

A quantitative detail that has emerged in recent years is worth mentioning. Assuming homogeneous vortex tangle [Barenghi & Skrbek (2006; 2007)], noting that vortices oriented parallel to the direction of propagation of the second sound do not contribute to the attenuation and taking proper account of the so-called "sine squared law" [Mathieu et al. (1984)], the line length L follows the expression

$$L = \frac{6\pi\Delta_0}{B\kappa}\left(\frac{a_0}{a} - 1\right) \quad \text{rather than} \quad L = \frac{16\Delta_0}{B\kappa}\left(\frac{a_0}{a} - 1\right), \tag{10.1}$$

where a and a_0 are, respectively, the amplitudes of the second sound standing-wave resonance response of Lorentzian form with and without vortices; its undisturbed linewidth is given by Δ_0, and B denotes the mutual friction coefficient that depends on the temperature and, weakly, also on the frequency [Donnelly & Barenghi (1998)]. The older expression for vortex line density [Stalp (1998)] did not take the line orientation into account properly (for details, see [Barenghi & Skrbek (2006); (2007)]). Thus, all values of L reported in the Oregon experiments [Smith *et al.* (1993); Stalp (1998); Stalp *et al.* (1999); Skrbek & Stalp (2000); Skrbek *et al.* (2000); Niemela *et al.* (2005); Niemela & Sreenivasan (2006)], as well as the early Prague experiments [Skrbek *et al.* (2003), Gordeev *et al.* (2005)], discussed later in this article, should be multiplied by a factor of $3\pi/8 \approx 1.2$; similarly, the effective kinematic viscosity (see Section 10.6) derived from the decay data should be multiplied by $(3\pi/8)^2 \approx 1.4$.

Since the second sound can be thought of as an anti-phase oscillation of the normal and superfluid components, this technique cannot be used in ^{4}He below about 1 K where the fraction of the normal fluid is negligible (see Figure 10.2), or in ^{3}He at any temperature, as second sound waves are overdamped due to the large kinematic viscosity of the normal fluid.

(2) Helium ions Helium ions have been successfully used by Milliken, Schwarz & Smith (1982) to investigate the decay of inhomogeneous QT created by ultrasonic transducers in He II at about 1.5 K. For measurements of the decaying QT below 1 K, an improved technique based on negative ions was recently introduced by Golov and coworkers in Manchester [Walmsley and Golov (2008); Walmsley *et al.* (2007; 2008); Eltsov *et al.* (2009)]. Negative ions (electron bubbles) are injected by a sharp field-emission tip and are manipulated by an applied electric field. Bare ions dominate for $T > 0.8$ K while, for $T < 0.7$ K, they become self-trapped on the core of quantized vortex rings of diameter about 1 μm (which are generated spontaneously while accelerating in an applied electric field). Short pulses of these probes are sent across the experimental cell. The relative reduction in their amplitude at the collector on the opposite side of the helium cell, due to the interaction of charged vortex rings with QT in the cell, is converted into vortex line density. An accurate *in situ* calibration is possible, as these experiments have been conducted on the rotating cryostat which provides a known density of rectilinear vortex lines corresponding to the rotation rate of the superfluid[1].

---

[1] Note that each vortex line carries one circulation quantum so the areal density of vortex lines constituting the vortex lattice is proportional to the angular velocity of the rotating apparatus [see, e.g. Tilley & Tilley (1986)].

(3) Andreev scattering ^{4}He is a bosonic superfluid. Several bosons can occupy the same quantum state and the wavefunction remains unchanged when two bosons are exchanged. The bosons obey Bose–Einstein statistics. On the other hand, ^{3}He-B is a fermionic substance. No two fermions occupy the same quantum state, resulting in a certain "stiffness" of fermionic matter, and the fermions obey Fermi-Dirac statistics. Given this fundamental difference between bosons and fermions, it is important to understand the decay of QT in ^{3}He-B as well. This has indeed been done. Due to very high viscosity of the normal fluid and strongly temperature-dependent mutual friction force, QT in ^{3}He-B occurs only at very low temperatures $T < 0.6\,T_c$, where, as already mentioned, the pressure-dependent transition temperature T_c is of the order of 1 mK [Finne et al. (2003)]. Much useful information has been obtained using the nuclear magnetic resonance technique [Eltsov et al. (2009)] but we shall be concerned here with the decay of QT in the zero temperature limit, for which the Lancaster ^{3}He group invented a new experimental technique based on Andreev scattering of quasiparticles caused by the velocity field of quantized vortices. The underlying physics of the technique is explained in detail by Fisher & Pickett (2009). Briefly, the equilibrium number of quasiparticles (and quasiholes, because ^{3}He-B is fermionic superfluid and excitations are thus thermally generated from below the Fermi level in pairs) can be sensed by a vibrating wire resonator. The drag force exerted by these quasiparticles is reduced if the vibrating wire is surrounded by a tangle of quantized vortex lines, as some of the incoming quasiparticles cannot reach it, being Andreev-reflected by the energy barrier of the superfluid velocity field near the vortex core. The fractional decrease in damping can then be converted into vortex line density of the tangle [Bradley et al. (2006; 2008)].

There are additional techniques such as the calorimetry combined with the towed grid, as well as existing data using ions in combination with oscillating grids at very low temperature in ^{4}He, obtained in Florida [Ihas et al. (2007)] and Lancaster [Davis et al. (2000)], but they have not yet produced usable quantitative data on QT. We will therefore not consider them further.

10.2.2 Quantitative data sets for comparison between HIT and QT

Conceptually, the zero temperature limit is the simplest case of QT because there is no normal fluid, and the vortex lines will not interact with normal turbulence as they do at higher temperatures. QT in He II below about 0.3 K turns out to be essentially of this form. Experimentally, it is much easier to access QT above about 1 K. In this range, the existence of turbulence in both components of the fluid and the coupling between them often leads to additional complications in comparison with conventional turbulence in classical viscous fluids. We will account for this mismatch between what is easier in experiment and what is easier in theory, as much as possible.

QT generated by classical methods In comparing decaying QT with its classical counterpart, it is useful to focus attention first on the QT data generated by methods of classical turbulence. At finite temperature, these methods force both the superfluid and the normal fluid in the same way and generate the so-called co-flow QT. The available experiments include: (1) the Oregon towed grid data in ^{4}He above about 1.1 K [Smith *et al.* (1993); Stalp (1998); Stalp *et al.* (1999); Skrbek & Stalp (2000); Skrbek *et al.* (2000)] (which we shall call *Oregon 1*, see an example included in Figure 10.3), obtained with the square grid of unconventional design consisting of six parallel 0.5 mm tines, 1.17 mm apart, joined from corner to corner by another tine – see Fig. 1b in Smith *et al.* (1993); (2) more recent data (*Oregon 2*) obtained with a new grid of "standard" fluid dynamical geometry (20 × 20 tines of square cross-section) [Niemela *et al.* (2005); Niemela & Sreenivasan (2006)]; (3) the Lancaster Andreev scattering oscillating grid data in ^{3}He-B in the zero temperature limit [Bradley *et al.* (2006)] (*Lancaster data*, see Figure 10.3); and (4) the Manchester ion spin down of a rotating cubical cell data in ^{4}He at temperatures below 1.6 K down to zero temperature limit [Walmsley and Golov (2008); Walmsley *et al.* (2007; 2008); Eltsov *et al.* (2009)] (*Manchester spin down data*).

All these data measure the temporal decay of the vortex line density, $L = L(t)$. The observables in these experiments are, respectively, the excess attenuation of the second sound, differential damping of the vibrating wire resonator and relative strength of the ion pulse at the collector, with the evaluation of L from the primary data being reasonably well established and understood. A discussion of possible caveats in the evaluation of L can be found in Skrbek & Sreenivasan (2012). The connection between L and the fluid dynamically interesting quantities such as vorticity will be discussed in Section 10.4.

Remarks on early investigations of decaying QT in counterflow
Thermal counterflow can be established by applying voltage to a resistive heater located at the closed end of a channel that is open to the helium bath at the other end. The heat flux is carried away from the heater by the normal fluid alone. By conservation of mass, a superfluid current arises in

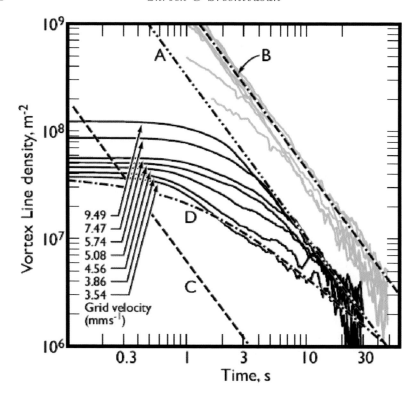

Figure 10.3 Solid black curves show the inferred vortex line density as a function of time after the cessation of grid motion in ^{3}He-B in the zero temperature limit for initial grid velocities indicated. Line A is the limiting behavior $L(t) \propto t^{-3/2}$ as discussed for the Kolmogorov QT in the text. The halftone data is that for the Oregon towed grid experiments of Stalp, Skrbek & Donnelly (1999) in He II, with line B showing the same power law late-time limiting behavior, while line C shows this behavior for the data assuming the classical dissipation law. Curve D illustrates the behavior for a random tangle (i.e., Vinen QT) in superfluid ^{3}He when the decay originates from the steady-state of lower vortex line density [Bradley et al. (2006)].

the opposite direction and a relative velocity, or counterflow

$$v_{\mathrm{ns}} = |\mathbf{v}_n - \mathbf{v}_s| = \frac{\dot{q}}{ST\rho_{\mathrm{s}}} \qquad (10.2)$$

is created along the channel. This velocity is proportional to the applied heat flux, $\dot{q} = \dot{Q}/A$, where A is the cross-sectional area of the flow channel, and S is the specific entropy of He II (i.e., its normal component, as the superfluid carries no entropy).

It is useful to remark briefly on the investigations of QT that began with Vinen's (1957) pioneering experiments on thermally induced counterflow in He II, and on their phenomenological description, even though the steady–

state counterflow has no direct classical counterpart. The remarks to follow are also helpful for introducing the concept of mutual friction, which applies to QT in co-flows as well. In counterflow, even for relatively small values of counterflow velocity $v_{\rm ns}$ the heat transport becomes affected by QT due to the appearance of a tangle of quantized vortex lines. The vortex line density, L, deduced from the experiment agrees with the steady-state solution of Vinen's phenomenological equation [Vinen (1957)], which expresses the dynamical balance between the generation and decay terms in the presence of a steady counterflow $v_{\rm ns}$:

$$\frac{\partial L}{\partial t} = \frac{\rho_{\rm n} B}{2\rho} \chi_1 v_{\rm ns} L^{3/2} - \frac{\kappa}{2\pi} \chi_2 L^2 + g(v_{\rm ns}). \tag{10.3}$$

Here χ_1 and χ_2 are undetermined dimensionless constants and B is the mutual friction parameter, tabulated by Donnelly & Barenghi (1998).[2] The unspecified function $g(v_{\rm ns})$ had to be included to comply with the experimental fact that there are no quantized vortices (and hence no mutual friction) below certain critical $v_{\rm ns}$. The assumptions that the vortex tangle is homogeneous, and that the quantity L characterizes it fully, seem to account for most of the observed phenomena in steady state counterflow turbulence – for a review on the early results, see Tough (1982). Neglecting the term $g(v_{\rm ns})$, the steady-state solution of Vinen's equation is

$$L^{1/2} = \frac{\pi \rho_{\rm n} B \chi_1}{\rho \kappa \chi_2} v_{\rm ns}, \tag{10.4}$$

which means that L is proportional to the square of the counterflow velocity, and that the measured (or numerically computed) steady-state value of L determines only the ratio of χ_1/χ_2. It was shown by Vinen (2005), for the simple case of the highly viscous quiescent normal fluid, that eddies larger than a critical size (given by the balance between the eddy size over its turnover time and the counterflow velocity) cannot exist. In that case Vinen's equation (10.3) should be applicable, allowing us to compute the decay rate. Indeed, by setting $v_{\rm ns} = 0$ (and neglecting the small term $g(v_{\rm ns})$) in (10.3), we obtain the differential equation

$$\frac{\partial L}{\partial t} = -\frac{\kappa}{2\pi} \chi_2 L^2. \tag{10.5}$$

---

[2] Equation (10.3) was later deduced by Schwarz (1988) using computer simulations of the laws of vortex dynamics under the local induction approximation. The local induction approximation neglects non-local interactions and is equivalent to neglecting advection and stretching processes which are essential in classical turbulence. The approximation describes the *steady* counterflow QT quantitatively because there are no large eddies (i.e., bundles of quantized vortices) in steady-state counterflow turbulence as the mutual friction destroys them faster than their turnover time.

Its solution leads to the decay $L(t) = L(0)(t+t^*)^{-1} \propto t^{-1}$ for long times. However, equation (10.5) is more general because it ought to describe the temporal decay of $L(t)$ of any unstructured tangle, called today by several authors as *Vinen turbulence*. As we shall see, this conclusion is confirmed by the *IBM ultrasonic* data and the recent *Manchester ion data* at low temperature. The inverse time decay was also clearly observed in ^{3}He-B *Lancaster data* for decaying $L(t)$, deduced by the Andreev scattering technique. Moreover, these data in the zero temperature limit clearly show that the form of the decay depends on the initial vortex line density. We shall discuss a crossover from Vinen to Kolmogorov QT, where the decay displays $L \sim t^{-3/2}$, in Subsection 10.3.5.

Further comments on decaying counterflow turbulence The main comment on the *Prague counterflow* data [for details, see Barenghi & Skrbek (2007)] and the relevant numerical simulations is that the decay of turbulence generated by a heat flow can be usefully divided into three stages. In the first stage, for $t < \tau$, say, He II is not yet isothermal. The time scale τ depends on the channel's length and cross-section and can be estimated using simple thermodynamic considerations. The initial decay of the counterflow turbulence is driven even when the heater is switched off, because of the excess heat present in the channel near the heater. Only in the second and third stages (with $t > \tau$) can the liquid be considered approximately isothermal and the decay free. The second stage is characterized by a slow decay rate. The vortex tangle induced by the counterflow is, in fact, quite polarized [Barenghi, Gordeev & Skrbek (2006)] by the normal fluid along the channel. This polarizing effect gradually relaxes when the heater is switched off, and vortex reconnections play an increasingly important role. In the third stage of the decay of counterflow turbulence ($t > t_{\text{sat}}$), the observed vortex line density acquires the $t^{-3/2}$ power law, similar to that described in Section 10.4 for the co-flow QT: see Figure 10.3.

QT generated by other means It is also useful to consider the following additional data on decaying vortex line density, chosen as representative of the many experiments performed over 50 years: (5) The data of Milliken, Schwarz & Smith (1982) on free decay of (spatially non-uniform) superfluid turbulence generated at 1.45 K in He II between the pair of ultrasonic transducers. The wavelength of ultrasound, at which QT was presumably forced, was of order 10^{-2} cm [Schwarz & Smith (1981)]. The decaying L and its spatial profile were deduced from the relative strength of ion pulses sent across the turbulent region (*IBM ultrasonic* data); (6) the Manchester

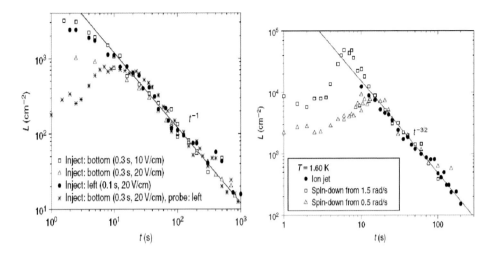

Figure 10.4 Deduced vortex line density plotted versus time for decaying QT generated by injecting ions in Manchester experiments [Walmsley and Golov (2008); Walmsley et al. (2007; 2008)] at $T = 0.15$ K (left) and $T = 1.6$ K (right). Solid lines with a slope of $-3/2$ and -1 illustrate two distinctly different behaviors of the late-time decay.

^{4}He decay data of QT generated by injecting ions (*Manchester ion* data – see Figure 10.4) in the same experimental cell as was used to obtain the spin-down data [Walmsley and Golov (2008); Walmsley et al. (2007; 2008)]; and (7) the Prague data using second sound in thermally generated decaying counterflow [Skrbek et al. (2003); Gordeev et al. (2005)] (*Prague counterflow* data); and (8) pure superflow turbulence data in He II [Chagovets & Skrbek (2008; 2008a)] (*Prague superflow* data). These experiments also measure decaying vortex line density, $L = L(t)$, though in configurations which cannot be connected to the grids directly.

10.3 Comparisons between QT and HIT: energy spectrum

10.3.1 *The energy spectrum in 3D classical turbulence*

The shape of the 3D energy spectrum for approximately homogeneous and isotropic turbulence in classical viscous fluids is well known, at least empirically. Neglecting intermittency corrections, we recall that Kolmogorov's dimensional reasoning yields that an inertial range of scales exists for high enough Reynolds numbers, within which the spectral energy density has the form

$$E(k) = C\epsilon^{2/3}k^{-5/3}, \qquad (10.6)$$

where the wavenumber $k = 2\pi/\ell$, ℓ being a physical scale, and the 3D Kolmogorov constant $C = 1.62 \pm 0.17$ [Sreenivasan (1995)]. At the low k end, the spectrum is naturally truncated by the size of the container $k_D = 2\pi/D$, reflecting the fact that eddies larger than the system, D, cannot exist. The low k end takes the form Ak^m; we shall discuss the value of the exponent m in Subsection 10.5.2. For increasing k, there is a broad maximum in $E(k)$ that corresponds to energy containing eddies. Most of the turbulent energy resides at scales around ℓ_e. In steady-state 3D turbulence, the energy is injected at the outer scale $\approx \ell_e$ which, by the Richardson cascade, is transmitted without loss in the k-space to the dissipation length scale $k_{\text{diss}} \approx (\epsilon/\nu^3)^{1/4}$, where it becomes dissipated by finite viscosity. Although the dissipation scale is a statistical quantity [leading to important effects (see, e.g., Sreenivasan (2004)], traditional dimensional reasoning uses its average value.

10.3.2 The measured spectral shape in the inertial range of QT

An important and informative experiment was performed by Maurer & Tabeling (1998), who measured with a Pitot tube the frequency spectrum of local pressure fluctuations in He I and He II by generating turbulence between counterrotating discs. The local pressure fluctuations were measured at a location of non-zero mean flow, whence the turbulent energy spectrum in the frequency domain was evaluated. In turn, via the (perhaps questionable) use of Taylor frozen hypothesis, it can be related to the energy spectral density $E(k)$. Three temperatures were used: 1.4 K at which liquid helium is largely superfluid, 2.08 K at which it is largely normal and 2.3 K at which it is entirely normal. The main, and striking, result of this experiment is that there are no detectable changes in the form of the spectrum for all these three temperatures, and an inertial range of wavenumbers of the classical Kolmogorov form was observed over the range of relevant wavenumbers detectable by this method. Moreover, the authors deduced the numerical value of the Kolmogorov constant well within the accepted range mentioned above; this situation was true even for the intermittency corrections (which we shall not discuss at any length here). The simplest explanation of this result is that, at scales to which the Maurer–Tabeling experiment had access, the two fluids move with the same velocity field that is identical to that expected in a classical turbulence at high Reynolds numbers.

There is no direct experimental evidence of the shape of 3D energy spectrum in QT as $T \to 0$, but significant support comes from computer simulations: Araki et al. (2002) (using the full Biot–Savart vortex filament approach) and Kobayashi & Tsubota (2005) (using the Gross-Pitaevski model). The inertial range of the Kolmogorov form was found in both works.

10.3.3 Cascade and dissipation of energy

It is not difficult to explain the experimental results described above, at least in the zero-temperature limit, in terms of reconnections of quantized vortices leading to the quasiclassical cascade. The support for the occurrence of reconnections in QT comes from the simulations of Koplik & Levine (1993) and the experiments of Bewley *et al.* (2006) and Paoletti *et al.* (2008). It appears that reconnections can occur also between two bundles of quantized vortices (closely mimicking classical vortex tubes) [Alamri *et al.* (2008)].

According to the current understanding, vortex reconnections cannot be the entire source of dissipation because they are probably not efficient enough [Paoletti *et al.* (2008)]. So the energy cascade towards higher k has to take another form, which is thought to be in the form of a Kelvin wave cascade. Kelvin waves on quantized vortices with well-known dispersion relation [see, e.g., Donnelly (1991)] transfer energy towards high enough frequencies[3], where acoustic emission of quasiparticles is efficient enough to balance the forcing at the outer scale. These emitted quasiparticles eventually thermalize and increase the temperature, effectively generating the normal fluid. Recent simulations also suggest that reconnections are accompanied by emissions of sound waves of wavelength of the order of the healing length [Leadbeater *et al.* (2001); Ogawa *et al.* (2002)], which may be the eventual source of energy decay in QT. It is currently a matter of intense discussion [see Kozik & Svistunov (2009) and references therein; L'vov, Nazarenko & Rudenko (2007)] as to whether Kelvin wave cascade joins the Kolmogorov cascade smoothly, or some form of a bottleneck exists, as in classical turbulence [Donzis & Sreenivasan (2010)]. These issues are not yet theoretically settled and the relevant experimental data are scarce [Eltsov *et al.* (2007)], so we will not discuss this aspect any further.

10.3.4 The role of the quantum lengthscale

In the zero-temperature limit, it is clear that there is only the superflow, up to smallest length scale given by the size of the vortex core. There is no viscosity, so the dissipation length scale cannot be defined in the traditional sense. It is common to introduce another physical length scale defined as

---

[3] For a single vortex line this process has been convincingly simulated by Vinen, Tsubota & Mitani (2003).

mean intervortex distance that could be estimated as $\ell_v \approx 1/\sqrt{L}$ [Vinen & Niemela (2002)]. Note, however, that the energy flux will have to be dissipated by some mechanism that cannot depend on viscosity (since it does not exist). It is then possible to introduce a characteristic quantum wavenumber as $k_Q \approx (\epsilon/\kappa^3)^{1/4}$, and the corresponding quantum length scale as $\ell_Q = 2\pi/k_Q$ [Skrbek & Sreenivasan (2012)].

The physical meaning of the quantum length scale is that it qualitatively divides the scales into large scales – for which quantum restrictions in the form of quantization of circulation do not play any role (or the "granularity" of QT does not matter) – and the small scales, where quantum restrictions are essential. For wavenumbers larger than k_Q, the spectral energy density ought to depend, besides $\epsilon = -dE/dt$ and k, also on κ. Dimensional argument similar to that of Kolmogorov [Vinen (2000); Skrbek & Sreenivasan (2012)] requires the spectral density in the wavenumber space k to take the general form

$$E(k) = C\epsilon^{2/3} k^{-5/3} f(\epsilon k^{-4} \kappa^{-3}) \,. \tag{10.7}$$

The quantum length scale plays another essential role in QT in so far as it separates the Kolmogorov (or quasiclassical) turbulence from the Vinen (or ultraquantum) QT. In the case of the Kolmogorov QT, the outer scale is larger than ℓ_Q and the 3D energy spectrum has a general form shown in Figure 10.5. In the Vinen turbulence, on the other hand, the quantum scale is larger than the scale at which energy is injected into the system and to quasiclassical cascade (and the Kolmogorov inertial range) cannot exist. A few brief comments on the Vinen QT may be useful, mostly in so far as they concern the crossover to the Kolmogorov turbulence.

10.3.5 Crossover from Vinen turbulence to Kolmogorov turbulence

Assuming the validity of equation (10.9) in from Subsection 10.4.1, we can estimate for the *Lancaster data* the quantum wavenumber as

$$k_Q = \left(\frac{\epsilon}{\kappa^3}\right)^{1/4} = \left(\frac{\nu_{\text{eff}}}{\kappa}\right)^{1/4} \frac{1}{\ell_v} \,, \tag{10.8}$$

where the intervortex distance as $\ell_v \approx 1/\sqrt{L}$. We shall see in Section 10.6 that, in He II, ν_{eff} is of order 10^{-2}–10^{-1} κ, so that ℓ_v and ℓ_Q are of the same order. Note that for these data, the crossover from Vinen to Kolmogorov QT corresponds to $L \approx 6 \times 10^3$ cm^{-2}, yielding an ℓ_v of about 100 μm,

10: How Similar is Quantum Turbulence to Classical Turbulence?

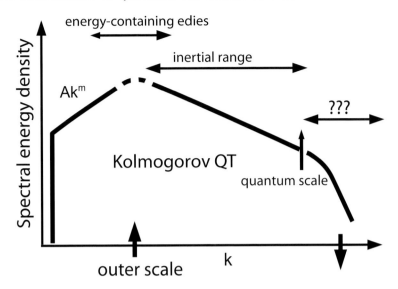

Figure 10.5 General form of the 3D QT energy spectrum, assuming that QT is forced at outer scale larger than ℓ_Q. The exact shape of the spectrum beyond k_Q cannot be predicted based on simple dimensional analysis.

similar to the ℓ_Q estimated above[4]. Similarly, the *IBM ultrasonic decay data* [Milliken, Schwarz & Smith (1982)] has an L whose largest magnitude is of order 10^4 cm^{-2}, from which we estimate ℓ_v (or ℓ_Q) to be about 10^{-2} cm, roughly equal to the wavelength representing the physical estimate for the outer scale of this QT. Thus, these two sets of data also represent Vinen QT.

The Vinen QT possesses quantum effects at all scales; neither stretching nor advection takes place, as the vortex tangle is random and there are simply no large eddies in the sense of classical turbulence. It is thus of no relevance to classical turbulence, but it is worth noting that the crossover in experiments from the Vinen to Kolmogorov turbulence behaves as expected from quasiclassical arguments. We can illustrate this for the *Lancaster data* on decaying QT, obtained in ^{3}He-B in the zero temperature limit[5]. The forc-

---

[4] The crossover state corresponds to a grid velocity of ≈ 5 mm/s, when the grid is forced by about 0.2 μN [Bradley et al. (2004)]. Multiplying these values, one gets the absolute energy decay rate in the stationary case, due to decaying turbulence some distance from the grid. Knowing the size of the grid, we estimate the QT decay to be of order 5×10^{-8} m^3. The density of ^{3}He is about 100 kg/m^3, so one gets the energy decay rate $\epsilon \approx 2 \times 10^{-4}$ m^2/s^3. With known circulation quantum in ^{3}He-B $\kappa \cong 0.066$ mm^2/s this results in $k_Q \approx (\epsilon/\kappa^3)^{1/4} \approx 30000$ 1/m. The corresponding quantum length scale is close to the mesh size of the grid, 50 μm.

[5] We believe that the described *Manchester ion data*, where QT is generated by ion jets displaying both $L(t) \propto 1/t$ and $L(t) \propto 1/t^{3/2}$ (see Figure 10.4), serve as another example for a crossover from Vinen to Kolmogorov QT regime.

ing is done by the same oscillating grid in all instances, so the scale of forcing is undoubtedly the same. By increasing the grid oscillating amplitude, more and more energy is supplied to the flow, all of which is being dissipated in the steady state. While the turbulence so generated is not homogeneous and isotropic, increasing the oscillating amplitude of the grid results in higher energy decay rate, ϵ; consequently, the quantum wavenumber $k_Q \propto \epsilon^{1/4}$ increases. This is completely consistent with the change in character of the decay of L (see Section 10.4).

10.4 Decaying vorticity

10.4.1 Basis for comparing classical turbulence with QT

The results of QT experiments are given in terms of decaying vortex line density, $L(t)$. To link this to energy decay, we should understand the role of quantized vortices broadly. The simplest system in which to do this is a rotating bucket of He II. In steady state the normal fluid is in solid body rotation, with vorticity $\boldsymbol{\omega}_n = \operatorname{curl} \mathbf{v}_n = 2\boldsymbol{\Omega}$ and He II is threaded with a rectilinear array of quantized vortex lines parallel to the axis of rotation. Although the superfluid vorticity is restricted to vortex lines, their areal density, n_0, evolves to match the vorticity of the fluid in solid body rotation. This results in $n_0 \kappa = \kappa L = 2\Omega$. Thus on length scales larger than the intervortex line distance, the superfluid is also in solid body rotation, with a velocity field equal to that of the normal fluid, i.e. $\langle \boldsymbol{\omega}_s \rangle = \boldsymbol{\omega}_n$. (Clearly, this picture does not apply for distances of the order of the vortex diameter.)

Even though QT is more complex to be fully understood by the paradigm of aligned vortices, it is helpful to keep the above observation in mind. In Kolmogorov-like QT the turbulent superflow at length scales exceeding ℓ_Q resembles classical flow possessing an effective kinematic viscosity ν_{eff}. Then the relationship between turbulent energy dissipation rate per unit volume and the mean square vorticity should apply, and we should have

$$\epsilon = -\frac{dE}{dt} = \nu_{\text{eff}}(\kappa L)^2 \ , \tag{10.9}$$

i.e., vorticity would be defined as κL in full analogy with the rotating bucket. This is the key relationship that enables broad comparisons of QT with its classical counterpart, and the basis for estimating ℓ_Q in Subsection 10.3.5.

The underlying physics of equation (10.9) has been discussed in detail by Vinen (2000); see also the review by Vinen & Niemela (2002). If equation (10.9) holds, the analysis of the decaying vortex line density in the *Kolmogorov-like QT* experiments, described earlier in Subsection 10.2.2, ought to be more or less straightforward, as it allows treating superfluids as a hypothetical classical fluid possessing an effective kinematic viscosity ν_{eff} (see Section 10.6).

The basic equation for turbulent energy decay of HIT is

$$-\frac{dE}{dt} = \epsilon = \nu\overline{\omega^2}, \qquad (10.10)$$

where ω is the root-mean-square vorticity.

If the outer scale of classical HIT is equal to the size of the system, the equations of decay can be closed and one can get explicit results for the decay. Equation (10.10) can be approximated as $\epsilon \approx E^{3/2}/D$, consistent with the familiar dimensional relation [Sreenivasan (1984)] $\epsilon = \epsilon_0 E^{3/2}/\ell_e$, where ℓ_e is replaced by the size of the system, D. It is immediately evident that the kinetic energy decays as $E(\tau) \propto D^2/(t+t^*)^2 = D^2/\tau^2$, where t^* is the virtual origin in time that depends on the starting value of energy as well as the transient processes that might have taken place at the start of decay. For decaying vorticity at late times, this corresponds to

$$\omega(t) \propto D/(t+t^*)^{3/2} \approx D/t^{3/2}. \qquad (10.11)$$

We emphasize that no assumptions on the shape of the 3D energy spectrum has been necessary in coming to this conclusion [Niemela & Sreenivasan (2006)].

Despite the inevitability of the above result, there is very little experimental data on decaying HIT that would confirm them convincingly. This is so because every grid-turbulence experiment has aimed to keep the length scale small compared to the size of the apparatus. On the other hand, numerical simulations such as those of Borue & Orszag (1995), Biferale et al. (2003), and especially those of Touil, Bertoglio & Shao (2002), appear closer. In particular, these authors investigated the decay of turbulence in a bounded domain without mean velocity using direct and large-eddy simulations, as well as the eddy-damped quasinormal Markovian (EDQNM) closure. The effect of the finite geometry of the domain is accounted for by introducing a low wavenumber cut-off into the energy spectrum of isotropic turbulence. It is found that, once the turbulent energy-containing length scale saturates, the root-mean-square vorticity follows a power law with the $-3/2$ exponent. This yields $L(t) \propto t^{-3/2}$, just as observed in *Oregon 1, Oregon 2, Lancaster data* and *Manchester spin down data* on Kolmogorov QT (see Figure 10.4 for an illustration).

Note, further, that equation (10.11) shows that data obtained in two geometrically similar channels must have the same exponent but a prefactor that shows a *linear dependence on the size of the channel*. As already mentioned, there are no data in classical HIT against which this prediction can be compared. In the *Prague counterflow* data, on the other hand, two geometrically similar channels, of square crossection 10 mm × 10 mm and 6 mm × 6 mm, yielded [see Figure 2 in Gordeev et al. (2005) or Figure 6 in Barenghi & Skrbek (2007)] not only the decay *exponent* of −3/2, but also the *linear dependence on the size of the channel*. This result illustrates how the QT data have already begun to provide new information applicable to classical decay of HIT.

10.4.2 Additional remarks

The only direct experimental evidence for the *energy* spectrum[6] in superfluid ^{4}He comes from the work of Maurer & Tabeling (1998), discussed already. No direct measurement of the *energy* spectra exists in ^{4}He below 1.4 K and in ^{3}He-B.

Roche and coworkers (2007) measured the fluctuations of the vortex line density in a turbulent superfluid ^{4}He at $T = 1.6$ K using a small second sound probe, which is a micromachined open-cavity resonator, and found the frequency power spectrum of L to possess a $-5/3$ roll-off. This temperature was within the range covered by Maurer & Tabeling (1998). Recently the Lancaster group [Bradley et al. (2008)] carried out a complementary analysis of the steady-state experimental data in ^{3}He-B in the zero temperature limit using Andreev reflection. The measured power spectra of vortex line density also displayed the $-5/3$ power.

These two results seem contrary to the notion that L is a measure *not* of the energy but of the mean superfluid vorticity, via the relationship $\omega(t) = \kappa L(t)$. Roche et al. (2007) argued that this contradiction disappears if the vortex line density field is decomposed into a polarized field (that carries most of the energy of the flow) and an isotropic field (which is mostly responsible for the observed spectrum); the thinking is that this larger component is swept passively along by the large scale flow.

---

[6] Although it follows from the work of Maurer & Tabeling (1998) that in QT the intermittent corrections most likely exist and are approximately the same as in classical HIT, we neglect them here: as shown by Skrbek & Stalp (2000), they do not change the decay exponent to the first approximation.

10.5 Decay of HIT when the shape of the energy spectra matters

10.5.1 The form of the energy spectrum

If the large scale of turbulence does not saturate, the problem of decay is not closed and the details depend on the precise form of the spectral density of turbulence energy. The spectral density in decaying HIT behaves differently in different scale ranges and obeys different forms of similarity, with appropriate crossovers between them. One can then use equation (10.10) to calculate the decay in terms of the dynamics of crossover scales and other spectral constants such as scaling exponents. This procedure was first used by Comte-Bellot & Corrsin (1966) and revised by Saffman (1967, 1967a). This approach was extended by Skrbek & Stalp (2000) and Skrbek, Niemela & Donnelly (2000), by taking account of ideas from Eyink & Thomson (2000), to develop a spectral decay model for classical HIT and related it to the available data on decaying turbulent energy. The outcome of the model was also compared with the *Oregon 1* data on the vortex line density $L(t)$. These authors identified four regimes of spectral decay, which we shall consider below.

One cannot account for all the fine details of the last stages of decay and to relate classical and quantum cases. The fundamental reason is that, the high k part of the energy spectra in Kolmogorov QT and HIT is quite different in character. For the simplest case of zero temperature where there is no normal fluid, no viscosity and no mutual friction, the kinetic energy cascading down the inertial range reaches the quantum length scale of the order of the intervortex distance, where quasi-classical description ceases to hold. It seems reasonably clear that the Kolmogorov cascades will yield place at that point to the Kelvin wave cascade in some fashion. In the Kelvin cascade, waves of very high frequency are generated on individual vortex lines, eventually leading to the dissipation in the form of phonon emission at very high k. We are reluctant to discuss this issue in detail here; in particular, we will not examine the relationship of the exact shape of the high k part of the QT spectrum on the temporal decay, or vice versa. Our reluctance is mainly the result of the fact that there are no relevant experimental data, and the *Lancaster data* and *Manchester spin down data* do not display significant late time departures from the $-3/2$ power. The situation at finite temperature is more complicated due to the two-fluid behaviour and the coupling by mutual friction of the normal and superfluid components making up the quantum working fluid.

In our subsequent discussion we therefore neglect the effects of the finite cutoff of the spectra at high k and write the latter in the simple form

$$\phi(k) = 0 \; ; \; k < 2\pi/D \qquad (10.12)$$
$$\phi(k) = Ak^m \; ; \; 2\pi/D \leq k_1(t) \qquad (10.13)$$
$$\phi(k) = C\epsilon^{2/3}k^{-5/3} \; ; \; k_2(t) \leq k, \qquad (10.14)$$

reflecting the fact that scales larger than the size of the turbulence box do not exist. In the vicinity of the energy containing scale $\ell_e = 2\pi/k_e(t)$, where $k_1(t) < k_e(t) < k_2(t)$, the spectral energy density displays a broad maximum whose shape does not have to be exactly specified. We shall now consider the first three stages of decay. Evaluating the total turbulent energy by integrating the 3D spectrum over all k leads to a differential equation for decaying turbulent energy; and applying $\epsilon = \nu\omega^2$ leads to a differential equation for decaying vorticity, this being the quantity of primary interest to us.

10.5.2 The four stages of decay of HIT

In the early stages of decay, the spectrum retains a self-similar shape. For $D \gg \ell_e$ the *first decay regime* corresponds to

$$E(t + t_{\text{vol}}) = E(\tau) \propto \tau^{-2\frac{m+1}{m+3}} \; ; \qquad (10.15)$$
$$\ell_e \propto \tau^{\frac{2}{m+3}} \; ; \; \omega \propto \tau^{-\frac{3m+5}{2m+6}} . \qquad (10.16)$$

Here t_{vol} is the virtual origin time. Assuming the validity of the Saffman invariant ($m = 2$), we obtain the *first regime* for decaying energy, $E \propto \tau^{-6/5}$ (i.e., we have recovered the original Saffman (1967, 1967a) result) and for vorticity, $\omega \propto \tau^{-11/10}$. On the other hand, assuming the validity of the Loitsianskii invariant ($m = 4$), we obtain the alternative expression for decaying energy, $E \propto \tau^{-10/7}$ (i.e., we have recovered the original Kolmogorov (1941) result) and for vorticity, $\omega \propto \tau^{-17/14}$.

As the turbulence decays further and ℓ_e grows, the lowest physically significant wave number becomes closer to the broad maximum around $2\pi/\ell_e$. The low wave number part of the spectrum can no longer be approximated as Ak^m. Instead, it can be characterized by an effective power m_{eff} that decreases as the turbulence decays, such that $0 < m_{\text{eff}} < m$. Equation (10.16) then shows that the decay rate slows down. As ℓ_e approaches D, m_{eff} essentially vanishes to zero and we arrive at the *second regime* of the decay $\omega \propto \tau^{-5/6}$.

At the saturation time, t_{sat}, the vorticity reaches its saturation value, ω_{sat}, and the growth of ℓ_e is completed. The *universal third regime* is predicted as

$$\omega(t + t_{vo2}) = \omega(\tau) = \frac{3\sqrt{3}D}{2\pi}\sqrt{\frac{C^3}{\nu}}\tau^{-3/2} \qquad (10.17)$$

with the different virtual origin time t_{vo2}. This regime is universal in the sense that a decaying system must sooner or later reach it, independent of the initial level of turbulence (provided it is high enough for viscosity corrections to be negligible). In practice, transitions from one decay regime to the next are gradual, and care must be taken while searching for the decay exponent from log-log plots of the decay data, as each of the decay regimes has its own virtual origin.

The last, or the fourth, stage of decay is strongly influenced by the high k end of the energy spectrum where, because of the quantization condition in superfluids, QT is expected not to correspond to classical turbulence. For this reason, we shall not consider this stage of decay further. In any case, in classical HIT – neglecting possible bottleneck effects [see, e.g., Donzis & Sreenivasan (2010)] – the high k end of the spectrum was modeled by Skrbek & Stalp (2000) by a sharp cutoff at an effective Kolmogorov wavenumber $k_\eta \approx \gamma(\epsilon/\nu^3)^{1/4}$, where the dimensionless parameter γ is of the order unity. It was shown there that, as the dissipation length grows during the decay of classical HIT, the decay rate gradually speeds up and cannot be expressed by a simple power law.

10.5.3 Stages of decay in $L(t)$ and comparisons with HIT

The pronounced feature of *Oregon 1* as well as *Oregon 2* data is that the four different decay regimes [Skrbek, Niemela & Donnelly (2000)] switch from one to the other successively as the spectrum evolves and the decay proceeds. We have already pointed out that the fourth stage of decay might in fact be different for classical HIT and QT due to the distinctive difference in the high k parts.

Focusing now on the first regime, we note that Skrbek, Niemela & Donnelly (2000) and Skrbek & Stalp (2000) based their work on the *Oregon 1* data and concluded that the decay exponents ($L(\tau) \propto \tau^{-11/10}$; $E(\tau) \propto \tau^{-6/5}$) are in agreement with the spectral decay model based on the validity of the Saffman invariant, which assumes that the total momentum of the flow is conserved and as a consequence the spectral energy density $E(k) \propto k^2$ at low k. If, however, one assumes that the angular momentum of an entire flow is conserved, as shown by Landau & Lifshitz (1987), this is equivalent to the validity of Loitsianskii invariant, as a consequence of which the spectral energy density $E(k) \propto k^4$ at low k. Though Proudman & Reid (1954) argued that the Loitsianskii invariant is not really an invariant, the issue appears more complicated. More recently, Yakhot (2004) found that a modified Loitsianskii integral is an invariant of the three-dimensional decaying HIT. In their recent work, Ishida, Davidson & Kaneda (2006) show that the Kolmogorov decay law $E(\tau) \propto \tau^{-7/10}$ holds numerically in a periodic domain whose dimensions are much larger than the integral scale. In view of this, we have critically re-analyzed the *Oregon* data and believe that, taking into account that in the first regime the decay exponents differ only slightly [$(L(\tau) \propto \tau^{-11/10}$ versus $L(\tau) \propto \tau^{-17/14}$], the experimental accuracy and the temporal resolution of the second sound technique does not allow firm conclusion to be drawn. As other available experimental sets on decaying QT appear no better, it is our opinion that no conclusive experimental answer exists at present. In this respect, measurements in QT can be very beneficial.

The second regime, which is more in the nature of a slow changeover, possesses a slower decay regime $(L(\tau) \propto \tau^{-5/6}; E(\tau) \propto \tau^{-2/3})$ that does not depend on the model (e.g., on the validity of the Loitsianskii or the Saffman invariant, as long as there is a broad maximum in spectral energy density); this state ought to occur in all experiments. Indeed this feature does exist in both *Oregon 1* and *Oregon 2* data, and is at least consistent with the *Manchester spin down* data (see, e.g., Fig. 3 of Walmsley *et al.* (2007)] and with the *Lancaster* ^{3}*He-B* data [see, e.g., Fig. 2 of Bradley *et al.* (2006)]. This second regime is followed by the third universal regime $L(\tau) \propto \tau^{-3/2}$.

It is unclear if a more careful quantitative analysis of the *Manchester spin down* data and/or *Lancaster* ^{3}*He-B* data could bring greater confidence in our reasoning, as the initial state of QT in these experiments is spatially non-uniform. Despite this uncertainty, it seems that the decaying QT is capable of experimentally resolving this issue. In the Oregon experiments, the channel dimension was 1 cm and saturation of ℓ_e took place too early to observe the first regime over sufficiently long time for extracting the power law exponent accurately enough. The conditions will probably be difficult to improve. On the other hand, increasing the channel size by an order of magnitude seems experimentally feasible. There is a clear call to design additional experiments on decaying QT in ^{4}He in order to contribute experimentally to the solution of the puzzle on Loitsianskii versus Saffman invariants. Experiments where both classical turbulence in He I ($Re \approx 10^7$) and QT in He II is generated in the same box between counterrotating discs with blades are under progress in Prague and Grenoble, and will be discussed elsewhere.

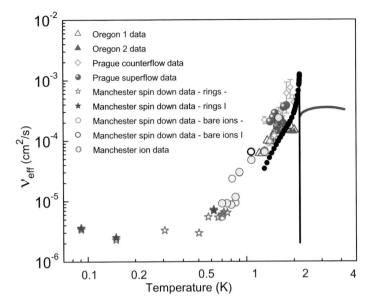

Figure 10.6 The observed temperature dependence of $\nu_{\text{eff}}(T)$ from various experiments as described in the text. The solid line above the λ-temperature is the kinematic viscosity of normal liquid He I [Donnelly & Barenghi (1998)]. The big black dots are our estimates $\nu_{\text{eff}} \approx \kappa q$, as described in the text.

10.6 Effective viscosity

Assuming that the Kolmogorov form of spectra holds for QT at all temperatures over length scales up to $\ell_e = D$, the classical decay model [Stalp, Skrbek & Donnelly (1999); Skrbek & Stalp (2000); Skrbek, Niemela & Donnelly (2000)] predicts for the decaying L not only the $-3/2$ power law but also the prefactor:

$$\kappa L(t) = \frac{D(3C)^{3/2}}{2\pi\sqrt{\nu_{\text{eff}}}}(t+t_{vo})^{-3/2} \cong \frac{D(3C)^{3/2}}{2\pi\sqrt{\nu_{\text{eff}}}}t^{-3/2} \ . \qquad (10.18)$$

When $L \propto D$, we can thus describe the decaying QT in quasi-classical manner by introducing the effective kinematic viscosity, $\nu_{\text{eff}}(T)$. At the same time this quasi-classical description of QT makes sense if, and only if, values of $\nu_{\text{eff}}(T)$ deduced from various data sets (especially among those measured by different techniques) agree with each other. So far this check has been possible only for QT in He II.

The first attempt to extract the temperature dependence of $\nu_{\text{eff}}(T)$ was made by Stalp *et al.* (2002), based on the *Oregon 1* data set. This effort was later refined by Niemela, Sreenivasan & Donnelly (2005) and Niemela

& Sreenivasan (2006) by adding the *Oregon 2* data set. Chagovets, Gordeev & Skrbek (2007) pointed out some corrections to these values of $\nu_{\text{eff}}(T)$ and added the values $\nu_{\text{eff}}^{cf}(T)$ determined from the *Prague counterflow* data. This last set of data is so far the only one that has allowed one to check the RHS of equation (10.18) with respect to the channel size, D. Walmsley and Golov (2008) and Walmsley *et al.* (2007; 2008), added their values of $\nu_{\text{eff}}(T)$, deduced from the *Manchester spin down* data, where $L(t)$ was determined by using bare ions and charged ion rings. In the overlapping temperature region all these sets of $\nu_{\text{eff}}(T)$ agree with each other within the experimental scatter. Moreover, recently Chagovets & Skrbek (2008; 2008a) deduced additional set of $\nu_{\text{eff}}(T)$, based on *Prague superflow* data set. Although these data are deduced from the exponential decay of L [for details, see recent papers by Chagovets & Skrbek (2008; 2008a)], they too agree with the earlier sets. The conclusion appears to be that a quasi-classical description of Kolmogorov QT in He II is physically plausible, provided conditions allow a coupling of the normal and superfluid eddies in co-flow turbulence. If so, it opens up possibilities of using He II as a working fluid with truly remarkable physical properties from the point of applied fluid dynamics.

10.6.1 Simple model for effective kinematic viscosity of turbulent He II

We now present a simple model for effective kinematic viscosity in QT [Skrbek (2010)], where two velocity fields exist, coupled by the mutual friction force. At all k below k_Q the two fields are locked together, but gradually become decoupled as k increases, due to combined action of the normal fluid viscosity, mutual friction force and quantum effects in the superflow.

Let us assume, for simplicity and order of magnitude estimates, that the decoupling takes place not gradually, but sharply at some characteristic k_d; that is, the normal and superfluid flows are fully coupled for $k < k_d$ and fully decoupled for $k > k_d$. Let us consider the flow for $k > k_d$ in the frame of reference of the normal fluid eddy of the size just $2\pi/k_d$, and neglect the possible motion of the normal fluid at higher k, i.e., we assume that for $k > k_d$ the normal fluid is at rest. Let us further assume that for $k > k_d$ there are enough number of quantized vortices in the flow arranged in such a way that the coarse-grained hydrodynamic equation [Finne *et al.* (2003); Volovik (2003)] holds:

$$\frac{\partial \mathbf{v}_s}{\partial t} + \nabla \mu = \mathbf{v}_s \times \boldsymbol{\omega}_s + q\hat{\boldsymbol{\omega}}_s \times (\boldsymbol{\omega}_s \times \mathbf{v}_s). \tag{10.19}$$

Here $q = \alpha/(1-\alpha')$ and μ denotes the chemical potential. The normal fluid thus provides a unique frame of reference and we have to deal only with the superfluid velocity $\mathbf{v}_s = \mathbf{v}$. As was first emphasized by Finne et al. (2003), this equation has remarkable properties that are distinct from those of the Navier–Stokes equation which can be written [see, e.g., Landau & Lifshitz (1987)] in a similar form:

$$\frac{\partial \mathbf{v}}{\partial t} + \nabla \mu = \mathbf{v} \times \omega + \nu \Delta \mathbf{v} \ . \tag{10.20}$$

Let us compare the dissipative terms on the RHS of these two equations. Taking into account that $\omega = \kappa L \approx \kappa/\ell_v^2$, where ℓ_v is the characteristic intervortex distance, we estimate the superfluid dissipation term as $q\kappa \mathbf{v}/\ell_v^2$. By analogy with $\Delta \mathbf{v} \approx \mathrm{v}/\ell_v^2$ we see that the order of magnitude of the effective kinematic viscosity must be given by $\nu_{\mathrm{eff}} \approx \kappa q$. We plot this quantity together with experimental data $\nu_{\mathrm{eff}}(T)$ in Figure 10.6. This model is not applicable in the $T \to 0$ limit, where the physical origin of dissipation is different, i.e., phonon emission by Kelvin waves, and not too close to the superfluid transition temperature, where equation (10.19) ceases to hold.

The agreement with the experimental data suggests that this simple model captures the underlying physics, at least approximately. Note that this model can be applied to ^{3}He-B, but there are no experimental data for comparison. A similar, more sophisticated, model that combines theoretical ideas of two coupled cascades with computer simulation has recently been published by Roche, Barenghi & Leveque (2009), displaying similar results.

10.7 Conclusions

Superfluid phases of liquid helium ^{4}He II and ^{3}He-B – quantum fluids of known and tunable properties – can be used to study superfluid dynamics and experimental studies of QT, in particular. Their special physical properties, together with contemporary advances in cryogenics, provide several independent methods for generating quantum flows experimentally.

QT in the limit of zero temperature takes the form of complex tangle of quantized vortices resulting in the stochastic nature of its velocity field and represents a simplest prototype of turbulence. In analogy with the Kolmogorov dissipation length scale in classical turbulence, a quantum length scale can be introduced on the basis of dimensional reasoning. Depending on whether or not the outer scale of QT exceeds this quantum scale, one can generate either the quasiclassical (or Kolmogorov) QT or the ultraquantum (or Vinen) QT. On small length scales, due to quantum mechanical restrictions, the classical and quantum flows are quite different. However, with or without the presence of the normal fluid, on length scales much greater than the quantum length scale or the average spacing between vortex lines ("inter-vortex spacing") in the vortex tangle, the flow of the superfluid component in Kolmogorov QT closely mimics that of a classical fluid.

We have particularly focused on comparisons in one instance: the case of decaying turbulence behind pull-through grids. In this instance, two major similarities seem to prevail. The first is that spectral density in an intermediate range of length scales in QT follows the classical Kolmogorov form with the $-5/3$ slope. The second one is the rate at which the energy rolls off with time. The most convincing case is that the energy decays in both QT and classical turbulence as -2 power of time (or the root-mean-square vorticity decays as the $-3/2$ power) in a certain range for which the large scale of turbulence becomes comparable to the size of the container and cannot grow any further. For other conditions, there exists very good correspondence between decay coefficients and the spectral shape.

The decay of Kolmogorov QT endows it the property of an effective kinematic viscosity, very similar to the effective viscosity in classical turbulence. Its (temperature dependent) magnitude in He II is of order 10^{-2}–$10^{-1}\kappa$, where κ is the quantum of circulation (approximately $10^{-3}\,\mathrm{cm}^2\,\mathrm{sec}^{-1}$). However, unlike the effective viscosity in classical turbulence which depends on the details of the flow geometry, the effective viscosity seems to be the same for all configurations examined so far. The universality of this conclusion is being explored currently. If it holds firm, one can imagine that new areas of inquiry in statistical mechanics will open up.

QT is an interesting and challenging subject in its own right. Moreover, complementary experiments on its decay have already provided a better understanding of its classical counterpart and might in turn lead to a better grasp of turbulence in general.

Acknowledgements We are grateful to many colleagues, especially C.F. Barenghi, R.J. Donnelly, A.I. Golov, M. Krusius, V.S. L'vov, P.V.E. McClintock, J.J. Niemela, M. Tsubota, W.F. Vinen and G.E. Volovik for stimulating discussions over the years. The research of LS was supported by GACR 202/08/0276.

References

Alamri, S.Z., Youd A.J. and Barenghi, C.F. (2008). Reconnection of superfluid vortex bundles, *Phys. Rev. Lett.* **101**, 215302.

Andronikashvili, E.L. (1946). A direct observation of two kinds of motion in helium II, *Journal of Physics (Moscow)* **10**, 201.

Araki, T., Tsubota, M. and Nemirovskii, S.K. (2002). Energy spectrum of superfluid turbulence with no normal-fluid component, *Phys. Rev. Lett.* **89**, 145301.

Barenghi, C.F., Gordeev, A.V. and Skrbek L. (2006). Depolarization of decaying counterflow turbulence in He II", *Phys. Rev.* **E74**, 026309.

Barenghi, C.F. and Skrbek, L. (2007). On decaying counterflow turbulence in He II, *J. Low Temp. Phys.* **146**, 5.

Bewley, G.P., Lathrop, D.P. and Sreenivasan, K.R. (2006). Superfluid helium – visualization of quantized vortices, Nature **441**, 588.

Biferale, L., Bofetta, G., Celani, A., Lanotte, A., Toschi, F. and Vergassola, M. (2003). The decay of homogeneous anisotropic turbulence, *Phys. Fluids* **15**, 2105.

Borue, V. and Orszag, S.A. (1995). Self-similar decay of three-dimensional homogeneous turbulence with hyperviscosity, *Phys. Rev* **E5**, R856.

Bradley, D.I., Clubb, D.O., Fisher, S.N., Guénault, A.M., Matthews, C.J. and Pickett, G.R. (2004). Vortex generation in superfluid He-3 by a vibrating grid, *J. Low Temp. Phys.* **134**, 381.

Bradley, D.I., Clubb, D.O., Fisher, S.N., Guénault, A.M., Haley, R.P., Matthews, C.J., Pickett, G.R., Tsepelin, V. and Zaki, K. (2006). Decay of pure quantum turbulence in superfluid He-3-B, *Phys. Rev. Lett.* **96**, 035301.

Bradley, D.I., Fisher, S.N., Guénault, A.M., Haley, R.P., O'Sullivan, S., Pickett, G.R. and Tsepelin, V. (2008). Fluctuations and correlations of pure quantum turbulence in superfluid He-3-B, *Phys. Rev. Lett.* **101**, 065302.

Chagovets, T.V., Gordeev, A.V. and Skrbek, L. (2007). Effective kinematic viscosity of turbulent He II, *Phys. Rev.* **E76**, 027301.

Chagovets, T.V. and Skrbek, L. (2008). Steady and decaying flow of He II in channels with ends blocked by superleaks, *Phys. Rev. Lett.* **100**, 215302.

Chagovets, T.V. and Skrbek, L. (2008a). On flow of He II in channels with ends blocked by superleaks, *J. Low Temp. Phys.* **153**, 162.

Comte-Bellot G. and Corrsin, S. (1966). The use of a contraction to improve the isotropy of grid-generated turbulence, *J. Fluid Mech.* **25**, 657.

Davis, S.I., Hendry, P.C. and McClintock P.V.E. (2000). Decay of quantized vorticity in superfluid He-4 at mK temperatures, *Physica* **B280**, 43.

Dobbs, E.R. (2000). *Helium Three* (Oxford Science Publications).

Donnelly, R.J. (1991). *Quantized Vortices in Helium II* (Cambridge University Press).

Donnelly, R.J. and Barenghi, C.F. (1998). The observed properties of liquid helium at the saturated vapor pressure, *J. Phys. Chem. Data* **27**, 1217.

Donzis, D.A. and Sreenivasan, K.R. (2010). The bottleneck effect and the Kolmogorov constant in isotropic turbulence. *J. Fluid Mech.* **657**, 171–188.

Eltsov, V.B., Golov, A., de Graaf, R., Hänninen, R., Krusius, M., L'vov, V. and Solntsev, R.E. (2007). Quantum turbulence in a propagating superfluid vortex front, *Phys. Rev. Lett.* **99**, 265301.

Eltsov, V.B., de Graaf, R., Hanninen, R., Krusius, M., Solntsev, R.E., L'vov, V.S., Golov, A.I. and Walmsley, P.M. (2009). Turbulent dynamics in rotating helium superfluids. In *Progress in Low Temperature Physics*, vol. XVI, Chapter 2, ed. by M. Tsubota and W.P. Halperin (Elsevier).

Eyink, G. L. and Thomson, D. J. (2000). Free decay of turbulence and breakdown of self-similarity, *Phys. Fluids* **12**, 477.

Feynman, R.P. (1955). Application of quantum mechanics to liquid helium. In *Progress in Low Temperature Physics*, vol. 1, ed. by C.J. Gorter (North Holland).

Finne, A.P., Araki, T., Blaauwgeers, R., Eltsov, V.B., Kopnin, N.B., Krusius, M., Skrbek, L., Tsubota, M. and Volovik, G.E. (2003). Observation of an intrinsic velocity-independent criterion for superfluid turbulence, *Nature* **424**, 1022.

Fisher, S.N. and Pickett, G.R. (2009). Quantum turbulence in superfluid ^{3}He at very low temperatures. In *Progress in Low Temperature Physics*, vol. XVI, Chapter 3, ed. by M. Tsubota and W.P. Halperin (Elsevier).

Gordeev, A.V., Chagovets, T.V., Soukup, F. and Skrbek, L. (2005). Decaying counterflow turbulence in He II, *J. Low Temp. Phys.* **138**, 549.

Ihas, G.G., Labbe, G., Liu, S.C. and Thomson, K. (2007). Preliminary measurements on grid turbulence in liquid ^{4}He, *J. Low Temp. Phys.* **150** 384.

Ishida, T., Davidson, P.A. and Kaneda, Y. (2006). On the decay of isotropic turbulence, *J. Fluid Mech.* **564**, 455.

von Kármán, T. and Howarth, L. (1938). On the statistical theory of isotropic turbulence, *Proc. R. Soc. London* **A164**, 192.

Kobayashi M. and Tsubota, M. (2005). Kolmogorov spectrum of superfluid turbulence: Numerical analysis of the Gross–Pitaevskii equation with a small-scale dissipation, *Phys. Rev. Lett.* **94**, 065302.

Kolmogorov, A. N. (1941). On degeneration (decay) of isotropic turbulence in an incompressible viscous fluid, *Dokl. Acad. Nauk U.S.S.R.* **31**, 538.

Koplik, J. and Levine, H. (1993). Vortex reconnection in superfluid-helium, *Phys. Rev. Lett.* **71**, 1375.

Kozik, E.V. and Svistunov, B.V. (2009). Theory of decay of superfluid turbulence in the low-temperature limit, *J. Low Temp. Phys.* **156**, 215 and references therein.

Landau, L.D., (1941). The theory of superfluidity of helium II, *J. Phys. (U.S.S.R.)* **5**, 356.

Landau, L.D., (1947). On the theory of superfluidity of helium II, *J. Phys. (U.S.S.R.)* **11**, 91.

Landau, L. D. and Lifshitz, E. M. (1987). *Fluid Mechanics* 2nd ed., (Pergamon Press).

Leadbeater, M., Winiecki, T., Samuels, D. C., Barenghi, C. F. and Adams, C. S. (2001). Sound emission due to superfluid vortex reconnections, *Phys. Rev. Lett.* **86**, 1410.

Lesieur, M. (1997). *Turbulence in Fluids* 3rd ed., (Kluwer).

Loitsianskii, L.G. (1939). Some basic laws for isotropic turbulent flow, *Trudy Tsentr. Aero.-Gidrodyn. Inst.* **3**, 33.

L'vov, V. S., Nazarenko, S. V. and Rudenko, O. (2007). Bottleneck crossover between classical and quantum superfluid turbulence, *Phys. Rev.* **B76**, 024520.

Mathieu, P., Placais, B. and Simon, Y. (1984). Spatial-distribution of vortices and anisotropy of mutual friction in rotating He II, *Phys. Rev.* **B29**, 2489.

Maurer, J. and Tabeling, P. (1998). Local investigation of superfluid turbulence, *Europhys. Lett.* **43**, 29.

Milliken, F.P., Schwarz, K.W. and Smith, C.W. (1982). Free decay of superfluid turbulence, *Phys. Rev. Lett.* **48**, 1204.

Niemela, J.J., Sreenivasan, K.R. and Donnelly, R.J. (2005). Grid generated turbulence in helium II, *J. Low Temp. Phys.* **138**, 537.

Niemela, J.J. and Sreenivasan, K.R. (2006). The use of cryogenic helium for classical turbulence: Promises and hurdles, *J. Low Temp. Phys.* **143**, 163.

Ogawa, S., Tsubota, M. and Hattori, Y. (2002). Study of reconnection and acoustic emission of quantized vortices in superfluid by the numerical analysis of the Gross–Pitaevskii equation, *J. Phys. Soc. Japan* **71**, 813.

Paoletti, M.S., Fisher, M.E., Sreenivasan, K.R. and Lathrop, D.P. (2008). Velocity statistics distinguish quantum turbulence from classical turbulence, *Phys. Rev. Lett.* **101**, 154501.

Proudman, I. and Reid, W.H. (1954). On the decay of a normally distributed and homogeneous turbulent velocity field, *Proc. Roy. Soc.* **A247**, 163.

Roche, P.E., Diribarne, P., Didelot, T., Francais, O., Rousseau, L. and Willaime, H. (2007). Vortex spectrum in superfluid turbulence: Interpretation of a recent experiment, *Europhys. Lett.* **81**, 36002.

Roche, P.E., Barenghi, C.F. and Leveque, E. (2009). Quantum turbulence at finite temperature: The two-fluids cascade, *Europhys. Lett.* **87**, 54006.

Saffman, P.G. (1967). The large-scale structure of homogeneous turbulence, *J. Fluid Mech.* **27**, 581.

Saffman, P.G. (1967a). Note on decay of homogeneous turbulence, *Phys. Fluids* **10**, 1349.

Schwarz, K.W. (1988). 3-dimensional vortex dynamics in superfluid He-4 – homogeneous superfluid turbulence, *Phys. Rev.* **B38**, 2398 and references therein.

Schwarz, K.W. and Smith, C.W. (1981). Pulsed ion study of ultrasonically generated turbulence, *Physics Lett.* **82A**, 251.

Smith, M.R., Donnelly, R.J., Goldenfeld, N. and Vinen, W.F. (1993). Decay of vorticity in homogeneous turbulence, *Phys. Rev. Lett.* **71**, 2583.

Skrbek, L. (2004). Turbulence in cryogenic helium, *Physica* **C404**, 354.

Skrbek, L. (2010). A simple phenomenological model for the effective kinematic viscosity of helium superfluids, *J. Low Temp. Phys.* **161**, 555–560.

Skrbek, L. and Sreenivasan K.R. (2012). Developed quantum turbulence, *Phys. Fluids* **24**, 011301 (48 pages).

Skrbek, L. and Stalp, S.R. (2000). On the decay of homogeneous isotropic turbulence, *Phys. Fluids* **12**, 1997.

Skrbek, L., Niemela, J.J. and Donnelly, R.J. (2000). Four regimes of decaying grid turbulence in a finite channel, *Phys. Rev. Lett.* **85**, 2973.

Skrbek, L., Niemela, J.J. and Sreenivasan K.R. (2001). Energy spectrum of grid-generated He II turbulence, *Phys. Rev.* **E64**, 067301.

Skrbek, L., Gordeev, A.V. and Soukup F. (2003). Decay of counterflow He II turbulence in a finite channel: Possibility of missing links between classical and quantum turbulence, *Phys. Rev.* **E67**, 047302.

Sreenivasan, K.R. (1984). On the scaling of the turbulence energy dissipation rate, *Phys. Fluids* **27**, 1048.

Sreenivasan, K.R. (1995). On the universality of the Kolmogorov constant, *Phys. Fluids* **7**, 27788.

Sreenivasan, K.R. (2004). Possible effects of small-scale intermittency in turbulent reacting flows, *Flow, Turb. Combust.* **72**, 115.

Stalp, S.R. (1998). Decay of grid turbulence in superfluid helium, University of Oregon, D. Phil. thesis.

Stalp, S.R., Skrbek, L. and Donnelly, R. J. (1999). Decay of grid turbulence in a finite channel, *Phys. Rev. Lett.* **82**, 4831.

Stalp, S.R., Niemela, J.J., Vinen, W.F. and Donnelly, R.J. (2002). Dissipation of grid turbulence in helium II, *Phys. Fluids* **14**, 1377.

Taylor, G.I. (1935). Statistical theory of turbulence, Parts I, II, III and IV. *Proc. Roy. Soc. Lond.* **151**, 421, 444, 455, 465.

Tilley, D.R. and Tilley, J. (1986). *Superfluidity and Superconductivity* (Adam Hilger).

Tisza, L. (1938). Transport phenomena in helium II, *Nature* **141**, 913.

Tough, J.T. (1982). Superfluid turbulence, *Progress in Low Temperature Physics*, vol. 8, ed. by D.F. Brewer (North Holland).

Touil, H., Bertoglio, J.P. and Shao, L. (2002). The decay of turbulence in a bounded domain, *J. Turb.* **3**, 049.

Vinen, W.F. (1957). Mutual friction in a heat current in liquid helium II, I. Experiments on steady heat currents *Proc. Roy. Soc.* **A240** 114; (1957). II. Experiments on transient effects, **A240**, 128; (1957). III. Theory of the mutual friction, **A242**, 493; (1958). IV. Critical heat currents in wide channels, **A243**, 400.

Vinen, W.F. (2000). Classical character of turbulence in a quantum liquid, *Phys. Rev.* **B61**, 1410.

Vinen, W.F. (2005). Theory of quantum grid turbulence in superfluid ^{3}He-B, *Phys. Rev.* **B71**, 024513.

Vinen, W.F. and Niemela, J.J. (2002). Quantum turbulence, *J. Low Temp. Phys.* **128**, 167.

Vinen, W.F., Tsubota, M. and Mitani A. (2003). Kelvin-wave cascade on a vortex in superfluid He-4 at very low temperatures, *Phys. Rev. Lett.* **91**, 135301.

Vollhardt, D. and Wölfle, P. (1992). *The Superfluid Phases of Helium 3* (Taylor and Francis).

Volovik, G.E. (2003). Classical and quantum regimes of superfluid turbulence, *JETP Letters* **78**, 533.

Walmsley, P.M. and Golov, A.I. (2008). Quantum and quasiclassical types of superfluid turbulence, *Phys. Rev. Lett.* **100**, 245301.

Walmsley, P.M., Golov, A.I., Hall, H.E., Levchenko, A.A. and Vinen, W.F. (2007). Dissipation of quantum turbulence in the zero temperature limit, *Phys. Rev. Lett.* **99**, 265302.

Walmsley, P.M., Golov, A.I., Hall, H.E., Vinen, W.F. and Levchenko, A.A. (2008). Decay of turbulence generated by spin-down to rest in superfluid He-4, *J. Low Temp. Phys.* **153**, 127.

Yakhot, V. (2004). Decay of three-dimensional turbulence at high Reynolds numbers, *J. Fluid Mech.* **505**, 87.

QA 913 .T463 2013

Ten chapters in turbulence